# Technopolis

Deog-Seong Oh · Fred Phillips
Editors

# Technopolis

## Best Practices for Science and Technology Cities

Springer

*Editors*
Deog-Seong Oh
World Technopolis Association
Chungnam National University
Daejeon
Republic of Korea

Fred Phillips
SUNY Stony Brook, New York

SUNY Korea, Incheon

and

General Informatics LLC
San Diego, CA
USA

ISBN 978-1-4471-5507-2 ISBN 978-1-4471-5508-9 (eBook)
DOI 10.1007/978-1-4471-5508-9
Springer London Heidelberg New York Dordrecht

Library of Congress Control Number: 2013953895



Printed on acid-free paper

Springer is part of Springer Science+Business Media (www.springer.com)

# Preface

A nation's global competitiveness comes from regions, and international cooperation is an essential part for regions to build their competitiveness.

The World Technopolis Association, a global network of local governments and actors of regional innovation including universities and technology parks was born in 1998 with an aim to share knowledge on science and technology and success stories of building science and technology parks (STPs) and innovation clusters as a growth engine that leads to sustainable regional development.

Developed nations build STPs to stay at the forefront in R&D and commercialization of cutting edge technologies. Developing nations also desire to develop STPs as a means of improving their capacity of technology-based economic growth and regional innovation which will eventually enhance their national competitiveness.

UNESCO and WTA have jointly carried out international cooperative projects to help developing countries' effort to build STPs by providing technical assistances, pilot projects consulting, and capacity building programs. These projects have produced many fruitful accomplishments and are well regarded by the international community.

In particular, the annual UNESCO-WTA International Training Workshop which started as the first joint project in 2005 has been very successful as a capacity building program for STP managers and policymakers. The Workshop also produced a number of quality research papers relating to STP development and management.

This publication is a collection of selected works that have been submitted to the Workshop. It includes papers on theories and concepts of STP development/management, best practices, and applications in three different chapters, and will be a great textbook or reference not only for researchers and students but also for those responsible for building and managing STPs.

I would like to thank Prof. Deog-Seong Oh, Prof. Fred Young Phillips, and authors for their invaluable contributions. I also thank all those who have helped in the preparation of this book.

Republic of Korea

Hong Chul Yum
Mayor, Daejeon Metropolitan City President,
World Technopolis Association

# Contents

# Editors' Introduction

At the center of the Technopolis/Science and Technology Parks (STPs) development is the creative process of taking a careful approach to fostering competitive regional growth based on science and technology. A frequently raised question is "how to measure the effectiveness of Technopolis/STP as an instrument of regional innovation policy to stimulate technology-led economic development and sustainable growth."

This volume presents highlights and representative works from eight international workshops conducted by the World Technopolis Association in cooperation with UNESCO on these topics. The workshops' purpose was (and continues to be, as the annual events continue) the sharing and dissemination of knowledge of STP, technopolis, and innovation cluster. The workshops devote special attention to knowledge that can be applied to participants' home countries, many of which are in the developing world.

## The World Technopolis Association

After the inaugural WTA General Assembly, which was held in September, 1998 in Daejeon, the WTA stepped forward as an international organization with 23 member cities from ten countries. In only 15 years since its establishment, the WTA has grown as an international organization with 84 members from 43 countries as of 2012.

Aiming to serve as the heart of a global network leading to sustainable STP/ technopolis development, WTA's triune mission is to:

- Support international exchanges between member cities and create ways to promote sustainable technopolis development by enhancing competitiveness in this era of simultaneous localization and globalization.
- Accelerate regional development and activate regional economies through practical exchanges of highly qualified human resources, information, and goods among member cities.
- Enhance technological competitiveness through international cooperation among businesses, research institutes, and universities.

WTA's mission is served by objectives including:

- Building a globally competitive technopolis.
- Connecting science and technology with regional innovation.
- Creating collaborative partnerships for development of STP.
- Promoting knowledge and technology exchange among technopolises.
- Hosting expositions and exhibits for new technologies and products.
- Assisting in the development of contracts between businesses.

To these ends, WTA uses a triangle approach, linking local governments, businesses, and universities. Significantly for this book, one of the triangle programs is an international training workshop each year for managers and researchers from five regions: Africa, Middle East, Asia, Latin America, and Eastern Europe. The participants of the workshop are members of a network for technopolis development in their region and will participate in the development of a regional pilot project. The workshops are held in Daejeon Metropolitan City, Republic of Korea, in cooperation with regional partners.

## UNESCO-WTA Cooperation

The cooperative project between WTA and UNESCO was designed as a novel approach to resolve the problem of technology exchanges that many countries, which require new technologies to implement their goals of sustainable development, are facing. In April 2006, WTA and UNESCO established an official relationship at the 174th Session of the UNESCO Executive Board. WTA can be henceforth described as an "NGO in operational relations with UNESCO" as WTA starts a 5-year cooperative project with UNESCO entitled for "Capacity Building and Technical Assistance on Technopolis Development." Both organizations express their intention to render assistance to each other with the goals of improving their service and functions in

- Capacity building activities for science cities' stakeholders.
- Provision of expertise and technical assistance in the planning, development, and management of science cities in developing countries.
- Promoting global networking among science cities around the world, including African, South American, and Middle Eastern countries.
- Promoting research and publication on the sustainable development of science cities and technopolises.

*The International Training Workshops on Science and Technology Park Governance*

Each of the workshops has centered around distinct themes. The themes were:

| Year | Theme |
|---|---|
| 2005 | Workshop on Science City Governance Science City Development and Management |
| 2006 | International Training Workshop "High-tech Clusters in Global Context" |
| 2007 | Towards Sustainable Growth of Science and Technology Park : Role of Technology Business Incubation in Techno Park |
| 2008 | Towards Creative Growth of Science Park and Innovative Cluster |
| 2009 | Green Growth based on the Science Park Initiatives |
| 2010 | Triple Helix Model of Innovation: Government-Academia-Business Cooperation |
| 2011 | Science and Technology Park—INNOVATION ECOSYSTEM |
| 2012 | Valorization: Tangible benefits from Science and Technology parks |

The thread linking all these themes has been sustainability. STPs must be sustainable socially, environmentally, creatively, and financially. Local culture and politics, then, are considerations as important to the STP as intellectual vitality and technological absorptive capacity. Each chapter in this book reflects the sustainability orientation.

Technopolis-building and Science and Technology park design are the most interdisciplinary of endeavors, requiring knowledge of architecture, city planning, sociology, finance, politics, geography, technology strategy, R&D management, technology transfer and commercialization, education, and entrepreneurship. This variety of considerations informs the chapters also.

Every country's leaders understand that innovation is the key to increased wealth and a better life for its people. They understand, moreover, that a nation's innovative activity happens in a place—or in just a few places. Thus the name *Technopolis*—from the Greek for technology city. When a part of a city is newly built or newly refocused for the purpose, innovation then occurs in an even tighter radius, called a science park, a technology park, or an innovation center.

We believe further that sharing stories of successes, challenges, best practices, and new methods of analysis significantly increase each participating nation's chances of finding the markets and the direct investment that lead to a sustainable technopolis hosting creative laboratories and viable business clusters.

Technopolis initiatives have to deal with naysayers and conservative segments of society opposing change. Citizens and legislators may oppose focusing investment in one area of the country for technology development purposes, when other districts remain needy. Refocusing educational institutions for innovation and entrepreneurship is not an easy task. Obstacles to technopolis abound.

We fully realize that not all technopolis projects will succeed. However, participants in WTA's international conferences and workshops agree: Failure is guaranteed if no effort is made, if no information is shared, and if no risks are taken. Risks are mitigated by sharing information. This is the function of the UNESCO-WTA workshops and of this book.

The editors are grateful to the workshop participants and speakers, to UNESCO, and especially Dr. Yoslan Nur of that organization's Science Policy and Sustainable Development Division, and to WTA President Hong Chul Yum, the WTA Board of Directors, and WTA staff[1] for making the workshops and this book possible[2].

March 2013

Deog-Seong Oh
Fred Phillips

[1] The help of Mrs. Hyonjung Kim was especially invaluable in compiling this book.

[2] This work was also supported in part by the MSIP (Ministry of Science, ICT and Future Planning), Korea, under the "IT Consilience Creative Program" (NIPA-2013-H0203-13-1001) supervised by the NIPA (National IT Industry Promotion Agency).

# Part I
# Theory and Concept

# A New Strategic Approach to Science Cities: Towards the Achievement of Sustainable and Balanced Spatial Development

**Gordon Dabinett**

**Abstract** This chapter addresses the new challenges facing the internationalisation of science city projects. It is predicated upon an assumption that such projects are underpinned by a desire to overcome and address the real or potential consequences of uneven spatial development. A further assumption underlying the arguments in this chapter acknowledges that technology and its development do not inevitably lead to a series of foregone or determined impacts. Rather the applications and uses of technology are mediated via social structures and influences, including spatial planning (Borja and Castells 1997). Spatial policies can shape and be shaped by the various ways in which the emergence of a knowledge economy might be mediated through specific localities and territorial characteristics. These specific characteristics of localities and the spatial outcomes of general policy ideas become elements for debate in any assessment seeking to understand such complex interactions. The chapter constructs a critical perspective of science cities in light of future scenarios generated by globalisation and sustainable development. These scenarios will frame future successful economic development and desires to increase prosperity and the quality of life in all cities and regions. Experiences and practices, largely from Europe, are examined with respect to the extent to which common approaches situated in markedly different socio-economic localities are also able to manage the transition from, initially, an industrial to post-industrial economic order, and currently, onwards to a knowledge-based regional economy. The chapter argues that a new strategic approach is required in order for science city projects to contribute to sustainable development in the future. A form of development that will create difficult trade-offs between economic, environmental and equity goals, based on new forms of indigenous development and territorial governance, and a new approach to the social construction and reproduction of innovation and learning. Such a new strategic approach finds expression in the practices of polycentric urban development, multi-level governance and integrated spatial planning. The chapter begins

G. Dabinett (✉)
Regional Studies, Research Exchange for the Social Sciences, Department of Town and Regional Planning, University of Sheffield, Sheffield S10 2TN, UK
e-mail: g.e.dabinett@sheffield.ac.uk

D.-S. Oh and F. Phillips (eds.), *Technopolis*,
DOI: 10.1007/978-1-4471-5508-9_1, © Springer-Verlag London 2014

with an account of the project-based development of science parks and associated developments, such as business and innovation centres and technology parks, through to the development of more strategic approaches to encouraging innovative milieu, such as the technopolis policy, science cities, regional innovation systems and 'smart' cities. This account is followed by an assessment of the regional development models and analyses that have underpinned the ideas and practice, an assessment that provides insights into the future role that such approaches and public policy might play in the future, and the limits to this in a new global order seeking greater sustainability. The final section begins to map out and give articulation to what a new strategic approach to science cities might entail.

## 1 Science Parks, Science Cities and Regional Development Practices

Komninos (1997) has suggested that the first phase of science and technology parks development occurred within Western Europe at beginning of the 1970s, and was based on a rather experimental approach, notably the science parks in Cambridge (England), Heriot-Watt (Scotland), and Sophia Antipolis (France). These initial schemes were often the outcome of specific local factors (Massey, Quintas and Wield 1992), but governments at local, regional and national scales were quick and numerous in identifying these as possible models for dealing with the then rapidly growing economic crises associated with de-industrialisation; and the need to address flagging national economic performances with respect to technology-based development, then dominated by the USA and Japanese-based production blocs. Komninos (1997) thus identifies a second phase of science and technology practice beginning in the 1980s, when it is possible to identify more than 100 parks being set up throughout Western Europe (see Table 1), or more specifically the European Union (EU). He suggests that these projects were often connected to wider political and economic frameworks, concerned with productive restructuring, the disintegration of productive capacities, the rise of small businesses and the new demands for R&D, innovation and producer services. Studies of this period of regional development all point to the significant variety in the purpose and scale of projects, varying from small business incubators such as business and innovation centres (BiCs) officially supported by EU financial instruments, to large physical and land-use developments seeking to significantly replace old productive systems or to create new economic spaces or growth poles. Common components have been found within these developments in practice (Monck et al. 1998):

**Table 1** Science and technology park developments in Europe (from Komninos 2002)

| Member state | No. of parks | Type of park | Origin stakeholder | Management | Objectives |
|---|---|---|---|---|---|
| Belgium | Eight large and small parks | Science park, incubator | Universities, federal ministry, non-profit associations | Administrative council | Technology transfer, housing of large companies |
| Denmark | Five small parks | Science parks | Universities, Ministry of research and technology | Business foundation | Technology transfer, housing of large companies |
| France | 40 small parks | Technology park, technopolis, technopolis networks | Municipalities, regional government, chambers | Associations of communes, mixed consortia | Development and attraction of high-tech activities, development of networks |
| Germany | 70 small parks | Innovation centre, technology park, business incubator | Local, Lander | Private company | Support start-ups, creation of skilled jobs, translation of R&D into practice |
| Greece | Five small parks, five incubators | Technology park, innovation centre | Ministry of research and technology, research institutes | SA | Support start-ups, technology transfer |
| Ireland | One large park, one small park | Technology park, incubator | Regional development agency | National Technological Park Company | Attraction of international investors |
| Italy | 11 parks | Science park, technopolis | Ministry of research, large companies, local authorities, associations | Associations, mixed consortia | Technology transfer, support SMEs, creation of business clusters |
| The Netherlands | Seven small parks | Science park, business incubator | State, local authorities, universities | Consortia of founding institutions | Location of firms near universities, technology transfer, industry-university cooperation |
| Portugal | Three large parks | Technology parks | Local government, private companies | | Development of local resources, growth of SMEs |
| Spain | Nine large and ten small parks | Technology park, business nursery | Autonomous communities, regional government, universities | Associations, public companies | Attract high-tech companies, industry-university cooperation, development of innovative firms, upgrade regional industry |
| UK | 40 small and large parks | Science park, technology park, innovation centre | Universities, local authorities, public agencies, private organisations | Management committee of associated founders | Property development, technology transfer |

1. University-production co-operation, which creates a technology and innovation environment open to firms;
2. Infrastructure which transfers technology and business services to SMEs or larger firms;
3. A number of innovative firms that create a pole for innovation capable of diffusing technology and know-how to the wider productive system round the park.

Research and policy analyses have shown that 'science parks' came to be used as a term to refer to quite different projects, and very different experiences. Perhaps unsurprisingly, the outcomes from such projects also varied: some succeeded, some failed, with those in faster-growing areas such as the South East of England and the South of France doing better than those in old industrial areas, such as Wales, Northern England, the former coalfields, etc. (Komninos 2002). Helpfully, Komninos (2002, p. 53) also outlines the five main types of projects that characterised this period of development. These still remain as elements of most regional development programmes in European-based practices:

*Science parks*—these are probably the simplest way to plan and develop new technology districts and industrial spaces. Local authorities, development agencies, European state governments and the European Commission have supported them, with the aim of creating environments favourable to technology transfer and technological development. The main focus of science parks is to boost the creation of technology-based firms. This means a firm whose strength and competitive edge are derived from the knowledge of its field, such as natural science, engineering or medicine, and the subsequent transformation of this know-how into products and services for a market. The definition includes not only manufacturing firms but also firms in industry-related services. To such firms, science parks offer a friendly environment to product development, cooperation with R&D organisations, support from technology transfer agencies, brand name and quality premises.

*Research parks*—are usually located close to one or more universities or similar academic and research institutions. Its emphasis is on research rather than development and the key is academic/research liaison at the leading edge of science and technology. Normally, production plants are precluded.

*Technology parks*—a development to accommodate companies engaged in the commercial application of high technology, with activities including R&D, production, sales and servicing. It is distinguished from science and research parks because of a greater emphasis on production. Academic involvement is also essential. Technology parks meet the specialised location requirements of high-technology companies, but they offer a higher proportion of non-production to production space. The emphasis is on the proximity of high-technology companies engaged in similar operations. There may be restrictions on tenants, and a requirement that they exhibit some high-tech activity.

*Innovation centres*—a facility catering for the needs of predominantly new businesses engaged in the development and marketing of new technological products and services. The purpose of an innovation centre is to promote the setting up of high-tech businesses with high market risk. The services provided include technical services and advice on finance, marketing and technology.

*Business incubators*—a place where newly created firms are located in a rather limited space. Its aim is to increase the chance of growth and rate of survival of these firms by providing them with modular building facilities, common technical facilities, and also managerial support and back-up services. The main emphasis of incubators is job creation and local development, but the technology orientation is often marginal.

At the beginning of the 1990s a new wave of policy schemes and approaches were to emerge within the practice of regional technological development in Europe. These new approaches were to have their roots in a number of factors, including the results of evaluations that revealed the limitations of solely relying on the science/technology park project-based approach in achieving job creation, business growth and industrial restructuring. Concerns were also being raised about the relationships between national innovation systems, EU scientific competitiveness and regional performance following the ascension of nation states such as Spain and Portugal in 1986 and East Germany in 1990. Similar fundamental debates are again currently resurfacing as the EU extends its territory and membership to include the 'new' eastern European states such as Poland, Hungary, Czech Republic, Estonia, Slovakia, etc.

This new strategic approach to regional development policy initially emerged through a series of experimental and selective schemes: STRIDE, SPRINT, RITTS, RTP and RIS. These were characterised by a quasi-activist approach, with emphasis on networking and institutional external economies to support technological innovation (Komninos 2002; Simmie 1997; Landabaso 1999). The European Commission (EC) first launched these approaches with STRIDE, a financial support initiative that sought to strengthen the European innovation and technology support services, such as science parks, innovation services, and networks of technology and innovation specialists. However, it also sought to go further than these existing projects by also seeking to (i) facilitate the diffusion of new technologies to firms by offering support to specific projects for technology transfer, support for innovation financing by smaller firms and inter-firm cooperation; and (ii) to improve the awareness and understanding of innovation, by supporting exchanges of knowledge and experiences between member states. Regional Innovation and Technology Transfer Strategies and Infrastructures (RITTS), Regional Technology Plans (RTP) and Regional Innovation Strategies (RIS) provided co-finance and guidance to regional governments to undertake an assessment of their regional innovation potential, and define strategies that promoted the cooperation and capabilities of the small firm sector, the research and technological community, and the public authorities. Overall, more than 60 initiatives were launched in the 1990s (Komninos 2002), and a European wide network ensures the continuity of these early schemes.

This shift in EU regional development policy is generally built upon (EC 1994):

1. A bottom-up approach, giving emphasis on the regional technology demand.
2. A regional approach, focusing on the development of a territorial entity on the basis of a consensus between the government, the private sector, the universities and the research centres.
3. A strategic approach, combining the analysis of the regional technological development and the definition of long-term priorities and short-term actions.
4. An integrated approach, linking the efforts of the public and private sectors towards the common goal of increasing regional productivity and competitiveness.
5. An international approach, considering the global market trends and enhancing international technology and economic cooperation.

In this approach the region is the key level of concrete action, and the elaboration of any strategy involves real actors and is meant to take into account the strengths and weaknesses of specific productive systems and the real capabilities of the research and academic community in the region (Komninos 2002). In this way, the development outcomes should be secured not only by projects such as science and technology parks, but also by the institutions in the region which can offer some guarantee of continuity and implementation. It is also acknowledged in this approach that there is no simple panacea for less developed regions, but rather the active searching to achieve these outcomes develops the institutional and cultural-based innovativeness of a region that in turn is necessary to trigger adaptation and flexibility required by the knowledge economy.

The relationship between regions and between regional and nation state policies on innovation and technology also underpinned the geographically wider study of technopoles by Castells and Hall (1994). A concern with wider systemic processes rather than individual projects is illustrated by their comment that: "What we are witnessing is the emergence of a new industrial space, defined by the location of the new industrial sectors and by the use of new technologies by all sectors…paradoxically, cities and regions are increasingly becoming critical agents of economic development, offer flexibility in adapting to changing conditions of markets, technology and culture." Against this context, Castells and Hall (1994) claim that: "technopoles in fact explicitly commemorate the reality that cities and regions are being profoundly modified in their structure, and conditioned in their growth dynamics by the interplay of three major, interrelated historical processes":

1. a technological revolution, mainly based on information technologies;
2. the formation of a global economy, that is, the structuring of all economic processes on a planetary scale, even if national boundaries and national governments remain essential elements and key actors in the strategies played out in international competition;
3. the emergence of a new form of economic production and management that can be termed informational. Characterised by the fact that productivity and competitiveness are increasingly based on the generation of new knowledge and on the access to, and processing of, appropriate information.

Whilst there are some examples of technopole development in Europe (Castells and Hall 1994), the key site for the development and implementation of this construct has been Japan and East Asia. In this process of generating new growth, it is argued that cities, regions and nation states compete with each other, but more often than not, such competition becomes a source of innovation, of efficiency, of a collective effort to create a better place to live and a more effective place to do business. This became expressed "within various attempts to plan and promote technologically innovative, industrial-related production within one concentrated area" (Castells and Hall 1994). An empirical typology based on international experience can explain the initiatives that had developed a focus on planned techno-industrial development:

1. Technology parks—attempts to induce new industrial growth by attracting high-technology manufacturing firms to a privileged space. Functions defined in economic development terms, deliberately established business area, resulting from government and/or university efforts.
2. Science cities—strictly scientific research complexes, with no direct territorial linkage to manufacturing. They are intended to reach a higher level of scientific excellence through the synergy they are supposed to generate in their secluded scientific milieu.
3. Technopolis programmes—localisation of national programmes, regional development and industrial decentralisation.

Within this typology science cities are seen as "new settlements, generally planned and built by governments, and aimed at generating scientific excellence and synergistic research activities, by concentrating a critical mass of research organisations and scientists within a high-quality urban space. What characterises science cities, in contrast with other types of technopoles, is their focus on science and research, independent of their impact on their immediate productive environment. They are generally conceived as supports to national scientific development, considered a positive aim in its own right, in the hope that better scientific research will progressively percolate through the entire economy and the whole social fabric. They are also often presented as tools of regional development, intended to assist the decentralisation of scientific research, with all the prestige that involves, to the national periphery or, failing that, the metropolitan periphery" (Castells and Hall 1994).

The concepts of science cities and the technopolis aimed to promote a new approach to regional development. The creation of secluded and privileged spaces, detached and independent from mundane concerns and their internal closure of space were supposed to spur the cohesion of intellectual networks that would support the emergence, consolidation and reproduction of a scientific milieu, with its own set of values and mechanisms to promote the collective advancement of scientific inquiry. They were to offer symbolic and material proof of the nation state's commitment to science and technology, and the spatial concentration was to give real as well as symbolic presence to national scientific resources. This concept was perceived to aim at promoting a new regional culture that differed from conventional regional development that centred on land utilisation and infrastructural improvements. Policies and practices in Japan in particular combined the elements of garden cities and Silicon Valley to create a vision featuring research universities, science centres, industrial research parks, joint R&D consortia, venture capital foundations, office complexes, international convention centres and residential new towns (Castells and Hall 1994).

Park (1997) argues that whilst these programmes may be regarded as successful against certain measures of national development, results varied heavily at local levels. The over-expectation on the technopolis plans and a drastic change in the macro-economic situation in the second half of the 1980s were two factors that consequently led to a re-assessment of the approach, which in order to reach the original goals of the programmes saw a further build-up of organisational infrastructure, adequate educational facilities and creative environments to be established in the areas (see Table 2).

The extension of support for technological development outside the confines of specific buildings (innovation centres), exclusive parklands (science parks) into

**Table 2** Overview of technopolis strategies (Source: Park 1997)

| First technopolis period | Second technopolis period |
|---|---|
| Conditions | Conditions |
| Industrial structural switch from heavy and large scaled fordist production to light and small-scaled flexible manufacturing systems | Increase of enterprises locating overseas along with the appreciation of the yen; progress of industrial structural adjustment |
| An approach run towards a technology nation | Progress of globalism |
| A period of financial reconstruction | A wave of technological innovation and international competition on technological development power |
| Demands for the revitalisation of local economies | Reinforced foundation of technological development power (maturing of first technopolis period) |
| Enlarging disparities of the technological capabilities between regions | New information technologies and spread of informationalisation into local economies |
| Decline of local industries, concentration of high-tech industries in large city areas, enlarging regional disparities | The advance of soft and service economies<br>Diversification and change of value and consciousness among the Japanese<br>Recovery of national financial power |
| Basic strategy of first period | Basic strategy of second period |
| Reinforce the foundation of technological development capabilities | Lay stress on endogenous modes of development |
| Transfer to high-tech local industry | Give prominence to individualism |
| Dig out local needs and make the most out of the local seeds | Formation of technopolis networks |
| Technological innovation from the grassroots | Accommodate to internationalisation (open policy, technology transfer, networks) |
| Respect independency of localities, attach to the soft-technology base | Implement a city making |

wider notions of the science city, the technopolis and the regional innovation system is being stretched even further with very recent initiatives that seek to directly confront the notion of an informational society (Castells 1989; Van Den Berg and Van Winden 2002). In these approaches, such as digital cities, 'Smart' cities, intelligent and creative cities, Komninos (2002) believes: "the typical city at the end of the twentieth century where the functions of industry, services, trade, housing, recreation, education and so on dominate is gradually conceding its position to a new urban system where the basic components (universities, exchange malls, industrial districts, science parks, recreation zones, etc.) operate on two levels, in the three-dimensional natural world and the virtual space of the Internet. The integration of these levels creates a reality much more complex than before but with increased capability for supporting individuals and organisations hosted by them. Intelligent cities see the merging of these two worlds." In practice one movement for transforming cities in relation to new knowledge and technology conditions in this way is the Smart Communities project in California, which focuses on promoting the philosophy and application of the information society at the level of cities and regions. A smart community is simply a community in which government, business and residents understand the potential of information technology and make a conscious decision to use that technology to transform life and work in their region in significant and positive ways. The rationale for the movement stands on the human interaction and social cooperation that are needed for making a smart community "Market forces may generate new technologies, but they do not give rise to smart communities; only people do—people with a vision, with a commitment to change, with a willingness to work together with others in their community to achieve a common purpose" (SMART Cities Guide Book 2002).

At the end of this review is it possible to draw out some overall and strategic lessons and reflections from these apparently disparate practices, especially with respect to the underlying concern of this chapter with uneven spatial development? Do general ideas and concepts underpin the models, and are there alternative strategic frameworks or approaches that emerge when ideas are mapped onto specific regions and cities in practice? It might be argued that the first attempts to establish science parks, technology parks and innovation centres with support of state policies were a phenomena giving expression to some form of localism. The projects were largely local initiatives, each in different ways seeking to capitalise on assets in their areas that were taking on new roles (such as universities) or offering the potential to create value in new ways (such as technology transfer mechanisms). In a period of severe economic restructuring these assets were often used to seek competitive advantage over other local areas or regions. Investments in land and property were used to create spaces to capture and retain tacit, codified and institutionalised knowledge. But as the model was adopted elsewhere, it also became perceived as a tool within policies to stimulate growth (new business start ups) and attract new assets (such as high-tech-based FDI) to the localities. Such a set of approaches were likely to help those areas with a strong asset base over the worst off and more traditional industrial areas. Therefore, when these localised initiatives were adopted within state policies at a larger scale, the effectiveness of the approach to tackle uneven spatial development was not fully or adequately considered.

The use of science cities, the technopolis programme and regional innovation strategies can therefore be seen as a significantly different approach within state policies towards uneven spatial development. They represented a shift in intervention logic, expressed in changes in desired governance structures, in conceptions of the relationship between the local and global, in the extension of interventions to labour, capital and institutional assets, and in the reproduction of scientific and technological knowledge. Although it would be rather simplistic to attempt to classify a series of complex and highly varied policy responses, two broad strategic approaches can be drawn out of these practices. First, the various initiatives might be construed as a means of regionalising or localising national or pan-national policy goals. In such logic, the support offered to science and technology developments, and more broadly innovation, was seeking to secure competitive advantage over other nation states (Japan) or trading blocs (European Union). Sub-national uneven spatial development may have been made explicit within such measures, but even when it was, it might be regarded as a secondary policy objective in the absence of any redistributive measures (such as national/EU scientific research programmes) or attempts to set up some form of control or regulation over high growth areas and mobile high-tech capital.

Second, the logic of these new forms of intervention might be seen as based on a new regionalism, where the promotion of competitive advantage might be criticised for creating a 'winner takes all' or a 'zero sum game', and instead should be replaced by initiatives and policies that seek to create regional comparative advantage. If these interventions are based on supporting endogenous growth, they might again have limited value in tackling uneven spatial development, but instead have potential to respect the diversity of regional economies and the possible plurality of values and outcomes that can arise from a similar diversity in technological developments. However, where structural factors persist, such as institutional weakness and path dependency, the policies may at best simply reproduce current patterns of uneven spatial development, and at worst, reinforce and further exaggerate economic inequalities.

## 2 New Challenges in Regional Development for Science Cities

Goldstein and Luger (1992) see the science city as a prominent instrument to establish an innovative milieu or core area of regional development. Oh and Masser (1995) broaden this notion by seeing science cities—incorporating the ideas of the technopolis, research park, science or technology park—as a way of linking high-tech industries with regional development through economic innovation. "In other words, a technopolis is a means to develop good places to live by attracting high-tech industries, universities, research labs, pleasant dwellings, and public service into a city which has a nice natural environment and convenient living conditions" (Oh 2005). Within this perspective, science cities are in fact given specific expression, an expression that requires careful and critical examination in light of the regional and national conditions within which the idea is implemented or seeks policy influence.

Oh (2005) proposes that science city development has a number of expressions or possible outcomes:

1. They provide a major location or space for business-science ventures;
2. They attract foreign investment;
3. They promote the transfer of new and emerging technologies;
4. They provide a major location and space for the development and commercialisation of knowledge-based products and services;
5. They provide the ability to adapt existing technologies to local conditions and needs;
6. They increase added value to exports;
7. They improve foreign exchange earning;
8. They provide the ability to purchase new technologies;
9. They improve environmental conditions;
10. They improve economic performance;
11. They usher the host country into the global economy; and
12. They reduce poverty.

The possibility therefore arises to utilise such a potentially powerful regional development instrument to enhance the capacities of developing countries in their management of science, technology and innovation. However, the foregoing review has strongly hinted at the extent to which the achievement of public policy goals through the implementation of such technology-based initiatives is highly varied. A variety in effectiveness and outcome which is deeply conditioned by a number of social, physical, cultural and governance factors as well as economic factors. Outcomes are not inevitable, nor do they follow supposedly inevitable pathways of successful development. Local conditions in specific spaces and places do condition important elements such as risk taking, entrepreneurship, the structure of capital, the motivation of labour, the form and nature of creativity. Perhaps most importantly, creating the abilities and assets to generate and capture added value that can overcome prevailing patterns of uneven spatial development is particularly problematic. The particular concern in this chapter is the extent to which different strategic policy frameworks can offer the possibility of addressing these issues in order to deliver balanced and sustainable development, and what characteristics might such a framework take or need to consider. In order to explore these practices further, this chapter now examines some of the critical assessments made of the technopolis concept, and then advances new frameworks of policy that can shape the key elements of successful regional technological development: growth models; spatial planning; socialisation of learning; and governance.

Critiques of the technopolis concept have taken a number of forms (see Park 1997; Sternberg 1997), but below are outlined some of the key arguments that provide insights into the challenges which the approach might face if it is to be the basis of future regional development policies.

Limit 1—the technopolis development remains as a 'satellite development', and fails to become fully integrated within the wider economy and spatial development processes. Further development increasingly becomes dependent on attracting new investment through the provision of land and skilled labour that is a cheaper offer than that in the major existing cities. Outcome could create a cost competitive model, highly dependent on wider economic conditions.

Limit 2—the technopolis is only able to significantly attract mobile FDI, creating a branch plant syndrome with little, if any, endogenous-based growth and embedding of capital, skills or technological knowledge. Productive and service tasks are largely routine, involving subcontracting and overspill functions decentralising on cost or capacity grounds from main core economies.

Limit 3—the technopolis fails to fully develop appropriate university–business links. There is a lack of key creative and entrepreneurial professionals in the local universities or businesses, or path-dependent social relationships continue as barriers to fully developed interactions. Failure or limits to such interaction can be reinforced by prevailing national state policies towards university-based education, research and entrepreneurial activities, and the lack of challenges to established professional and social relationships which discourage new relationships and entrepreneurship.

Limit 4—the technopolis is unable to develop sufficient soft infrastructure to support research and technological development, in particular no forms of venture capital and international knowledge exchange. Further growth becomes more dependent on increasing the hard infrastructure of the area—roads, airports, buildings—thus reinforcing satellite and dependent development trajectory.

Limit 5—failure or limits to the relocation of high-level, internationally competitive R&D activities, furthering the spatial division of high order scientific and technical labour.

Limit 6—the technopolis is characterised by a lack of inter-industry linkages and voluntary collaboration, and a lack of spin-offs from universities, laboratories and research centres. Business and entrepreneurial cultures remain embedded in previous cultures of trust, reward and dependency.

Limit 7—technopolis development is limited and frustrated by unproductive rivalry and competition between local governments and development agencies. This can lead to ineffectiveness and inefficiency, but can also increase the fiscal burden.

Limit 8—technopolis success is severely limited by the failure for national and regional policy development objectives to be appropriately integrated. Contradictions occur between sector goals and spatial outcomes.

There is therefore, within the science city/technopolis model, a potential threat that practices will too easily develop an emphasis on the property and physical dimensions of urbanisation and regional development. Emphasis can shift towards marketing and image building by the city management in order to attract tenants to the properties, and away from nurturing networks and new cultures. Expectations and reward systems largely shaped by historical patterns or spatially benign systems can reinforce low institutional links, in particular those that require the involvement of higher education and research institutions. In this way, despite considerable investment and extensive programmes of action, there is a threat that the area will not break away from embedded path dependency, and simply reconfigure localities, by attempting to make spatial assets more attractive to the next round of capital investment, rather than achieving transformative change, new trajectories and a wider modernisation of the region.

Furthermore, as the Limits 7 and 8 explore above, the nature of governance and critically the role of national state structures, processes, norms and goals can significantly determine the final outcome of science city/technopolis development. Komninos (2002) reflects that there "is no doubt that the state policy for innovation is a key factor. It would seem easier to create the norms and conventions of trust and association on a national scale, taking advantage of the cultural and political consistency of the nation, and use them for action on the global and local scales". And continues to argue that it is the region that "is the level of concrete action", the elaboration of strategy must involve real actors and take into account the strengths and weaknesses of specific productive systems and the real capabilities of the research and academic community. Furthermore, "Two issues that should be stressed are the differences in the organisation of innovation systems in different countries, and the difficulty of defining models. The institutional framework for innovation appropriate for a nation is shaped by complex political, economic and cultural factors forged over a long period of time. On the other hand, in many countries a major restructuring in the state research, technology, education, vocational training and finance institutions is necessary. This involves both organisational and cultural changes, and the building of institutions capable to substitute rules with conventions of trust and cooperation" (Komninos 2002).

'Science city' based urban development is thus a highly normative construct. It has on many occasions been advanced as a way of replicating Silicon Valley, an expression of a particular form of growth and production of wealth and power, which itself has not been reproduced throughout the USA. The assumptions and goals in this model are inherently linked to certain patterns and outcomes of economic growth. First, it advocates a high growth outcome, both in scale and rate. Very often this is also seen to support any national state objectives to improve economic performance, but as outlined above, is also often rationalised through the desire to overcome uneven spatial development, especially in peripheral areas, or localities peripheral to main urban cores (Tokyo, Seoul, Paris, London, Madrid). Second, it advances the notion of high value added growth with a view that this can increase prosperity and counter-balance alternative outcomes such as low-waged economies or cost-based competitiveness. Finally, the model places great emphasis on spatial proximity, agglomeration effects, the creation and promotion of milieu and clustering of activities—"New technological knowledge is usually in such a tacit form that accessibility is bounded by geographic proximity and/or by the nature and extent of interactions among actors in the innovation system" (Acs 2002).

Such normative constructions would appear to be able to address the challenge of the new global knowledge economy, where the new geography of economies is supposedly "not the result of natural endowments of land, labour and capital, as economists have long thought. Rather it is powered by innovation and entrepreneurship; and this in turn is the product of real people acting in real places. In other words, the factors that really matter are the ones we create ourselves. We do this by recombining the knowledge and other resources in new and novel ways. What is more, some places are better than others at doing this. That is because they are able to attract, mobilize and connect the factors that really matter—innovative people and creative entrepreneurs" Richard Florida (Florida and (2001) Foreword in Acs 2002). The attractiveness of the normative set of values and associations inherent in this model are exhibited in the way that 'science city' has also become used as a way of branding existing cities, either those already knowledge economy rich (e.g. Science City York, in the north of England) or those undergoing restructuring and that use the notion as a way to organise and promote urban assets (e.g. Newcastle Science City, in the north east of England). In both these cases the knowledge base largely resides within higher education institutions. As Castells and Hall (1997) remind us, the first type of technopole consisted of industrial complexes of high-technology firms that were built on the basis of innovative milieu. They assert that these complexes, linking R&D and manufacturing, were the true command centres of the new industrial space, and arose without deliberate planning, though governments and universities did play a crucial role in their development.

Despite such imperative claims for supporting and encouraging certain normative constructs of economic growth, UNESCO highlights, "there can be few more pressing and critical goals for the future of humankind than to ensure steady improvement in the quality of life for this and future generations, in a way that respects our common heritage—the planet we live on. As people we seek positive change for ourselves, our children and grandchildren; we must do it in ways that respect the right of all to do so. To do this we must learn constantly—about ourselves, our potential, our limitations, our relationships, our society, our environment, our world. Achieving sustainable development is a life-wide and lifelong endeavour which challenges individuals, institutions and societies to view tomorrow as a day that belongs to all of us, or it will not belong to anyone" (UNESCO Decade for Education for Sustainable Development 2005–2014).

In the United Kingdom (HMG 2005) these broad goals have been translated into eight core areas of activity in an attempt to ensure the future planning of 'sustainable communities':

1. Governance—well run, with effective and inclusive participation, representation and leadership.
2. Connectivity—well connected, with good transport services and communication linking people to jobs, schools, health and other services.

3. Services—well served, with public, private, community and voluntary services which are appropriate to people's needs and accessible to all.
4. Environmental—environmentally sensitive, providing places for people to live that are considerate of the environment.
5. Equity—fair for everyone, including those in other communities, now and in the future.
6. Economy—thriving, with a flourishing, diverse and innovative local economy.
7. Housing and the built environment—well designed and built, featuring quality built and natural environment.
8. Social and cultural—active, inclusive and safe, fair tolerant and cohesive with a strong local culture and other shared community activities.

It might be envisaged that the science city concept might lend itself to the achievement of these broad goals in two main ways: by being sustainable and by promoting sustainable development. Future planning of new developments might be seen to contribute to environmentally sensitive places by, for example: actively seeking to minimise climate change, including through energy efficiency and the use of renewables; protecting the environment, by minimising pollution on land, in water and in the air; minimising waste and dispose of it in accordance with current good practice; making efficient use of natural resources, encouraging sustainable production and consumption; protecting and improving bio-diversity (e.g. wildlife habitats); enabling a lifestyle that minimises negative environmental impact and enhances positive impacts; and creating cleaner, safer and greener spaces. The promotion of sustainable development might take the form of eco-science parks, or initiatives such as Wise@CAT, that recognises the importance of linking science-based research, technological development, learning and basic education in changing the norms and values that underpin currently unsustainable behaviours. The later project creates a 'micro-campus' for the transmission and development of environmental education and research.

A central proposition in this chapter is that the notion of science cities or technopolis is a normative policy construction, based on spatial organising ideas—that is, it prescribes how the process of urbanisation should occur, in order to achieve certain high growth technological/economic outcomes. The ability for these ideas to be delivered in reality, and for implementation to match these broad goals, is conditioned by a number of factors. Thus, science cities are socially constructed as much as they are the product of bricks and mortar, financial and agency instruments. It is argued that the prevailing strategic planning frameworks are one of the factors that can, in part, determine these outcomes. Critical to these are the role seen for the state (e.g. highly interventionist, enabling, regulatory, etc.). Above it is also argued that science cities face two major challenges as a spatial organising idea in the future planning of urban development: responding to the still emergent global knowledge economy which is creating further uneven spatial development; and to the demands to achieve sustainable development, which will certainly increase over time and territories. The future internationalisation of the science city idea needs to be considered within the strategic spatial frameworks of polycentric urban development; integrated spatial planning; and multi-level governance.

*Polycentric urban development*—Serious concerns are emerging in different contexts that balanced urban development will not result from current trends in urbanisation. The existing urban order sees mega-cities and global cities attracting further investment, wealth and power, and they themselves are becoming highly spatially segregated between rich and poor populations (Parr 2004). This raises serious questions about the role of smaller and less well located urban areas, a situation seemingly furthered by inter-urban competition with its associated fiscal and political costs and 'zero-sum game' outcomes from the promotion of endogenous growth theory. The normative application of the polycentric urban development idea sees urban areas collaborating in order to combine and maximise from their complementary and distinctive functions and roles (Davoudi 2003). Clearly, competition still occurs in some areas, but within a spatial development framework and an inter-urban governance arrangement that allows strategic collaborative decision making where appropriate. All urban areas are inter-connected through the flows of modern capital and the informational society, and spatial organising ideas must reflect this, and civic leaders recognise this. Science cities would need to look to their external relationships carefully, to realise where benefits could accrue to the constituent elements of their wider urban systems from their specific and specialised form of development.

*Multi-level governance*—This is seen as necessary (Gualini 2004) since checks and counter-balances are needed within governance systems to gain benefit from top-down approaches to science city planning (technopolis programme) and bottom-up approaches to the development of embedded regional innovation development (EU RIS). Higher levels of government are necessary to formulate appropriate policies towards uneven spatial development (re-distribution of resources), scientific excellence and international competitiveness (R&D programmes, frameworks for higher education) whilst regional and local levels need devolved powers and control to ensure the localisation of benefits, and embedding of cultural and social shifts associated with new modes of innovation and entrepreneurialism.

- Multi-level governance offers pluralistic processes, which envisages governance as an open and chaotic pattern of events—no single actor has sufficient potential to dominate.
- Bargaining is not simply between two or even three levels, but encompasses multiple levels, many of which lie outside the control of state executives, such as development agencies, business representative bodies, etc.

*Integrated spatial planning*—In its most essential form, spatial planning can be regarded as a set of procedures, processes and practices concerned with what is built, where and when. The notion of a 'built environment' is given a very broad meaning in this interpretation of spatial planning, incorporating many forms of 'development', at various scales (e.g. a housing estate and a city), and expressed through many investments (e.g. a building, a public open space, infrastructure). It is an activity that inherently involves 'government', but also many other actors, stakeholders, interests and populations.

There is widespread acceptance of the need for integrated approaches to spatial planning to deal with these complex problems and the challenges arising from changes in our cities and regions (Echenique and Saint 2001; Jenks and Dempsey 2005). Despite extensive agreement on the need for integrated approaches, there remain concerns about how to design and implement such approaches in particular localities, and to deliver a variety of spatial planning outcomes. In practice, integration may be achieved through:

*Linking functions*—The successful planning of city areas is more likely to be sustainable if planning is able to pursue actions in all relevant domains of urban activity and make explicit links between these in any specific project. Thus, housing actions should be linked to employment measures, which need to be integrated with transport interventions that in turn need to seek social and environmental outcomes. The increasing complexity in urban life means such integration is not simple or straightforward.

*Linking organisations*—Urban planning is facing more complex governance relations, expressed in terms of multi-level governance, partnerships, devolution, capacity building, regionalism, inter-professional teams and inter-disciplinary working. Successful and sustainable development requires spatial planning practice to be able to make links between different levels and functions of government, between different types of government agencies, between government and other interests such as community groups, business interests and other non-governmental organisations.

*Linking spaces*—An integrated approach to planning also needs to ensure that any impacts are considered and realised within a variety of spatial scales—connecting the local with the global. Major investment projects are often inward looking, seeking to provide solutions for an area and within that area. This could extend beyond the city/adjoining urban area relationship to incorporate cross-border and possibly trans-national relationships.

*Linking citizens*—Integrated spatial planning by its nature has its roots within democratic and participative practices, but it also has to face up to the challenges being created by fundamental changes in modern democracies. These shifts in civil society underpin ideas that integrated spatial planning should change from a state-initiated and top-down engagement with citizens to a more two-way, interactive and deliberative set of practices that take account of the many groups, individuals and organisations that have an interest in, or may be affected by the outcomes of spatial policy.

*Linking through WTA*—Evidence suggests that learning in a trans-national context is problematic because of significant national differences, which make it difficult for participants to understand both the context and content of policy, and to readily apply experience gained from other countries. However, learning does take place in networks and through trans-national collaboration where partners work together to solve common problems in a goal-oriented fashion, and where concrete and practical outcomes are more likely.

## 3 Concluding Remarks

This chapter has attempted to construct a critical perspective on the practice of science cities in light of future scenarios generated by globalisation and sustainable development. These scenarios will frame future successful urban economic development and desires to increase prosperity and the quality of life in all cities and regions. Experiences and practices have been situated in markedly different socio-economic localities, and over time have sought to manage the transition to a post-industrial economic order, and currently, onwards to a knowledge-based regional economy.

The chapter argues that a new strategic approach is required in order for science city projects to contribute to sustainable development in the future. A form of development that will create difficult trade-offs between economic, environmental and equity goals, based on new forms of indigenous development and territorial governance, and a new approach to the social construction and reproduction of innovation and learning. It is suggested, mainly from experience and reflections from a Western European perspective, that such a new strategic approach can find expression in the practices of polycentric urban development, multi-level governance and integrated spatial planning.

## References

Acs, Z. (2002). *Innovation and the growth of cities*. Cheltenham: Edward Elgar.

Borja, J., & Castells, M. (1997). *Local and global: Management of cities in the information age*. London: Earthscan Publications.

Castells, M. (1989). *The informational city*. Oxford: Blackwell.

Castells, M., & Hall, P. (1994). *Technopoles of the world: The making of 21st century industrial complexes*. London: Routledge.

Cooke, P., & Morgan, K. (1997). *The associational economy: Firms, regions and innovation*. Oxford: Oxford University Press.

Davoudi, S. (2003). Polycentricity in European spatial planning: From an analytical tool to a normative agenda. *European Planning Studies, 11*(8), 979–1000.

EC. (1994). *Regional technology plans guide book*. Brussels: European Commission DGXVI/DGXII.

Echenique, M., & Saint, A. (Eds.). (2001). *Cities for the new millennium*. London: Spon Press.

Florida, R. (2001). Foreward in Acs, Z. (2002). *Innovation and the growth of cities*. Cheltenham: Edward Elgar.

Graham, S., & Marvin, S. (1996). *Telecommunications and the city*. London: Routledge.

Goldstein, H., & Luger, M. (1992). Theory and practice in high-tech economic development. In R. Bingham, et al. (Eds.), *Theories of local economic development*. CA: Sage.

Gualini, E. (2004). *Multi-level governance and institutional change*. Aldershot: Ashgate.

Harvey, D. (1989). *The condition of postmodernity*. London: Basil Blackwell.

Harvey, D. (1998). *The urban experience*. Oxford: Basil Blackwell.

Hassink, R. (1993). Regional innovation policies compared. *Urban Studies, 30*(6), 326–335.

HMG. (2005). *Securing the future: Government's sustainable development strategy*. London: The Stationary Office.

Jenks, M., & Dempsey, N. (Eds.). (2005). *Future forms and design for sustainable cities*. London: Architectural Press.
Komninos, N. (1997). After technopoles: Diffused strategies for innovation and technology transfer. In J. Simmie (Ed.), *Innovation, networks and learning regions* (pp. 181–195). London: Jessica Kingsley.
Komninos, N. (2002). *Intelligent cities: Innovation, knowledge systems and digital spaces*. London: Spon Press.
Landabaso, M. (1999). EU policy on innovation and regional development. In F. Boekema, K. Morgan, S. Bakkers & R. Rutten (Eds.), *Knowledge, innovation and economic growth: The theory and practice of learning regions*. London: Edward Elgar.
Masser, I. (1990). Technology and regional development policy: A review of Japan's technopolis programme. *Regional Studies, 24*(1), 41–53.
Massey, D., & Wield, D. (1992). Science parks: A concept in science, society and space. *Environmernt and Planning D: Society and Space, 10*, 411–422.
Massey, D., Quintas, P., & wierld, D. (1992). *High tech fantasies: Science parks in society, science and space*. London: Routledge, Chapman and Hall.
Monck, C., Quintas, P., Porter, R., Storey, D., & Wynarczyk, P. (1998). *Science parks and the growth of high technology firms*. London: Croom Helm.
Morgan, K. (1997). The learning region: Institutions, innovation and regional renewal. *Regional Studies, 31*, 491–503.
Oh, D. S., & Masser, I. (1995). High-tech centres and regional innovation: Some case studies in the UK, Gerrmany, Japan and Korea. In C. Betugila et al. (Eds.), *Technological change, economic development and space*. Berlin: Springer.
Oh, D. S. (2005). *Development strategies of WTA in global context*. UNESCO-WTA Cooperative Workshop, 1–7 November 2005.
Park, S.-C. (1997). The Japanese technopolis strategy. In J. Simmie (Ed.), *Innovation, networks and learning regions* (pp. 137–158). London: Jessica Kingsley.
Parr, J. (2004). The polycentric urban region: A closer inspection. *Regional Studies, 38*(3), 231–240.
Porter, M. (2003). The economic performance of regions. *Regional Studies, 37*(6/7), 549–578.
Sassen, S. (1997). Electronic space and power. *Journal of Urban Technology, 4*(1), 1–18.
Simmie, J. (Ed.). (1997). *Innovation, networks and learning regions*. London: Jessica Kingsley.
Sternberg, R. (1997). New industrial spaces and national technology policies: The case of Kyushu and the Japanese 'technopolis strategy'. In J. Simmie (Ed.), *Innovation, networks and learning regions* (pp. 159–188). London: Jessica Kingsley.
Van Den Berg, L., & Van Winden, W. (2002). *Information and communications technology as potential catalyst for sustainable urban development*. Aldershot: Ashgate.

# Regional Innovation and Cooperation among Industries, Universities, R&D Institutes, and Governments

**Eberhard Becker, Bettina Burger and Thorsten Hülsmann**

**Abstract** Globalization and regionalization seem like contradictory concepts. However, they present the framework for economic activity, innovation, and cooperation in our time. Under these circumstances, regions are, to an increasing extent, in an entrepreneurial position. They must compete among each other for: enterprises, investments, highly skilled personnel, and jobs. The extent to which a region is competitive depends on several factors, the most important of which are knowledge, social capital, networks, and supporting structures. These factors may be self-organized or government-based. In reality, there is no single recipe known that guarantees success in all cases. However, in view of the relevance of knowledge and learning, the cluster approach can offer an explanation as to why some regions manage to develop their knowledge-based industries more successfully than others. When we talk about globalization we have a new economic setting in mind, one in which profit-oriented multinational strategies of large business corporations lead to an international division of labor. The resulting post-Fordist concept of flexible production is comprised of a formation of independent units within large corporations. These units are integrated into regional networks, instead of being hierarchically sub-ordered into the centralized corporation. To local communities, this situation poses a challenge but also provides new opportunities concerning the task of directing their development to the advantage of the region. M. E. Porter (1990, 1998, 2000) is one of the prominent writers to draw attention to the relevance of regions in a globalized economy. The first part of this chapter gives

E. Becker (✉)
TU Dortmund University, Fakultät für Mathematik, D 44221 Dortmund, Germany
e-mail: eberhard.becker@tu-dortmund.de

B. Burger
TUMentrepreneurship, Technische Universität München, Arcisstr.
21 D 80333 München, Germany
e-mail: bettina.burger@tum.de

T. Hülsmann
EffizienzClusterLogistik Ruhr, Kölner Str. 80-82, D 45481 Mühlheim an der Ruhr, Germany
e-mail: Thorsten.huelsmann@effizienzcluster.de

D.-S. Oh and F. Phillips (eds.), *Technopolis*,
DOI: 10.1007/978-1-4471-5508-9_2, 

greater insight into the concept of regional cluster development, its characteristics, and its meaning for regional policy. Very often, Science or Technology Parks are seen as a suitable means to form clusters of modern high technological industries. The second part deals with the example of the City of Dortmund, Germany, since the recent history of this city is a perfect example for successful regional development and the key ingredients involved. Special attention is given to the role of research and higher education within the structural policy that helped Dortmund to overcome a severe crisis. The technology park in Dortmund demonstrates how active cluster development can help improve the economic development of a region, thus improving its competitiveness and the life of its population.

## 1 The Concept of Regional Competitiveness

Globalization and regionalization do not contradict each other, but are mutually dependent. Global economic activity increases while national trading borders decrease. To the same extent, local and regional networks and conditions for production gain in importance. One reason scholars decided to look at the regional level was the fact that in the 1970s some cities and regions were able to deal with the economic crises in a much more successful way than others, even though the national surrounding conditions were the same. E.g., special attention was given to the development of the high-technology cluster "Silicon Valley" in California, USA.

The observation that in recent decades certain regional economies managed to develop into success stories while others decreased is a fact. It becomes clear that in times of globalizing markets and regional specialization of production systems, there is not only international competition between enterprises, but also between regions. The question is whether favorable conditions for successful regional development can be created and if so, which factors are of importance.

Cities and regions are increasingly aware of this growing competition. As a consequence, they try to increase their regional competitiveness. As a first thought in this context, the individual economic power of the enterprises located in a region comes to mind. It seems obvious that the economic performance of the regional enterprises is related to the prosperity of the region. However, the concept of regional competitiveness includes more than just the added competitiveness of the individual enterprises in one region. Moreover, it refers to synergetic effects that are created through interaction and cooperation among those enterprises, as well as certain framework conditions specific to the territory under consideration. Success in regional competitiveness can be measured by a few parameters:

- Economic growth
- Prosperity
- Quality of Life.

Regional competitiveness and the competitiveness of enterprises are therefore mutually dependent. Successful enterprises usually lead to the well-being of the region in which they are located. However, the success of an enterprise is closely related to the quality of regional conditions for production. A very individual, regionally special mix of location factors is the prerequisite for the success of enterprises. In the following section we will give more insight into the factors that promote regional competitiveness. However, it can already be stated that economically successful regions are usually more likely to create and offer those conditions.

## 2 Factors for Regional Competitiveness

Competitiveness is the driving force for development. When regions find themselves in competition, the different factions within the region are more likely to cooperate in order to achieve mutual benefits. The common aim is to improve the regions comparative advantages with respect to other, competing regions. The diverse, loosely knit networks and interactions seem to play an important role for the development of dynamic economic regions. Interactions are usually based on knowledge exchange and spill-over between enterprises and their surroundings. The following section deals with factors that influence the level of regional competitiveness. They are: The availability of knowledge and the ability to learn, the presence of social capital and networks, and the form of organizational structures.

### *2.1 Knowledge and Learning*

Knowledge and learning have a fundamental meaning for regional prosperity and growth:

> "Knowledge is the fundamental resource in our contemporary economy and learning is the most important process" (Johnson & Lundvall 1994, p. 24).

Knowledge, on the one hand, is nowadays globally available. Through modern communication technology it can spread globally within seconds: it is extremely mobile. One consequence is the global activity of not only networks of enterprises, but also of informal networks. On the other hand, knowledge tends to accumulate in certain areas and places. It clusters wherever conditions for generating and managing knowledge are best. Favorable framework conditions include the presence of R&D-adequate infrastructure and the availability of specialized personnel.

Moreover, social factors like networks play an important role in the development of knowledge. Proximity, social interaction, and face-to-face-contacts foster the evolution of what we call implicit or tacit knowledge. As opposed to explicit or codified knowledge, tacit knowledge is not available in books, rules, or formulas. However, it is crucial for innovations to evolve as a factor for competitiveness.

As a consequence, the availability and access to knowledge, i.e., the presence of institutions of education, research and development, are an important advantage for regions.

## 2.2 Social Capital/Networks

Social capital can be defined as a set of norms and values which govern interactions between people, the institutions where they are incorporated, relationship networks set up among various social actors and the overall cohesion of society (Camagni 2003). Social capital evolves through the interplay of individual and organizational learning. Thus, it is reproduced at the local and regional level because here, people live and interact. The global perspective rules the markets, but social capital as a comparative advantage is closely related to the regional level. Social capital is a resource that comes to life through the relations between different actors.

A network describes a type of cooperation that stands in between the (vertical) formal relations, i.e., supplier contracts, and the (horizontal) informal platform of the market. Networks are characterized by the cooperation of competitors. A network is strongly based on personal, informal relationships between actors from politics, business, institutes of education and research, and cultural and social institutions. In some ways, networks substitute institutionalized structures of decision-making through cooperation on the personal level.

Therefore, networks offer the conditions for social capital to evolve. Social capital can only be used in a reasonable way within a limited size of entity. Within that framework, social networks develop that endow a common identity. Certainty is increased, and therefore, cost for economic transactions between the members of a network decrease. The cooperation of a group according to informal principles causes higher prosperity and growth among the individual members of the group. However, the members produce positive external effects, thus increasing growth and prosperity for the entire region.

## 2.3 Self-Organized Structures

What we have in mind when we talk about self-organized structures in the context of regional development is a form of organizational system that is, to a certain extent, self-dependent and flexible. Self-organized structures have the ability to organize their activities and processes of development as well as the process of learning. Their goals are self-determined, and they pursue targeted activities to achieve them. The ability to act self-organized is one of the important factors that enable regions to be competitive.

Evolving self-organized structures is a reaction to the fact that traditional forms of regional governments failed, to an increasing extent, in the presence of today's global challenges to regional development. Self-organizing structures have the ability to involve a more diverse range of regional actors in the process of regional development. Institutions of higher education and research institutes have great potentials to participate in these activities. On the one hand, they offer human resources that create social capital for the region. On the other hand, they are important hubs for regional networks, especially to connect regional enterprises with the scientific community. With the help of politics and administration, these positive impacts can be further strengthened.

## *2.4 Cooperation of Actors*

From the preceding sections it should be clear that networks of relevant actors account for the competitiveness of regions. Therefore, it is necessary to take a closer look at the actors who have to interact Fig. 1.

Regarding economic success, companies are the main drivers. The ability to produce products people asks for and which are competitive in a national or international environment is vital to the well-being of a region. To this end they rely on a qualified work force that has to be trained in institutions of education, universities, and research institutes. A broad range of tasks has to be performed: these institutions altogether should be able to cope equally with a huge diversity of talents and at the same time produce highly qualified research-oriented experts. In particular, modern high technology companies require graduates from universities and benefit from close contact with research institutes.

Observations and studies show that regions with high-level universities and research institutes enjoy an advantage, at least as far as modern technological products are concerned. Both sides, institutions of education and research on the one hand and companies on the other hand, are asked for new forms of interaction traditionally not established in most cases.

**Fig. 1** Actors for regional development

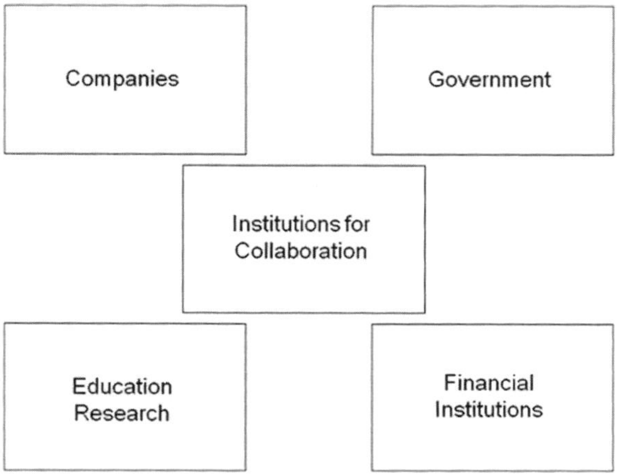

Institutions and procedures for cooperation between various partners have to be developed. This applies not only to companies and their partners from the educational and research sector, this task concerns much more "agents" of regional development: the financial institutions, the various chambers of commerce and the local as well as regional government. They all contribute their share to enable the region, the companies, and their employees to achieve a high level of competitiveness.

The degree to which governments at all levels are interfering in the economic processes may differ from country to country. But whatever impact is maintained, the role of governments is very important. They set the frame for and encourage promising activities, they have to be concerned with attractive living conditions and they are instrumental in guaranteeing the sound and stable prerequisites that economic activities need for success.

In conclusion, regional competitiveness does not only refer to the environment a territory offers to the individual enterprises located within this certain spatial entity. Rather, processes of collective learning and socialized growth of knowledge play a crucial role for regional competitiveness. Important factors are the conditions and processes of knowledge accumulation and the development of interpretative codes, as well as models of cooperation and decision-making on which the innovative progress of local companies is based. Competitiveness is created through:

- Specific advantages strategically created by single firms.
- Territorial synergies and cooperation capability enhanced by an imaginative and pro-active public administration.
- Externalities provided by local and national governments.
- The specificities historically built by a territorial culture (Camagni 2003).

At this point it is appropriate to conclude by quoting one of the most prominent researchers in this area, Michael E. Porter, who keeps on communicating that

> "(…) the enduring competitive advantages in a global economy lie increasingly in local things—knowledge, relationships, motivation—that distant rivals cannot match" (Porter 1998, p. 78).

## 3 Theories of Regional Development: The Cluster Concept

Several approaches exist to explain the development of economic activity in regions (Fuerst 2001). Whereas the concept of "Innovative Milieus" (Frommhold-Eisebith 1999) lays its focus on the meaning of networks based on common cultures, the concept of "Learning Regions" (Johnson & Lundvall 1994) concentrates on the meaning of knowledge and learning, along side the ideas of networking and implementation. An even more holistic approach to regional development strategies offers the concept of cluster development. In the following, we will present main features of the cluster concept in order to describe in some detail a certain type of regional economic development that is based on the factors of regional competitiveness discussed above with special emphasis on innovation and cooperation.

## *3.1 Introduction to the Cluster Concept*

A cluster is composed of regional agglomerations of enterprises belonging to a certain sector of economic activity. They are closely linked through supplier connections. These connections offer a high potential for rationalization by the specialization of single enterprises in a certain element of production. However, those interactions are not limited to the functional organization of the production process, but explicitly include communication in the fields of research and development, marketing, and even common strategic development measures. Regional actors and institutions work through the informal exchange of knowledge and experiences on a common problem. Apart from businesses, institutions of education, research and development institutes, think tanks and technology centers play an important role, as well as NGOs and cultural institutions. Their spatial proximity and their regional context of social interaction is the basic prerequisite for successful clusters.

Michael Porter, through extensive research, came to the conclusion that economic success does not occur in single industrial branches, but in clusters (Porter 1990, 1998, 2000). These clusters are made up of actors from related branches and subsectors. They are linked through vertical and horizontal, formal and informal connections. Once a cluster has evolved, intensive exchange within the cluster starts. This interaction leads through the diffusion of new technologies, increased availability of trained specialists, rapid dissemination of innovations through the channels of subcontractors and clients, to the discovery of new possibilities for competitiveness and new market opportunities, and more. In an economically successful region, established enterprises compete with each other. Thus, a reciprocal strengthening system is created. The dynamics of the cluster on the one hand sustains itself, but also provides prosperity and growth for the region.

## *3.2 Characteristics of Clusters*

A cluster is not simply an accumulation of different actors of a certain economic sector. Special conditions and frameworks help a successful cluster to evolve. They are all related to the regional dimension:

- An intensive local rivalry for gains and image; this leads to a constant improvement and diversification in products.
- The emergence of new enterprises on the market, for example spin-offs, enhances the dynamic competition.

- An extensive culture of cooperation exists within the different actors of the cluster. It is fostered by the chambers of commerce, the departments for economic promotion, and institutions for collaboration.
- Clusters are based on networks of individuals, brought to life through informal interaction.
- Connections to related branches are available.
- Highly skilled personnel are available.
- New technologies are available.
- Markets and demand are close by (Ketels et al. 2003, p. 18).

Therefore, the entire value creation chain for a certain sector is present in a cluster. Networks are strongly related to the economic sector of the cluster. Knowledge as a factor is relevant in its regional production, availability, and exploitation. In a well-functioning cluster, the value creation chain is made up of regional demand for and regional supply of knowledge.

Thus, the research community includes the most important actors in a cluster, apart from the enterprises. Due to the crucial role of social capital for successful regional development, institutions of research and education which are part of the value creation chain of the respective cluster are the hubs for cluster networks. They hold a central position for knowledge spill over, for individual interaction and thus for cooperation and innovation. Interactions among cluster participants are fostered through institutions for collaboration. They create specialized platforms that allow knowledge to flow more easily and enable the cluster to organize collective activities. Thus, they operate as a mediator for cluster activities.

## *3.3 Development of Clusters*

It has been argued repeatedly that lower costs and new technologies for communication will cause a "de-clustering" of the economy or even the creation of virtual clusters that supersede the need for physical proximity. However, empirical findings suggest that so far this development has not occurred. In fact, many of the cluster effects arise only because of proximity, for example the spill overs from unplanned meetings (Ketels 2003, p. 12). Physical proximity will most likely remain the crucial factor for cluster development.

Not only economic factors, but also the so-called "soft" location factors are relevant. Most soft location factors are regionally bounded and therefore belong to the innate assets of a region. Soft location factors include for example:

- Social climate
- Modes of conduct of public administration
- Image of the location
- Presence of R&D institutions
- Institutions for education and qualification

- Cultural institutions
- Infrastructure
- Natural environment
- Plus many more.

They are predominantly immobile locational factors. Consequently, public policy has the aim to improve the immobile factors in order to attract the mobile factors: Investments, jobs, and highly skilled workers. The role of public policy concentrates on promoting the kind of resources that are relevant for innovation: Human capital, innovative capital, the competence to interact, and coordinate and entrepreneurial performance. Successful clusters are located in places where people can acquire and share tacit knowledge.

Clusters can take a long time to develop. For many clusters, historical reasons can be found for their existence today. For example, natural factors like resources or the location at a major trading route or river can be the root for evolution of a cluster. Also, the existence of an initial institution, such as a company or a university can be the root for cluster development. They can over time act as an anchor for the cluster by spinning-off new businesses and attracting investment from companies outside the region. The evolution of a cluster can take years or decades. However, the literature indicates that some clusters have developed much faster "because of the determined action of regional leaders who had spotted the potential of their region for the cluster" (Ketels 2003, p. 6).

## *3.4 Clusters and Economic Performance*

Clusters create economic benefits. These benefits have three dimensions:

- Companies can operate with a higher level of efficiency, drawing on more specialized assets and suppliers with shorter reaction times than they could in isolation.
- Companies and research institutions can achieve higher levels of innovation. Knowledge spill overs and close interaction with customers and other companies create more new ideas and provide intensive pressure to innovate while the cluster environment lowers the costs of experimenting.
- The level of business formation tends to be higher in clusters. Start-ups find external suppliers and partners on whom to rely. Clusters also reduce the cost of failure, as entrepreneurs can fall back on local employment opportunities at many other companies in the same field (Ketels 2003, p. 7).

Empirical findings suggest that clusters improve the economic performance of individual participants of the clusters and as a consequence the entire region benefits. There is a positive and significant relationship between the existence of strong clusters and higher overall wages. Strong employment in a cluster category in which a region is strongly specialized leads to higher overall wages in the region. Moreover, if regions increase their concentration in employment across clusters over time, their wage growth is even higher. A strong economic activity in some fields seems to be more effective for regional prosperity than having a presence in all fields (Ketels 2003, p. 13). Apparently, the economic success of a cluster also has positive effects on the economic sections of the region that do not participate in the cluster. Overall prosperity increases.

## *3.5 Cluster Initiatives*

Empirical findings already support a complex concept of cluster development. However, the research on clusters is not only interested in analyzing a phenomenon. In addition, a new approach for economic policy might be developed from the cluster concept. The underlying rationale is that there are certain externalities that sustain the development of clusters. These externalities do not necessarily occur automatically, but can be triggered or strengthened through purposeful action. Also, policy can influence the time it takes for a region to develop a cluster.

The conceptual knowledge about clusters and their strategic implications for regional development policies has already been applied in so-called cluster initiatives. Although empirical findings are still limited, they already provide some interesting insight (Ketels 2003, p. 17): The introductory step to forming a cluster initiative comes in almost equal shares from government and business, or a combination of both. Companies are usually engaged in the set-up and governance of the initiative, whereas governments tend to be important in terms of financing and securing some level of organizational support.

Cluster initiatives are involved in a broad range of activities. Activity areas are:

- Research and networking
- Policy lobbying
- Commercial cooperation
- Education and training
- Innovation and technology
- Investment attraction (Ketels 2003, p. 17).

Empirical data suggest that there are several drivers of success for cluster initiatives. They seem to be more successful if the cluster on which they focus is already strong and located within a good business environment. They are also more likely to have success if they are part of a broader strategy to improve the business environment in the region. Another important factor is the basis of a shared conceptual framework of competitiveness. The lack of a broad consensus turns out to be the factor most likely to cause failure of the cluster initiative. Last but not least, a small operational budget to finance an office with a dedicated cluster facilitator seems to be necessary to sustain the initiative over time. The second part of this paper will demonstrate a very successful example of cluster development, namely the case of Dortmund.

## 3.6 Regional Structural Policy as Cluster Policy

Insights in the concept of clusters as a principle of regional development are of great consequence for public policy. Today, public policy must bear the challenge to facilitate economic change and the adoption of innovations, while at the same time enhancing local assets which is also a key factor in growth (Konvitz 2000, p. 657).

Interregional competition is increasing, and a neglect of cooperation within regions causes soaring costs. Thus, different actors on the regional level have to come and work together, to pool their resources and cooperate. This necessity also calls for a new focus in public policy, the focus on governance. A regionally coherent system and non-governmental structures necessary to promote clusters are not a mere product of the market. Instead, they are based on political activity: Mutual learning, negotiations, and compromises between crucial actors of the region. Especially, important is that the regional economy and the scientific community create favorable conditions for successful clusters. This is where policy, administration, education, research, and business come together. The common aim is to pursue a policy that ensures the availability of resources and shapes external restrictions that comply with the common goals.

Regional government in Germany used to be centralized and hierarchical. Municipal authorities had the objective to work for predetermined targets, to make use of concessions for infrastructural development and to oversee distribution of social welfare. Structural policy at the municipal level was largely ignored. With the change from the Fordist-era to Post-Fordism, cities and regions have begun to assume their roles as entrepreneurs. Their task is to strengthen and develop their economic bases. For local governments, a new management system has developed. Instead of holding a central function, local governments today participate in networks that manage structural policy through a process of consensus.

Ketels sums up the objectives for regional development:

> "Regions need to activate their clusters, address cross-cutting weaknesses in their general business beyond the life cycle of specific administrations, and define an overall understanding of the unique value they intend to provide relative to other locations" (Ketels 2003, p. 19).

Against the background of the concept of clusters as a strategy for regional development, regions have to evaluate in which sectors they have the potential to develop clusters. Regional as well as national strengths have to be taken into account. There has to be a regional consensus on which sectors should be promoted as clusters. With a common goal in mind, public and, if possible, private means will be directed on sector-related development, for example for physical infrastructure, the establishment of new institutions for education and research, the financing of new firms, marketing, and the setting up of networks.

> "The integration of local educational institutions with regional development objectives is crucial; without it, the region will not be able to supply the specialized workforce on which any cluster is dependent" (Gerszewski/Krieger 2002, p. 107).

At the end of this section where we were conveying a positive view of cluster development it is only fair to state that critics of cluster policy may be found in the literature: E.g., the cluster concept might not be generally transferable to any industry and any region (Alsleben 2005; Alecke et al. 2006; Audretsch 2003). In (Huber 2012) certain seemingly well-grounded statements about synergies in clusters are questioned. Therefore, further research on the cluster concept is needed.

However, the case of the City of Dortmund, presented in the sequel, provides a convincing example for regional cooperation which resulted in resolving a severe economic crisis of the 1980s. In Dortmund, all relevant actors including institutes of research and education, enterprises, local and regional government, and financial institutions worked together (and continue to do so) in a vibrant network in order to promote regional innovation and economic development. In this case the cluster concept proved to be a promising approach to promote economic development of knowledge-based industries, thus improving the regional competitiveness and overall economic situation of the region concerned.

# 4 The City of Dortmund

## *4.1 Location of Dortmund*

The City of Dortmund is located in the federal state of North Rhine-Westphalia (NRW) in the western part of Germany. This state is Germany's largest with respect to population (17 million inhabitants). The metropolitan area of the Ruhr, a region of around 4,500 $km^2$, 53 communes, and a population of 5.2 million is situated in this state. Located at the eastern edge of this area, Dortmund is one of these communes–with about 600,000 inhabitants the seventh-largest city in Germany and a regional metropolis Fig. 2.

## *4.2 Economic History*

For a period of more than 100 years the economy of Dortmund was based on coal mining and the production of steel. In addition, Dortmund was famous for its beer industry. One may say that Dortmund and its region essentially contributed to Germany's industrial productivity and wealth for decades. However, starting in the late 1970s of the last century these industrial sectors began to decline, as could be seen in many others places world-wide. This decline affected the fortunes of Dortmund severely. In 1997 the two biggest steel producing companies in Germany "Thyssen" and "Krupp Hoesch" decided to merge and closed steel production in the City of Dortmund. They then transferred production to the city of Duisburg, located at the other edge of the Ruhr area, about 100 km away.

**Fig. 2** Geographic location of Dortmund in Germany. *Source* Institute of Spatial Planning, University of Dortmund

Steel production in Dortmund came to a definite end due to the withdrawal of ThyssenKrupp.

To date, the successful period of the coal, steel, and beer industries is history. In a short time Dortmund had suffered the loss of 80,000 jobs. The community and the economic sector faced a nearly hopeless task of creating a huge number of jobs in a short period of time.

Two main areas of counter actions will be described in the following. They both serve as informative samples for the principle of regional development displayed in the earlier sections of this chapter:

- The development of the Technology Center Dortmund and the Technology Park Dortmund (started in 1985, "TechnologieZentrumDortmund" and "TechnologieParkDortmund" in German).
- The development and implementation of the city development model Dortmund-Project (started in 2000) with its cluster initiatives.

## *4.3 The Technology Center Dortmund and Technology Park Dortmund*

In the early 1980s when the inescapable process of declining traditional industries was at its beginning the city urgently needed a new basis for economic development. It was then that various stakeholders in the city of Dortmund acted in an amazingly fast manner. They reached a consensus regarding analysis of the situation and concepts suitable for future recovery of the presently decaying traditional industry. This consensus has been referred to as the "Dortmund Consensus" ever since and formed the basis, shared by all, for a remarkable transition from old to new economic structures. This consensus can be considered as an informal but influential new form of government.

Cooperation between the local players is evident in the cases of the Technology Center Dortmund and the adjacent Technology Park. While searching for innovative ways to promote the restructuring of the region, a Dortmund initiative led to the establishment of a technology center. The founding fathers, so to say, were

- Dortmund's institutions of higher education
- The City of Dortmund
- The local Chambers of Commerce
- Dortmund's financial institutions.

The opportunity was grasped quickly. In 1985, 16 months after the preliminary negotiations, the first buildings were completed. Here Dortmund was among the pioneers in Germany, second only to Aix-la-Chapelle and Berlin.

The idea itself and the prompt implementation of the project were attributable to the advantages that the university and the city undoubtedly possessed. The University of Dortmund had deliberately developed strength in the fields of engineering, computer science, and the natural sciences, and therefore the potential for innovation was high. In addition, the University is situated on a large green field site that offered enough space for an industrial periphery.

The 'Dortmund Consensus' had become well known throughout the region; it signified a reliable partnership between the public sector (politics, administration, and education) and the private sector (industry and enterprise). Therefore, the project was the result of successful networking:

- The Rector of the University promoted the idea and served as a mediating influence among academia, politics, and industry; the infrastructure and fields of excellence of the university defined the focus of a local technology center.

- The City of Dortmund purchased the land and, financially supported by the state of North Rhine-Westphalia and the European Union, had the buildings constructed.
- The City, Chamber of Commerce, Chamber of Handicrafts, and the more influential financial institutions set up a managing company to run the center–or, more precisely, to lease the office space, to sublet it to start-up and existing companies, and to offer various services.

Still today, this successful networking is reflected in the legal structure of the technology center which is a private company with the following shareholders:

- The two universities in Dortmund
- The Chambers of Commerce and Handicrafts, respectively
- The City of Dortmund
- Various financial institutions.

The outline and structure of Dortmund's Technology Center became a model for the German innovation centers. It has concentrated both on technological development and spin-off companies; in other words, it has supported the development and application of technological products as well as helping and encouraging young entrepreneurs wishing to set up a business.

Since that time the Technology Center has developed into one of the most successful and largest centers in Germany. In 10 building complexes with floor space of 120,000 $m^2$ it accommodates about 190 companies with more then 1,500 employees. The Dortmund technology center has maintained its orientation toward the main fields of research pursued in Dortmund, more details will be presented below in Sect. 4.5.

### 4.3.1 Technology Park Dortmund

The Technology Park was founded in the same year as the technology center, 1985. It is located on an area of 40 ha next to the joint campus of the University of Dortmund and the technology center Dortmund. The park offers space for construction of their own premises to those interested in pursuing early stage projects and enterprises which have hatched out of the incubator at Technology Center. It is likewise an excellent location for enterprises seeking contact with scientific institutions but have no need for the kind of support provided in the technology center.

The Technology Center has grown fast and today accommodates more than 250 companies with over 8,000 employees.

A more detailed analysis of the concept of Dortmund's Technology Center and Park as well as the mode of its operations will appear in (Becker and Herrmann 2013) Fig. 3.

**Fig. 3** Dortmund campus with Technology Center and Park. *Source* University of Dortmund

## *4.4 The Dortmund-Project*

To explain the establishment of the initiative Dortmund-Project (Dortmund-Project 2000) one has to go back to the year 1997 and the decision of the two biggest steel producing companies in Germany "Thyssen" and "Krupp Hoesch" to merge and to close steel production in the City of Dortmund. Within a Memorandum of understanding both companies agreed upon certain activities to substitute and replace the lost jobs, including the establishment of an automotive supply factory in Dortmund. After some counseling, the ThyssenKrupp AG engaged the consulting firm McKinsey & Company to development a 10 year strategy for the City of Dortmund to halve the unemployment rate.

In June 2000, as a first step of implementation, the Dortmund City Council voted to establish the Dortmund-Project under the direct supervision of the Lord Mayor and reserved budgets of approx. €5 Mio. p.a. up to the year 2010. From the very beginning the Dortmund-Project was supported by ThyssenKrupp AG and the Economic and Employment Promotion Dortmund. The overall strategy was described in six goals:

- Set up new anchor industries in Dortmund.
- Strengthen companies resident in Dortmund.
- Expand training programs, skills upgrading plans, and R&D of an international standard.
- Turn the City of Dortmund into a modern business city with a high quality of life and unrivaled leisure amenities.

- Expedite planning and approvals procedures: one-stop shopping for start-up and/or relocating companies.
- Substantially boost the level of employment.

## *4.5 The Current Situation*

Without any doubt, the Technology Center, the Technology Park, and the Dortmund-Project constitute key elements of Dortmund's successful restructuring and recovery in economic terms after the collapse of traditional industries in the last 30 years. However, it has to be stressed again and again that it worked out so well due to various actors from different sectors of the society who worked together in a coherent manner:

- The City
- Financial institutions
- Chambers of Commerce
- Universities and Research Institutes.

No single actor can achieve what has been achieved so far. However, a few additional remarks are in order.

Technology center and park are focusing on areas and companies of modern high technology. This requires close contact to research entities which are available in Dortmund such as:

- TU Dortmund University, the University of Applied Science.
- Max-Planck Institute for Molecular Physiology.
- Two Fraunhofer Institutes for Logistics and Information Technologies, respectively.
- Two Leibniz Institutes for Analytical Science and, Working Environments, and Human Factors, respectively.

Various financing schemes have been established and are still being used. Venture capital and seed funds were made available by local institutions. In addition, a combination of investments from the city of Dortmund, the regional state of North Rhine-Westphalia and European Structure Funds is being applied to secure the financial basis.

Technology Center and the Dortmund-Project, in cooperation with the above scientific partners have successfully initiated several so-called "Competence Centres" which are medium-sized high-tech clusters.

The BioMedizinZentrumDortmund, in English—the Bio Medical Center Dortmund, is a recent Biotech-Cluster that is located on the campus of TU Dortmund University which also offers a well-known academic program in the field of biotechnology. About 30 biotech companies enjoy proximity to research institutes of world-wide reputation, to name the Max-Planck Institute for

Molecular Physiology as a typical example. This center was built in three steps between 2002 and 2009 with an investment of nearly €60 Mio and provides space of 15 000 $m^2$ in total with 8,000 $m^2$ designed as laboratories.

In 2005 and 2007 the MST.factory dortmund was opened, offering 10,000 $m^2$ total and absorbed investments of approximately €50 Mio. This center is supplemented by a Center for production technologies, constructed in 2008 with half of the previous investment. It is interesting to note that both centers are located on so-called "brown fields," previously the site of the last Dortmund steel mill. Building the new economy on the sites of an old industry is a contribution to urban development. Moreover, it vividly displays the rise of a new economy in the city and region.

It is important to note the Technology Center, is successful despite the fact that it is run as a private company. Tenants in the Technology Center and other Centers enjoy the provisions of high-tech infrastructures and the general assistance from the centers, even though they have to pay high rents. The fact that this concept is working successfully is a proof of the quality of the services offered, and of the perceived advantages it brings to those companies that make use of them.

## *4.6 Economics Effects*

> "Governmental investments into the future of a region–pushing economy in the right direction or wasting tax money into a black hole"

Questions of this type are raised frequently, and they have to be raised in view of the huge spending pouring out of state sources (Becker and Herrmann 2013). In the case of the Dortmund situation a recent study (Gundeland and Luttmann 2008) has identified and measured the economic effects that can be related to the Technology Center and Park Dortmund (TPDO for short). At this moment, a balance of governmental investments versus fiscal effects is unknown to the authors. Nevertheless, taking into account also the non-fiscal effects we cannot but state that an overall positive outcome is beyond serious doubt, at least in the case of Dortmund.

In detail, the study shows:

- The gross national annual income linked to TPDO is €900 Mio where
- About €600 Mio stays inside the region
- About 16,000 jobs nation-wide are related to TPDO where
- About 11,500 of them are located regionally
- Tax income on the side of region amount to 20 Mio. p.a.
- Innovation in the region has greatly improved, 35 % of the companies in TPDO are filing patents, more than 20 % go international
- Academically trained personnel in TPDO amount to 70 %
- TPDO has positive impact on the entire region in regard to innovation, networks, ability to attract highly qualified personnel.

These figures and facts may justify high investments and show that the transition of old industrial structures, gone forever in Dortmund, into modern industrial and business structures can be described as successful to some extent.

However, there are old and new challenges to address as well, by measures still to be worked out. A few final words are due.

The rate of unemployment is still high. So far, the loss of so many jobs in the past has not been fully compensated. While the traditional industries were able to offer decent employment opportunities to a large number of low- to semi-qualified employees, technology-based enterprises predominantly employ a highly skilled work force. In a way, the shift of economic sectors leads to a substitution of work force, but not a substitution of employment opportunities. Also, high-tech companies are usually more specialized and smaller. They do not necessarily lead to mass employment. Other forms of business or industrial production also have to be developed in order to sustain a demographically diverse region such as Dortmund.

In addition, the huge threat of demographic decline in Germany, as well as in many western countries, poses questions not yet understood, let alone answered. Companies will be facing a fast growing difficulty to find the skilled work force they need to be competitive. At first glance, this seems to be a problem that enterprises have to resolve on their own. However, resuming the holistic view adopted in this paper, this problem will be of the greatest concern to all players who are involved in the economic process.

In the past 30 years in the City of Dortmund the enterprises, financial as well as educational and scientific institutions managed to develop strategies to cope successfully with the big economic crisis caused by the collapse of traditional industries. The authors of this paper are convinced that the previous arsenal of creative tools and actions will be suitably augmented to address the new challenges.

## References


Alecke, B., et al. (2006). Are there really high-tech clusters? The geographic concentration of German manufacturing industries and its determinants. *The Annals of Regional Science, 40*, 19–42.

Alsleben, C. (2005). The downside of knowledge spillovers: An explanation for the dispersion of high-tech industries. *Journal of Economics*, *84*(3), 217–248.

Audretsch, D.B. (2003, December): *Specialization or Diversification of Regional Economic Structure: Empirical Findings and Strategic Orientation*. Prepared for Presentation at the Conference on Clustermanagement in der Strukturpolitik–Internationale Erfahrungen und Konsequenzen fuer NRW, Duisburg.

Becker, E., Herrmann, S. (2013). Changing dark coal into illuminating high-tech, Ways out of an economic crisis–The case of Dortmund. *World Technopolis Review*, 1(4), 276-286

Camagni, R. (2003, December). *Regional Clusters, Regional Competencies and Regional Competition.* Paper delivered at the International Conference on "Cluster management in structural policy"–International experiences and consequences for Northrhine-Westphalia, Duisburg.

Dortmund-Project (2000), A new jobs package for Dortmund. Dortmund.
Fromhold-Eisebith, M. (1999): Das „kreative Milieu–Nur theoretisches Konzept oder Instrument der Regionalentwicklung? In: Raumforschung und Raumordnung, 57 (2/3).
Fuerst, D. (2001). Regional governance–ein neues Paradigma der Regionalwissenschaften? *Raumforschung und Raumordnung, 59*, 370–380.
Gerszewski, S., & Krieger, F. (2002). The city and University of Dortmund: From coexistence to partnership. *Industry & Higher Education*. London: IP Publishing Ltd.
Gundel, S., & Luttmann, J. (2008). *Die regionalwirtschaftlichen Effekte von TechnologieZentrum und TechnologiePark Dortmund*. Materialien zum Siedlungs- und Wohnungswesen 42, Westfälische Wilhelms-Universität Münster
Johnson, B. & Lundvall, B. (1994). The learning economy. *Journal of Industry Studies,* (1)2.
Huber, Franz. (2012). Do clusters really matter for innovation practices in Information Technology? Questioning the significance of technological spillovers. *Journal of Economic Geography, 12*(1), 107–126.
Ketels, C. (2004): Cluster sind kein Selbstzweck, interview given in G.I.B. info 4.04.
Ketels, C. (2003, December). *The Development of the cluster concept–present experiences and further developments*. Paper prepared for NRW conference on clusters, Duisburg, Germany.
Ketels, C., Lindquist, G., & Sölvell, Ö. (2003). *The Cluster Initiative Greenbook*. Stockholm: Ivory Tower AB.
Konvitz, J. W. (2000): Changing Economies: The Territorial Dimension, in: Informationen zur Raumentwicklung 11/12.2000, BBR Bonn.
Porter, M. E. (1990). *The Competitive advantage of nations*. New York: The Free Press.
Porter, M. E. (1998). Clusters and the new economics of competition. In: *Harvard Business Review*, *6*(76), 77–90.
Porter, M. E. (2000). Location, competition and economic development: Local clusters in a global economy. *Economic Development Quarterly, 14*, 15–34.

# Regional Innovation Support Systems and Technopoles

**Robert Hassink and Su-Hyun Berg**

**Abstract** In the 1990s, technopoles, a land and property-led technology policy concept which aims at spatially clustering high-tech firms and R&D organisations, have been very popular among local, regional and national policymakers to boost regional economic growth. No matter how they are called, be it science parks, technopoles, high-tech centres, incubator centres, technology parks, technoparks, science cities or innopolis, they have given hopes to policymakers in many countries to boost regional technology transfer, innovativeness and hence competitiveness.

## 1 Introduction

In the 1990s, technopoles, a land and property-led technology policy concept which aims at spatially clustering high-tech firms and R&D organisations, have been very popular among local, regional and national policymakers to boost regional economic growth. No matter how they are called, be it science parks, technopoles, high-tech centres, incubator centres, technology parks, technoparks, science cities or innopolis, they have given hopes to policymakers in many countries to boost regional technology transfer, innovativeness and hence competitiveness. Many detailed studies have been done both on technopoles in individual countries such as the USA (Luger and Goldstein 1991), Japan (Bass 1998) and Germany (Sternberg et al. 1996) and on technopoles in an international comparative perspective in order to find some lessons that could be learned from successes and failures (Castells and Hall 1994). Now at the beginning of the twenty-first century, the concept seems to have reached some point of saturation, particularly in industrialised countries such as the USA, Western Europe and Japan. Moreover, this mainly property-led policy

R. Hassink (✉) · S.-H. Berg
Department of Geography, University of Kiel, Kiel, Germany
e-mail: hassink@geographie.uni-kiel.de

D.-S. Oh and F. Phillips (eds.), *Technopolis*,
DOI: 10.1007/978-1-4471-5508-9_3, © Springer-Verlag London 2014

tool has been criticised for not sufficiently supporting regional networking and technology transfer to regional firms.

At the same time there have been arguments in favour of putting technopoles in a broader perspective of regional innovation support systems (OECD 2011). The aim of this Chapter is therefore to analyse the contribution of regional innovation support systems to the development of technopoles. Based on some empirical examples of regional innovation support systems in Western Europe, that is Great Britain, the Netherlands and Germany, the Chapter uses a typology of these systems, consisting of grassroots, networked and dirigiste kind of systems (Cooke 2004). In the next two sections we will further introduce technopoles (Sect. 2) and regional innovation support systems (Sect. 3). In Sect. 4 we will then describe the empirical examples, whereas conclusions and lessons will be drawn in Sect. 5.

## 2 Technopoles

According to Castells and Hall (1994), technopoles are cities, suburbs or even rural areas whose existence is dominated by the presence of high technology in the form of research, development, manufacturing or some combination of all three. They may be planned or unplanned, privately financed, publicly financed or set up by public–private partnerships. They come in several varieties and have many different names. The French word 'technopole' is used to describe them all generically. They are also called science cities and in this Chapter both terms will be used synonymously. Technopoles are located all over the world, but mostly in North America, Europe and East Asia. The most famous technopole, Silicon Valley, came into existence in the 1960s in what was then basically a rural area. Others, such as Route 128 in Greater Boston, are found in the suburbs of huge metropolitan areas. Successful technopoles are characterised by synergy between industry, academia and government. Although any region can have higher education institutes, public research establishments, high-tech firms or even all three at once, that alone does not guarantee the special kind of relationship that generates synergy.

The success of unplanned technopoles, such as Silicon Valley and Los Angeles, inspired other countries to deliberately create their own (Castells and Hall 1994). This Chapter will focus on these technopoles as planned policy concepts. In fact, many scholars regard just these planned concepts as technopoles. Lagendijk and Charles (1998:16), for instance, define technopoles as "a land and property-based form of technology policy, geared towards the establishment of a spatial agglomeration of hi-tech businesses and organisations". The first examples of these planned technopoles are, among others, Sophia-Antipolis in France, *Daedeok* Innopolis (formerly known as *Daedeok* Science Town) in South Korea, Tsukuba Science City in Japan and Akademgorodok in the former Soviet Union (Castells and Hall 1994). There are large variations between the characteristics and scopes of technopoles between different countries. In Europe, for instance, where technopoles were established in two waves (1969–1973 and 1983–1993) (Komninos

1997), large *attraction-led* technopoles aiming at attracting inward investment of high-tech companies and R&D departments of multinationals can be found in France and Spain. In countries such as the UK, Germany, the Netherlands and Belgium, on the other hand, "technopoles" are nothing more than small *incubator-led* parks aiming at boosting spin-offs from higher education institutes and public research establishments and other small high-tech firms.

Although there are large variations among science and technology parks in different countries (Anttiroiko 2004), broadly speaking, technopoles aim at achieving three goals. First, the most obvious goal is to foster economic development. High-tech and innovation-led growth is regarded as absolutely necessary for maintaining and increasing competitiveness of firms, regions and nations. Second, in some countries, particularly those with over-populated and congested urban areas, building an out-of-the-way technopole in the countryside is often seen as a way to reduce regional economic inequalities. This line of reasoning was behind the creation of Tsukuba in Japan, Daedeok in South Korea and technopoles in France, where economic planners hoped to draw research and development out of the over-burdened capital cities by relocating their national research facilities and universities to new sites in the country and by luring firms with incentives to follow them there. This kind of 'dirigiste' or mission-oriented regional policy is based on the growth pole concept and can only be found in countries with powerful central governments (Cooke and Morgan 1998). Third, technopoles aim at creating synergy between higher education institutes, public research establishments and firms in order to foster technology transfer, innovation and hence competitiveness. In the end, this should lead to creating an environment geared toward innovation. Capital and resources are naturally drawn to such a "milieu of innovation", rather than having to be relocated through central planning (Castells and Hall 1994). Synergy is crucial in the long-run. Even technopoles based on branch plants, controlled from distant locations, should have synergy, as source of innovation, as their long-term objective.

Although creating synergy and fostering technology transfer and networking both between firms and between firms and higher education institutes and public research establishments in the region is an important goal of technopoles, the failure to achieve this goal is at the same time one of their largest weaknesses. The main reasons for this lack of technology transfer and networking are the following. First, because technopoles are a property-based initiative, a great deal of their management is property-related and puts much emphasis on marketing and image instead of promoting networking on the spot. According to Lagendijk and Charles (1998:19) "The most important question… is whether other instruments not based on property or on attracting investments to particular places might be more effective to facilitate technology transfer". Second, since the technology-push philosophy (linear model of innovation) prevails at many technopoles, they often lack explicit technology transfer instruments. Third, in some larger technopoles in peripheral areas, externally controlled branch plants dominate, which have few links with local suppliers and lack the R&D base necessary for collaboration with local higher education institutes and public research establishments. According to

Castells and Hall (1994) it is not sufficient to simply provide the networks in a physical sense at technopoles, it is also necessary to take definite steps to open up the social networks and break down barriers to networking.

One of the main weaknesses, particularly with technopoles planned by central government, is that technopole plans are often over-ambitious and focus on several, often the same, technologies and industries. Planners often forget that different technopole policies are appropriate to different levels of regional development (Castells and Hall 1994). At lower development levels, relatively modest technology parks will be perfectly appropriate. Particularly in the latter case, it is necessary to concentrate on one or two target areas or niches that are best adapted to local needs and facilities such as regional higher education institutes, public research establishments, industrial traditions, entrepreneurial capacities and political leadership in the region.

Since many technopoles are specialised in the same fields, such as in microelectronics or in biotechnology, recently voices have become louder stressing the need for diversification (Hassink and Hu 2012). Interesting theoretical concepts in that respect are regional innovation platforms (Harmaakorpi et al. 2011) and regional branching (Boschma and Frenken 2011), which could be applied in order to promote diversification of techopoles into related and unrelated technology fields.

In most advanced industrialised countries, such as the USA, Germany and to some extent Japan, the technopole concept seems to have reached some form of saturation. In these countries, therefore, technopoles are certainly in the maturity or even declining phase of their life cycle. With regard to science parks in the USA, Luger and Goldstein (1991) detected some degree of saturation already in 1991, as they stated that any new research park will have more difficulties, especially if it is not linked to a higher education institute/public research establishment and/or located in peripheral areas. Similar conclusions were drawn by Sternberg et al. (1996) with regard to Germany. In a European context Komninos (1997:193) clearly sees technopoles at the end of their life cycle, particularly if they are put in the wider range of innovation support initiatives: "... the spread of new effective tools for technology transfer, based on networks, institutions and services, questions the established character of technopolitan development. The novel feature of these tools is that they operate without property or spatially polarised dimensions". Annerstedt (2006, 279), however, has a contrasting view, as he states that "over the past 10 years, with the promotion by policymakers of specialised, local 'clusters' of firms and supporting institutions as strategic means for industrial policy, science parks have come back into the limelight of the centre stage for industrial policy deliberations" (see also OECD 2011).

In this chapter we will argue that technopoles need to be well integrated in regional innovation support systems in order to tackle the problem of lacking technology transfer and networking. In the following, we will therefore analyse regional innovation support systems in more detail, both theoretically in the next section and empirically in Sect. 4. We will take an evolutionary stance in doing this and will link the development stage of national economies with the life cycle of technopoles and the related regional innovation support systems.

## 3 Regional Innovation Support Systems: A Theoretical Framework

Technopoles should clearly be understood as part of a larger array of regional innovation policy initiatives. Cooke (2001, 22): "... there is emerging recognition that science parks are a valuable element but not the only or main objective of a localised or regionalised innovation strategy". The position of technopoles in these wider regional innovation policies, however, differs from country to country, depending on the relative importance of technopoles compared with the other elements of these policies that is technological financial-aid schemes, the innovation support infrastructure and cluster support initiatives and a range of other measures (see OECD 2011). In Japan and South Korea, where we can find large-scale technopoles devised and partly financed by central government, technopoles seem to have a more prominent position in regional innovation policies than in Germany and many other countries in Europe. In Europe, technopoles are not frequently mentioned neither in support programmes of the European Union nor in theoretical development concepts (Moulaert and Sekia 2003). Technopoles, however, can and should be integrated in these policy programmes and development concepts. However, no matter what position technopoles have in wider regional policies, in all industrialised countries the dispersed networks of demand-oriented innovation support agencies (software) and the spatially constrained, property-led, supply oriented technopoles (hardware) seem to be quite separated from each other.

The popularity of the concept of regional innovation systems is closely related to the surge in regional innovation policies in many industrialised countries of the world. This is due to the fact that the importance of the regional level is increasing with regard to diffusion-oriented innovation support policies (Lagendijk 2011; Asheim et al. 2003, 2011; Fritsch and Stephan 2005). Central governments, however, keep their key role in supporting basic, pre-competitive technologies, which have spill-over effects that go far beyond the borders of regions (Storper 1995). Partly supported by national and supranational support programmes and encouraged by strong institutional set-ups found in successful regional economies such as Baden-Württemberg in Germany and Emilia-Romagna in Italy, many regions in industrialised countries have been setting up science parks, technopoles, technological financial-aid schemes, innovation support agencies, community colleges and initiatives to support clustering of industries since the second half of the 1980s. The central aim of these policies is to support regional endogenous potential by encouraging the diffusion of new technologies both from universities and public research establishments to small and medium-sized enterprises (SMEs), between SMEs and large enterprises (vertical co-operation) and between SMEs themselves (horizontal co-operation).

This increasing importance of regions for innovation policy can be considered as the outcome of a converging of regional and technology policy since the early 1980s (Fritsch and Stephan 2005). These two policy fields converged into regional innovation policies since their aim became partly the same, namely supporting the innovative capabilities and thus competitiveness of SMEs. It also fits into what Amin (1999) observed as a shift from a firm-centred, incentive-based, state-driven and standardised regional economic development policies to bottom-up, region-specific, longer term and plural-actor policies. These policy trends cannot only be seen in European countries, but also in North America and some countries in Asia (OECD 2011). Although we can therefore speak of a general phenomenon, there are of course large differences between individual regions and countries concerning the extent to which these trends take place. Generally, contributive factors to regional innovation policies are a federal political system, decentralisation, strong regional institutions and governance, a strong industrial specialisation in the region, socio-cultural homogeneity and thus relationships of trust, large economic restructuring problems and a strong commitment of regional political leaders.

One of the main strengths of the regional level for innovation support has been called the "garden argument" (Paquet 1998): if the economy is regarded as a garden with all kinds of trees and plants, for the gardener (government) there is no simple rule likely to apply to all plants. Growth is therefore best orchestrated from its sources at the level of cities and regions. At this level, rather than at the national level, policymakers can better tailor policy in relation to demand (Nauwelaers and Wintjes 2003; Lagendijk 2011). Regionalisation, therefore, allows for differentiation in policies, which is necessary because of differing regional economic conditions and institutional settings, thus different support needs of industries and firms. Regionalisation also raises the enthusiasm and motivation of regional policymakers, as they are now able to devise "their own" policies. Moreover, because of the large variety of institutional set-ups and initiatives in Europe and North America, these laboratories of experimentation offer both national and regional policymakers plenty of institutional learning opportunities (Hassink and Hülz 2010).

Closely related to this "garden argument" is the positive relationship between institutional embeddedness in regions, entrepreneurial learning processes and competitiveness (Lorenzen 2001). For their competitiveness firms depend on innovation processes. In order to come to such innovation processes firms have to exchange information and reproduce this information into knowledge, in other words they have to learn. Due to an increasing cut-throat competition and shorter product life cycles, firms, particularly SMEs, are increasingly dependent on information and knowledge sources that are only available outside the firm. Firm innovation processes therefore increasingly take place in interaction with other organisations, be it with other business partners, such as customers, suppliers or competitors or with public research establishments, higher education institutes, technology transfer agencies and regional development agencies. Innovation processes hardly ever take place in isolation any more. Innovations can thus be understood as manifest results of cumulative learning processes of firms. The spatial environment provides different institutional contexts for interactive learning. These contexts differ not only nationally, but also regionally and locally from each other. Firms are therefore institutionally embedded in different contexts for interactive learning. Spatial proximity stimulates communicative interaction between actors. However, it is not a sufficient condition. In order to achieve this interaction social proximity (equal or similar characteristics such as age, vocation, language and equal or similar views on values and norms) and organisational proximity (concern structure, intra- and inter-firm network structures) are necessary factors as well (Boschma 2005). The knowledge form determines to what extent proximity is necessary for learning by interacting. Typically, innovation-relevant information is not a publicly available, codified good, but private tacit knowledge—those parts of personal knowledge as well as personal skills that cannot be communicated in an impersonal way. Only through personal, communicative interaction between actors there are possibilities to exchange, understand and to apply this kind of information. In order to communicate, tacit, and to a lesser extent codified knowledge 'code keys' are needed, which are only understandable if (social) coherence and proximity are available. Thus, institutional embeddedness in regions positively affects the communication of tacit knowledge in particular and learning by interacting in general, which in turn is positive for competitiveness. Collective learning processes and a collective tacit knowledge are linked to the location because of the coinciding of social, cultural and spatial proximity (Boschma 2005). At the same time, however, Bathelt (2003, 772) stressed that one should not forget the role of the non-local for competitiveness: "In addition to mobilising internal resources, regional policies should also support agents in developing linkages and networks with external agents and markets … Caution should … be exercised in prioritising the local capabilities over non-local opportunities".

Since regional innovation policies have been emerging starting in the mid-1980s, several academics have started to develop theoretical and conceptual ideas on regional innovation strategies since the mid-1990s (Moulaert and Sekia 2003). These concepts, which form an important part of the so-called family of territorial innovation models, that is regional innovation systems (Cooke 2004; Mothe and Paquet 1998; Asheim et al. 2011), the learning region (Morgan 1997) and clusters (Porter 2000; OECD 2007) have been partly developed for policy reasons, namely as a response to organisational and strategic weaknesses of regions. Scholars also wanted to derive conceptual policy lessons from successful regional economies and to clarify why the regional level is an important level as a source for learning and innovation.

Of the developed concepts, the regional innovation systems concept is most widely dealt with in direct combination with regional innovation policy, both in a conceptual way (Cooke 2004; Mothe and Paquet 1998) and concerning empirical case-studies, including North American and Asian ones (Cooke et al. 2004; Mothe and Paquet 1998). The learning region, on the other hand, clearly lost its importance (Hassink and Klaerding 2012). Cooke et al. (1998:1581) define regional innovation systems as systems "in which firms and other organisations [such as research institutes, universities, innovation support agencies, chambers of commerce, banks, government departments] are systematically engaged in interactive

**Table 1** Typology of regional innovation support systems

| | Grassroots | Network | Dirigiste |
|---|---|---|---|
| Initiation | local | multi-level | central government |
| Funding | local agencies | diverse | national agencies |
| Research & support | applied/near-market | mixed | basic |
| Specialisation | low | mixed | high |
| Intra-regional co-operation | high | fair | low |
| Co-ordination | low | potentially high | potentially high, but often low |

*Source* adapted by the authors after Cooke 2004

learning through an institutional milieu characterised by embeddedness". The aim of regional innovation systems is to integrate traditional, context-linked, regional knowledge and codified, world-wide available knowledge in order to stimulate regional endogenous potentials.

A typology of regional innovation support systems helps to apply the concept to a broad range of regions and to clarify the 'scale' of involvement of public policy, which is from mainly national to mainly local. It also clarifies the relationship between national and regional innovation support systems. Such a typology consists of grassroots systems, network systems and dirigiste systems (Table 1) (for other typologies in relation to regional innovation policies see Cooke 2007; Nauwelaers and Wintjes 2003).

In grassroot systems, the initiation of technology transfer action is locally organised—for instance, at district or town level. Funding is supplied by local banking, local government, and local chamber of commerce. The competence of the research and support agencies of the region focuses on near-market and highly applied fields. However, the technical specialisation is likely to be low, thus it has the tendency to concentrate on general problem-solving. The level of supralocal co-ordination is low, because the nature of initiation is local-based.

In network systems, the initiation includes local, regional, federal and supranational levels. Moreover, the competence of research and support is multilevel, since funding is supplied by agreement between banks, government agencies and firms. In the research subsystem, a mixture of basic and high applied research is more likely. The co-ordination efforts between institutions of different spatial levels are potentially high, because there are many stakeholders involved. In this type of system, specialisation is mixed, which is due to a wide range of demand, going SMEs to multinationals (Cooke 2004).

These two systems show similarities to what Amin (1999) has labelled bottom-up, region-specific, longer term and plural-actor kind of regional economic development policies.

In nationally initiated dirigiste systems, on the other hand, intraregional institutional embeddedness and "systemness" tend to be weaker. They come close to the firm-centred, incentive-based, state-driven and standardised kind of regional economic development policies (Amin 1999). Therefore, the effort to build a more effective networking system is needed with intensive cooperation between innovation actors in the region.

In Fig. 1 the central idea of systems of innovation is presented in a graphical way. This model makes clear that firm innovation is influenced by mainly four factors: internal factors to the firm (A), the firm's production environment, which includes the firm's most significant external sources for innovation, namely customers and suppliers (B), R&D infrastructure and policies (C) and the innovation support system (D). Factors A and B form the production system, whereas C and D make up the institutional system. The spatial level of impact (local, regional, national or international) plays a role in factors B, C and D. Depending on the role local/regional authorities play compared to national/international authorities when it comes to affecting the institutional system, we can speak of a grassroots, network or dirigiste system (Table 1). Which factors to include in such a model is, of course, debatable and goes beyond the scope of this Chapter. What is important to

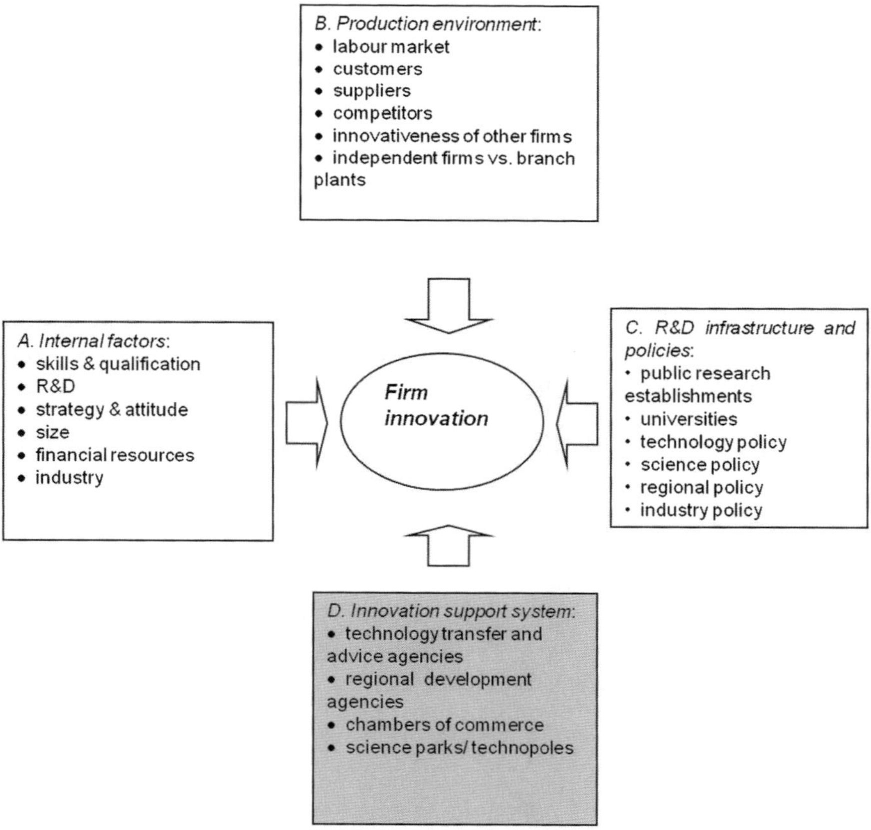


**Fig. 1** The position of regional innovation support systems in a broader framework of innovation systems

know here, is that the innovation support system (D) is part of a system of innovation and that it can be set up, controlled and financed by authorities at several spatial levels.

An innovation support system consists of all agencies found in three support stages, namely the provision of general information, technological advice and joint R&D projects, between firms (of which technology-following SMEs are the main group) and universities and public research establishments. Agencies found in these stages try to help to solve innovation problems mainly of technology-following SMEs by either giving them advice themselves or by referring them to other agencies in a further stage of support. If it concerns a highly complicated technological problem, the SMEs might finally be referred to a university or public research establishment. The agencies can be mainly nationally initiated or regionally initiated.

In the next section, this analytical concept of an innovation support system, consisting of three support stages, will be used to identify the agencies found in a region, to find out whether they are co-operating with each other (system or infrastructure?), whether overlap can be observed and to what extent they have been set up, controlled and financed by local, regional, national or international authorities.

## 4 Case-Studies of Regional Innovation Support Systems from Western Europe

On the basis of the theoretical framework and typology presented in the previous section, it will be analysed what kind of institutionally embedded innovation support systems can be found in the case of Baden-Württemberg, Germany, the North East of England and Utrecht, the Netherlands, in this section in order to draw some general lessons on how regional innovation support systems are related to technopoles in Sect. 5.

### *4.1 Baden-Württemberg, Germany*

In Germany, which has a long history in supporting science, technology and innovation, the two federal ministries are mainly responsible for implementing the policies and providing funding. The Federal Ministry of Education and Research (BMBF), which had an annual budget of over €12.7 billion in 2010 (BMBF 2012), provides funding for education, research and development. This includes institutional funding for Germany's research organisations (jointly funded by the Federal Government and the Länder (regional state governments), contributions to university large infrastructure investments, priority research programmes in key sectors, and international subscriptions.

The Federal Ministry of Economics and Technology (BMWi) allocates an annual €450 million to innovation in the form of support programmes for innovative SMEs, industrial collaborative research and priority projects in the area of energy and civil aeronautics. The federal and Länder governments have joint responsibility for a number of policies, including forward planning in education, the expansion of existing and construction of new universities and major infrastructure equipment. They jointly support Germany's research organisations, including the German Research Association (Deutsche Forschungsgemeinschaft—DFG), the Max Planck and Fraunhofer Societies and Germany's Academies of Science. They are also responsible for harmonising the regulations and syllabuses for further education and vocational training. In general, technology policy in Germany is marked by a high degree of sectoral selectivity which results from the dominance of direct project support (Klodt 1998). Its second main feature is the persistence of public research institutions, although technological priorities have significantly changed over time.

Due to its *co-operative federalism* that is characterised by a strong interdependence between the federal government and the Länder (regional states), many government tasks in Germany are performed by the Länder (Berg 2011). Higher education and technology policy are areas in which the Länder have their own responsibilities. The regionalisation of technology policy soared particularly in West Germany after Baden-Württemberg successfully started regional technology consultancy centres in the 1980s (Cooke and Morgan 1998). Generally, each Land has selected a wide range of regional technology policy measures to assist enterprises in developing their innovative potentials, to build up a technology and science infrastructure, to transfer information, to train employees technologically and to promote business start-ups. In Germany, we can therefore find a system which lies between grassroots and network systems in the old established Länder. In the new Länder in eastern Germany, however, we can find a system which has, due to strong support of the federal government, more dirigiste characteristics (Koschatzky 2000). The latter, however, are clearly moving towards network support systems, as the federal government is slowly decreasing its support.

Not only do Länder have a relatively strong position in innovation support in Germany (innovation-oriented regional policy), also the central government has discovered regions as an implementation platform of its innovation and technology policy, which has been labelled regionalised national innovation policy (Koschatzky 2000). One prominent example of this latest trend of regionalised national innovation policy is the BioRegio contest, an initiative of the central government to boost Germany's competitiveness in biotechnology. In a competitive procedure three regions were selected for support, namely Munich, the Rhine-Neckar Triangle and the Rhineland. According to Koschatzky (2000, 17) the programme contributed to increasing Germany's competitiveness in biotechnology to a considerable extent.

Other initiatives with similar characteristics are the EXIST-University-based start-ups and InnoRegio contests. EXIST focuses on supporting regional concepts for co-operation between higher education institutes, companies and other partners. The InnoRegio contest promotes regional innovation strategies in a broad sense in the new Länder of Germany. These three contests mark a “change of paradigm in the Germany technology and innovation policy”, as national technology policy for the first time regards the region as a relevant platform of support programmes (Koschatzky 2000, 21). In order to participate in these contests, regions need to have co-ordinative power and abilities to submit a sound proposal. These contests can lead to increasing regional inequalities, as only those regions that have both the demand for these innovation policies and the co-ordinative power can benefit from these programmes.

The High-Tech Strategy for Germany, adopted in August 2006, is one of the latest national innovation strategies that consolidates and further develops existing scientific-technical competences of Germany. The High-Tech Strategy consists of five fields: health/nutrition, climate/energy, mobility, communication and security. The Federal Government’s funding of the High-Tech Strategy for Germany provided €291 million along with 13 private investors in 2011 (BMBF 2012).

The economy of *Baden-Württemberg*, which has about 10 million inhabitants and is the most south-western of the old Länder in western Germany, has long been considered to be one of the most prosperous of Germany and even of Western Europe. Unemployment rates have been the lowest in Germany since the early 1970s. In the year of 2009, Baden-Württemberg had one of the lowest unemployment rates (5.1 %, End of 2009) of any Länder in Germany. Also other indicators, such as export rates (42.5 % in 2010), the gross domestic product (€362 Mio. in 2010), the development of the number of employees (5.3 million in 2007), economic growth and, the proportion of R&D expenditure in relation to GDP (4.8 % in 2009) confess the stable and strong economic position in the hierarchy of regions in Germany.

Industrial policy has a long tradition in Baden-Württemberg. Ferdinand von Steinbeis already supported many small craft firms in Württemberg with regard to technological knowledge, export and training in the nineteenth century, whereas Heinrich Meidinger was active in these fields in Baden at the same time. Also, since the mid-1970s policymakers in Baden-Württemberg have been active and innovative with regard to technology policies. Baden-Württemberg developed its own technology programme in 1976 as the first state of West Germany. Embedded in the framework of the federal and European technology policy, Baden-Württemberg’s technology policy measures are developed in fields in which organisational and spatial proximity are essential. Therefore, the support of SMEs and technology transfer are of main importance (Sturm 2002).

Since 1987, technology policy in Baden-Württemberg, on which Lothar Späth, prime minister from 1978 until 1991, had large impact, has always been based on four features: supporting the public research infrastructure, technology transfer, technological aid schemes focused on individual firms and technology centres and business start-up support. The government of Baden-Württemberg declared technology transfer as the core area of its technology policy. Technology transfer is seen as of paramount importance, as the economy of Baden-Württemberg is dependent on the diffusion of incentives from core technologies rather than on development of core technologies themselves. The most important ministries conducting technology policy in Baden-Württemberg are on the one hand the Ministry of Finance and Economic Affairs, and on the other hand the Ministry of Science, Research and Arts.

An important change, which partly occurred due to the economic recession in the 1990s, was the shift from a supply oriented technology policy towards more demand-orientated. In order to become more oriented towards industrial demand, the so-called Joint Initiative Economy and Politics ('Gemeinschaftsinitiave Wirtschaft und Politik') was introduced (Hassink 1996). This initiative aims at gathering the state government, firms, trade unions and intermediaries at one conference table. Due to this initiative, private involvement in technology policy initiatives has increased. Firms, albeit mainly large firms, are participating in the discussion meetings held in the framework of the Joint Initiative Economy and Politics. The initiative is meant to speed up the innovative developments that already took place on the market. In addition to discussing, the parties mentioned above also develop, implement and carry out concrete projects, such as the support of co-operation between suppliers and customers in the car industry, measures to secure competitiveness in mechanical engineering, and the multimedia pilot project. The initiative clearly strengthened the already existing institutional embeddedness in the region.

When it comes to general advice on aid schemes, business support and the referring of firms to technological experts, the main economic intermediaries are the Chambers of Commerce and to a lesser extent the Chambers of Crafts which serve small craft firms. Other centres try to solve technical problems themselves or carry out R&D-projects with firms. These technology transfer centres, which have close links with higher education institutes and public research establishments, can be divided in two groups. First, all universities and public research establishments, mainly concentrated in Stuttgart, Karlsruhe, Heidelberg, Mannheim and Freiburg, have their own transfer facilities trying to solve problems of firms in the state. Second, particularly SMEs are well served by a dense infrastructure of 400 Transfer Centres of the Steinbeis Foundation for Economic Promotion, which was founded in 1971 to promote technology transfer between polytechnics and firms (Beise et al. 1995).

By attaching these centres to polytechnics the state of Baden-Württemberg hoped to reach particularly SMEs, since practice-oriented polytechnics can overcome the initial resistance of SMEs to discuss R&D problems with research institutes. Since in many cases polytechnic professors are directors and since the centres are specialised in the technologies that prevail in the regional production structure, there are close social ties between the directors of the centres and the managing directors in the regions. This structure, therefore, guarantees a high institutional embeddedness in the region.

Steinbeis Transfer Centres can be divided in two main groups. First, a core group of traditional Transfer Centres for Technology Consultancy, which have no subject orientation, provide SMEs with general advice on technological issues and refer them to other experts. Second, a much larger group of subject-oriented

Transfer Centres have been set up besides the existing centres, since the problems of firms became more specific and complex. The main activities of all Steinbeis Transfer Centres are general consultancy services, technology and marketing consultancy, R&D, and further training (workshops, seminars, conferences). The total staff of the Steinbeis Foundation increased from 830 in 1983 to 5,000 in 2011, and the budget grew from DM 8.3 million to €134 million (B–W 2000; STW 2012). The Foundation is nearly able to pay itself (92 % of its income is earned by own activities), although one has to keep in mind that the state of Baden-Württemberg is paying the salaries of all professors at polytechnics. Although the headquarters of all individual centres in Stuttgart are supposed to co-ordinate and organise the system in order to realise a state-wide division of labour, the networking reality of the Steinbeis Foundation can be doubted, since the pressure to earn revenue induces a great deal of rivalry between the individual centres.

Finally, the state governments have been supporting Gründer- und Technologiezentren since the end of the 1980s. This can be considered as buildings, which provide technology-oriented business start-ups with relatively inexpensive office space including services, such as a secretary, meeting rooms, etc. There are now about 41 of these centres in Baden-Württemberg and the latest centres have been focusing on specific industries. There are for instance some software centres and biotechnology parks (B–W 2000). Furthermore, it is one of the few states in Germany which set up a science city: the Ulm Science City or 'Wissenschaftsstadt Ulm' was developed in 1985. Four organisations initiated the development of this concept: the DaimlerChrysler concern, which wanted to extend links with science and other large companies, the state government of Baden-Württemberg and the city of Ulm, which wanted to boost the structurally weak region of Ost-Württemberg and the University of Ulm, which wanted to extend its science part. The whole Ulm Science City complex comprises four components. First, the university and polytechnic created new research subjects with a high industrial applicability, such as electrical engineering, computer science, energy technology, medical engineering, life sciences, information and communication technology, bio and nano materials. Second, the Daimler-Benz concern wants to concentrate nearly all basic research departments of its divisions into one research centre in the Ulm Science City. The main research fields of this centre are information technology, energy technology, material and production technology and technology assessment. Third, several 'An'-institutes were set up on the site of the science city. These mixed public–private funded research institutes are meant to bridge the university–industry gap. Fourth, a science park on the campus attracted small, technology-orientated firms and spin-offs.

All in all, the Baden-Württemberg regional innovation support system can be considered as between a grassroots and network type of system, with a relatively strong leeway for regionally supporting and steering the system. Both the only real technopole, Ulm Science City, as well as many smaller incubator centres are relatively strongly embedded in the regional innovation support systems, which is characterised by long-term stability.

## 4.2 The North East of England

The North East of England used to be one of the main industrial centres of Great Britain during a long time. Industry was mainly based on coal mining and shipbuilding in Tyne and Wear (around the cities Newcastle and Sunderland) and steel production in Teesside (Middlesbrough) (Hassink 1993). Since the mid-1970s governmental financial-aid for these traditional industries has been declining in a rapid pace, so that the already falling number of employees in shipbuilding and coal mining has been decreasing dramatically from then on. Regional policy reacted very early on the decline of traditional industries, since the North East became one of the first 'depressed areas' in the country in the early 1930s. Measures were from then on aimed at attracting inward investment. The loss of jobs in traditional industries was partly compensated by the attraction of branch plants. In addition, the service sector could compensate part of the male job loss, although particularly women were employed in this sector. Furthermore, the service activities consist mainly of government administration and retailing, which created particularly low-wage and part-time employment (Pike et al. 2006).

With the decline of traditional industries, the region lost economic control, since the attracted branch plants are externally controlled. On the other hand, the previous monostructured production structure diversified as several new industries were attracted, but the size of the firms remained relatively large, which impedes the economic flexibility of the region. Further, because of the increasing number of branch plants the North East lost research and development activities and thus innovative potential orientated towards product innovations (Pike et al. 2006). The loss of economic control was also responsible for the lagging behind of the North East concerning high-grade production orientated services, such as financial and business consultancy. The North East possesses only few potential customers for these kinds of services. In addition, the regional economy of the North East is disintegrated, because of the over-representation of externally controlled branch plants and the under-representation of innovative SMEs. Firms have few material links with other firms in the region and thus a network economy does not exist at all. Furthermore, they are faced with a peripheral location, the lack of public research centres in the region, the brain-drain of graduates and the dearth of an enterprise culture, as the relatively small number of business start-ups shows.

The North East of England has a long tradition of solving economic problems by attracting inward investment. As traditional industries suffered early from a decrease in demand, action was needed during the heyday of the traditional regional policy (the 1960s). The aim was to reduce regional economic differences by bringing work to the people. Since the mid-1970s expenditure for regional policy grants has been cut, but some other activities of central government and institutions in the North East emerged instead.

First, central government increased its efforts in improving the situation in inner cities. Urban policy was equipped with more financial back-up and Urban Development Corporations were installed in the cities with the severest manifestations of urban decline, such as Tyne and Wear. These corporations aim at the promotion of physical development of inner cities through subsidies for property developers. They are much more focused on housing, retailing and leisure activities than on rejuvenating the local and regional manufacturing industry. In Labour-controlled areas such as the North East, these centrally led corporations often clash with local authorities.

Second, some new organisations in the North East were set up during the late-1980s, which were partly supported by the central government (quangos; quasi-autonomous, non-governmental organisations). The Northern Development Company used to be one of the most important institutions within the "economic development industry" (Hassink 1993). This Company, which started its work in 1986, was joined by trade unions, employers' organisations and local authorities. The Company tried to attract inward investment by improving the image of the North. Additionally, it had an important lobby function and started to devise initiatives to support indigenous firms and authorities. Until the end of the 1990s, the Company received much less financial backing from London than the Scottish and Welsh Agencies and the Tyne and Wear Development Corporation. Hence, it was not able to co-ordinate regional economic development initiatives very well. This situation improved in 1999 when the NDC became part the of the larger organisation called One NorthEast (ONE), one of the nine English regional development agencies established by the UK government. Due to devolved powers and finance from the central government to this organisation, it was able to steer and co-ordinate regional innovation policy. Unfortunately, however, the regional innovation support system weakened again in 2011 when One North East, together with other regional development agencies, was abolished and replaced by much weaker Local Enterprise Partnerships (Pugalis and Townsend 2012).

Third, technology political organisations and measures compensated for the loss of regional political aid. The Enterprise Initiative of the Department of Trade and Industry tries to support innovative SMEs by providing financial investment grants and financial support for business and technological consultancy. Further, a country-wide network of Regional Technology Centres has to foster technology transfer between institutions of higher education and firms in the regions. These centres advice SMEs on technological problems of all kind and often refer firms with specific questions to professional organisations. One of the most successful initiatives taken by the Regional Technology Centre North is Knowledge House, a project aimed at overcoming the barriers between SMEs and universities. Financial aid for technology transfer, though, has been short-term orientated. Hence, the approach towards firms in the region is reactive instead of proactive. They tend to focus on innovative and financially strong firms instead of helping weak and small companies. One of the main initiatives of One NorthEast was the establishment of five Centres of Excellence (Coenen 2007).

There have also been a few attempts to set up science parks or technology centres in the North East. As the examples of the Sunderland Technology Park and the Offshore Technology park show, there are difficulties to fill these parks with innovative SMEs. Another example of a science park is the North East Technology

Park at the University of Durham. More recently, Newcastle, the largest city in the region, has received "science city status", which means that it is earmarked for support to develop science and technology facilities.

In short, both the emerged 'economic development industry' and technology transfer offices in the North East of England consist on the one hand of centrally led organisations and on the other hand of spontaneously appeared regional organisations, which often lack enough governmental aid. The multitude of organisations and the lack of a regional elected body bear the dangers of too much overlap and too little co-ordination and transparency (Duke et al. 2006; Coenen 2007), which also leads to a vulnerable position of technopoles in the overall regional innovation support system.

## 4.3 Utrecht, the Netherlands

Although the province of Utrecht is seen as a relatively strong economic region if employment and income indicators are used, it has weaknesses concerning the innovativeness of SMEs and business start-up rates (IPO 2004; Technopolis 2000). Looking at the sectoral composition, a relatively large number of people is working in services, whereas manufacturing industry is clearly under-represented. Apart from publishing and printing, all manufacturing industries in Utrecht are under-represented, including medium-tech industries, such as the car industry and chemical industries. Within service industries, financial services and business consulting have the highest location quotients.

Syntens is not only considered as the main actor concerning innovation support in Utrecht, it is in fact regarded as the main actor in the Netherlands (Hassink 1997). This dense network of 18 centres, which was set up in 1988, is financed by the central government and co-ordinated by a national clearing house. The Syntens Centres are anchored in the region as their boards consist of regional decision-makers from chambers of commerce, higher education institutes, public research establishments and companies. Regional embeddedness of the individual centres is seen as of great importance, since consultants of the individual centres should know the local, mental and cultural characteristics of SMEs in order to be able to build up trust relationships. The aim of the Syntens Centres is to support SMEs in finding and applying technological knowledge that should help them to renew production processes, products and services. The centres aim to link practical problems of SMEs with the supply of technological knowledge. The centres do not carry out any R&D themselves. The 150 consultants of the 15 centres, who all have business experience, first try to solve problems they encounter in randomly visited SMEs themselves. They do not approach clients with new technology or new product ideas, but deliberately come "empty-handed". If consultants lack the required expertise to help the firms themselves, they will refer to experts in other Syntens Centres, companies, higher education institutes, public research establishments, management and engineering consultancies. Furthermore, the centres try to stimulate networking among SMEs in the region by bringing together managers of SMEs at workshops, business meetings or presentations of innovation projects. The regional centres also advice firms about national aid schemes. The effectiveness of Syntens has been extensively evaluated in the first half of the 1990s and, generally, the results of these studies are positive (Technopolis 2000).

In Utrecht, Syntens has good connections with the Provincial Government, the Chamber of Commerce and the Faculty of Economics and Management at the University of Professional Education (Technopolis 2000). It, has, however, very few connections with the other colleges and universities, even though Syntens have tried on several occasions to establish better rapport. The main agency for developing innovation in the region (Syntens), therefore, is not well enough connected to the largest potential supply (higher education institutes and public research establishments in the region). Moreover, the transfer office of the university focuses its support on advising its own researchers instead of companies in the region.

In addition to Syntens, the Chamber of Commerce is the dominant organisation when it comes to providing all firms with general business information and referring them to other agencies. Syntens can be regarded as the main agency when it comes to providing support to innovative SMEs in the region. Utrecht, therefore, seems to have a 'Two Stop Shop' system, with the Chamber of Commerce and Syntens as the main agencies. Moreover, some of the industry associations in the region, Metaalunie/PKM in particular, provide their members with useful business information and technological and innovation advice. Since the industry associations are national organisations in the first place, with offices in the regions which cover areas much larger than provinces, they are not much linked to the provincial institutional set-up.

Compared to famous university regions abroad and less famous ones in the Netherlands, such as Leiden and Twente, which have been successful in stimulating spin-offs from university, Utrecht University has been relatively passive until recently. It established a building with office space for spin-offs, the Matthias van Geuns building (Van Weesep and Wever 1996), when other universities had done so already for a long time. The building has not only been built very late, offices are also mainly rented by university institutes and the centre lacks any kind of technological, financial or business advisory support for start-up businesses. During many years there have been talks and negotiations to set up an incubator centre for biomedical business start-ups at the university campus, but until recently nothing was realised.

Finally, some five years ago a couple of innovation partners, such as the province, the city government, the university and the newly created coordinating body, Taskforce Innovatie Regio Utrecht, have pushed the establishment of Utrecht Science Park, which aims at both attracting inward investment in R&D and fostering knowledge transfer and spin-offs. The latter is particularly facilitated by UtrechtInc as the main incubator in Utrecht Science Park. Arguably due to the weak role of the provinces in innovation policy, there has been little local and regional pressure to set up a technopole or incubator centres for a relatively long time.

During a long time, regional innovation support systems did not exist in the Netherlands, mainly due to the weak political and financial power of the provinces. Recently, as the case of Utrecht shows, provinces are getting more active in innovation policy, particularly as coordinators of policies and institutes devised and financed at several spatial levels (local, national, EU) (IPO 2004). Interestingly, in the case of Utrecht the Regional Innovation and Technology Transfer Strategy (RITTS) supported by the EU set a process of regional coordination in motion, first by the Utrecht Network for Innovation and Technology (Unité) and most recently by on the Taskforce Innovatie Regio Utrecht (IPO 2004; TFI 2009).

## 5 Lessons for Science City Governance

To conclude, the above described regional innovation support systems and their relations with technopoles in different countries of Western Europe differ strongly from each other (Table 2). Furthermore, the innovation support systems also change through time, and most of the time they develop in the direction of network type of regional innovation support systems. Cooke (2007) also made this point and ascribes this to the strong influence of EU programmes such as RITTS on regional innovation support systems, particularly in politically more centralised countries, such as the UK and the Netherlands. Arguably one tend to find smaller, but better regionally embedded and integrated technopoles in countries with a decentralised political-administrative structure, such as Germany.

As stated above, one of the main problems of many technopoles is the lack of regional technology transfer and networking that take place both between firms and between firms and higher education institutes and public research establishments in the region. To solve this problem, technopoles need to be integrated in wider regional innovation policies and strategies. These strategies should be based both on thorough and in-depth studies of the strengths, weaknesses, opportunities

**Table 2** Regional innovation support systems compared

| | Baden-Württemberg (Germany) | The North East of England | Utrecht (the Netherlands) |
|---|---|---|---|
| Initiation | mainly regional government | regional and central government | central and increasingly provincial government |
| Funding | regional and national agencies | regional and national agencies | mainly national agencies |
| Research & support | mixed | mixed | mainly basic |
| Specialisation | mixed | mixed | mixed |
| Intra-regional co-operation | high | fair | fair |
| Co-ordination | high | fair | fair |
| Type of system | between grassroots and network | between dirigiste and network | between dirigiste and network |

and threats of the regional economy and firms' production environment and on the national and supranational institutional framework. Only if the establishment of technopoles are based on these kind of studies, they are able to focus both on the demand for technologies among local firms and on niches in the regional production structure. In this way the development of supply oriented technopoles based on the linear model of innovation can be avoided. Therefore, this Chapter strongly endorses the argument for a differentiated regional innovation policy put forward by Tödtling and Trippl (2005). They (2005, 1203) state that "there is no "ideal model" for innovation policy as innovation activities differ strongly between central, peripheral and old industrial areas". Careful design of innovation support systems guarantees a good fit between the development stage of technopoles and the innovation support system in which they are embedded. Since each country is at a different development stage, technopoles will accordingly be at different positions in their policy life cycle: at the end of their life cycle in Japan and Germany, at earlier stages in emerging economies. Technopoles in industrialised countries, therefore, need to be embedded in other regional innovation support systems than technopoles in emerging economies.

In addition to its role of boosting technology transfer and networking in the region, there are two other advantages of differentiated, coherent regional innovation strategies. First, over-ambitious technopole planning focused on a too broad range of technologies can be avoided with the help of regional innovation strategies. Second, regional innovation strategies help to place technopoles in the proper regional innovation policy context and by doing that they help to coordinate all innovation-oriented measures relevant to the region, which might be devised both at local, regional, national and supranational level. This will help to avoid overlap and duplication of policy measures and thus to foster transparency and efficiency of the innovation support system. Moreover, to develop technopoles and to reap benefits from them for the regional economy is a long-term process, certainly longer than the political election cycle. Carefully devised regional innovation support systems can help regions to yield benefits of technopoles sooner than without them.

All in all, regional innovation support systems develop in an evolutionary manner and they can become particularly useful in supporting technopoles if they develop from a dirigiste to a network kind of character. The development of technopoles can be seriously harmed, if there are mismatches between the developmental stage of technopoles and the broader regional innovation support system in which they are embedded.

## References

Amin, A. (1999). An institutional perspective on regional economic development. *International Journal of Urban and Regional Research, 23*, 365–378.

Annerstedt, J. (2006). Science parks and high-tech clustering. In P. Bianchi & S. Labory (Eds.), *International handbook of industrial policy* (pp. 279–297). Cheltenham: Edward Elgar.

Anttiroiko, A.-V. (2004). Editorial: Global competition of high-tech centres. *International Journal of Technology Management, 28*, 289–323.

Asheim, B., Isaksen, A., Nauwelaers, C., & Tödtling, F. (Eds.). (2003). *Regional innovation policy for small-medium enterprises*. Cheltenham: Edward Elgar.

Asheim, B. T., Lawton Smith, H., & Oughton, C. (2011). Regional innovation systems: Theory, empirics and policy. *Regional Studies 45*, 875–891.

Bass, S. J. (1998). Japanese research parks: National policy and local development. *Regional Studies, 32*, 391–403.

Bathelt, H. (2003). Geographies of production: Growth regimes in spatial perspective 1—innovation, institutions and social systems. *Progress in Human Geography, 27*, 763–778.

Beise, M., G. Licht, G., & Spielkamp, A. (1995) *Technologietransfer and kleine und mittlere Unternehmen: Analysen und Perspektiven für Baden-Württemberg*. Baden–Baden: Nomos Verlagsgesellschaft (Schriftenreihe des ZEW No. 3).

Berg, S.-H. (2011) Centrally steered or self-organizing? regional innovation systems in South Korea and Germany. Unpublished Master Thesis at the University of Flensburg—Syddansk Universitet.

BMBF. (2012). *Bundesbericht Forschung und Innovation 2012*. Berlin: Bundesministerium für Bildung und Forschung.

Boschma, R. A. (2005). Proximity and innovation: A critical assessment. *Regional Studies, 39*, 61–74.

Boschma, R., & Frenken, K. (2011). Technological relatedness, related variety and economic geography. In P. Cooke, B. Asheim, R. Boschma, R. Martin, D. Schwartz, & F. Tödtling (Eds.), *The handbook on regional innovation and growth* (pp. 187–197). Cheltenham: Edward Elgar.

B-W, W. M. (2000). *Innovationssystem Baden-Württemberg: Innovations- und Technologieförderung als wirtschaftspolitische Aufgabe*. Stuttgart: Wirtschaftsministerium Baden-Württemberg.

Castells, M., & Hall, P. (1994). *Technopoles of the world*. London: Routledge.

CBS. (1999). *Innovatie en provincie: Regionale innovatieprofielen van het MKB in Nederland*. Voorburg: Centraal Bureau voor de Statistiek.

Coenen, L. (2007). The role of universities in the regional innovation systems of the North East of England and Scania, Sweden: Providing missing links? *Environment and Planning C, 25*, 803–821.

Cooke, P. (2001). From technopoles to regional innovation systems: The evolution of localised technology development policy. *Canadian Journal of Regional Science, 24*, 21–40.

Cooke, P. (2004). Introduction. In P. Cooke, M. Heidenreich & H-J. Braczyk (Eds.), *Regional innovation systems: The role of governances in a globalized world* (pp. 1–18). London: Routledge.

Cooke, P. (2007). Regional innovation, entrepreneurship and talent systems. *International Journal of Entrepreneurship and Innovation Management, 7*, 117–139.

Cooke, P., Heidenreich, M., & Braczyk, H.-J. (Eds.). (2004). *Regional innovation systems: The role of governances in a globalized world*. London: Routledge.

Cooke, P., & Morgan, K. (1998). *The associational economy: Firms, regions, and innovation*. Oxford: Oxford University Press.

Cooke, P., Uranga, M. G., & Etxebarria, G. (1998). Regional systems of innovation: An evolutionary perspective. *Environment and Planning A, 30*, 1563–1584.

Duke, C., Hassink, R., Powell, J., & Puukka, J. (2006). *Supporting the contribution of higher education institutions to regional development; peer review report: North East of England*. Paris: OECD.

Fritsch, M., & Stephan, A. (2005). Regionalization of innovation policy—introduction to the special issue. *Research Policy, 34*, 1123–1127.

Harmaakorpi, V., Tura, T., & Melkas, H. (2011). Regional innovation platforms. In P. Cooke, B. Asheim, R. Boschma, R. Martin, D. Schwartz, & F. Tödtling (Eds.), *The handbook on regional innovation and growth* (pp. 556–572). Cheltenham: Edward Elgar.

Hassink, R. (1993). Regional innovation policies compared. *Urban Studies, 30*, 1009–1024.
Hassink, R. (1996). Regional technology policies in the old and new länder of Germany: Case-studies from Baden-Württemberg and Thuringia. *European Urban & Regional Studies, 3*, 287–303.
Hassink, R. (1997). Technology transfer infrastructures: Some lessons from experiences in Europe, the US and Japan. *European Planning Studies, 5*, 351–370.
Hassink, R., & Hu, X. (2012). From specialisation to diversification in science and technology parks. *World Technopolis Review, 1*, 6–15.
Hassink, R., & Hülz, M. (2010). Editorial: Debating the dilemmas of interregional institutional learning processes. *Regional Studies, 44*, 495–498.
Hassink, R., & Klaerding, C. (2012). The end of the learning region as we knew it; towards learning in space. *Regional Studies, 46*, 1055–1066.
IPO. (2004). *De rol van provincies op het gebied van innovatie en kenniseconomie*. Amersfoort: Interprovinciaal Overleg.
Klodt, H. (1998). German technology policy: Institutions, objectives and economic efficiency. *Zeitschrift für Wirtschaftspolitik, 47*, 142–163.
Komninos, N. (1997). After technopoles: Diffused strategies for innovation and technology transfer. In J. Simmie (Ed.), *Innovation, networks and learning regions?* (pp. 181–195). London: Jessica Kingsley Publishers.
Koschatzky, K. (2000) *The regionalisation of innovation policy in Germany—theoretical foundations and recent experience*. Karlsruhe: Fraunhofer Institute for Systems and Innovation Research (FhG-ISI) (Working Papers Firms and Region No. R1/2000).
Lagendijk, A. (2011). Regional innovation policy between theory and practice. In P. Cooke, B. Asheim, R. Boschma, R. Martin, D. Schwartz, & F. Tödtling (Eds.), *The handbook on regional innovation and growth* (pp. 597–608). Cheltenham: Edward Elgar.
Lagendijk, A., & Charles, D. (1998). *Regional institutions facilitating technology transfer to SMEs: A review paper for OECD*. CURDS: Newcastle upon Tyne (unpublished manuscript).
Lorenzen, M. (2001). Localized learning and policy: Academic advice on enhancing regional competitiveness through learning. *European Planning Studies, 9*, 163–185.
Luger, M. I. & Goldstein, H. A. (1991). *Technology in the garden: Research parks and regional economic development*. Chapel Hill: University of North Carolina Press.
Morgan, K. (1997). The learning region: Institutions, innovation and regional renewal. *Regional Studies, 31*, 491–503.
Mothe, J. De, & Paquet, G. (Eds.). (1998). *Local and regional systems of innovation*. Boston: Kluwer Academic Publishers.
Moulaert, F., & Sekia, F. (2003). Territorial innovation models: A critical survey. *Regional Studies, 37*, 289–302.
Nauwelaers, C., & Wintjes, R. (2003). Towards a new paradigm for innovation policy? In B. Asheim, A. Isaksen, C. Nauwelaers, & F. Tödtling (Eds.), *Regional innovation policy for small-medium enterprises* (pp. 193–220). Cheltenham: Edward Elgar.
OECD. (2007). *Competitive regional clusters*. Paris: OECD.
OECD. (2011). *Regions and innovation policy*. Paris: OECD.
Paquet, G. (1998). Techno-nationalism and meso innovation systems. In R. Anderson, et al. (Eds.), *Innovation systems in a global context: The North American Experience* (pp. 58–75). Montreal: McGill-Queen's University Press.
Pike, A., Rodríguez-Pose, A., & Tomaney, J. (2006). *Local and regional development*. London: Routledge.
Porter, M. E. (2000). Location, competition, and economic development: local clusters in a global economy. *Economic Development Quarterly, 14*, 15–34.
Pugalis, L., & Townsend, A. (2012). Rescaling of planning and its interface with economic development. *Planning Practice & Research, 5*, 157–174.
Sternberg, R., Behrendt, H., Seeger, H., Tamásy, C. (1996). Bilanz eines Booms. Wirkungs-analyse von Technologie- und Gründerzentren in Deutschland. Ergebnisse aus 108 Zentren und 1021 Unternehmen. Dortmund: Dortmunder Vertrieb fuer Bau- und Planungsliteratur.

Storper, M. (1995). Regional technology coalitions: An essential dimension of national technology policy. *Research Policy, 24*, 895–911.

Sturm, R. (2002). Neben Brüssel und Berlin: Wirtschaftspolitik in Baden-Württemberg. In H. Cost & M. Körber-Weik (Eds.), *Die Wirtschaft von Baden-Württemberg im Umbruch* (pp. 280–296). Stuttgart: Landeszentrale für politische Bildung.

STW (2012). Steinbeis-stiftung, www.stw.de (Accessed 31 Aug 2012).

Technopolis (2000). *The Utrecht RITTS project report on supply side analysis*. Brighton: Technopolis Ltd (unpublished manuscript).

TFI. (2009). *Motor voor Duurzame Innovatie; TFI 2009–2013*. Utrecht: Taskforce Innovatie Regio Utrecht.

Tödtling, F., & Trippl, M. (2005). One size fits all? towards a differentiated regional innovation policy. *Research Policy, 34*, 1203–1219.

Weesep, J. Van, & Wever, E. (1996). *Een universiteit in Utrecht: De banden tussen Universiteit Utrecht, stad en regio*. Utrecht: Universiteit Utrecht.

# The Triple Helix: International Cases and Critical Summary

**Fred Phillips, Sukaina Alarakhia and Pornpimol "Joy" Limprayoon**

**Abstract** This chapter will use five cases to examine the role of government–industry–academic cooperation on the development of technology clusters, research parks, and metropolitan-area technopoleis like Austin or Silicon Valley. We find that triple helix cooperation is necessary in deliberate cluster initiatives (as opposed to spontaneously formed clusters) but is not a sufficient concept in itself. A fourth strand, consisting of non-profit enterprises, NGOs, and voluntary associations, is necessary.

## 1 Introduction: University–Industry–Government Cooperation

In a presentation at last year's WTA-UNESCO workshop, Phillips (2009) displayed Table 4.1. For this chapter, we have added emphasis to the phrase "especially with cross-sectoral links."

Cross-sectoral links are central to the idea of the triple helix. This briefing will use five cases—in many of which one or more of the present authors have been involved—to examine the role of government–industry–academic cooperation on

F. Phillips (✉)
Department of Technology and Society, College of Engineering and Applied Sciences, Stony Brook University – State University of New York, c/o SUNY Korea, 187 Songdo-dong, Yeonsu-gu, Incheon 406-840, Korea
e-mail: fred.phillips@stonybrook.edu

S. Alarakhia
General Informatics LLC, San Diego, CA, USA

P. "Joy"Limprayoon
General Informatics LLC and University of Redlands, Coleman University, California, San Diego, CA, USA

D.-S. Oh and F. Phillips (eds.), *Technopolis*, 
DOI: 10.1007/978-1-4471-5508-9_4, 

**Table 1** Technopolis success factors

| |
|---|
| Embracing change |
| Social capital, especially with cross-sectoral links |
| Cluster strategies that target specific company groups for collocation |
| Visionary and persistent leadership |
| The will to action |
| Action |
| Constant selling |
| Self-investment in infrastructure |
| Outreach and networking |

*Source* Phillips (2006)

the development of technology clusters, research parks, and metropolitan-area technopoleis like Austin or Silicon Valley.

Tech clusters and technopoleis are nonlinear phenomena. The accident of one influential person moving to a city, or the accident of one company spinning out from a university—the "butterfly effect" of complex system theory—can set the cluster/technopolis on a trajectory of growth or one of stagnation. A cluster/technopolis on a growth trajectory is subject to positive feedback, or lock-in. This means the benefits of locating in the city increase faster than the number of companies present. Lock-in implies that each new or relocated company, university, or government office makes the city more attractive.

It also implies that once located in the city, companies will be slow to leave, even if business conditions change. Locales lucky enough to experience this lock-in will enjoy lower recruiting costs, and fewer problems with "clawing back" incentives extended to companies that do not stay in the region for the contracted amount of time (or create the contracted number of jobs).

The triple helix hypothesis is that cross-sector cooperation creates a second, reinforcing lock-in cycle. The hypothesis also notes that cross-sector cooperation makes an already complex system even more complex, thus increasing the opportunities for disaster (Leydesdorff 2000; Etkowitz 2002; Benner and Sandström 2000; Cho 2008). Even if disaster is avoided, saturation will eventually set in. Silicon Valley's success led to congested highways and unaffordable housing—and as a result many experienced entrepreneurs and investors moved to Austin, Texas, to start leaner, cleaner companies.

Most of the cases in this briefing involve locales that interacted with Austin's $IC^2$ Institute in developing their own technology-based growth strategy. It will help set the scene for the following cases to quote one of the Institute's clients (UTEN 2009) about $IC^2$:

> The $IC^2$ Institute has a 30-year track record of working with emerging, developing, and developed regions worldwide on how to effectively structure industry-science-academic relationships to transfer and commercialize innovative and creative knowledge/technology to build wealth and high quality jobs while providing for a sustainable quality of life. Austin, Texas, is known internationally as having leveraged academic, business, and government collaboration to transform a mid-sized central Texas government and

> university town into a globally competitive technology center that successfully educates, attracts, and retains scientific and entrepreneurial talent from leading technology regions in the US and worldwide.
>
> Based on many national and international rankings, Austin is judged one of the top US cities in terms of entrepreneurship, economic growth, and quality of life and [its success] is often referred to internationally as the "Austin Model"... Key to Austin's successful technology-based growth is the fact that the city and The University of Texas at Austin are able to educate, attract, and retain key US and international talent. This talent has been crucial to the establishment of globally competitive clusters in semiconductors, software and IT, and computers and peripherals as well as emerging clusters in biosciences, nanotechnology, digital media, clean energy and wireless technology.

## 2 University Technology Enterprise Network UTEN

In the mid-1990 s, Professor Manuel Heitor of the Instituto Superior Técnico (the engineering school of the Technical University of Lisbon) visited the $IC^2$ Institute and launched a program of student and faculty exchange between Austin and Lisbon. He became Deputy-President of IST in 1993, and was named a Fellow of the $IC^2$ Institute. He also joined the editorial board of Technological Forecasting & Social Change. His talent and energy led to his being appointed Secretary of state for Science, Technology, and Higher Education for Portugal. In this capacity, he expanded the Portuguese government's interactions with American universities, focusing especially on the University of Texas at Austin, Carnegie-Mellon University, and MIT. With UT-Austin's $IC^2$ Institute, Heitor conceived and structured the University Technology Enterprise Network (UTEN).

This initiative followed a six-month assessment of the match between the capabilities of the $IC^2$ Institute and its partners and Portugal's needs and challenges in the areas of technology development and commercialization. A 5-year agreement was signed in 2007. It involved 15 Portuguese universities, technology parks, and research centers. UTEN's goal is to build a competitive and sustainable S&T transfer network and infrastructure in Portugal. Objectives include strengthening Portuguese technology transfer practices, and building academic–science–business cooperative networks (Pfotenhauer 2010).

Under UTEN the $IC^2$ Institute has worked with Portuguese academic, government, and businesses. In 2008, the Institute received visitors from Portuguese universities and incubators, and from the Ministry of Science, Technology, and Higher Education. Visits by UTEN partners in Texas included (in Austin) the Austin Technology Incubator, UT-Austin's Office of Technology Commercialization, the Greater Austin Chamber of Commerce; (in Dallas) UT Dallas' Institute for Innovation and Entrepreneurship, Office of Technology Commercialization, and Arts & Technology Institute; and (in San Antonio) UT San Antonio's Center for Innovation, Technology Entrepreneurship, INCELL Corporation, and TEKSA.

Later in 2008, "a biotechnology and medical technology expert team representing UTEN Austin visited with entrepreneurs, universities, incubators, research parks, and other institutions throughout Portugal" (UTEN 2009). Under UTEN one of the present authors (Phillips) has gone to Lisbon each summer with US colleagues to evaluate proposals from Portuguese researchers to the Fundação para a Ciência e a Tecnologia, the Portuguese national science foundation.

> In Austin, the team attended training on market-based entrepreneurship, learned about government-sponsored and private funding opportunities and models… and visited with entrepreneurs, venture capitalists, business angels, IP lawyers, and professors from the $IC^2$ Institute's Master of Science in Technology Commercialization program (UTEN 2009).

Portugal's government has provided more than € 50 million for UTEN-related initiatives within Portugal and with the three US universities. Business-sector partners contributed a few million additional euros in 2009 (Fischer 2010). UTEN's missions are to:

- "Overcome a historic gulf in Portugal between academe and industry, an insular higher education system, and a business climate with little tolerance for risk" (Fischer 2010).
- Encourage Portuguese universities to work together on common research problems. "Individually, many of the country's 13 public universities lack the capacity to tackle large-scale research projects" (ibid).
- Internationalize the educational experiences of Portuguese students.

Fischer (2010) adds,

> …the Portuguese strategy is, at least in its conception, top-down, driven by government initiative. In Portugal, however, those government officials are also academics—both Mariano Gago, the longtime minister of science, technology, and higher education, and Mr. Heitor, the secretary of state, or deputy minister, are former engineering professors.

UTEN's annual report (UTEN 2009) lists a dozen viable companies in each stage of growth (start-up, growth, mature) that have benefited from the program, and notes these spin-off benefits are additional to significant progress on the academic and entrepreneurial culture-change objectives.

## 3 Kansas City's Life Sciences Cluster

The making of life science clusters primarily require three main ingredients: (i) proprietary science that can support a wide array of commercially viable products, (ii) risk capital, and (iii) science and entrepreneurial talents. While all key ingredients are indispensable, the research centers of excellence are central. All life science clusters emerge around prominent universities and research institutions, which provide the intellectual base both in terms of highly trained human capital and proprietary technologies. These three key ingredients are prerequisites for the creation of a life science cluster but for it to thrive, a confluence of other factors is needed—environmental, infrastructural, and cultural.

## 3.1 Government Support

A Life Science industry needs to operate in a favorable environment, with well-maintained infrastructure. A government's supportive tax and regulatory policies imparts an enabling climate, fostering the commercialization practices and attracting more basic ingredients to the cluster. Many aspects of the infrastructure are public in nature which only the government is in a position to provide. Some of these include, but are not limited to:

- A legal framework that protects intellectual property rights of inventors.
- Mechanisms to license these rights, transfer them from public to private domain, commercialize academic discoveries, and ultimately reward innovation.
- Education infrastructure to train large pools of labor, preparing human capital for the industry.

The crucial role of the government in the creation of a life science industry existed long before its birth. For instance, in North Carolina, behind the excellence of universities was decades of government support.[1] Government funding continues to pour into research institutions' labs, in tens of billions of dollars every year.

## 3.2 Commercial Infrastructure

There are other infrastructural elements which generally fall in the private sector. These include "professional service providers such as patent agents, intellectual property attorneys, consultants and accountants; wet-lab facilities and bio supplies and equipment; specialized suppliers such as contract manufactures and clinical research organizations; and related industries such as medical devices and information technology" (Song 2004), all of which aid in facilitating the creation and growth of a cluster.

## 3.3 Collaboration

The life science industry is probably more collaborative than any other industry where deal-making is a way of life. Extensive and intensive collaboration pervades the entire cluster.

- Academic collaboration

It is rather common for faculty members from different institutions to collaborate on the same research project, and their institutions jointly own the IP rights of resulting inventions.

[1] See Song (2004)

- Intra-industry collaboration

Companies form alliances with each other to achieve synergy in R&D efforts, expand product pipelines, and leverage each other's expertise and resources. These alliances take a number of forms, including one-way licenses, cross-licenses, R&D collaborations, and commercial and sales partnerships, in a wide range of product modalities such as proteins, peptides, and small molecules (see Song 2004).

- Academic-industry partnership

In successful life science clusters, academic institutions and companies are mutually engaged. Through technology transfer, institutions license their government-funded technologies to companies for commercial development. Institutions may also conduct private research sponsored by companies. Institutions' research expertise helps companies' business; companies provide institutions with strategic directions of commercializing new ideas.

- Academic–industry–public partnerships

In leading life science clusters, academic institutions and companies are also involved with governments and local communities to shape public policy, and push the economic agenda forward. In the late 1980 s to early 1990 s when public understanding of biosciences was foggy, the U.C. campuses and Stanford University's public seminars and papers took a lead role in addressing public concerns in the State of California (Song 2004).

The intensity of these collaborations is an essential stimulant for research innovation and creativity. It helps cluster entities remain current on the industry's rapidly evolving technical aspects, and gain access to information, financial resources, and alliance opportunities. Previous studies of the biotechnology industry showed that the number of network ties that a biotech firm or institution has within a cluster, as well as its position and centrality in relation to other firms in the cluster, can affect performance. "In the short term, firms lacking in alliances will be slower to generate research discoveries, obtain patents, and turn scientific results into marketable products. In the long run, firms that learn to manage diverse portfolios of collaboration, involving multiple projects at different stages of development, are less likely to fail."[2] The energy of these interactions also contributes to the vitality of the cluster as a whole.

[2] Walter W. Powell et al., "Network Position and Firm Performance: Organizational Returns to Collaboration in the Biotechnology Industry".

## 3.4 Critical Mass

Critical mass appears to be an attractive element in these clusters. Larger population of entities means higher availability of meaningful collaboration opportunities. Critical mass, which often implies recognition and reputation, is significant in many other ways too. A critical mass of talent attracts more talent. A critical mass of healthy companies attracts more capital. Critical mass allows economy of scale in time allocation, procurement of resources and build-out of specialized infrastructure such as lab or manufacturing facilities.[3] It lowers market risk for specialized services and vendors, therefore encouraging their supply. A critical mass of companies gives managers and employees a sense of option, that even if their own company fails there are other potential employers nearby.

Critical mass is not only desirable, it is a necessity (Song 2004). Smaller locales, lacking critical mass, face a challenge in attracting external talent as well as capital. When asked whether they would invest if an opportunity is promising in itself but from a less established geographic area, most venture capitalists say that they would consider investing, however, they would have the company locate to a major cluster, because the odds of success are much lower in regions without the support network.

## 3.5 The Nature of Life Science Clusters

With the onset of a mature life science industry, the industry tends to continue to gravitate toward established clusters where there is critical mass, provided those clusters can meet the growing needs at sustainable cost.

Even when the emergence of a new traditional cluster may not seem to be in sight, however, a combination of compelling force conspires to create a new type of cluster—i.e., specialized subclusters, or small clusters specialized in one or two niche strengths of a region. An article in *Genetic Engineering News*, dated September 2004, reported future trends in life science cluster development.[4] "Tomorrow's company in Seattle can have manufacturing in Research Triangle Park, marketing offices in New Jersey, and clinical trials in Kansas. It is much more of a hub-and-node scenario now."

Hence, as regions and companies realize the impracticality of developing full-scale clusters today partly due to the huge cost investment, they are instead adapting their strategy to develop in niches they have competitive advantage in. Also reported in the Genetic Engineering News article mentioned earlier, "In the past, regions simply wanted to duplicate San Diego or Boston in their backyard."

---

[3] "Texas Biotechnology and Life Science Cluster Report", State of Texas, August 2005.

[4] "Novel Model for Biotech Cluster Development", David G. Jensen, Volume 24, Number 16, September 15, 2004.

The author quotes an interviewee. "In the future, the smart states will be trying to find out what their niches are. Then, they can make plans to develop their market opportunities and build business and clusters around these strengths." Another economic development expert concurred, "The biotech industry will grow more reliant upon an outsourced business model much like the 60 % of the global pharmaceutical industry that relies upon outsourcing...."

## *3.6 The Life Science Cluster of Kansas City, Missouri and Kansas City, Kansas*

Kansas City, MO-KS, lies within the bi-state region of the Greater Kansas City metropolitan area (Fig. 1).

According to the 2009 Life Sciences R&D census report, released May 24 2010 by the Kansas City Area Life Sciences Institute (KCALSI), despite the struggling economy, the region's life sciences research and development industry has grown slowly but steadily over the past 3 years. "The demand for health sciences innovation remains high, and we are fortunate to have a strong presence in both the

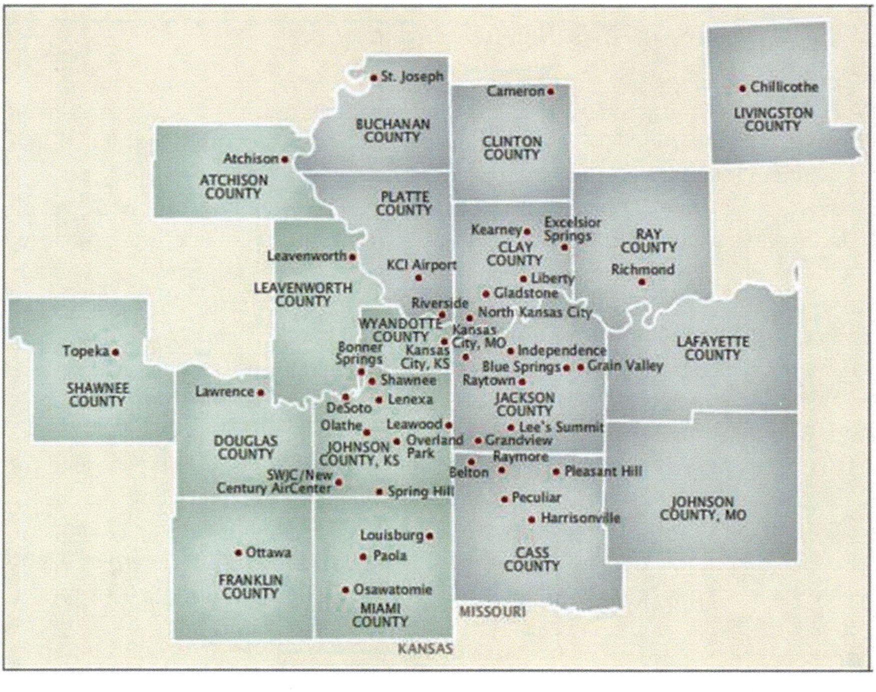


**Fig. 1** Greater Kansas City Metropolitan Area, Credit: Kansas City Area Development Council (KCADC)

human health and animal health industries," said Dr. Dan Getman, president of KCALSI.

A key element in the growth of the life science sector in Kansas City is that both Kansas City, MO and Kansas City, KS are home to leading universities, research centers, and institutions and life science organizations, all of which support a strong research base responsible for uncovering new discoveries and innovation, leading to intellectual property, and ultimately developed into products and therapies.

The University of Missouri Kansas City (UMKC) located in Kansas City, MO is home to a large life science program. The number patents filed over the last 4 years has increased with 65 patents filed in 2006 and 110 patents filed in 2009. Licenses and options signed also increased during the same period with 17 licenses signed in 2006 to 77 licenses signed in 2009, bringing a licensing income of $10.4 million in 2009 up from $2.4 million in 2006. Their annual licensing income goal for 2014 is $50 million. The approximate amount of funding awarded to faculty to help bring their ideas closer to market was $600,000 (Fig. 2).

Total externally sponsored research expenditures in the University of Missouri system-wide for the year 2009 was $308,296,926 million, of which the University of Missouri Kansas City (UKMC) research expenditure was $34,792,030 million.

- UMKC's commitment to Kansas City's life sciences initiative has been strengthened by the opening of its new $50.2 million Health Sciences Building in Fall 2007. It also has plans for a second building—the Center for Health Sciences Research—to accommodate the expansion of interdisciplinary and translational research (Table 2).
- The UKMC School of Biological Sciences (SBS) houses more than 30 active research laboratories with faculty members engaging in cutting-edge research, including structural biology, biochemistry, microbiology, molecular cell biology, and developmental biology.

A goal of the SBS is to create an infrastructure and environment where research programs exploring fundamental questions in basic biology can be developed and addressed. Annual research funding for faculty research projects exceeds $4 million.

The UMKC School of Medicine has a diversity of biomedical research strengths enabling it to provide leadership in research at UMKC. Since medicine is at the very heart of all biomedical research efforts, the School's role is critical for UMKC to achieve maximal productivity from its life sciences research. It has also jointly recruited 18 Endowed Chairs in collaboration with its affiliated hospitals. These Chairs provide an established nest for highly productive translational research, which has led to an overall increase in Federal funding, as well as foundation and Industry sponsored research.

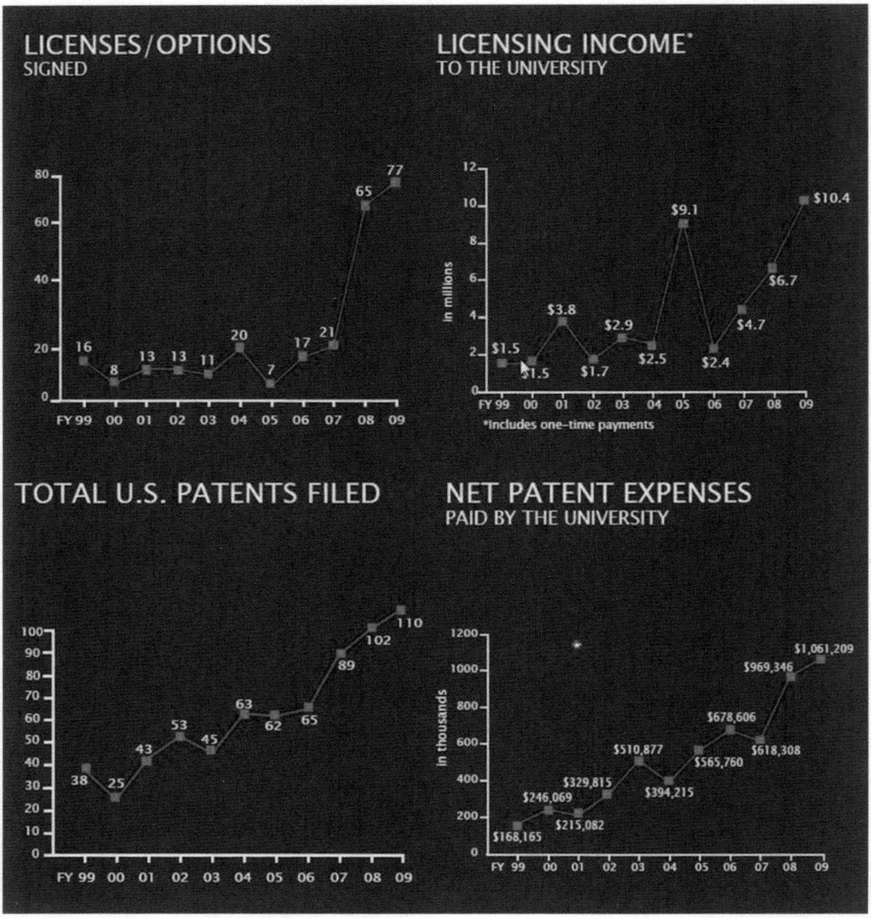


**Fig. 2** Credit: Unversity of Missouri Kansas City (UKMC)

**Table 2** Kansas City Research Centers of Excellence that received NIH Funding in Fiscal Year 2009

| | Number of 2009 awards | 2009 NIH Funds |
|---|---|---|
| Research Universities<br>University of Missouri Kansas City | 30 | $11,451,966 |
| Research Institutions and Hospitals<br>Children's Mercy Hospital | 4 | $1,557,269 |
| Kansas City Clinical Oncology Program | 1 | $520,000 |
| Stowers Institute for Medical Research | 15 | $4,154,415 |

*Source* NIH

## *3.7 Health and Life Science Partnerships*

At UMKC, the foundation for their strength in life science lies in interdisciplinary partnerships and intercollegiate collaborations. Two primary examples of interdisciplinary collaborative research at UMKC are located in the Center of Excellence in the Study of Dental and Musculoskeletal Tissues (CEMT) and the Vision Research Center (VRC).

The CEMT includes investigators from the UMKC Schools of Dentistry, Medicine, Nursing, Biological Sciences, and Computing and Engineering, while the VRC is a synergistic collaboration between the UMKC Schools of Medicine and Pharmacy, Truman Medical Center, as well as other disciplines and schools, combining clinical and research. The VRC positions Kansas City to become a national center for eye research where ocular diseases will be studied by top researchers in the country resulting in the seamless transfer of findings into practical use with patients.

As a founding stakeholder and supporter of the Kansas City Area Life Science Institute (KCALSI) and its initiatives, UMKC has forged new partnerships with community institutions and strengthened existing relationships to expand basic and applied research capacity. The vast majority of UMKC's faculty in the life and health sciences have formal affiliations or joint appointments with a partner institution.

The UMKC School of Medicine has established partnerships with several clinical affiliates. These include:

- **St. Luke's Hospital of Kansas City**: Serves as a center of medical education and a primary teaching hospital for the UMKC School of Medicine. It delivers care to 11 hospitals and related health services in the Kansas City area and surrounding region. Saint Luke draws sponsorship from the National Institutes of Health, the American Heart Association, the Saint Luke's Hospital Foundation, and others. Their medical researchers are conducting more than 200 research projects and participating in more than 250 clinical trials.
- **Truman Medical Center**: A not-for-profit two acute-care hospital health system in Kansas City, which also serves as a primary teaching hospital for the UMKC Schools of Health Sciences.
- **Children's Mercy Hospital**: A center of excellence for children's clinical care, medical education, and research.

The UMKC clinical affiliates provide knowledgeable and productive clinical faculty to enhance research programs, as well as access to patients from diverse demographic backgrounds presenting a wide variety of clinical conditions. In addition, the Department of Basic Medical Science has developed strong collaborative relationships with UMKC's clinical partners providing opportunity for translational research. "Collectively, these relationships enable the School of Medicine to study basic mechanisms of disease in the laboratory, contribute to the development of specific modalities, and evaluate the efficacy of these therapies in clinical settings." (*Source*: UMKC Department of Basic Medical Science)

Partner institutions in life science also include:

**Midwest Research Institute (MRI)**:

MRI is an independent, not-for-profit research laboratory for applied scientific research and technology development. It includes life sciences research among its five research programs.

MRI's life sciences clients are government agencies, including the Centers for Disease Control (CDC), U.S. Department of Agriculture (USDA), U.S. Food and Drug Administration (FDA), and many other health-related agencies at the federal, state, and municipal level. In addition, MRI serves industry, covering the pharmaceutical, veterinary medicine, and biotechnology business sectors; and academia, including many of the nation's leading universities.

With its alliance with Kansas State University (K-State) in Manhattan, KS, it promotes collaboration on bioscience research.

MRI also formed a for-profit sub-subsidiary, MRI Ventures, which handles the commercialization of intellectual property and new technologies that are developed either at MRI or through collaborative efforts involving the Institute. "The strategies used in translational research—bringing research developments out of the lab and into the marketplace—include all stages of new product development, licensing, technology assessment, market forecasting, investment management, strategy development, and business planning." (*Source: MRI*)

**Stowers Institute for Medical Research**:

Stowers Institute is a 501 (c)(3) charitable organization structured as a Medical Research Organization under U.S. Treasury statutes. Since it is privately funded, it gives Stowers researchers freedom to investigate promising ideas before there is enough evidence to garner outside financing—an arrangement that gives Institute researchers opportunities to initiate groundbreaking research.

It also houses the headquarters of BioMed Valley Discoveries, a for-profit translational R&D organization, whose role is to develop basic biomedical discoveries into applications to improve human health.

**Kansas City University of Medicine and Biosciences**:

The oldest medical school in Kansas City, Missouri, it is known for its excellence in academic medicine including education, research, and patient care. Through their affiliation with the Stowers Institute and the Kansas City Area Life Sciences Institute, they maintain an active research development and recruitment program to attract and retain top-notch researchers.

**University of Kansas Medical Center**:

Located in Kansas City, KS, it advances the discovery of new knowledge through its Schools of Allied Health, Medicine, Nursing, Pharmacy, and Graduate Studies.

It works in partnership with The University of Kansas Hospital which offers students and residents opportunities in patient care.

**Kansas City Area Life Sciences Institute (KCALSI)**:

Founded by the Civic Council of Greater Kansas City and the Kansas City Area Development Council. The organization works to advance the life sciences research, commercialization, and workforce development in Kansas.

Their role is to foster constructive relationships between the academic and private sectors; assist scientific collaborative research efforts; oversee and manage fundraising and marketing activities; advocate for the life sciences at the local, state, and national levels; and provide support to economic development and technology transfer & commercialization organizations.

**Kansas City Life Sciences Fund**:

Established at the greater Kansas City Community Foundation serves to support the region's life science initiatives. The fund received an initial contribution of $1.5 million from the Ewing Marion Kauffman Foundation located in Kansas City, MO which fosters entrepreneurship. The Kansas City Life Sciences Fund enables donors to help Kansas City recruit top medical research talent focusing especially on cures for cancer and diabetes.

**Kansas Bio**:

A non-profit servicing the bioscience community in Kansas, composed of private sector companies, and partnerships with public and academic institutions.

Kansas City, MO-KS benefit from organizations such as the Kansas Bio whose roles include:

- Maintaining the funding for state agencies, programs, and incentives that provide support for economic development activities within the bioscience industry in Kansas, such as: Kansas Bioscience Authority, Kansas Technology Enterprise Corporation, Angel Investor Credit Tax credits, and programs of the Department of Commerce related to industrial attraction, expansion, and retention, as well as programs that address talent development and recruitment in the bioscience industry.
- Develop guidelines for advantaging Kansas' partners related to intellectual property (IP) protection for bioscience companies and Kansas's research universities who wish to partner and engage in research projects.
- Evaluate the use of Research & Development tax credits in Kansas.
- Ensure the bioscience industry is engaged and allowed to give feedback in legislative-led tax incentive comparison discussions.
- Strengthen Kansas' workforce development activities that advantage the biosciences.
- Strengthen education programs and expand curricula which enhance bioscience education in K-12.
- Enhance Kansas' image as a globally recognized leader in the biosciences and as a preferred place to do business and start businesses involving advanced technologies.
- Support legislative and/or regulatory policies to promote and incentivize energy efficiency and sustainability measures led by KansasBio member companies.

(*Source*: Kansas Bio)

## *3.8 State Efforts in Life Sciences*

Both Missouri and Kansas have been committed to the development of a comprehensive network to support research, facilitate commercialization, and to promote the adoption of new technologies.

### 3.8.1 Missouri

In order to induce existing businesses to increase their research efforts, businesses are permitted to claim a tax credit equal to 6.5 % of the excess of qualified research expenses during the tax year, over the average amount of qualified research expenses incurred in Missouri during the preceding three tax years. The credit may be carried forward for up to five additional years.

New Enterprise Creation Act

The New Enterprise Creation Act is intended to generate investment for new, startup Missouri businesses that have not developed to the point where they can successfully attract conventional financing or significant venture capital from later stage funds.

Certified Capital Companies

A Certified Capital Company (CAPCO) may invest in an eligible business, which is in need of venture capital and cannot obtain conventional financing. The eligible businesses must derive their revenue primarily from manufacturing, processing or assembling or products, conducting research and development, or, service businesses, which can demonstrate that more than 33 % of its revenue would be from outside the state of Missouri.

### 3.8.2 Kansas

In Kansas, two state organizations are instrumental in the development of new technology: The Kansas Technology Enterprise Corporation, which works with start-up technology ventures and the Kansas Bioscience Authority, which works with bioscience companies investing in Kansas.

The Kansas Bioscience Authority was created by the Kansas Economic Growth Act of 2004 which established a funding engine that will generate more than

$580 million over the next 10–15 years, which will be fueled in research, commercialization, and workforce development programs that support the growth of the state's life science presence.

### Seed and Venture Capital Funds

A credit for a portion of a taxpayer's investment in Kansas Venture Capital, Inc., Sunflower Technology Venture, LP, or a certified private venture capital company or local seed capital pool may be claimed against its Kansas income tax liability.

The amount of the credit will be 25 % of the total amount of cash investment. The amount of credit exceeding the taxpayer's liability in any one taxable year may be carried forward until the total amount of credit is used.

### Angel Investor Tax Credit Program

Tax credits are offered against Kansas income tax liability for accredited investors making investments in seed and early stage capital financing for emerging Kansas businesses engaged in the development, implementation, and commercialization of innovative technologies, products, and agencies.

- The credit is 50 % of the investor's cash investment in the qualified business.
- If the amount of credit exceeds the investor's tax liability in any one taxable year, the remaining portion of the credit may be carried forward until the total amount of the credit is used.
- There is a $50,000 tax credit limit per company per year. Accredited angel investors can receive a total of $250,000 tax credits per calendar year.

Further incentives that make Kansas City, MO-KS a magnet for investment include:

- Its combination of big-city business amenities and small-market ease of living
- One of the fastest growing major job markets in the Midwest
- Lower business and lifestyle costs than most major metros
- A well-educated, extremely productive workforce
- The most geographically central major metro in the country.

## 3.9 Challenges and Prospects

Compared to other top life science clusters, Kansas City's needs larger life science companies, other than its current sole dominance of research institutions and hospitals. Having a presence of larger life science companies would give the region more name recognition and gravity in attracting skilled workers and companies. Both the states of Kansas and Missouri have provided great incentives to both relocating companies and existing companies which offer great cost-effective advantages over other life science clusters.

Unlike California life science clusters, Kansas City's life science cluster falls short of entrepreneurs and venture capital. At a recent Greater Kansas City Chamber of Commerce's Innovation Conference held in July 2010, Nicholas Franano, president and founder of Lenexa-based Novita Therapeutics asked a rhetorical question, "Can anyone name a very prominent medical device or biotech firm that has started in Kansas City in the last three years?" His point was that Kansas City has had "a pretty long drought" when it comes to new life science companies.

One of the key reasons cited during the panel discussion was a lack of venture capital, which all seemed to be in the West Coast or East Coast. Steven St. Peter, managing director of MPM Capital Group explained, "It's not that people hate the Midwest, it's just that that's where the venture capital is formed." The panel also added that it is not that area companies are not capturing venture capitalists' attention—9 of the 20 companies that received investments from Enterprise Center of Johnson County's (ECJC) angel investor networks were in the life science areas.

According to St. Peter, if Kansas City wants to attract more venture capital money, area companies need to be aware of where that money is coming from because financing plays a critical role in the process and success of drug development. Moreover, the process of getting a drug from the lab to the shelf takes more time and money than some venture capitalists willing to gamble. The average drug discovered in a university takes between 10 and 17 years and costs $1.2 billion and upwards before it gets to market, according to Scot Weir, director of the Institution of Medical Innovation at University of Kansas Medical Center. He added that KU is making strides to curb that cost and time commitment. Weir cited a recent instance in which the KU Cancer Center and the Leukemia and Lymphoma Society were able to get a novel drug into a clinic in 13 months for $1.5 million.

The average drug developer needs to be ready to spend between $50 and $150 million before the drug reaches the market. However, academia still faces challenges in turning drug discoveries into drug development as pointed out by Weir:

- Universities are holding onto technologies longer, which increases patent costs
- Innovative, advance technologies are more expensive
- There is a disconnect between intellectual property investment and faculty accountability
- There are large human capital investments associated with startup ventures.

## 3.10 KC Life Science Cluster-Related Websites and Publications


Children's Mercy Hospitals and Clinics http://www.childrensmercy.org/mobile/

Ernst & Young Beyond Borders. Global Biotechnology Report 2009 http://www.ey.com/Publication/vwLUAssets/Beyond_borders_2009/File/Beyond_borders_2009.pdf

Ewing Marion Kauffman Foundation: Kansas City Life Science Fund http://www.kauffman.org/advancing-innovation/kansas-city-life-sciences-fund.aspx

Jensen, D.G., "Novel Model for Biotech Cluster Development", Volume 24, Number 16, September 15, 2004

Kansas Bio http://www.kansasbio.org/about/documents/IAP_Directory_2_20090125.pdf

Kansas City Area Development Council http://www.thinkkc.com/SiteLocation/GreaterKCProfile/GKCProfile_main.php

Kansas City Area Life Sciences Industry Census: 2009 Census http://www.kclifesciences.org/pdf/Newsfeeds/Life%20Sciences%20Census%20press%20release.pdf

Kansas City Area Life Sciences Institute http://www.kclifesciences.org/index.php

Kansas City Business Journal: Panelists Tackle Reasons For Kansas City's Medical Device 'Drought', July 29 2010 http://kansascity.bizjournals.com/kansascity/blog/2010/07/panelists_tackle_reasons_for_kansas_citys_medical_device_drought.html

Kansas City University of Medicine and Biosciences http://www.kcumb.edu/

Midwest Research Institute http://www.mriresearch.org/BusinessPartners/Pages/default.aspx

Milken Institute. The Greater Philadelphia Life Sciences Cluster 2009: An Economic and Comparative Assessment.
http://www.milkeninstitute.org/pdf/PhillyLifeSciencesRprt_ex.pdf

Missouri Biotechnology Association http://www.mobio.org/industry_resources/

Missouri North America's Business Center http://www.missouridevelopment.org/index.html

Powell, W.W., Koput, K.W., Smith-Doerr, L., & Owen-Smith, J. "Network Position and Firm Performance: Organizational Returns to Collaboration in the Biotechnology Industry"

Saint Luke's Health System https://www.saintlukeshealthsystem.org/

Song, B., U.S Biotech Industry Report and An analysis on Bio Cluster Development. A report prepared for the Asia Bio Cluster Study, 2004

St. Louis Business Journal: Nixon pushes for biotechnology incentive fund, December 16 2009 http://www.bizjournals.com/stlouis/stories/2009/12/14/daily46.html

Stowers Institute for Medical Research http://www.stowers-institute.org/MediaCenter/docs/FactSheet.pdf

"Texas Biotechnology and Life Science Cluster Report", State of Texas, August 2005 http://www.texasindustryprofiles.com/PDF/twcClusterReports/TexasBiotechnologyandLifeSciencesCluster.pdf

Truman Medical Centers http://www.trumed.org/

UMKC Health and Life Sciences http://lifesciences.umkc.edu/

UMKC School of Medicine http://www.med.umkc.edu/research/default.html

University of Kansas Medical Center http://www.kumc.edu/

University of Missouri System. Office of Research and Economic Development, Annual Report 2009 http://www.umsystem.edu/ums/departments/ed/documents/2010annualreport.pdf

U.S Department of Health and Human Services. NIH Awards by Location and Organization http://report.nih.gov/award/organizations.cfm?ot=&fy=2009&state=MO&ic=&fm=&orgid=&view=statedetail

## 4 Northwest Education Cluster

The *New York Times*[5] recently gave one of us (Phillips) credit for being godfather to the Northwest Education Cluster. It is gratifying to watch this group's progress. While piles of politics, coveys of consultants, and bales of bucks have failed to produce robust entrepreneurial clusters in the industries that the State of Oregon officially targets, NWEC took off and has thrived for 10 years—at the cost of a few pizzas.

NWEC now comprises more than 40 companies and 260 individual participants. Pearson has paid a half billion dollars to acquire one of the companies (eCollege), and others are raising out-of-town capital at an impressive pace. To our knowledge there are no other clusters in the education space anywhere else in the US.

NWEC was born when Fred Phillips met investment banker Kelvin Ng, through the BudoDojo and the Harvard Club. In August 2003, Kelvin wanted to get Portland's education companies together and asked Drs. Niki Steckler and Phillips

[5] http://www.nytimes.com/2007/06/21/business/smallbusiness/21edge.html?_r=2&adxnnl=1&oref=slogin&ref=education&adxnnlx=1182492498-4uabtxWFL9zq2Uaigw6f/g

to emcee. Phillips arranged a room at the Oregon Graduate Institute, added his education industry contacts to Kelvin's, and ordered some pizzas. Kelvin brought along Jim Snyder, who has been leading the group ever since.

Phillips mapped a way forward in the event that the group found value in meeting with each other. Phillips also suggested the right questions to ask, to determine whether that value was there. (See the flow chart below.) It was Kelvin's idea to call the meeting, "just to see what will happen." So, Kelvin's initiative and Phillips' complemented each other nicely. We decided not to force matters; if sparks flew, we would host more meetings. If no sparks, we would enjoy the group's company for an evening, and then forget about it.

As it turned out, there were sparks. Subsequent developments played out actually much as the flow chart prescribed.

NWEC received no direct government support, though some state education-related agencies participated. It is thus a good example of a decentralized, spontaneous, networked initiative for knowledge-based industry, and economic development. Our European and Asian colleagues are sometimes surprised that this can happen; in their countries, most such things are instigated by governments. But it is possible because when it comes to cluster formation, knowledge, and a shared sense of possibility and empowerment are currencies every bit as valuable as money (Fig. 3).

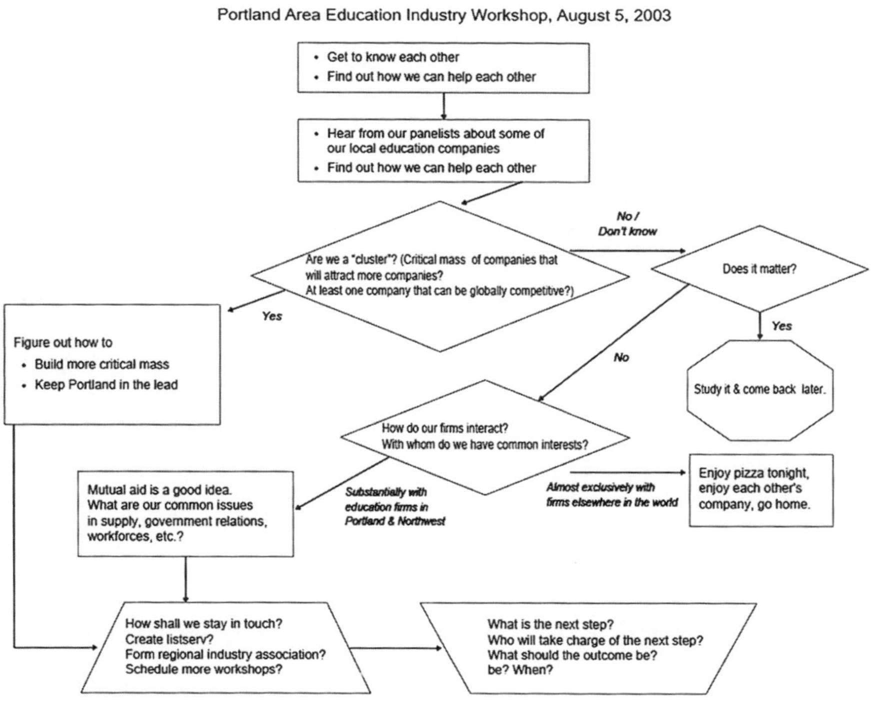


**Fig. 3** Why are we here, and what can we accomplish?

Oregon continues to attract educated people who are passionate about education. They are obvious employment targets for education companies, and obvious candidates to become education-related entrepreneurs. They do not want to leave Oregon due to the high quality of life locally, so the companies have to come here and/or stay here.

NWEC does what clusters do: meet for knowledge sharing and social networking, and connect with universities, governments, and other cluster initiatives in the region. Technically, NWEC is an industry association, representing companies in a nascent cluster (Phillips 2008).

Where VCs used to shun companies with "assets that could walk out the door tomorrow," that is now changing. This means an enterprise with smart, committed, creative, relatively immobile people can now attract venture capital. A number of these companies also have proprietary code assets, but educational software is still not a mass market, nor one that commands high markups. Perhaps slow but sure growth, without the pressure for investor liquidity, is a success factor for NWEC.

NWEC, then, started without government assistance. Its connection with academe is straightforward, as all of its members are concerned with academics from Kindergarten level through graduate school level. However, as government is a key player in education at all those levels, connections between the cluster and governments were desirable and inevitable.

Jim Snyder writes (Snyder 2008, and personal communication 2010):

> At [a] December 2007 meeting the cluster held a roundtable where the Oregon State Superintendent of Schools (Susan Castillo), the Oregon Senate Majority Leader (Senator Richard Devlin) and others from the teacher's union and Governors' Office all sat together to talk about working towards building 21st Century schools. The businesses themselves could not have had these people at the same table except for the leverage of the cluster. Over 70 attended the meeting and ideas percolated.
>
> As a follow up to the roundtable summit, the cluster had its quarterly meeting where the Portland Schools Foundation and the Chalkboard Project presented what they were up to and how we could work together. We are getting some pro-bono work from a local PR firm that is interested in what the cluster is doing.
>
> How can the cluster support the Governor's workforce strategy? One idea is to connect with very established manufacturing cluster here in Oregon to push for an educational/workforce reform agenda.
>
> Saltare Software did get an NSF grant in 2007 for $1.5 mllion in conjunction with the work they do with CASIO.
>
> The cluster does work with the Oregon Department of Education and the State in a few ways. ODE sends panel speakers to the cluster's meetings. The international trade person from Oregon State is a cluster member and is eager to know details of our member companies' international trade. A few years back we cooperated with the Willamette Educational Service District on rolling out their new project for students called Accelerate Oregon, a statewide initiative that leverages public and private commitment to Oregon's K12 education system to improve teaching and learning through the integration of technology.
>
> Part of what we are doing now is working with the Oregon Business Plan and Oregon Cluster Network to help define our cluster economic analysis and build a strategy report. I have a business student creating a first draft of a written cluster economic impact study (Tables 3, 4).

**Table 3** Northwest Education Cluster—participants by sector

| Government | Academe | Foundations/NGOs | Industry |
|---|---|---|---|
| Oregon State Superintendent of Schools | Oregon Health & Science University | Oregon Business Plan | eCollege |
| Oregon Governors' Office | Portland State University | Chalkboard Project | Cenquest |
| Oregon Department of Education | University of Oregon | Portland Schools Foundation | Saltare Software |
| Willamette Educational Service District | Various public and charter schools | Oregon Cluster Network | Local PR firm |
| | University of Portland | Teachers' union | Other firms |

# 5 Russia and China

## *5.1 Russia*

In the mid-1990s, Prof. Nikolai Rogalev of Moscow's Institute of Power Machinery and Mechanics traveled to the $IC^2$ Institute to learn the techniques of incubation and commercialization. He built Russia's first incubators and started his institute's commercialization and spin-off programs. 15 years later, Rogalev's pioneering work is being replicated elsewhere in post-Soviet Russia because of that country's government's efforts to foster closer ties between Russian academe and industry.

Project Eureca (Enhancing University Research and Entrepreneurial Capacity) joins a number of Russia's national research universities with four American universities for training and action in technology transfer and academic–industry cooperation. Private foundations from the US and Russia have pledged up to $2.5-million a year for Eureca's initial 2 years of support (Table 5).

According to the *Chronicle of Higher Education* (Blumenstyk 2010) and those the paper interviewed, Eureca's goals are to:

- Familiarize Russian governments and universities with the (to them) new ideas of technology transfer and university–industry partnerships.
- Increase Russian knowledge of intellectual property best practices.

**Table 4** Northwest Education Cluster—what each sector contributes

| Government | Academe | Foundations/NGOs | Industry |
|---|---|---|---|
| Policy direction | Research on learning | Philanthropy | Advanced products |
| Access to policy makers | Markets for educational products | Social entrepreneurship | Jobs |
| Export assistance | Course content | Research on educational needs | Input on policy |

**Table 5** Eureca participants

| Government | Russian Universities | US Universities | Companies | NGOs/Foundations |
|---|---|---|---|---|
| Russia | Lobachevsky State University of Nizhni Novgorod | U. Maryland | American venture-capital firms | New Eurasia Foundation |
| | | Purdue | Russian university spin-off companies | Association of University Research Parks |
| | St. Petersburg State University of Information Technology, Mechanics, and Optics | UCLA | US entrepreneurs | U.S.-Russian Foundation for Economic Advancement and the Rule of Law |
| | | U. Washington | | American Councils for International Education |
| | | | | National Council for Eurasian and East European Research |

- Educate on the effects of corruption on business growth.
- Build a culture of academic entrepreneurship.
- Build a knowledge economy that will be less dependent on oil and mining revenues.
- Contribute to the success of Skolkovo, a city near Moscow that is intended to be the Silicon Valley of Russia.
- Integrate Russian universities into local and regional economies.

The US universities will train Russian academics as trainers. They in turn will diffuse the knowledge of tech transfer to other Russian researchers and institutions. At the same time, the project will concentrate on the transfer of real technologies, with an emphasis on nanotechnology and computer science.

The Russian government has provided the needed directives and permissions, and has also passed a Russian version of the US Bayh-Dole Act, which allows universities to keep revenues from licensing of their own inventions. The partnership is expected to attract US entrepreneurs who will seek new technologies as well as funding from investors recently made rich by Russian oil and gas revenues. US venture capitalists will also be seeking investment opportunities.

## *5.2 China: TusPark*

Started 16 years ago but given a new boost by China's 2007 15-year plan, Tsinghua University's technology park emphasizes social capital, combining Chinese and Silicon Valley styles. Life at TusPark includes interacting with local executives from Microsoft, Google, and Sun, as well as with the entrepreneurs and service providers residing in the park. TusPark aims for a distinct culture, different

**Table 6** What each sector contributes

| Government | Academe | Associations/NGOs | Industry |
|---|---|---|---|
| Policy direction | TT theories/frameworks | Philanthropy | Licensees |
| Supportive laws | Educated knowledge workforce | Jobs | Jobs |
| Idea leadership | Idea leadership | Social entrepreneurship | Idea leadership |
| Export assistance | Leading-edge curriculum | Licensees | Taxes |
| | Interns | Networking | Philanthropy |
| Funding. Subsidies for young but essential future industries. | Enhanced local quality of life | Economic development initiatives | Investment |
| Invention | Invention | Invention | Invention |

from surrounding Chinese culture and more tolerant of Western styles of investment and entrepreneurship (Young 2010).

The park has had successes, including attracting 200 companies and the participation of the US' Purdue University. However, some tenant companies are headed by sons of top Communist party officials. This implies that strict meritocracy has not been adopted. The park's companies have been accused of violating WTO rules because excessive central government support drives down the effective price of exports. Then too, lax enforcement of intellectual property laws makes entrepreneurs reluctant to start companies even within the protection of TusPark (Blumenstyk 2010).

# 6 Lessons

Immediate lessons from the cases are:

- NGOs may play a role as important as governments, universities, or industry. It is really a quadruple helix, though that phrase lacks the poetic rhythm of "triple helix."
- Movement of people among the sectors facilitates inter-sectoral cooperation and flexibility. (There is the danger that the "revolving door" will reinforce the status quo.)
- Each strand of the triple helix may involve more than one university, more than one government agency, more than one company or industry.

As a first stage of analysis and as a conclusion to this chapter, let us examine what each sector brings to the table (Table 6).

## References

Benner, Mats, & Sandström, Ulf. (2000). Institutionalizing the triple helix: research funding and norms in the academic system. *Research Policy, 29*(2), 291–301.

Blumenstyk, G. In New Project, Russian Universities Tap American Expertise in Tech Transfer. *Chronicle of Higher Education*. Retrieved September 30, 2010, from http://chronicle.com/article/In-New-Project-Russian/124657/.

Cho, M.-H. (2008). Corporate helix model: The industry and triple helix networks. *International Journal of Technology and Globalisation, 4*(2), 103–120.

Etzkowitz, H. (2002). The Triple Helix of University–Industry–Government: Implications for policy and evaluation. Working paper 2002·11, Science Policy Institute. Institutet för studier av utbildning och forskning, Drottning Kristinas väg 33D, SE-114 28 Stockholm.

Fischer, K. (2010). Portugal aims to modernize with help from the U.S.: Effort to align academe and industry relies on campus partnerships. *Chronicle of Higher Education*. Retrieved September 12, 2010, from http://chronicle.com/article/Portuguese-Universities-Turn/124364/.

Leydesdorff, L. (2000). The triple helix: an evolutionary model of innovations. *Research Policy* 29 2000 243–255

Northwest Education Cluster. http://www.portlandedcluster.com.

Pfotenhauer, S. M. (2010). Portuguese partnerships seek to tackle European higher-education reform challenges. *Chronicle of Higher Education*. Retrieved June 24, 2010, from http://chronicle.com/article/Portuguese-Partnerships-Seek/66043/.

Phillips, F. (2006). *Social culture and high tech economic development: The technopolis columns*. London: Palgrave Macmillan.

Phillips, F. (2008). Cluster Bucks. Retrieved January 13, 2008, from http://consciousmanager.blogspot.com/2008/01/cluster-bucks.html.

Phillips, F. (2009). Toward a sustainable technopolis. 2009 UNESCO-WTA International Training Workshop "Green Growth based on the Science Park Initiatives".

Snyder, J. (2008). Consider the Northwest Education Cluster. Retrieved January 25, 2008, from http://blogs.edweek.org/edweek/edbizbuzz/2008/01/friday_guest_column_consider_t.html.

Song, B. (2004). U.S Biotech Industry Report and an Analysis on Bio Cluster Development. A report prepared for the Asia Bio Cluster Study.

UTEN Portugal2008–2009 Annual Report. http://utenportugal.org/wp-content/uploads/uten-annual-report-2008-2009.pdf.

Young, J. R. (2010). Chinese research park incubates hope for scholarly spinoffs. *Chronicle of Higher Education*. Retrieved September 14, 2010, from http://chronicle.com/article/Chinese-Research-Park/124420/.

# Sustainable Development of Technopolis: Case Study of Daedeok Science Town/Innopolis in Korea

**Deog-Seong Oh**

**Abstract** This chapter attempts to draw out transferable lessons for sustainable technopolis development by analyzing the Daedeok Valley (DV) in Korea, the attempt to redesign a technopolis by restructuring a pure science city (DST). This chapter would discuss the definitions and key concepts of technopolis and sustainability, and analyses the DV development progress toward sustainability through in-depth analysis of the structure, industrial activities, and network buildings. The major principles of sustainability (environment, equity, and futurity) are used as the evaluation criteria for these development trends. The results indicate that the DV development could achieve a higher level of sustainability progress in most of the evaluation criteria when compared to former model of pure science city (DST).

## 1 Introduction

Technopolis is a regional innovation strategy that generates sustained and propulsive economic activity through the creation and commercialization of new knowledge (Preer 1992). A technopolis is not merely a concentration of high-technology firms or research and development organizations, but also a comprehensive urban development strategy with sustainable context. At the center of the technopolis is the creative process of developing new technologies, translating them into commercial products or processes, and taking a careful approach to fostering competitive regional growth in terms of economic, social, and environmental sustainability. A frequently raised question is "how to measure the effectiveness of technopolis as an instrument of regional innovation policy to stimulate technology-led economic development and sustainable growth".

D.-S. Oh (✉)
Chungnam National University, Daejeon, Korea
e-mail: ds_oh@cnu.ac.kr

D.-S. Oh and F. Phillips (eds.), *Technopolis*,
DOI: 10.1007/978-1-4471-5508-9_5, 

Korean experience is interesting in this respect not only because of the extent to which it reflects regional innovation and economic growth but also, because of the way in which policy trends have been incorporated into regional conditions and tasks of national development (Ko 1998). Recently, in Korea, regional innovation policy based on technopolis has been shown to be a critical factor for achieving regional economic growth and sustainable development through networked collaboration between HEI's, research institutes, and industries. It is expected that technopolis could be a regional platform for a comprehensive approach of technological-based regional development in sustainable context. This is particularly apparent in Daedeok Science Town (hereafter referred to as DST/Innopolis uses technology-led development as a means of promoting regional innovation for sustainable development in Daejeon Metropolitan City (Oh 2002). DST/Innopolis is also a good example of explaining the impact of the creation of a technopolis on regional development. Particularly, there is clear evidences of strong expansion of high-tech venture businesses due to the successful development of the DST/Innopolis shift from pure science city to technopolis.

This research attempts to draw out transferable lessons for future development and assist sustainable technopolis growth in Korea by analyzing the DST/Innopolis development in Daejeon City, Korea. This chapter begins by identifying key definitions and principles of sustainability and technopolis through a literature review. It then explains the key research methodology and evaluation framework adopted. The major research interest is focused on the question "what kinds of shift have been made on the changing process from pure science city to technopolis (DST/Innopolis) for the last 30 years in terms of three main aspects of technopolis development like structure, activities, and network in the sustainable context." The last part of the chapter evaluates the effectiveness of technopolis policy and discusses some key issues for sustainable development.

## 2 Literature Review : Concept and Major Principles

### *2.1 Technopolis*

The technopolis concept emphasizes the need for a balanced approach to high-technological development. Instead of only focusing on technology it involves the creation of new settlements, complete with research parks, new universities, technology centers, housing and cultural facilities (Tatsuno 1986). Masser (1991) has pointed out that technopolises are larger in scale and often linked to the development of infrastructure and facilities on the new town model, whereas science parks are more limited in scope. Technopolises also tend to be more production oriented than science parks and have both national and regional objectives. The national and technological objectives are to offer to high-tech industries adequate industrial land and an environment suitable for creative research. These resources have become scarce in the major metropolitan areas. Consequently, the regional and technological objective is to promote technological development in less developed areas. For this purpose, physical, scientific, and institutional infrastructure is developed in a decentralized pattern by a combination of measures taken at the local and regional levels and by national government (Stöhr and Poeninghaus 1992). A useful distinction can be made between a technopolis and a science city. Whereas new settlements with many high-tech production firms but relatively few basic research institutes are referred to as technopolises, "science cities are areas dominated by basic research institutes, with relatively few high-tech production firms."[1]

## 2.2 Sustainability: Concept and Major Principles

The term "sustainable development" has been described in various ways without universal consensus. To extract a common definition, many researchers (e.g., Pezzey 1989; Pearce 1989; Trzyna 1995, etc.) have attempted to analyze various definitions of sustainable development.

Trzyna (1995) argued that at least seventy definitions of sustainable development are in circulation. Despite extensive efforts to extract a common definition, there is still no commonly accepted single definition of sustainable development (Haughton and Hunter 1994). However, the most widely used definitions of sustainable development are those in Our Common Future (1987) and Caring for the Earth (1991) (Trzyna 1995). In addition, it becomes apparent that there are three key domains in the conceptualization of sustainable development:

1. Environmental sustainability: Goodland (1995) argued that to achieve environmental sustainability: (i) in the 'source site' harvest rates of renewable sources must be kept within the regeneration rates; (ii) in the 'sink site' waste emissions must be controlled within the assimilative capacity of the environment so as not to impair it (Khan 1995). Environmental scientists argue that environmental sustainability is most essential for the continued well being of human society since, only within ecosystems, can vital life-support processes take place.
2. Social sustainability: this implies that human/social capital (e.g., tradition, equality, accessibility, empowerment, participation/sharing, cultural identity, etc.) must be preserved or improved to maintain sustainable development. Social scientists (e.g., Ruttan 1991) strongly argue that there is a close link between social condition (i.e., poverty) and environmental decay. It is argued

[1] These two terms and the types of communities they represent are not mutually exclusive. For example, Silicon Valley in the USA is well established as both a science city and technopolis (Rosers and Dearing, 1990). The term 'science city' applies best to Tsukuba science city in Japan, which was consciously planned as a basic research city.

that sustained poverty leads to sustained colonization and exploitation of natural resources (Khan 1995). Therefore, social scientists contend that pursuing social sustainability (e.g., increasing equality in resource distribution, accessibility to services, decision-making opportunities, poverty alleviation, etc.) is the key to preservation of environmental sustainability.

3. Economic sustainability: this implies that human-made capital, and its surrogate (money) must be maintained or improved for sustainable development. This traditional view of economic sustainability places undue reliance on the capacity of the market to encourage the efficient allocation of resources (Khan 1995). The assumptions are: (1) the supply of natural resources is unlimited, and (2) the power of technology will help replenish whatever natural resources are destroyed in the process.

This research uses these three key domains (environmental, social, and economic sustainability) of sustainability as an evaluation framework for assessing the level of DST's progress toward sustainability.

## 3 Research Methodology and Evaluation Framework

This study offers an analyzing framework to examine the experience of DST in-depth. This highlights the ways in which the technopolis policy has provided a framework for local efforts to promote regional innovation and economic growth in the sustainable context. The framework contains the aspects of technopolis development (structure, activity, and network) and evaluation dimensions of their sustainability (environment, equity, and futurity). Figure 1 presents an evaluation framework, which was devised and applied throughout the case study analysis.

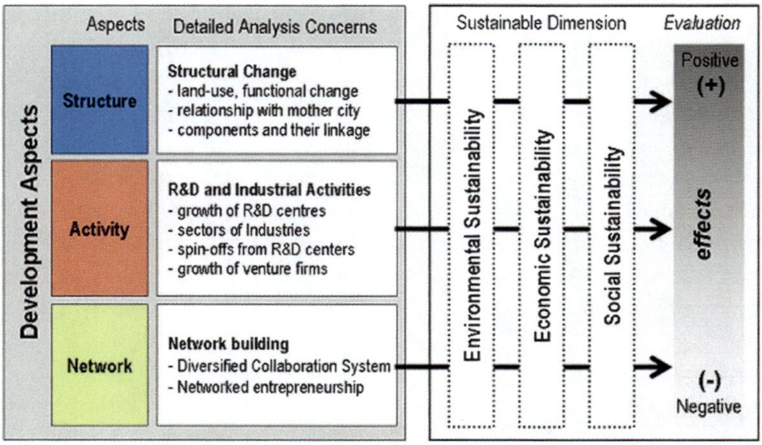


**Fig. 1** Analysis framework

When we analyze the technopolis development of DST in the sustainable context, the detailed factors and contents are as follows: First, in terms of structure, there are four factors of analysis related to technopolis development such as the relationship with mother city, land use, major components, and their linkage. Second, in the part of activity, growth of R&D centers, sectors of industries, spin-offs from R&D, growth of venture firms are analyzed in detail. Because significant value of the technopolis lies in its potential for achieving a synergistic rate of technological innovation through assisting the development of dense communication networks among heterogeneous R&D activities (Rogers and Dearing 1990). Third, research interests are focused on network, which is a strategic linkage among the R&D centers, entrepreneurs, and universities for regional development. In these respects, we analyze network building in terms of a diversified collaboration system and the process of the networked entrepreneurship in DST.

After analyzing the above mentioned aspects of DST development, we try to evaluate its sustainability in terms of three major principles (hereafter sustainable dimension) like environment, equity, and futurity, which were identified through theoretical review. When we take a cross check between development aspects and sustainable dimension, the evaluation of DST development can be undertaken.

# 4 Case Study: Development Pattern of Daedeok Science Town/Innopolis in Terms of Sustainability

## *4.1 Overview*

DST is located in the center of national territory, approximately 150 km from Seoul, the capital of Korea (see Fig. 2). At a distance of about 1 h from major cities including Seoul, Daegu, Gwangju, exchages with other regional government officials, industry leaders, and researchers is convenient. Total area of DST is almost 3,205 ha and is occupied by 71 R&D institutes, 6 universities, and approximately 800 enterprises. The site is surrounded with a good quality natural environment including ample green areas around Daejeon City, where people are not allowed to build houses or factories by specialized law.

There are 2 development stages in DST development. At initial stage, Daedeok Science Town (DST) established in 1973 within the area of Daejeon Metropolitan City is the unique science town in Korea which has been developed as a mecca of science and technology with strong research workforce. DST was intentionally created as an engine of enhancing national competitiveness of high-technology R&D Capability through the agglomeration of research institutes bringing together many national and regional development policy efforts from the last 30 years. DST, research complex has been built with the investment of $3.16 billion for the past 3 decades to better respond to the economic demands of the nation. As mentioned before, it has been well developed with the balance among research

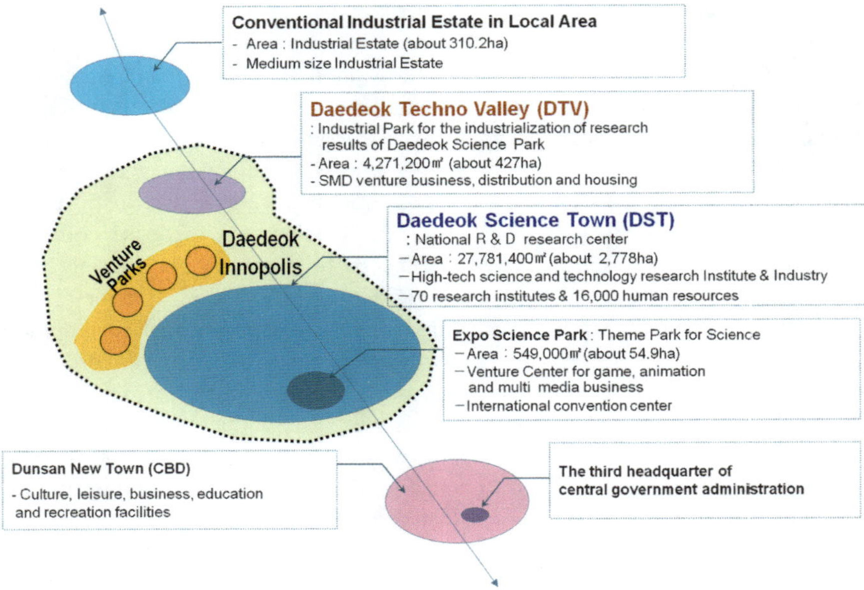


**Fig. 2** Composition of Daedeok valley and growth axis

institutes, academic institutes, industries, and public sectors where they possess the model of next "Silicon Valley" in Korea. Recently, the changed model of technopolis development can be seen in DST. Very active developments of venture park are being established to cover the huge number of high-tech SME's based on DST development (see Fig. 2).

Figure 2 shows the growth of DST containing high-tech industrial park (DST venture parks and DTV), conventional industrial estates, and CBD.

### 4.1.1 Former Stage of Development: A Model of Pure Science City

The first development phase began in the 1970s when the National Science Town was established at Daedeok, in Daejeon. The master plan represents a concrete attempt by central government to create a science city outside the capital region. The main objectives of DST development include: (1) creating a foundation for joining the ranks of advanced countries. The twenty-first century based on science and technology development, (2) fostering closer links among research institutes, academia, and industry through the effective placement of government-supported research institutes, universities, and private research institutes, and (3) establishing a pollution-free science garden city with cultural facilities. Especially, it was expected that DST can enhance research efficiency through a systematic and comprehensive R&D investment to each institute, create an optimal research environment by mutual exchange and application of a variety of information, knowledge and know-how, and promote investment effectiveness by the shared use of facilities, manpower, and projects. In order to achieve these aims, the development of DST in the initial stage was centered on public sector research institutes and nationally funded universities (Oh 1995). The dominant role of the government is particularly evident in the initial stage of DST development. In the 1980s, the function of DST was to establish a R&D center for government supported institutes and industries through the relocation policy of central government. There was a change in the expectations of the role of DST due to changing economic and social climate. Applied research and development that are directly applicable to commercialization have become an expectation. Therefore, DST has been criticized for being unable to fulfill such needs until mid of 1990s (Shin and Sul 1999).

### 4.1.2 Later Stage of Development: Technopolis Model with the High-Tech Cluster (Daedeok Innopolis)

Since the mid-1990s, efforts are being made to enable high-tech venture firms to be established in DST with mechanisms to support the commercialization of R&D results achieved in DST. The main components of this mechanism are Technological Business Incubator (TBI), Post-TBI, and venture park where start-up firms can direct their efforts in commercializing their new technology. In this consideration, Daejeon Metropolitan City Administration mapped out a scheme to develop a high-tech industrial district, Daedeok Techno Valley (DTV). Daejeon City establishes a citywide industry development structural plan in 2001 linking DST, DTV, existing conventional industrial estate, and Expo Science oriented amusement park. Within this structural plan, DTV has key role for developing later stage of DST development as an advanced science and technology venture town and connecting the City's existing industrial estates and former stage of DST development.

In the mid-2000s, central government had designated DST and its vicinity as Daedeok Innopolis for the further development of technology commercialization based on R&D. This variety of high-technology based development activities enable the DST to be re-designated, the first National R&D Special Zone National R&D Special Zone[2]: to carry out the pilot project of regional innovation and technology-led economic development with strong support from central government. Late stage of DST's development provided important opportunities for regional development.

DST/Innopolis is a place where people, technology, and nature harmoniously coexist as well as where business and research activities are conducted efficiently

[2] National R&D Special Zone : all accommodating within the Daejeon Metropolitan City—Daedeok Science Town (DST), Daedeok Techno-Valley (DTV), Industrial Complex, Yuseong Tourist zone.

and conveniently. But much more than that, it is a world-class innovation cluster where creativity openly blossoms and breathes. It is now becoming the high-technology clusters where the applications of intellectual property based on R&D are made very active. It is, also, the only place for highly educated workforce of R&D with 10 % of PhDs in entire Korea. It consists of the most convenient facilities for the technology commercialization of specialized industrial sectors like IT, BT, and NT with the research institutes and universities together.

## *4.2 Analysis of DST/Innopolis Development in the Sustainable Context*

### 4.2.1 Structure and Environment

Linkage with Mother City and Land Use

When DST was incorporated administratively into Daejeon in 1983, DST was merely a secondary urban center. Since the end of 2000, DST has faced a new turning point in terms of its role for urban structure. DTV is a bridge between pure science city and technopolis toward regional strategy in order to generate a maximum synergistic effect by linking these critical areas functionally and spatially. DST can be understood as the comprehensive approaching policy to contain R&D function, technology commercialization of DTV, and mass production by local industry (Table 1).

In 1981, 38.5 % of the DST site was designated as a research and education area so that research facilities could be established there. This research and education area was again increased to 3,032 acres in 1985, 46.6 % of the whole site.

The change from DST to Daedeok Innopolis brought about the change of land use at three times (Oh 2001). In the initial stage (1986–2001), land use of DST focused on the site for R&D and education. Then, the total area of 5,221 ac consisted of academic and research area, residential area and green belt area, and

**Table 1** Land use allocation according to the development stage

| Land use (ac) | Former stage: DST | | Later stage: Daedeok Innopolis(DST + DTV) |
|---|---|---|---|
| | 1st master plan (1981–1985) | 2nd master plan (1986–2001) | 2002-present |
| Academic research | 2,633 (38.5 %) | 3,203 (46.6 %) | 3,203 (40.5 %) |
| Residential | 577 (8.4 %) | 577 (8.4 %) | 761 (9.6 %) |
| Greenbelt | 1,127 (16.4 %) | 1,048 (15.3 %) | 1,179 (14.9 %) |
| Industry (venture business/light production) | – | – | 368 (4.7 %) |
| Other uses | 884 (36.7 %) | 2,036 (29.7 %) | 2,396 (30.3 %) |
| Total | 5,221 | 6,684 | 7,907 |

area for other uses. The academic and research area occupied 38.5 % of the present area, it was designated to a few public institutions which were moved into DST by the central government. In 1985, the master plan of DST was changed for the first time. About 46.6 % of the site was designated for research and education purposes so that research and universities facilities could be built there. Thereafter, when central government encouraged public institutions to move into the DST development the concept changed from pursuit of a self-sufficient city to conurbation with its mother city (satellite town). The total area of DST expanded to 6,684 ac. The expansion of academic and research area was for public institutes and private R&D centers moving into DST. And mature stage (2002-present). In particular, the allocation of greenbelt is decreased, although the huge expansion of technopolis has been made. The designation of a national R&D zone enabled DST to expand its territory of DTV and then to establish the Daedeok Innopolis site. A significant change occurred in the land use of DST caused by feverous venture inauguration. About 368 ac (4.7 % of total area) for light production and venture business which allocate the R&D and industrial use zone is expanded in DST because the spin-offs from many research institutions secured their site within the area to stay in close proximity with their mother organizations. In fact, most spin-offs at an initial stage used existing space of their mother organizations. Several institutions launched their own TBI business. Thereafter, Daejeon City established the project of Daedeok Techno-Valley (DTV) as high-tech park near the DST. Because R&D and manufacturing activities of industries and venture firms are inseparably related to each other, industrial sites for light production are designated here in DTV comprehensively (Oh 2005). On the contrary, the percentage of greenbelt zone has been decreased gradually from DST to Daedeok Innopolis.

## Functional Change

In terms of functional change, DST functioned as a pure science city in the initial stage. It focused on R&D and education to enhance the capability of high-technology development. Above all, it concentrated on pure R&D stressing on basic research and restricting production and other industrial activities. Since manufacturing was allowed with restrictions in DST, this model was not appropriate for the regional innovation on the basis of technology transfer. In addition, Daedeok Innopolis development could facilitate the formation and growth of new businesses in the region because it is going to be developed as a high-tech venture business center for domestic and multi-corporations commercializing advance science and technology.

However, as many private R&D centers settled in DST, it began to have another function gradually developing into a multi-functional technopolis. Private R&D centers had aptitudes to directly link R&D to industrial activities because they had their attributes to pursue profits like an enterprise. Accordingly, they caused the acceleration of technological commercialization there—and venture incubation began to grow remarkably. Under these circumstances, a demand for industrial function was raised not from outside but inside Daedeok Innopolis. It was Daedeok Techno-Valley project created near DST in 2003. Since then, the development process of DST, as a multifunctional technopolis, has been accelerated to mature stage (See Fig. 3).

In particular, we can see the critical change of functional structure from pure science city to technopolis for complex model of research and business development. The role of Daedeok Valley in the region is activity in connection with the promotion of regional innovation activities based on the collaboration among universities, research institutes, and industries. The main concerns of RIS center in DV put on the technology-based regional development through the support of the organization of production system, management support, and the commercialization of technology by R&D activities. Moreover, the role of local authorities varies substantially between these three major components of technopolis development. As the collaboration among industry, universities, and research institutes becomes increasingly critical and the support role of local government within the DV

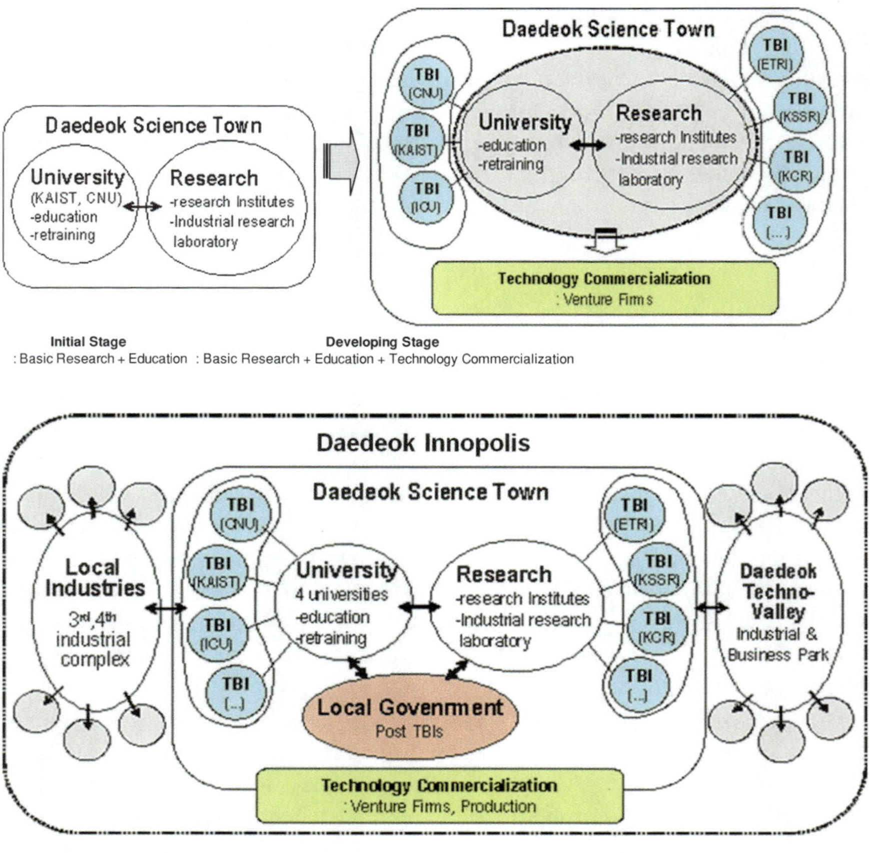


**Fig. 3** Functional change; from science city to technopolis

responds by extending their roles. Accordingly, the structure of pure science city has developed into that of techopolis of regional innovation cluster among the agents of innovation in RIS (see Fig. 3).

Evaluation: Environmental Sustainability

In terms of sustainability, we can evaluate the development features shifted from DST to Daedeok Innopolis in the following aspects. Integration with mother city, which has been made through the structural change of Science City enables technopolis development and its mother city, Daejeon, for the future. The problem of limited development of a high-tech sector in DST could be resolved by establishment of a new high-tech industrial park called DTV. This is very positive for economic sustainability. However, the huge expansion of technopolis site including venture parks near DST could give negative impact on environmental dimension of sustainability. The expansion of self-help establishment of small venture parks near DST must be carefully controlled to protect the environmental sustainability. However, the negative influence can be solved through the strategies of compact city development, which will target regeneration projects of inner city areas of conventional industrial parks with eco-industrial park planning concepts. In addition, the supply of housing and urban facilities in the mother city can save further expansion of technopolis site in the current greenbelt area (Table 2).

In conclusion, the structural change of DST gave the positive effect for sustainable development not only for science city but also its mother city. Only the expansion of venture parks around DST must be controlled by careful land use strategy to protect the environmental sustainability in Daedeok Innopolis.

### 4.2.2 Activity: R&D and Industrial Activities

R&D

DV is a host to a concentration of 21 Government Research Institutes including ETRI, 10 public research institutes, and 40 private research Institutes.[3]

[3] ETRI led the world's first ever commercialization of CDMA and has been instrumental in developing DRAM technologies. More recently, ETRI successfully developed WiBro (Wireless Broadband Internet) and DMB (Digital Mobile Broadcasting) which are fast proving to be next generation IT industries. Other prominent research institutes including the Korea Atomic Energy Research Institute, which successfully produced localized atomic fuel, are part of our complex as well. Daedeok Innopolis is also home to the Korea Aerospace Research Institute which developed KITSAT satellites 1–4 which are widely recognized as embodying globally advanced technologies. (DAEDEOK INNOPOLIS Management Office 2006).

**Table 2** Sustainability evaluation in terms of *structure and environment*

| Factors and contents | Development features shifted from DST to Daedeok Innopolis | | | Sustainability evaluation |
|---|---|---|---|---|
| | Former stage | Later stage | | |
| *Linkage with mother* | *Satellite town* (unique *Science city*) | – *Integration with its mother city* (Multi-functional *Technopolis* as RIS center) | | *Positive* for economic sustainability<br>– sharing the DV's development potential with mother city area<br>– compact development |
| *Functional change* Component and their Linkage | *Basic* | | | *Research + Education(Univ.)*<br>– pure R&D<br>: High-tech development oriented<br>: restricting production |
| *Basic* | *Research + Education (Univ.) + Technology Commercialization* (venture firms) | *Basic Research + Education (Univ.) + Technology Commercialization* (local industries, industrial & business park)<br>→ *Regional Innovation* | | |
| *Land use* | R&D use area *(38.5 %)*<br>: for public Institutes | R&D use area *Expansion (46.6 %)*<br>: for Public institutes and Private R&D centers | *National R&D Special Zone (70.2* $km^2$*)*<br>– DST, DV, DTV<br>– R&D, TBI business, Industrial & Venture business, manufacturing etc | *Negative* for environmental sustainability (demand for careful land use strategy with respect to ecological planning) |

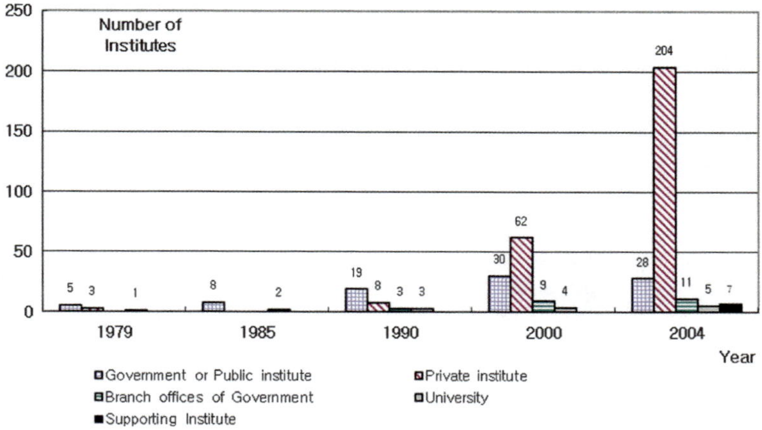


**Fig. 4** Growth of R&D centers and related institutions in DV. *Source* DAEDEOK INNOPOLIS Management Office, 1979–2004 (http://ddinnopolis.or.kr/)

There are typically two levels of R&D activities in DST. One is corporate R&D labs, which are engaged in basic research; the other is division-level labs, which are used for research directly related to the business and manufacturing needs of their particular division-level activities, and also focus on the commercialization of their products for their business. The characteristics of R&D activities in DST have changed according to the changes in the main body of R&D activities. For example, the research institutes of private firms in Daedeok are varied in their sectoral pattern of activities, while many government institutes are biased toward basic research and development rather than for production and commercialization (Fig. 4).

At an initial stage, DST focused mainly on pure basic research, because most of the institutes in DST were public organizations. In 1988, about 42 % of all government research institutes were located and this percentage had increased until year 2004. From the late 1980s, private sector research institutes began to relocate from the Capital Region into DST, which resulted from the changed objective of government that promoted applied research by private R&D center. Private sector substantially has been higher than public sector in the R&D centers since 1992. Table 3 shows the increasing development of R&D centers and their employees.

Since private institutes carried out R&D for their mother enterprise, they had more interests in the technological innovation and profits of mother enterprise than public benefit. These attributes of private institutes made R&D activities in DST more subdivided and varied. This variety of activities is important, because it suggests the potential for future development in DST as well Daejeon, as it can be attractive to a wide range of research-focused technologically advanced and innovative organizations. For example, the R&D fields which were conducted in private R&D labs in 2005 were more subdivided than those of 1992. There are 32 private institutes in 12 fields related to particular industrial sectors like IT, BT, NT,

**Table 3** Sectoral of R&D centers; number (employee)

| Description | Public sector | Private sector | Total |
|---|---|---|---|
| Comprehensive R&D | 4 (1,230) | 1 (83) | 5 (1,313) |
| Biotechnology | 4 (482) | 5 (455) | 9 (937) |
| Information technology | 3 (2,655) | 5 (848) | 8 (3,503) |
| Precision chemical | 1 (403) | 9 (1,672) | 10 (2,075) |
| New material (including high molecule) | | 5 (549) | 5 (549) |
| Meccatronics (including marine science) | 2 (571) | 2 (305) | 4 (876) |
| Resource, Energy development | 5 (2,173) | 2 (76) | 7 (2,249) |
| Astronomy, Aerospace, Astronautics | 3 (806) | 1 (109) | 4 (915) |
| The others | 6 (2,483) | 2 (112) | 8 (2,595) |
| Total | 28 (10,803) | 32 (4,209) | 60 (15,012) |

etc. As R&D activities of private institutes became subdivided and varied, the potentiality of DST for technology commercialization has increased.

### Industrial Activities: Growth of Venture business and SME's

The industrial activities of DST have focused on venture business from research institutes' spin-off firms. There is a concentration of highly educated and skilled research manpower in public and private R&D center. This means that there is a high possibility of commercialization of research results. It is argued that the high potentiality is because there are about 2000 doctorate degree holders. Recently, active spin-offs are increasing in the research institutes and universities and significant sources of technology-oriented new ventures are being generated. Actually, 7.1 % of total venture firms in Korea is located in Daejeon Metropolitan City and Chungnam Province (6 % in Daejeon, 1.1 % in Chungnam Province). It is the highest concentration outside the Seoul Capital Region.

The start-up activities in DST began to be observed from the time when institutions launched their TBIs. As research institutes and universities including ETRI, KAIST, and CNU participated in TBI business, the number of venture firms suddenly increased. This enabled venture start-up to be accepted as an essential aspect and activity area rather than be understood as a mere part of R&D activity. When venture firms from the major 10 R&D centers in DST were surveyed in 2004, there were only 2 venture firms in 1990, 130 firms in 2002, and an increase to 219 firms in 2004. If job creation is analyzed for the same years, 35 persons were employed in 1990, 2,212 employees in 2002, and 3,237 persons were employed in 2004 (See Fig. 5) ETRI, KAIST, and Korea Standard Science Research Institute are the main bodies, which actively contributed to technology commercialization. And, the venture start-ups extended to all area of Daejeon since late 1990s. The venture inauguration set out from DST/Innopolis began a full-scale relation with regional territory not only to Daejeon Metropolitan City but also to its vicinity.

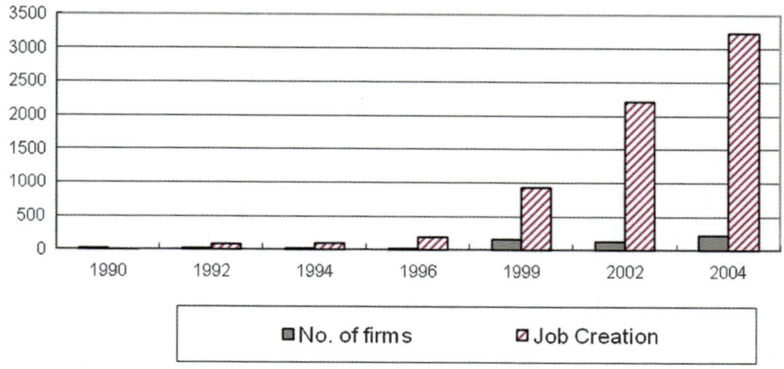


| Year | 1990 | 1992 | 1994 | 1996 | 1999 | 2002 | 2004 (DV total) |
|---|---|---|---|---|---|---|---|
| No. of firms | 2 | 6 | 7 | 20 | 154 | 130 | 219 ( 824) |
| Job Creation | 35 | 84 | 96 | 187 | 924 | 2,212 | 3,237 (22,395) |

**Fig. 5** Growth of venture firms from major R&D centers in DST/Innopolis. *Source* DAEDEOK INNOPOLIS Management Office 2006 (http://ddinnopolis.or.kr/)

According to the report of Daedeok Innopolis Foundation (DIF) in 2006, within 10 years DST/Innopolis will have obtained immense growth as well as the number of our occupant companies and R&D institutes and the number of international patent registrations will rise. Especially, the companies are estimated to be 3,000 units in 2015 and the number of foreign R&D institutes is expected to be 20 in 2015 up from 2 in 2004. In addition, DIF is predicting the international patent registration will be increased about 10 times more than the number in 2004 and sales by technology commercialization are expected to reach $31,296 million in 2015 (Fig. 6).

Figure 7 shows the current status of DST/Innopolis development including high-tech industries, HEI's, and R&D institutes. Growth of venture business is shown in sales increases. According to the investigation of Korea Information ServiceINC (2002), the amount of turnover by venture businesses in Daejeon area was ₩382.5 billion ($ 375 million) in 2000. It is about double that of 1999, and corresponds to about 5.6 % of all output of manufacturing industries in Daejeon area. If those figures are classified by scale, venture firms which had less than ₩30 million showed the highest ratio by 38 % of all firms in 1998. However in year 2000, venture firms which had sale over ₩500 million showed the highest ratio of 35 %. Venture firms in Daejeon made great strides. Average sales value per company was shown to have grown 36 times, from ₩ 25.9 million in 1998 to ₩ 928 million in 2000 in relative terms. Daejeon expects that the scale of venture firms will be increased up to ₩ 6 trillion in 2005 if sales of venture businesses in Daejeon grow at current trend. It corresponds to about 56 % of whole sales from manufacturing industries in Daejeon and 33 % of GRDP from the region.

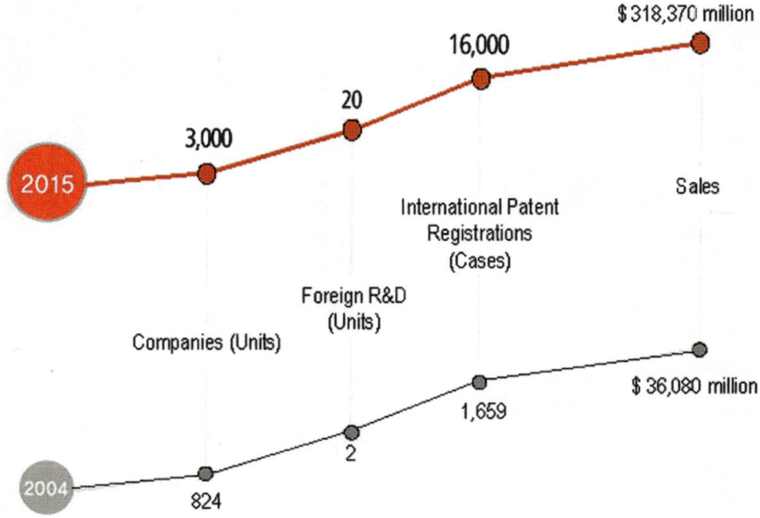


**Fig. 6** Estimation for the growth of DV from 2004 to 2015. *Source* DAEDEOK INNOPOLIS Management Office 2006 (http://ddinnopolis.or.kr/

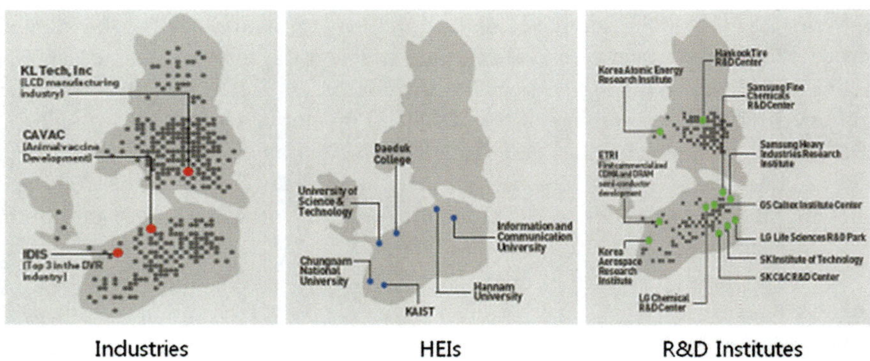


**Fig. 7** Current status of DST/Daedeok Innopolis (2006). *Source* DAEDEOK INNOPOLIS Management Office 2006 (http://ddinnopolis.or.kr/

An impact of DST on regional economic growth is related to industrial restructuring. Until 1993, the industrial structure of Daejeon city did not change much in spite of the establishment of DST. However, since its function was changed in 1993, Daejeon Metropolitan City has made efforts to enable high-tech venture firms to move into DST by providing support mechanism for the commercialization and production of R&D results in DST. The main components of these mechanisms are TBI, post-TBI, and venture industrial parks where start-up firms can concentrate on commercialization of their technology. To support the venture firms and SME's in the high-tech industries in DST, we offer various tax

**Table 4** Sustainability evaluation in terms of activity

| Factors and contents | Development features shifted from DST to DV | | | Sustainability evaluation |
|---|---|---|---|---|
| | Former stage | Later stage | | |
| Growth of R&D center | Pure basic research (5 public institutes, 3 private sectors, 1 national Univ.) –focused on public R&D and education | Applied R&D (30 public institutes,4 universities): Private sector(62) increased and varied | Comprehensive development of high-tech industry (71 R&D institutes, 6 universities) | Very positive for economic sustainability (base for technology-led regional development) |
| Industrial activities (spin-offs and growth of high-tech SME's) | None | Spin-offs from Public R&D center (launched by their TBIs only) : KAIST, ETRI, KSSRI (Total 255 including 219 spin-offs firms) | Spin-offs from various sector of private and public R&D centers(TBIs, BIs): active and strong growth of venture firms and their settlement in the city (219 in DV, 824 in Daejeon) | Positive for economic sustainability (base for future economic growth) |

benefits including exemption or reduction in acquisition taxes, registration taxes, and property taxes. In addition, a DAEDEOK Special Fund has been formed to support stable venture capital investments. Also, full efforts are being exerted in the expansion of occupancy areas & facilities. We offer further support in the form of marketing activities so that product development can be transformed into profits. As a result, venture businesses grew rapidly in the Daejeon area.

Development of high-tech industry due to the growth of venture business gave rise to changes in the industrial structure of Daejeon. According to the data of National Statistical Office (2002), the industrial structure of Daejeon shows that the ratio of manufacturing industry has increased from 12.9 % in 1999 to 13.6 % in 2000. The changes in industrial structure appear in the business feature of the transferred firms. Daejeon City, for example, expects to change the composition rate of industries between traditional industries and IT industries from current 70 : 30 to 40 : 60 in the near future.

Evaluation: Economic Sustainability

The growth of R&D centers in DST is a very important base for future development in Daejeon City, the representative high-tech region in Korea. The active spin-offs and high-tech SME's have had a very positive impact on its economic sustainability. These developments also had good effects for regional economic growth and industrial restructuring. In particular, this trend is spreading over the whole city. This trend is very positive for the sustainable urban growth in terms of equity to share the development potential in DST with the mother city. The expected effects of high-tech industrial growth enable local high-tech manpower from HEI's and R&D centers to sustain their workplaces in Daejeon in the future. In particular, possible growth of specialized high-tech industry is a positive symbol for the future competitiveness of technopolis. This is the evidence of positive evaluation of DST's role for future regional development (Table 4).

### 4.2.3 Network and Clustering

Networking: Industry-Academia Collaboration Network

At the initial stage of DST's development, research institutions formed an organization, that is called 'Daedeok Science Town Association (DSTA)'. DSTA is an organization promoting mutual friendship and contributing to the development of DV through regular meetings of chief personnel of research institutes.[4] The

---

[4] The chief managers of 52 institutions in DST are its members, of which about 84% are public institutions. DSTA has major jobs such as activation of DV, policy formulation for DV development, formation of research atmosphere, building up international cooperation. Enhancing cooperation among member institutions, and building ties with local communities are also its major tasks.

collaborative activities among research institutions in DV was gradually increased and subdivided as time progressed. KAIST,[5] CNU, and ETRI individually began to match the collaborating project with industrial sectors. The different way of collaboration between R&D institution and industrial sector in DST can be made in technology transfer activities. In particular, it was KAIST and CNU that have led this collaboration since the initial time. They facilitated knowledge creation, human capital formation, technology transfer provision of regional leadership and knowledge infrastructure, etc.

Furthermore, regarding collaboration between university and industry, there are also BI (Business Incubator), Regional Research Center (RRC), and Rental Research Institutes (RRI). ETRI has been also active in the collaboration with the industrial sector as a research institute in DV. It has two major roles in the collaboration with the industrial sector: one is venture incubation through TBI; the other is the management of higher education institution called ICU.[6] Figure 8 explains how the collaborative network in DV works in the initial time.

Networking among venture firms in DV can be classified into two types. One is an active and direct networking by organizing community, and the other is a passive and indirect networking by information infrastructure. 'Daedeok 21C Venture Family' is an example of networking by organizing community. It is an early model, in which venture firms in DST were networked and confined to organizations which were made up of spin-offs only from Public research institutions.[7]

However, networking alone is not enough to make organizations but to maintain close relationships for active technological commercialization and project building. Therefore, it is necessary to develop cooperative projects for members to keep them in close cooperation with a particular project.

---

[5] KAIST concentrates its efforts in the field of new technology, particularly in the R&D of high-tech industry, and supporting venture start-ups so that it could create a base for international competition in DST. KAIST has a Technology Innovation Center (TIC) and a Consulting Center for Technology Transfer which is to accelerate technology innovation by providing research results from KAIST to opening firms and support venture business firms by the establishment of the national technology utilization system.

[6] ETRI is the second research institution which launched its own TBI business in DST. The TBI of ETRI is the core venture incubator in DST, as it has been playing a critical role in incubating many venture firms in the field of IT. Moreover, ETRI established Information and Communication University (ICU) to provide bachelor, master and doctorate courses for training professional manpower in the field of IT.

[7] The aim of 'Daedeok 21C Venture Family' is to accelerate information exchange and technology innovation among venture firms in DST and to develop a venture 'ecosystem'. It had changed its name to '21C Venture Family' in 2000 and expanded its number up to 280 firms from Daejeon and Chungnam province.

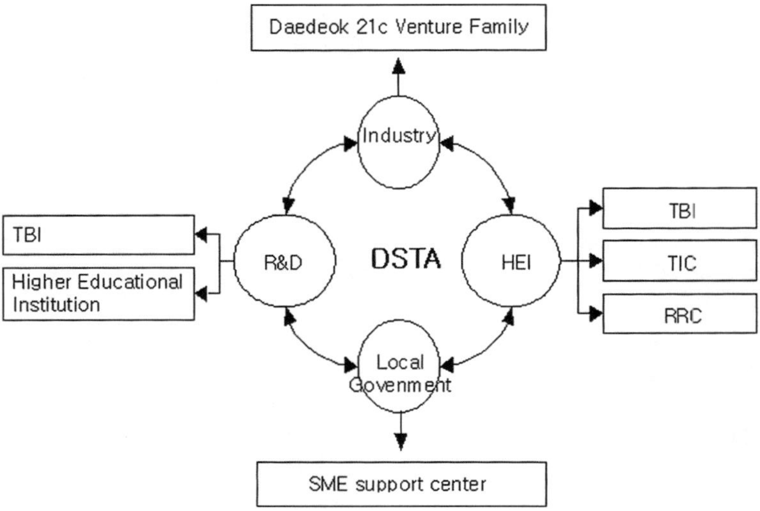


**Fig. 8** Collaborative network between R&D and industrial sector in DV

## Regional Cluster of Specialized Industries

The several specialized industries in Daedeok like IT, BT, NT are now characterized by many high-tech SME's concentrated in regional clusters and engaged in symbiotic relationships with multi-international firms. The networked communities consisted of the firms of these specialized industries and are present at mature stage of DST development. A survey of these patterns (Oh 2002) in detail has indeed confirmed that life-style reasons are the main factor in the initial location decision, while this might suggest some magnetic effect as the basis for all parts of DST and Daejeon City area. In other words, high-tech clusters start by accident due to the presence of focal organizations (R&D centers, universities) that generate spin-offs. These spin-offs, lead to continuous links, which sustain clusters of several specialized sectors of industry in the long run. There are several good examples such as 'Daedeok IT-net', and 'Daedeok Bio Valley'.

Daedeok IT-net in Daedeok Valley is a university—R&D center partnership in field of IT industry, whose purpose is to create a new business model for clustering. Daedeok IT-net is a cluster that promotes joint ownerships for gathering market information and know-how and does marketing by members together in the field of IT. Especially, this cluster keeps on doing the closer collaboration with the mother organizations of spin-offs like ETRI, KT, and KAIST. The commercialization from IT-net is involved with SME, venture firms, and also large enterprise through the collaboration. The supporting institute, Daejeon High-tech Industry Promotion Foundation plays an important role for high-tech clusters in the field of IT industry by supporting networking among universities, R&D center, and industry in the region (Fig. 9).

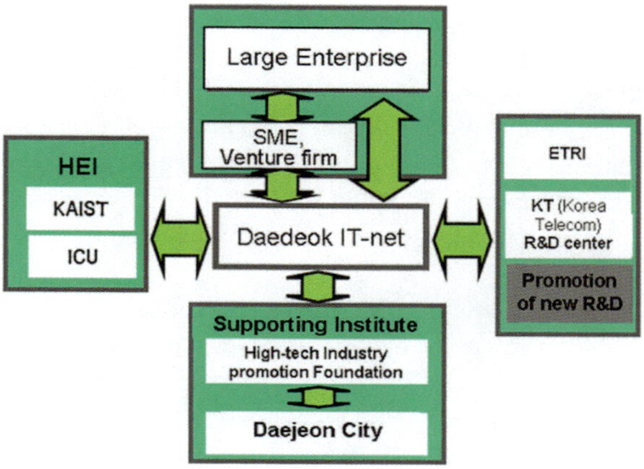


**Fig. 9** Daedeok IT-net Clustering in the Field of Semiconductor

Daedeok Bio-Community consisted of 14 venture firms. It is a group whose purpose is developing competitiveness through co-work, information exchange, and joint marketing. The local agency of regional innovation gave its full support for the establishment of Bio-cluster with the close collaboration of national Biotech Institute in Daedeok Innopolis, because this sector of industry is the driving force for future development of Daedeok Innopolis in the twenty first century. This pattern of clustered development, through which venture firms are collaborating with one another, is very meaningful in terms of an alternative business model. It can facilitate the SME's competitiveness in the market of specialized industrial sectors (Fig. 10).

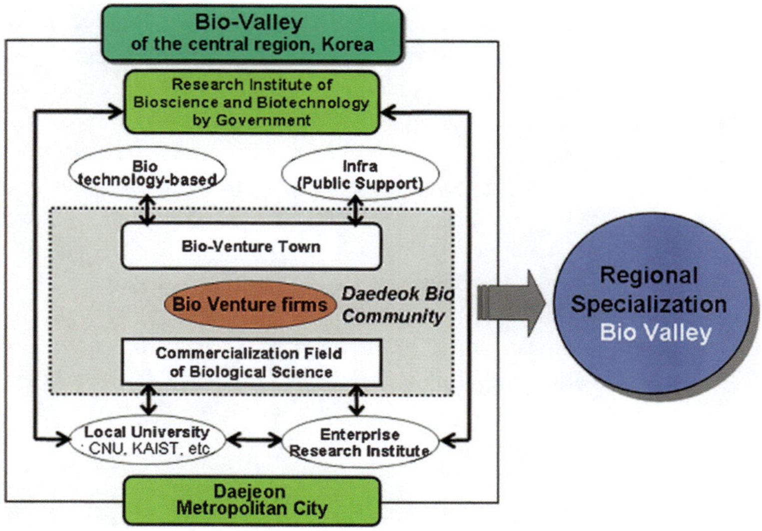


**Fig. 10** Daedeok Bio-Valley cluster

The other networking type among venture firms in DST/Daedeok Innopolis is collaboration based on the information infrastructure. It is networking that utilizes cyber space offered by Internet. For instance, there are two firms that manage online collaboration systems such as Hallow DD and DVI News in DV. These include serve information, DB for business, publicity (information) activities, and consulting through online network. Collaboration by information infrastructure has some advantages because it is not restricted by space and is able to connect in real time.

High-tech cluster in RIS has been made on the basis of the close relationships between research institutes, universities, and industrial sectors including venture businesses in DST/Daedeok Innopolis. It has been developed in three steps; network building for the collaboration between R&D center and industry, diversified collaboration and networked entrepreneurship of venture businesses and regional cluster of several specialized high-tech industries. The three different ways of network development have been aided by self-help meetings and information exchanges at initial stage and networked community building for developing competitiveness of industrial activities with the indirect support of local government for RIS system.

## Evaluation: Social Sustainability in Terms of Networking

The changing process from formal network to regional clustering is meaningful for the regional technopolis development. The network building called DSTA was a formal association only to play a limited role for public interest. It opened a possibility of collaboration among research institutes in DST at the first stage. The second stage of development initiated by major research university (KAIST) and research institute (ETRI) in 1990s was the starting point for the close cooperation of technology transfer and commercialization. In particular, the regional cluster of specialized industrial sectors (IT, BT, etc.) is the most important development. It can accelerate the regional capability of technology-based development and regional innovation through closer collaboration among high-tech SME's which were spin-offs from universities and research institutes and keep them closely linked. This kind of process enables the city to form RIS system to integrate these three stages of the sustainable development for regional innovation. The high-tech cluster in RIS can be a very important base for the economic sustainability of DST/Daedeok Innopois and Daejeon Metropolitan City. Clustered development enables the local parents of industrial sector to join the stage of regional economic growth through close collaboration with high-tech sectors including R&D centers, HEIs in DV. Furthermore, the regional cluster of IT, BT, can give the potential of high-tech industrial growth in the future (Table 5).

**Table 5** Sustainability Evaluation in Terms of Network

| Factors and contents | Development features shifted from DST to Daedeok Innopolis | | | Sustainability evaluation |
|---|---|---|---|---|
| | Former stage | Later stage | | |
| Networking | Simple network among research institution for minor communication(DSTA) | Network building for the collaboration among universities, research institutes and industries (total 255 including 219 spin–off firms) | Networked entreprenurship<br>– venture community<br>– Internet community | Positive for social sustainability (balanced growth of local industry in high–tech sectors)<br>Very positive for economic sustainability (potential base for future leading high–tech industry development) |
| Cluster | – | – | Regional cluster the specialized industrial sectors<br>– clustered development | |

## 5 Conclusion

This chapter tries to identify the effectiveness of technopolis policy to sustainable development with particular reference to Daedeok Science Town/Innopolis in Korea. Through the case study, it is identified that development from pure science town (DST) to technopolis (Daedeok Innopolis) has had positive impacts on sustainable development based on the strategy for the structural change, growth of high-tech firms based on R&D, and a collaborative network for regional cluster. Results show us positive changes for sustainable development in the following three aspects.

**First,** structure of Daedeok has been changed from science city to multi-functional technopolis which can facilitate technology commercialization and production. Structural change through the designation of Daedeok Innopolis gave further potential of facilitating the growth of high-tech venture business based on DST's R&D. There is an interesting give and take strategy between Daedeok and mother city. While Daedeok provides city with the potential for high-tech industrial growth, mother city, Daejeon provides it the city amenities such as good residence and cultural services. It means that DV and other city areas can benefit each other for the sustainable development of the region.

**Second, there** is a change of activities from pure R&D to industrial development through huge growth of venture business in Daedeok. Private research institutes and venture businesses, started from spin-offs, are the driving force of this change. Spontaneous entrepreneurship of venture firms and their cooperation with R&D centers and local HEI are also very important factors that can make Daedeok actually contribute to the regional economic growth and innovation in Daejeon Metropolitan City, Korea. Daedeok is also a good example to show us that private R&D centers can lead the technopolis development through their profit-oriented projects and also subdivided researches on technology transfer. They are another potential for endogenous development. These economical benefits from industrial activities are related with futurity which is a concept of sustainability. The benefits can create economic positive-effects without compromising the ability of future generations. Technopolis can enrich resources to meet future generations' needs through the growth of high-tech sectors including HEIs. Thus, it can be argued that technopolis is very positive for economic and social sustainability.

**Third,** network building for collaboration among universities, research institutes, and industries is a critical factor that enables technopolis to have the ability of sustainable development. The regional cluster of specialized industrial sectors at third stage is the most important development. It can accelerate the regional capability of technology-based development and regional innovation through closer collaboration among high-tech SME's which were spin-offs from universities and research institutes and also maintain the close link with them. Network building construction, regional cluster, and networked entrepreneurship enable resource sharing for the profitable growth of high-tech sectors. This sharing allows technopolis to have a social equity which is a concept of sustainability and a positive effect for sound entrepreneurship, and a closer linkage between academia and industry and formation of high-tech cluster.

In summary, technopolis can take more advantageous structure for sustainable development. In the environmental aspect of sustainability, technopolis has similar effects with the mixed-use development because technopolis has multi-function while science city has a single-function. In addition, technopolis could create economic benefits without compromising the resources of future generations. Furthermore, technopolis could support social sustainability via building a new regional cluster and networked entrepreneurship through information and resource sharing.

## References


DAEDEOK INNOPOLIS Management Office (2006) http://ddinnopolis.or.kr/.

Goodland, R. (1995). The Concept of Environmental Sustainability. *Annual Review of Ecology and systematics*, *26*, 1–26

Haughton, G. & Hunter, C. (1994). Urban development and the environment. In *Sustainable cities. Regional Policy and Development Series 7*. London: Jessica Kingsley Publishers, 9–29

Khan, A. (1995). Sustainable development: The key concepts, issues and implications. In *Sustainable Development*, 63–69

KO, S. C. (1998). A study on the determinants of imbalanced regional development : An application of regression model for a bias due to heterogeneity across region, *Regional Study*, *14*(Supply. 2)

Korea Information Service INC (2002), Management performance analysis for foreign-invested enterprises

Masser, I. (1991). By accident or design: Some lessons from technology led local economic development initiatives. *Review of Urban and Regional Development Studies, 3*, 78–93.

Oh, D.S. (2002) Technology-based regional development policy: Case study Daedeok science town, Daejeon Metropolitan City, Korea, *Habitat International*, (Vol.26, pp. 213–228). UK: Elsevier Science.

Oh, D.S., Kim, K.B., Jeong, S.Y. (2005) Eco-Industrial park design: A Daedeok Technovalley case study, *Habitat International*, (Vol.29, pp.269–284). UK: Elsevier Science

Oh, D.S. & Masser (1995). High-tech centers and regional innovation—some case studies in the U.K, Germany, Japan and Korea, in Bertuglia. C.S (eds.), *Technological change, Economic development and space*. Germany: Springer

Oh, D.S., et al. (2001). Daejeon science park development plan. Daejeon: Daedeok Technovelly Co.

Pearce, D. W., Markandya, A., & Barbier, E. B. (1989). *Blueprint for a green economy*, London: Earthscan

Pezzey, J. (1989). *Economic analysis of sustainable growth and sustainable development*, Environment Department Working Paper, No. 15, 12–23, The World Bank, Washington, DC

Preer, R. W. (1992). *The emergence of technopolis: Knowledge-Intensive technologies and regional development*, New York: Praeger

Ruttan, V. W. (1991). Constraints on sustainable growth in agricultural production: Into the 21st Century, *Canadian Journal of Agricultural Economics. 39*, 567–580

Shin, Dong-ho, et al. (1999). *Long-term development strategies for the Taedok science town*. STEPI: Korea.

Stöhr W., Poeninghaus R. (1992). Towards a date-based evaluation of the Japanese technopolis policy,*Regional Studies*, *26*(7), 605–618

Tatsuno, S. (1986). The technopolis strategy: Japan, *High technology and the control of the 21st Century*, New York: Prentice Hall

Trzyna, Thaddeus C. (1995). Linking values and policy for sustainable development: An international strategy to build the sustainability ethic into decision-making. Sacramento, California: Commission on Environmental Strategy and Planning, IUCN. IUCN/CESP Working Paper No. 6.

# Support Mechanism in Technopolis Toward Green Growth

**Herbert Chen**

**Abstract** The phrase "Green Economy" was first mentioned in '*Green Economy Blue Book*' by the British economist Pierre published in 1989. Green Economy promotes economic growth, instead of blocking it in the name of protecting the environment. It advocates changing extensive economic growth with the features of big investment, huge consumption, and serious pollution into intensive economic growth with the features of high efficiency, less resource-consuming, and less waste discharging and calls for harmony between economic and social growth and the proper load that nature can bear. As a new economic model aiming at harmonious development of economy and environment, Green Economy can fully satisfy the requirements of the scientific outlook on development of harmony and people first with energy saving and environmental protection as its goal.

## 1 The Concept and Significance of Green Growth

### *1.1 The Basic Concept of Green Economy*

The phrase "Green Economy" was first mentioned in '*Green Economy Blue Book*' by the British economist Pierre published in 1989. Green Economy promotes economic growth, instead of blocking it in the name of protecting the environment. It advocates changing extensive economic growth with the features of big investment, huge consumption, and serious pollution into intensive economic growth with the features of high efficiency, less resource-consuming, and less waste discharging and calls for harmony between economic and social growth and the proper load that nature can bear. As a new economic model aiming at

H. Chen (✉)
Tsinghua University Science Park(TusPark), Beijing, China
e-mail: chenhongbo@tuspark.com

D.-S. Oh and F. Phillips (eds.), *Technopolis*,
DOI: 10.1007/978-1-4471-5508-9_6, 

harmonious development of economy and environment, Green Economy can fully satisfy the requirements of the scientific outlook on development of harmony and people first with energy saving and environmental protection as its goal.

## *1.2 The Practical Significance of Green Growth*

### 1.2.1 Green Growth is a Crucial Step for the World to Take to Get Out of the Economic Crisis

The international crisis this time is actually an economic crisis. It is the result of violation of sustainable development laws and violation of economic and social development laws, warning us that the traditional economic development model is in doubt. It is the requirement of sustainable development to persist in Green Growth, which brings to the world harmonious development among economy, society, population, resources, and environmental protection, co-progress of virtual and real economy, coordination macro-control of the government and market allocation of resources. Thus Green Growth would help the world to form new industries, new resources, new products, new demands, and new employment, promote domestic demand, break out of the haze of global financial crisis, and step into an overall economic revival.

### 1.2.2 Green Growth is the Correct Path to be Taken for Future Global Economic Growth

The development of a global economy, especially the development of modern and current industrial economies, have passed through a zigzag history of huge energy consumption, material-consumption, waste discharging, pollution in return of not always good profitability, and sometimes, even negative benefit. Countries, regions, units, and enterprises worldwide shall consider Green Growth as their common goal and pursuit, drawing up the roadmap for global green development and develop a plan tailor-made for each region, unit and enterprise, coordinating various policies and activities, and fulfilling their due responsibilities.

### 1.2.3 Green Growth Shall be the Orientation for Governments to Develop Economic Policies

Governments from all over the world, especially from developed and developing countries shall speed up technological innovation and application, promote the development of green industries, and advocate the construction of green cities via various control policies. Systems including GDP KPI, Green Growth appraisal system, green product appraisal system for the manufacturing industry, as well as the standards and certification systems for green food and medicines will be completed in the future to push Green Growth into a scientific, regulated, and standardized track.

#### 1.2.4 Green Growth is the Path of Hope Leading us to Sustainable Development

To create the miracle of Green Growth, there should be a role model, a Green Growth region, and city to propagate and promote the idea and methodology of Green Growth, advocate the green method of production, initiate the green life-style, and so as to finally realize sustainable development for the region and city.

## 2 The Definition and Basic Function of Technopolis

### *2.1 The Definition of Technopolis*

Technopolis is an important media for promoting the progress of industrial technologies and the development of regional economies, as well as a place packed with various resources including government, enterprises, strong learning facilities, and research, financial, and intermediary institutions. In the Technopolis, we can witness integration and the emergence of innovative factors, new industries, and high-tech industries, accelerated upgrading of traditional industries, and the speedy development of eco-industries. Technopolis has become a major contributor to the development of the city, and the mainstay for industrial restructuring and upgrading of the region.

### *2.2 Basic Functions of Technopolis*

Create friendly working and living environment to attract high-tech enterprises and capable talents.

Support the growth of pioneering enterprises, foster entrepreneurs, and incubate innovative companies.

Attract large-size enterprises to provide them with value-added services and to create positions based on knowledge.

Build an effective platform for cooperation between politics, industry, academic achievements, research, finance, and so on to facilitate the transfer of scientific and technological results.

# 3 Several Ways for the Technopolis to Support Green Growth

The definition and basic functions of Technopolis tell that it supports and influence Green Growth from the following aspects:

## *3.1 The Development of the Technopolis Shall Satisfy the Requirements of Green Growth*

The construction and development of Technopolis have a growing impact on the region, with some of them becoming a major part of regional, or even national economic growth. Take a high-tech zone in China for an example: in 2006, there were 11 high-tech zones with a gross output value that contributed 20 % to that of the city, 33 high-tech zones were with an industrial added value contributing 20 % to that of the city, with 22 of them over 30 %. Therefore, whether the development of the Technopolis supports the concept of Green Growth is deciding the possibility of Green Growth in its city or region.

### 3.1.1 Case: Foxconn Science and Technology Park in Yingkou, Liaoning, China

Located in the coastal industrial site of Yingkou, Foxconn Science and Technology Park emphasizes "green, science and technology, and culture," bearing in mind the concept of sustainable development, aiming at establishing a park with completed functions, flexible development, and intelligent and green environment. Its features include:

*Create an attractive environment for talents*
Provide an ecological and landscape environment, meaning to create green landscape systems among functional zones and buildings, dotted with running water, so as to realize the ideal harmony between the architecture and the environment with the park boasting the features of a real park, with tranquil greens, singing followers, fragrant flowers, and fresh air.

*Plan for the big picture with by-phase implementation*
Overall planning shall be given to the road and piping in the functional area. The park can be constructed by phase according to the priority of the production of the factories so as to minimize the impact on the environment.

*Take full advantage of environmental-protection technology*
Reduce room temperature in the summer with the design of ventilation lane ways so as to save air-con cost. Besides, a recycling system of rain is available in the park, collecting the rain from the house and the ground in the reservoir for the use of water scenery, greens, and car washing, so as to reduce water consumption.

*Take advantage of energy-saving technology*
To present the harmony between science and technology and nature in the park, the renewable resources including solar energy and wind are fully utilized. Advanced solar photovoltaic panels are applied on both the inside and outside of the workshops in order to achieve passive solar energy heating and a solar energy in-door hot-water system. Wind power generators are installed on both sides of the main roads to supply electricity for the factories.

*Emphasize on both science and technology and leisure*
The fencing system outside the architecture includes Winping energy-saving glass and a double-layer glass wall, which not only represent a sense of modern architecture, but also contribute to better performance for ventilation, heat insulation, sound insulation, frost-proofing, etc. In addition, the park has adopted a ground-energy-powered air-con system, that generates a warm environment in winter and a cool environment in summer, as well as providing hot water by using groundwater and soil.

## *3.2 The Park Shall Train Experts for Green Growth*

Cristina Fernández, the person in charge of employment in Southeast Asia and Technology and Energy Strategic Plan of the Organization for Economic Cooperation and Development, pointed out on Sustainable Development and New Industrialization, sub-forum of Caifeidian Forum, the two most important factors for Green Growth: one is employment, the other is green skills. To better adapt to a low-carbon economy, the labor market is in need of an intensified education and training system, so as to achieve new green skills, which shall be far more advanced than current skills, and fulfill the needs of future industry. In 2020, jobs requiring medium- and high-skills will take up over 80 % of the job market, the trend started to show its potential in 2008. Meanwhile, the chances for low-end skill bearers will be limited to less than 20 % of the job market by 2014.

As the bridge between academic institutions and the commercial environment, Technopolis carries out a close cooperation with research institutes, boasting abundant resources of experts and scholars. On the other hand, the Technopolis has attracted a great number of young high-tech talents. The Technopolis can invite experts and scholars to give technical and skill trainings on Green Growth to its technicians, and teach them the concept of Green Growth and practical skills.

### 3.2.1 Case: Green Growth Overseas Training Organized by Beijing Zhongguancun

To achieve the goal proposed in the 11th Five-year Plan is to "establish an energy-saving and environmental-friendly society," which requires exchanges between government officials and the overseas energy-saving and waste reduction experts so as to carry out the energy-saving and waste reduction project with scientific and technological input, and implement the measures advocated by the state.

Therefore, in 2009, Beijing Zhongguancun organized a training in America on "energy-saving, waste reduction, and Green Growth" for institutions including governments of all levels, science and technology parks, high-tech zones, economic development zones, industrial parks, and college science and technology parks.

The main contents of the training include:

1. Introduction of status and related policies of energy saving and waste reduction in America.
2. Research and application of renewable energy technologies and energy saving regarding technology "energy-saving" and "waste reduction"

   – Biomass biogas technology
   – Biomass gasification and pyrolysis technology
   – Biomass liquid fuel technology
   – Wind energy technology
   – Solar energy technology
   – Car exhaust control technology (including hydrogen cell technology).

Institutions involved in the training in American include:

- California Energy Commission
- California Energy Commission (CEC) established in 1979, CEC is a key energy administration in California
- California Air Resources Board
- CARB short for California Air Resources Board belongs to California Environmental Protection Agency.

  – University of California—Davis
  – University of California—Berkeley
  – CEC, CARB, and EPA
  – California Biomass Association
  – California Wind Energy Association
  – Auckland Household Garbage Treatment Center.

## *3.3 The Park Shall Promote the Development and Utilization of Green Growth Technologies*

Green Growth poses high requirements on a technological level, and the application of modern science and technology is a crucial precondition for Green Growth. Modern science and technologies include new energy and renewable energy, low-carbon energy (solar energy, wind energy, geothermal energy, etc.), waste-water treatment, application of LED, green transportation systems (expand low-carbon railway network, increase light rail cars and trains, and reduce usage of cars), use substitution for water, etc.

The Technopolis is full of high-tech enterprises, which makes it the place with the most integrated high technologies in the region. Therefore, Technopolis shall take advantage of its resources and promote development and utilization of Green Growth technologies and related products such as: industry association, industry alliance, and research alliance.

### 3.3.1 Case: Beijing Zhongguancun Science and Technology Park

The Technopolis is full of high-tech enterprises, which makes it the place with the most integrated high technologies in the region. Therefore, Technopolis shall take advantage of its resources and promote development and utilization of Green Growth technologies and related products such as: industry association, industry alliance, and research alliance.

Promote energy saving and waste reduction via industry alliances in order to integrate government resources, enterprises and research institutes to a realized $1 + 1 > 3$. An example is the New Energy (heat pump) Application Alliance. Presently, this alliance is carrying out two projects: one is the Recycled Water Heat Pump Cooling and Heating Project in the Olympic Park which can heat or cool a building area of 413,000 sqm. This project generates from recycled water an amount of energy equal to that of 4,900 t of standard coal or 3,350,000 $m^3$ of natural gas with a goal of reducing the discharge of carbon dioxide by 4,300 t, sulfur dioxide and nitrogen oxides of 213 t, dust of 49 t. The other project is "Recycled Water Heat Pump Model Project in Qinghebaoshengli Resident Compound." With this project, Qinghebaoshengli Resident Compound will become the first resident compound in Beijing, or even in China, with a cooling and heating system powered by waste water.

The industry alliances in Zhongguancun Science and Technology Park are just one of the models of energy-saving and waste reduction industry parks in China. It has become a new mode in the industry parks to promote regional Green Growth via energy saving and waste reduction supported by system innovation.

## *3.4 Technopolis Shall Support Green Enterprises' Growth*

At the early stage when the Technopolis is established, a huge number of high-tech enterprises are expected to move in, among which those featuring technologies or products that are green growth-oriented shall be given priority for acceptance so that an atmosphere of encouraging green growth among enterprises will be formed at the very beginning. After the enterprises move in, the Technopolis shall make more efforts to utilize the innovation service system to help such enterprises, including knowing their difficulties with regard to capital, technology, human resources, patent protection, providing relative solutions, enlarging enterprise scale, improving their technologies and products, promoting sales, and so on; so that green growth will be ultimately realized in concerned fields and regions.

### 3.4.1 Case: Beijing Yizhuang Technopolis Supports Green Growth-Oriented Enterprises

In the first half of 2008, Yizhuang Economic and Technology Development Zone achieved 34 % growth of total industrial value and 60 % taxation, while the newly added consumption of land only increased by less than 10 %, water 11.9 %, and power 25 %. The anticipated full-year industrial value is up to 200 billion RMB and taxation has reached 12 billion RMB. Rapid economic growth supported by low-energy consumption could never have been realized without the system of "selective merchant recruitment" and systematic measures for water saving. It has been a major task for the Yizhuang Development Zone, over recent years, to set up a scientific standard of "selective merchant recruitment" for attracting high-tech industries that are technology-intensive with low energy consumption and pollutant discharge.

For that purpose, the Development Zone has rejected investment valuing over 6 billion RMB over the past 2 years. In addition, systematic water-saving measures have also been an innovative action taken by Yizhuang Development Zone to drive the mission of energy saving and emission reduction. By the end of May in last year, 140 institutions in Yizhuang had received water balance test and the authority has formulated the *Comprehensive Water Resource Plan*, which aims to realize integrated utilization of "five types of water", i.e., rain water, polluted water, tap water, surface water, and recycled water, so that the water resource can be intensively utilized in a recyclable way.

## *3.5 The Science and Technology Park Shall Try to Gain Policy Support from Local Government for Green Growth*

Development of Technopolis can never proceed without the support of government policies. Vice versa, to formulate policies and implement them, the government also needs to refer to the experience and ideas of Technopolis.

Development of the science and technology park is one of the driving forces for government to make relevant policies. The Technopolis represents the most advanced productive force in the region. It is said that "The economic base determines the superstructure". Therefore, to continuously develop the productive force, it is imperative for the government to renew old policies and develop new ones to adapt to new situations and needs. The Technopolis has a great group of enterprises and talents, with whom the government shall consult regarding new policies; also, new policies shall be implemented in the Technopolis for trial for a certain period of time prior to official implementation of the modified version.

The Technopolis shall give full play to its significant role in influencing government to conduct trials and providing feedback regarding policies and completely report about the needed support on the government's part for improving the idea of green growth plus realistic approaches during implementation. In that way, the Technopolis will be able to promote green growth in the region hand in hand with the government.

### 3.5.1 Case: Enterprises in High-Tech Industrial Park of Tianjin Take Measures to Reduce Emission

To encourage and promote green growth, the High-tech Industrial Park of Tianjin gave full play to its role of influencing the government in policy-making, and finally through discussing and negotiating with the concerned governmental section time and again, *Measures to Encourage High-tech Industrial Park of Tianjin to Reduce Emission* was formulated, mainly including:

Set up "Special Fund for Emission Reduction" to support actions for emission reduction in the park. Principally, the annual budget shall be ten million RMB.

Following projects shall be particularly stressed in the park: enterprises shall build their own system of polluted water processing and recycling; environment-friendly and energy-saving equipment shall be installed; clean or renewable energy shall be used for power generation, lighting, cooling, and heating. Comprehensive utilization of resources such as solid wastes (including domestic wastes, electronic wastes, etc.) should also be promoted.

Meanwhile, the Municipal Government of Tianjin offers an annual fund of 200 million RMB for supporting the utilization of clean and renewable energy in the purpose of ensuring the good effect of promoting green growth in Tianjin.

## *3.6 The Science and Technology Park Shall Create a Favorable Environment for Green Growth*

Development and construction of the science and technology park shall coordinate closely with its surrounding environment. It is particularly important that when a new Technopolis is being built its development shall be properly integrated with the existing city. Construction of a new Technopolis is always accompanied by huge amounts of investment, moving-in of large number of enterprises, and talents and attraction of various service institutions and innovative organizations, which all impact on the development trend of the existing city to a great extent.

Establishers of Technopolis should make full use of the opportunity to greatly promote and support the idea of green growth and the proposing and implementing of various measures. Through the construction of new Technopolis, original industries in old cities will be upgraded while the technologies can be improved, which in return will further facilitate green growth.

### 3.6.1 Case: Songhu Lake Technopolis in Dongguan, Guangdong Province

For the purpose of providing support for overall transformation and innovation of old Dongguan city through the construction of Songshan Lake Technopolis, the original plan placed the focus on developing the new Technopolis. Songshan Lake Science and Technology Park invited the China Academy of Urban Planning to compile the *Overall Planning for Songshan Lake Science and Technology Park of Dongguan.* According to the functional shortage of old Dongguan city, it is clearly proposed that: Songshan Lake area must aim to develop itself into a new Technopolis and follow the idea of promoting coexistence of humans and nature as well as coordinated development of industry and nature. Approaches such as "center around ecological balance", "develop with resource reservation and cultivation" and "harmonious co-existence of technology and nature" have also been proposed. The Technopolis is also expected to develop simultaneously with the old city while balancing each other in terms of space expansion and functional improvement; in that way, the Technopolis will become an important part of the old Dongguan city and strengthen the overall function of the major urban area of Dongguan.

Driven by Songshan Lake Science and Technology Park, seven functional sections have been constructed in 5 years that respectively feature: administration, commerce and residence, research and development, education, culture, industry and tourism. High-end supporting facilities for education, culture, medical care, commerce and residence have been built to meet the professional and life needs of management and staff in the region. Major public supporting facilities include: Dongguan Science and Technology College, Guangdong Medical College, Songshan Lake School, Northern Industrial City Primary School, kindergarten, Affiliated Hospital of Guangzhou Medical College, Songshan Lake Library, Songshan Lake Academic Exchange Center, Administrative Office of Management Commission, Hyatt Hotel, Songshan Lake Villa No.1, Northern Industrial City Mansion, Bachelor Apartment, staff dormitory and large shopping center. The goal of upgrading old city through Technopolis has been reached, which in return formed a sustainable environment for the development of the Technopolis.

## 3.7 Technopolis Shall Encourage Internal Enterprises to Strengthen International Exchanges

With advantageous resources, it's relatively easy and convenient for the Technopolis to communicate with overseas industrial fields related to green growth. The Technopolis shall make full use of the advantage to seek cooperation and exchanges with enterprises related to green growth in foreign countries

### 3.7.1 Case: Nanjing University: Drum Tower State University Technopolis

In May, 2009, the first enterprise-based international technology transfer center, i.e., China (Nanjing)—Finland Energy-saving and Emission Reduction Technology Transfer Center was established in Nanjing Shengnuo Heat Pipe Co., Ltd. The center was co-founded by Finland-China Cooperation Center and Nanjing Shengnuo Heat Pipe Co. Ltd, aiming to promote science and technology exchanges and cooperation between Chinese (Nanjing) and Finnish enterprises and institutions. The emphasis will be placed on new environment protection fields that can facilitate green growth, where the two parties will conduct staff exchange, technical discussion, collaborative research, etc. This will be a platform to organize international cooperation between academy, industry and research institutions, promote transfer of new technology, commercialization of research findings and industrialization of high technologies.

Over the past 6 months since the center was set up, there have been continual communications between the two parties, while a workshop has been organized to discuss the feasibility of applying the Finnish technology of geothermal energy and energy-saving technology in civil architecture to the heat pipe on Nanjing's part. For the next step, the two parties will expand their cooperation field, elevate cooperation level and seek realistic action. Nanjing Shennuo Heat Pipe Co., Ltd is planning to cooperate with related Finnish universities or institutions and set up in Finland an oversea R&D institution, which will help promote utilization of the heat pipe technology for civil needs in Finland according the local cultural environment and realistic needs. Through this platform, the mature technology and equipment of Nanjing can also be exported, promoting technology transfer and commercialization of research findings between the two countries.

*Achievements Scored in China by Supporting Green Growth*
Remarkable progresses have been made in energy saving and emission reduction. Energy consumption per GDP unit has been decreasing by season (year): decreased by 1.79 % in 2006, 4.04 % in 2007, 4.59 % in 2008, adding to an aggregate lowering of 10.1 % over 3 years; the energy amount saved is about 270 million t coal equivalent. Nationally, discharge of sulfur dioxide and COD have

kept decreasing, with each respectively reduced by 4.66 and 3.14 % in 2007, 5.95 and 4.42 % in 2008, while the accumulative lowering rate were 8.95 and 6.61 % over the first 3 years of the "11th 5-year plan".

## 4 Existing Problems

Because a unified valuation system for green growth is lacking, it is difficult to quantify contributions made by the Park to promote green growth.

In terms of state policies, current support for technologies and products that are green growth-oriented is still inadequate and the stimulation system is also incomplete

The monitoring system has yet to be completed for it is unable to detect and/or solve emerging problems for promoting green growth over time.

Also, there exist great conflicts between meeting basic living standards and the promotion of green growth in some regions, which makes it difficult for government to make a choice.

## 5 Solutions and Approaches

1. Technopolis shall make its own contribution to setting up a global valuation system for green growth. Currently, the UN has not developed a satisfactory standard to evaluate the performance of governments across the world according to the needs of sustainable development. We hope that the UN and relative institutions will offer such a standard, while governments shall also have their own standard to evaluate sustainable development. By this standard, behavior of government, enterprises, and people will be guided, and it will also serve to give the programming of economic and social development proper guidance. In key fields or implementing key projects, the fundamental requirement of sustainable development should be taken into consideration.
2. Based on a valuation system, the Technopolis shall organize research to study the approaches and measures taken by various countries and regions to valuate actions taken for energy saving and emission reduction, their execution and relative stimulation and punishment.
3. The Technopolis shall make full use of its intensive resources and collaborate with other Technopolis to accelerate the development and promotion of green growth across the world. Centering around fields such as energy, resource, and environment, the Technopolis shall setup and complete a series of international cooperation centers, try to make breakthrough on issues such as high-efficient power generation, clean production in heavy-pollutant industries, energy saving in architectures, etc., and conquer a number of critical and generic technologies.

4. Technopolises and Science and Technology Parks in different countries shall strengthen cooperation and exchanges, seek stronger supervision on energy saving and emission reduction while improving the approach toward international supervision. Technopolis shall also promote a system for calculating, monitoring, and evaluating energy saving and emission reduction. Monitoring and supervision on energy saving and emission reduction in different regions shall be strengthened, while effective measures can be taken to punish heavy waste of energy and resource, severe damage on environment, etc.

Technopolises and Science and Technology Park shall be built in larger scale for only green growth of economy can ultimately eliminate poverty and ensure people the basic living standard. We shall clearly realize that, all the problems shall be solved on the premise of development rather than to solve problems prior to development. Some policies and investment may not show its effect immediately; for proposing such policies and making such investment, we need convincing perseverance, a forward-thinking vision, and precise judgment.

## 6 Conclusion

Technopolis plays an important role for promoting green growth across today's world.

In order to become an important power in enhancing green growth, the construction and development of Technopolis must follow the principle of green growth.

Technopolis shall promote green growth based according to the its own characteristics by training talents, supporting relevant technologies and enterprises, accelerating the formulation and execution of government policies, creating certain social environments, which are helpful for green growth and strengthening international communications and cooperation.

There are still many difficulties on the pathway to promoting green growth. However, we believe that they will be solved during the process of sustainable development as the idea of green growth wins more and more recognition from the general population.

# Arrangement of Actors in the Triple Helix Innovation

**José Alberto Sampaio Aranha**

## 1 Introduction

Increased productivity/competitiveness of countries and cities depend more and more on public policies established by local government. The potential of a site no longer depends on much on its location, climate, or natural resources, but on their willingness, ability, energy, values, and human organization (Kother 1993).

The competitiveness of these places in the knowledge era is in the speed with which innovation is generated; this relationship was one of the critical success factors for population's quality of life. The more it generates innovation, more economic development we have, and therefore more resources for investment in quality of life of the population that adds further knowledge to the process—what can generate more technological changes, that could lead to more innovation, forming a virtuous circle as shown in the UNDP's development human report.[1]

And how is it possible to make the country more innovative? What public policies are needed to stimulate and organize people to innovate? How to explain that Brazil is the thirteenth country (2008) in indexed scientific articles,[2] 28th in patents[3] and 68th (2009) in innovation?[4]

First it is important to identify that these indicators depend on different public policies. They can help themselves, but are not necessarily correlated or ordered.

---

[1] Human Development Report 2001—United Nations Development Program—http://www.undp.org/hdr2001/

[2] http://acessolivrebrasil.wordpress.com/2009/05/08/noticia-da-agencia-brasil-relata-nova-colocacao-do-brasil-no-ranking-de-numero-de-artigos-publicados/

[3] http://www.telecentros.desenvolvimento.gov.br/sitio/destaques/destaque.php?sq_conteudo=3840

[4] Global innovation index 2010—3rd edition of Insead report, with the Confederation of Indian Industry (CII)—http://www.gii.networkedreadiness.com/main/home.cfm.

J. A. S. Aranha (✉)
Genesis Institute of PUC-Rio, Rua Marquês de Sao Vicente, 225 Edificio Dom Jaime de, Barros Câmara Gavea, 22451-900 Rio de Janeiro, Brazil
e-mail: aranha@puc-rio.br

D.-S. Oh and F. Phillips (eds.), *Technopolis*,
DOI: 10.1007/978-1-4471-5508-9_7, 

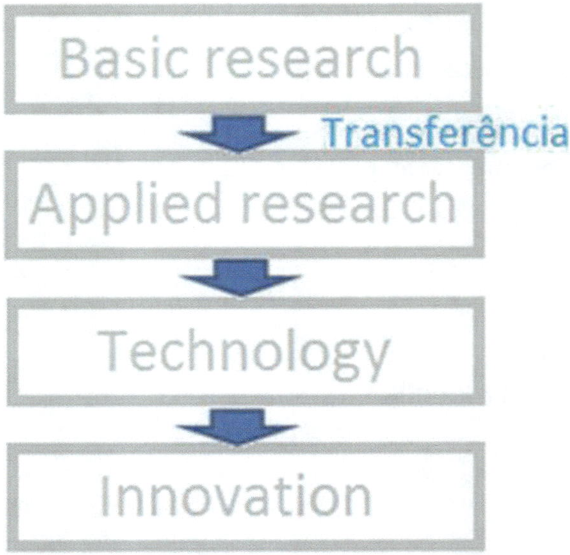


**Fig. 1** Linear model-transfer of basic research in innovation

That is, we need a specific public policy toward innovation, in addition to those already used for the generation of knowledge and technology Fig. 1.

All these indicators belong to what we call innovation system, working with the generation and use of knowledge, but with different goals; only the innovation itself directly brings economic development, competitiveness, and a possible social and environmental development Fig. 2.

One of the factors that affect an innovation system is its transfer. The transfer is the action of diffusion of innovation among all elements of the system. To develop its full potential, one of the adopted models is linear, in which basic research becomes innovation through two intermediate steps: the conversion of basic research in applied research and the transformation of this technology.[5]

The innovation really takes place by the time this technology is introduced into the market (Oslo Manual).[6] An innovation is the consolidation in the market of a new product or service. This way, it is called "transfer process" the translation of the results of basic and applied research in technology and its subsequent use by the society through goods, services, and processes Fig. 3.

[5] II Encuentro Internacional de Rectores—Innovación y transferencia del conocimiento—Debate general—Reflexiones sobre el análisis de un sistema de innovación—Guadalajara, Mexico—08/03/2010—http://iytc.universiablogs.net/2010/03/08/reflexiones-sobre-analisi-sistema-innovacion/.

[6] http://www.finep.gov.br/dcom/brasil_inovador/arquivos/manual_de_oslo/cap3_02_inovacao.html

**Fig. 2** Systemic approach

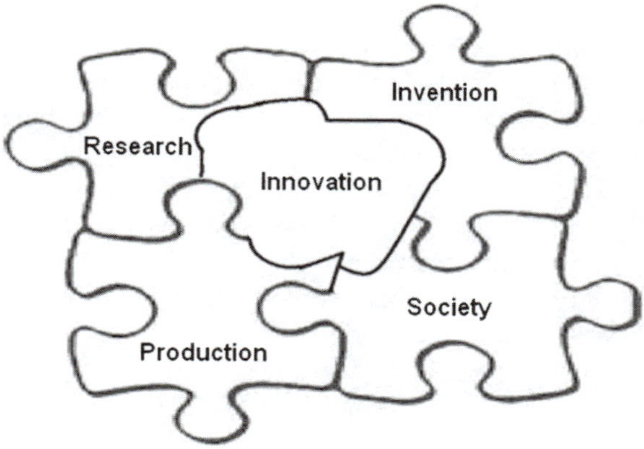


**Fig. 3** Annual evolution in the number of incubators and technology parks in Brazil *source* Portfólio de Parques Tecnológicos no Brasil—ANPROTEC—2008

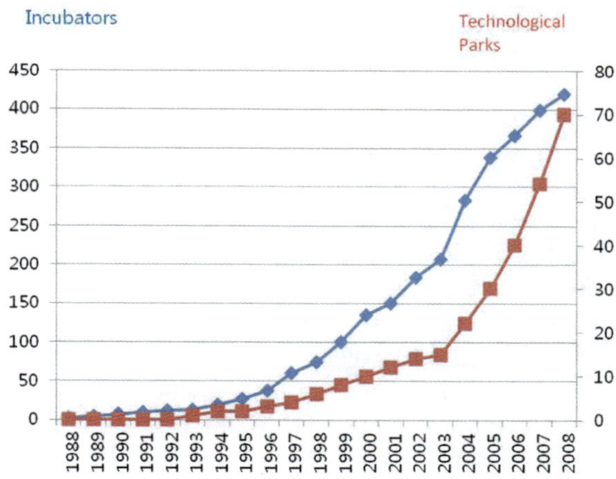


## 2 Actors from the Innovation System

In basic research, the major *scientific discoveries* (usually called theories for consisting of a set of scientifically proven explanations about natural phenomena) in most cases involve a large number of scientists who contribute with parts of the theory, either through basic and experimental research work or by drafting laws (Sabbatini 2000).

Unlike the discoveries, inventions usually belong to a single person or a small number of employees who work in the same project. They enhance the technological potential of the country and are a way of transforming the knowledge generated through basic and/or applied research in projects and prototypes that can be patented.[7]

---

[7] ANI—Associação Nacional dos Inventores—http://www.inventores.com.br/sistema/home/quem_somos.aspx.

The *innovation*—word derived from the Latin term *innovatio*—refers to an idea, method or object that is created and little resembles previous patterns. Currently, the word "innovation" is being considered as the invention that reached the market. So that innovation is consolidated, it requires different actors, researchers or scientists, inventors, and producers or businessmen.

This relationship of continuity was established at the advent of the Industrial Revolution, involving scientific progress-invention-innovation, so that the innovative process represented the end of a chain where technological practice was linked with other social systems, causing resistance and remeaning the various institutions (Thales 2006).

However, the current trend of formation of a technoscience significantly alters this panorama. The connection between components of technological advancement is organized as follows: invention-innovation-growth. The innovation becomes as the Schumpeterian perspective—means and instrument for the effectuation of economic growth. This implies in an instrumentalization of innovative practice aimed solely to growth, making it contained and programmable (Thales 2006).

In this context, one of the competitiveness factors of companies and nations is the time required to transform knowledge into products or services, that is, it takes time to introduce them into the market before others do, i.e., before changes occur.

We could then say that the future of innovation lies in reducing the time of knowledge generation and of its use by society (Aranha 2009). The linear historical sequence shown above where innovation is born according to different views—on academic research or in the company—tends to be modified so that it is effectively born on the market.

The presence of civil society in innovation has a key role in the viability of projects, which makes the management of the innovative process better consider this group in the set of actors in innovation environments. The society then stands in the center of the triple helix (Aranha 2005).

My view is that one should treat knowledge generation with a focus on solving problems of society and on the welfare of people, and set the proper use of knowledge generated by the society itself as a measuring parameter of this generation quality.

Furthermore, it is necessary to integrate the research laboratories of one region in a steady stream of preparation of professionals and production of knowledge, through the creation of competitive enterprises directed toward a particular industry which can also provide local economic and social development.

The biggest efficiency of this process will occur when there is complete utilization of knowledge by the society. A good example of this process was the mobilization of Pasteur so that farms implanted minimum standards of hygiene—when he developed the process of "pasteurization." That is, he extended his laboratory to the farms.

The incentive for knowledge generation, reducing the distance between the generation, accumulation, and its use by society, will result in shortening the time between the product or service creation and its use in a greater interaction Industry-University, actors of this innovation system.

A social system is a plurality of individuals who develop interactions according to shared standards and cultural meanings. Human relationships are the actions and attitudes developed by individuals and groups. The behavior of people is highly influenced by the environment and the existing informal attitudes and rules that exist in groups and are founded on individual processes, based on the interactions and relationships between people. These produce approach—cooperation, accommodation, assimilation (or expulsion), competition, and conflict (Aranha 2009).

The man builds his individuality in a contradictory way, because he is supported and constrained by singularizing (Vigostski 1929). "We become ourselves through others." The personality is made by the society or in social life, in a process that involves the internal workings of the human being to unify to others and distinguish from them, assuming a role more or less different from those performed by other group members (Góes 2000).

In this context, we must consider that the innovation process is composed of different social groups (researchers, inventors, producers, innovative entrepreneurs, and civil society), each one with a specific character, and who needs to assume a group consciousness with the same goals when treated as a macro group of innovation. The approach, rather than linear, must be seen then by the systemic point of view (Aranha 2010).

The term Social Capital refers to social networks based on trust, cooperation, and innovation, developed by individuals within and outside an organization, facilitating access to information and knowledge. Such networks can adopt a formal character (determined by the hierarchical links typical from the formal organization chart); but above all are informal in nature, involving horizontal links (between pairs) and diagonal (between and stakeholders collaborators from different areas) (Garcia 2010).

The Social Capital is the glue that interconnects the various forms of human capital, creating the most valuable intangible asset of organizations: the human working networks. The innovative entrepreneur is the actor who can deliver the benefits of Social Capital Management and create the necessary environments for innovation.

## 2.1 Innovation Environments

The innovation environments take into consideration factors arising from cultural heritage and the particular creativity of a social group, which traditionally are not recognized as components of innovation.

My perception is that creativity includes three different aspects related to entrepreneurship (theory and practice): *People*, as innovators, entrepreneurs, and articulators who are at the heart of creative production and function as agents of transformation; *Culture*, which helps to give people motivation and create a value system for communities and a sustainable creative culture; and *Environment*, where innovation occurs through the appropriation of knowledge and use of technology as productive factors to encourage more creativity (Miranda 2009).

The difference between creativity and innovation is the same as that between thinking about achieving goals and executing them. "Creativity imagines new things, Innovation does new things" (Levitt 1974). We might say then that the innovation process is to think new things, create new things, and distribute new things. In these steps we need to: research and learn the thinking of new things; make and build in the creation of new things; and implement and commercialize in the distribution of new products, processes, and services.

This is the business environment today. It requires more skill to manage change than strict controls to obtain results. Dynamic and integrated teams help more than a rigid structure with defined hierarchical levels. The alignment of mission, vision, values, and social and ethic responsibility among people in the company, research institutions, and society are more important than the balance sheet.

This mechanism is composed of work teams built through continuous organizational learning, or acquisition and socialization of knowledge through individuals, and their subsequent transformation into collective standards of performance. The main goal of these organizations is the human being.

Therefore, managing becomes structuring the abilities of people in the formation of communities able to learn and safely act. Malvezzi (1999) calls these communities in the company "communities of action," where the important thing is not having individuals who learn, but groups (human communities of work) that absorb knowledge, incorporate it, and turn it into collective behavior.

## 3 Smart Cities

This is the characteristic of intelligent cities, localities of the triple helix, communities that apprehend and are creative (Eleutheriou 2010). The creative cities are linked to urban transformation, as an intergenerational collective that brings together professionals from various fields to think and propose creative solutions that have art and culture as drivers of urban and social transformation.[8]

Smart City[9] is a strategy for economic and social development of regions, promoting quality of life for its citizens.[10] These cities use innovative technologies based on open platforms, providing Internet services for the development of innovation ecosystems—which creates opportunity for new sustainable and high quality services for citizens and businesses.

Smart City is different from a "digital city" when environmental and social capital play an important role. Smart Cities can be identified (and classified) along six main axes or dimensions:

---

[8] http://www.cidadecriativa.org/

[9] Ranking of European medium-sized cities. Universidad Tecnológica de Viena, Universidad de Ljubljana y Universidad Tecnológica de Delft. 2007.

[10] Vision of Parque de Innovacion de Servicios para las Personas da La Salle de Madrid—http://www.lasalleparquedeinnovacion.es/.

A *smart economy*, that focused in stimulating competitiveness, innovative enterprises, productivity and international relations;

The *smart mobility*, that includes a secure, sustainable, and innovative transport system for accessibility;

The smart environment, that stimulates the attractiveness of natural conditions, with environmental protection and sustainable management of resources, working technology, innovation, and creativity to improve quality of life on the planet, targeting a global, sustainable, and productive growth.

*Smart people*, citizens as co-producers and consumers of content and services, engaged earlier in the innovation process and being respected in their patterns of behavior in their new behaviors. People encouraged to increase their skills with ongoing training in services and entrepreneurship, highlighting the social and ethnic diversity, the creativity, and the participation in public life.

*Smart life*, quality of life with healthy conditions, cultural facilities, safety, educational facilities, attractive tourism, and social cohesion. Participatory urban planning and co-design. One of the roles of culture in this context is the social integration, which disrupts the distances between social groups through "fostering creativity, the rescue of population self-esteem, the rescue of traditional values and, through them, the socio-cultural identity." (Vetrale 2000).

*Smart governance*, which involves the participation of citizens in decision making and encourages public–private partnerships. Participatory governance, responsible, transparent, receptive, efficient, equitable, and inclusive. Develop long-term plans, creating a strategic vision that meets the needs of future generations. Foster innovation in education and learning. We need smart people: more than infrastructure legacy of the big events, we have to think about the knowledge legacy for people—in the human capital.

## 4 Innovation Mechanisms in Cities

Entrepreneurship in Brazil has grown significantly in recent years. Much of this growth is underpinned by a model where the innovation habitats (incubators and parks), are supported or backed by fostering institutions where the government has a significant role.

The incubator movement had a sharp increase between the years 1995–2008 and the technology parks have shown a sharp increase from 2003 on.

Technology and Science Parks are directly related to the local and regional socio-economic development, implemented from structuring programs. The state and local governments realize that it comes to strategies to stimulate growth and direct the development of their regions.

These environments cause impacts with social reflect in their surroundings and, as a whole, in trade, services, and real estate sectors, which will feel more acutely the effects of its operation. The existence of a technology park in a region (states and districts) tends to generate a change in behavior of the private and academic sector; among other impacts is the demand of professionals training with knowledge excellence.

When well planned and structured, the impact of the Technology Park in local and regional economy, traditionally associated with a productive sector, promotes a considerable boost in the region. It is noteworthy that the impacts generated by the implementation of a Technology Park should be measured over the long-term.

It serves as the bridge between research and production in the process of transforming the local economy, playing a decisive role to leverage the state's development by introducing in academic field, the demands of society and productive sector and, at the same time, responding to these demands (Aranha 2010).

In a city that has no more physical space for the installation of large industries, a Technology Park is an alternative to diversify the local economy and prevent its stagnation.

Brazil's position on the current world scenario and the supply of resources that characterizes the current moment of national economy, however, should be seen as a friendly warning that calls for the construction of a new model. In this new model, entrepreneurs, managers of innovation habitats, fund managers, and investors themselves need to reexamine their performance.

A possible strategy is to stimulate self-sustainable incubators and parks, which are those who are financially viable, besides the effectiveness to achieve their goals by optimizing resources and results. Only self-sustainable parks or incubators can be considered effective and with good prospects of becoming perennial.

Self-sustainable incubators and parks should be managed and financed as private companies. They are born of a conjunction between entrepreneurs, venture capital and corporations, and must have a manager or director who participates as a venture partner, sharing the enterprise risk: if the business fails, these actors are subject to loss of investment, but if succeed, will participate in the obtained financial return (Frick 2010).

## 5 A New Profile for Innovation Actors

An initiative of SEBRAE and ANPROTEC developed between the years 2008 and 2009 allowed tostart the building process of a performance model for the incubators, what received the name of *Centro de Referência para Apoio a Novos Empreendedores—CERNE* (reference center to support new entrepreneurs). The main objective was defined as to provide an important change in quantity and quality of their efforts in support of entrepreneurship.[11]

The implementation of the project led to the conjunct definition along with the incubators themselves, of a reference model containing "the key systems,

[11] Modelo de Referência para apoio a novos empreendimentos. Publicação (s.d) da ANPROTEC e SEBRAE.

components and practices that an incubator must implement to systematically generate a growing number of successful innovative businesses."[12]

The model led to the identification of four levels of maturity to be achieved in future by the incubators, being the goals of each, from minor to major, the following ones:

CERNE 1: To professionalize the generation process of innovative entrepreneurships.
CERNE 2: To improve internal processes of prospection for demand and supply of innovative projects, strategic planning, and attendance to associate companies.
CERNE 3: To strengthen the partners' network in order to increase the insertion into the regional system, implement a system of distance incubation, and establishing a network of experts.
CERNE 4: To implement an innovation management system, to be supplemented with other information security systems, environmental management, and social responsibility.

The CERNE is still under discussion by the various actors in all habitats of innovation. And it is extremely important that occurs as an effective implementation at each stage as an initial step toward the consolidation of institutions capable of promoting the rapprochement of venture capital (venture capital and private equity).

The table below lists some of the activities envisaged under CERNE according to investment funds requirements (Castro 2010).

Table 1 summarizes the support measures of general nature. It is also important to complement with the discussion of other actions to be implemented by incubators, accelerators, and managers of technology parks. The ultimate goal is to reach the concept of profitable or self-sustaining incubators.

[12] Modelo de Referência para apoio a novos empreendimentos. Publicação (s.d) da ANPROTEC e SEBRAE.

**Table 1** Some support actions from incubators

| Criteria | How is the CERNE model able to help? |
|---|---|
| Favorable position | The incubator must provide guidance to entrepreneurs in *developing the Technology Plan*. In addition, the incubator needs to assist entrepreneurs in the development of aspects related to products, services, and technologies, so as to make them more competitive. The incubator can assist in finding the best way to create value based on their competitive differential |
| Committed entrepreneur | During the selection process, the incubator should have *evaluation criteria of the entrepreneur profile*. In addition, the incubator must develop behavioral aspects and the entrepreneur profile, always thinking in his life included in the entrepreneurship (taking account of the possible difficulties in his personal life during the development of business). It is also important to support the entrepreneur in the relationship with your staff and partners, especially when internal disagreements occur between collaborators and partners |
| Business model | The incubator must *develop aspects related to the management of commercial sector*, including at least topics on organizing and motivating sales staff, marketing strategies, and sales strategies. Consultancies and target audience must be defined to be undertaken. For all that to be developed, the incubator needs to evaluate and structure, where necessary, the business model of the enterprise |
| Business plan | The incubator should have a Business Plan model so that *entrepreneurs can present their entrepreneurship proposals and update them throughout the process of incubation*. The incubators can also assist entrepreneurs in developing the executive summary, raising the most important issues in the business. In addition, the incubator can help the entrepreneur to develop managerial skills for the management of critical functions and processes of the enterprise |
| Investment structure | *The incubator guides entrepreneurs in preparing a formal document outlining the planning of short-, medium-, and long-term* for the economic and financial development of the enterprise, considering the need of working capital to support its activities and identifying the needs for capital investment. In addition, the incubator develops the aspects related to resource management, investor relations, and risk analysis, among others |

## 6 Proposal for an Actors' Interface in the Triple Helix

Innovation takes place through a set of institutions—research centers, technological, productive, and society. These in turn are represented by different actors—researchers, inventors, producers, investors, and consumers. Sometimes the same person can take some of these roles, but either way the relations between the actors, known as social capital, will be a decisive factor in the innovation process.

To catalyze these relations we recommend the innovative entrepreneur, a role that needs to be prepared to articulate the networks that produce innovation. Innovation must be focused on the well-being of people and stimulate economic, social, and environmental development with international scope. That is, it must be sustainable.

Since we will depend on the interaction ability of the entrepreneur—a bridge between the generation and use of knowledge—he has to be analyzed in its relation with the "social context" or the surrounding environment. This innovation environment within the Schumpeterian view is connected to the competitiveness of nations, as mentioned above, and should be part of public policy in providing conditions for competitiveness to the cities which will absorb these products in the market.

But data from more advanced countries show that the innovative capacity of a company or a nation does not depend simply of their economic capacity to invest in new technologies, or of their leader's, to develop appropriate economic strategies; but the social, cultural, and political capacity to productively apply and socially enjoy the tangible and intangible assets available (Maciel 2001).

The constructivist sociology says that the choice of specific technologies and the refusal of others are not based on purely economic or rational criteria, but rather in compatibilization involving beliefs and interests of various strategic groups and sectors that are in technological activity. In this sense, the economic interests of the triple helix follow up, but do not determine the route of innovation. Not only are the scientific and economic agents able to determine the technological practice, but also the social actors.

## References


Aranha, J. A. (2005). *Local socio-economic development micro cluster in 5th triple helix congress,* Itália, Maio.

Aranha, J. A. (2009). The future of innovation is the approximation between knowledge generation and the market – publication. *The Future of Innovation.* – http://thefutureofinnovation.org/, Article © 2009.

Aranha, J. A. (2009). *Interfaces: A chave para compreender as pessoas e suas relações em um ambiente de inovação.* Editora Saraiva (SP) .

Aranha, J. A. (2010). *Da visão de bem estar da sociedade as ações de realização local – latin america and the caribbean regional workshop on development of science parks and technology business incubators.* UNESCO: Costa Rica.

Aranha, J. A. (2010). As relações humanas no processo de inovação – XI encontro de reitores do grupo tordesillas -seminario: Fomentando el espíritu emprendedor. – Universidade Mackenzie (SP).

Andrade, Thales Novaes De. (2006). Aspectos sociais e tecnológicos das atividades de inovação-Lua Nova. *São Paulo, 66*, 139–166.

Castro, P. O. (2010). Material para publicação de venture capital e sua relação com os habitats de inovação – publicação. – BMF&Bovespa (SP).

De Goes, M. C. R. (2000). A formação do indivíduo nas relações sociais: Contribuições teóricas de Lev Vigotski e Pierre. Janet Educ. Soc., Campinas, v. 21, n.71. http://www.scielo.br/.

Eleutheriou, V., Aranha, J. A., E Zardo, J. B. (2010). Smart cities – 1o. workshop do programa 2014-Bis – Edição Finep – RJ, Out.

Frick, O., Aranha, J. A. (2010). Minicurso: O *Venture Capital* e sua relação com os habitats de inovação – XX Seminário Nacional de Parques Tecnológicos e Incubadoras de Empresas. – MTS.

García, I. (2010). Do capital humano ao capital social. A nova ciência das redes organizacionais na gestão de pessoas – site 14/06/2010. – http://www.rh.com.br/Portal/Mudanca/Artigo/6627/do-capital-humano-ao-capital-social-a-nova-ciencia-das-redes-organizacionais-na-gestao-de-pessoas.html.

Kotler, P., Haider, D., & Rein, I. (1993) *Marketing Places*. New York:The Free Press.

Levitt, H. T. (1974). *Marketing for business growth* (2nd ed.). NewYork: McGraw-Hill.

Maciel, M. L. (2001). Hélices, sistemas, ambientes e modelos: Os desafios à sociologia da inovação. Sociologias, Porto Alegre, ano, 3, n 6, jul/dez, p. 18–29.

Malvezzi, S. (1999). O Agente Econômico Reflexivo - Um novo desafio para a psicologia organizacional. - (EAESP-FGV & IP/USP) - http://www.ufba.br/~conpsi/conpsi1999/F009.html.

Miranda, P. C., Aranha, J. A., E Zardo, J. (2009). IASP world conference – Barcelona.

Sabbatini, R. (2000). As maiores descobertas científicas do milênio – Jornal Correio Popular, Campinas, 5/3/2000. http://www.sabbatini.com/renato/correio/ciencia/cp000305.htm.

Vetrale, S. (Coord.). (2000). Estudio internacional sobre políticas culturales urbanas. Montevideo.

Vygotsky, L. S. (1929). Concrete human psychology. In *Soviet psychology* 27(2), 53–77- 1986.

# Part II
# Bridging Theory and Practice

# Business Incubation Strategy of High-Tech Venture Firms in a Science Park

**Deog-Seong Oh**

**Abstract** In this chapter, the infrastructure and support mechanisms for technology commercialization in a science park has been evaluated with particular reference to business incubation of high-tech venture firms. The critical question is how science parks should develop in order to support the technology commercialization and business promotion, and benefit the regional economy as well as the local industrial structure. We have taken four findings from science park development experience; in a science park, there is a need to create business incubators (BIs) and technological and financial support mechanisms in order to maximize opportunities for technology transfer between academic and research facilities and private firms; science parks should include four major thrusts: a small BI, a technology commercialization center, a venture capital, and a cooperative linking mechanism for active R&D venture development. The four major thrusts will provide a focal point for the creation of new venture businesses and may serve as magnets to attract professionals in high technology fields and venture capital into the region; science park should develop a strong linking mechanism by network building which is largely composed of action programs that combine each resource of research institutes, universities, and a community; their efforts should focus on attracting high-tech industries and establishing promotional organizations. In particular, system approach of supporting entrepreneurship of high-tech venture firms is very crucial in terms of sustainable business incubation strategy. Start-up firms, after having successfully matured in the incubation center can then enter the productive-oriented industrial park. The combination of these three sectors (research park, high-tech industrial park, and incubation center) can be a kind of complex model of sustainable business incubation for high-tech venture firms in a science park.

D.-S. Oh (✉)
World Technopolis Association, College of Engineering Chungnam National University, Daejeon, Korea
e-mail: ds_oh@cnu.ac.kr

D.-S. Oh and F. Phillips (eds.), *Technopolis*,
DOI: 10.1007/978-1-4471-5508-9_8, 

# 1 Introduction

As part of a goal of regional economic development, venture businesses are of primary importance because of their potential for job creation in the areas of high technology and new products in the knowledge-based society of the twenty-first century. In a process of regional economic development and innovation, high-tech venture business is a basic ingredient for developmental success in terms of long-term commitment and persistence. Venture business means a firm which some of the few creative business challengers establish in order to commercialize their innovative technology using highly advanced technical power and technological knowledge. These businesses may take high risks because of poor capital even with high success possibilities, but also have high success potential on the basis of unique technical knowledge and new advanced technology. An increasingly popular tool for creating high technology-based venture businesses is the science park. Hence, many universities and cities promote science parks, since small- and medium-sized venture businesses from science parks create employment and aid municipal tax bases. These are the reasons for nurturing small businesses and science parks as an effective method of local economic development. As a matter of fact, among the strategic resources of a business, the cultural climate is the most important factor. Strategic resources of a business do not lie only in capital or technology, but also human capacities and adaptability.

In addition, when some of the venture firms from science parks grow over time, employment increases and, capital inflow and other development outcomes have a noticeable impact. These are the reasons for nurturing venture businesses as an effective way of economic development and regional innovation.

With these considerations in mind, this chapter aims to find a way of supporting technological commercialization and high-tech venture business promotion in science parks.

The chapter begins with a theoretical review of science parks and technology business incubation as a whole. The second part of the chapter describes model of business incubation in a science park. The third section discusses key issues for the application of this model in Daedeok and suggests development strategies for the future.

# 2 Theoretical Review

## *2.1 Science Park and Technology Business Incubation*

An increasingly popular tool for creating high technology-based venture businesses is the science park. Hence, many universities and cities promote science parks since small- and medium-sized venture businesses from science parks create employment and aid municipal tax bases. These are the reasons for nurturing small businesses and science parks as an effective method of local economic development. As a matter of fact, among strategic resources of a business, the cultural climate is the most important factor. Strategic resources of a business do not lie only in capital or technology, but also in human capacities and adaptability. A science park is a property-based initiative which (UKSPA 2006) has formal links with a university and/or other higher educational and research institutions (HEI); is designed to encourage the formation and growth of knowledge-based businesses and other organizations normally residing on site; has a management function which is actively engaged in the transfer of technology and business skills to the organizations on site. Within this definition, it is also possible to identify several sub-forms like technology park, technopolis, and so on, which complement other initiatives designed to stimulate a more productive relationship between industry and academia.

At the core of the science park phenomenon is a view about how technologies are created. Science parks constitute a channel by which academic science is linked to commerce. This then, is a highly particular model of scientific research and industrial innovation. Fundamentally, it is a linear model, in which there is a chain of successive, interrelated activities. These begin with basic scientific research and pass through applied and more developmental research activities, the development of new product and process ideas, the evolution and testing of prototypes, to commercial production and finally to diffusion (Massey et al. 1992).

According to Massey et al. (1992, pp. 58–60), two major policy questions have historically emerged from the linear model. The first is how to increase the supply of basic research ideas available for development. The second major question has been how to quicken the development and commercialization of basic ideas. The science park model fits as one possible means of solving the second problem. The purpose of the science park model is to nurture high technology-based small- and medium-sized firms for local economic development as spin-offs from science parks. Instead of one process of innovation from research to commercialization, however, they have also suggested an interactive model in which new ideas are generated and developed at all stages of innovation, including the production stage.[1] Small business has become increasingly attractive as a focus for local economic developmental attention. Given the relatively high failure rate of new small businesses on the one hand, and the public sector's expectations for them as the bulwark of the local economy on the other, it is only logical that a portion of most local economic development efforts is aimed at the survival and success of small firms (Allen 1985, pp. 16–17).

---

[1] Suggesting the ten major city examples of science park plans in Korea, Park noted that most of the science parks are under an initial planning stage and R&D units or firms are not actually located in the parks. He continued to argue that there are no significant differences in the selection of key industries among the proposed science parks according to the two models (Park 1992, p. 242–248).

## *2.2 Critical Success Factors for Business Incubation of High-Tech Venture Firms*

According to Fache (1992), the success of the innovation process in science parks is basically a cultural one. Among all identifiable factors of success (good planning, management, location, link with universities, aggressive marketing, international networks, and a realistic financial approach), he indicated that the cultural factor is probably the least considered because it is the most difficult to measure and to implement; it has to do with human attitudes, management of time, and social and cultural life. Business-oriented culture will create conditions for the rise of a daring entrepreneurial spirit. Starting a venture business requires a daring spirit and a strong character capable of implementing ideas.

Since use of the concept of entrepreneurship, there has been no precise definition of the word. Managers often describe entrepreneurs as being innovative, flexible, dynamic, risk taking, creative, and growth oriented. The popular press, on the other hand, often associates entrepreneurship with starting and operating new ventures (Feeser 1987). In the present context, this describes the process of actually putting an innovation into the marketplace; the innovator assumes the risks inherent in the process (Gibson 1992). The critical point in creating venture business is the interaction of knowledge with a recognized social need or want. Entrepreneurship bridges the gap between science and the marketplaces. Silicon Valley is a good example of entrepreneurship.

According to Larsen and Rogers (1988), key factors in Silicon Valley are the availability of technical expertise, infrastructure, venture, job mobility, information-exchange networks, and learning entrepreneurial fever from local role models. Among the factors, the major ones to stimulate entrepreneurship for hatching venture business are venture capital and other infrastructure. A venture capital firm serves as an intermediary between investors looking for high returns for their money and entrepreneurs in search of needed capital for their start-ups.[2]

A new firm is also dependent on a lot of infrastructure including universities, research institutes, governments, non-profit organizations, suppliers, financiers, markets, and others. Venture businesses are more likely to occur, and more likely to succeed, where the necessary infrastructure exists. There are also important roles for local government and universities to play in their relationships with the business incubators (BIs). They should build more BIs in response to their demand and consider building more of these facilities into the economic development strategies for effectuating their comprehensive plans. Thought should be given to linking incubators with other local development efforts in a complementary manner, in order to achieve an economic development synergy. Care should be taken to avoid the "zero sum game" of hatching new venture businesses that merely compete with and ultimately displace existing firms (Oh and Kang 1997) (Fig. 1).

---

[2] The literature on venture capital suggests that many venture capital firms like to be in close proximity to the entrepreneurs with whom they work (Lay Gibson 1992, pp. 39–41).

**Fig. 1** Technology-based venture growth path, supporting "entrepreneurship" of high-tech venture firms

## *2.3 Tools and Mechanism: Business Incubator/Technology Commercialization Center*

Business incubator is one of the major instruments for stimulating entrepreneurship and venture business development in the science park. Small BIs are facilities that support new and small firms by providing affordable space, shared office services, financial services, and management assistance. According to Allen (1985), the incubator concept must include four dimensions, such as a local network, multitenant space, shared services, and management consulting assistance. Sponsorship of small BIs falls into five major categories (1) public entities; (2) nonprofit organization; (3) universities; (4) private corporations; and (5) public private partnerships (MacDonald 1987). Differing sponsorship has tended to mean variation in incubator objectives and administration. Public sector and nonprofit organization sponsors seek to create jobs, diversify the economy, and expand the tax base. Universities become involved in incubator sponsorship as a means of creating a working laboratory for students or to provide a vehicle for marketing the products of faculty research. This causes them to be a bit more relaxed in their policies regarding tenant admission. Like publicly sponsored incubators, they are more likely to establish a time limit for individual tenant occupancy due to a concern that they might otherwise be seen as providing an inequitable advantage to private entities. University incubators tend to support high technology-oriented tenants over other industries (Oh 2002).

A Technology Commercialization Center is also one of the major instruments for stimulating entrepreneurship and venture business development in the science park.

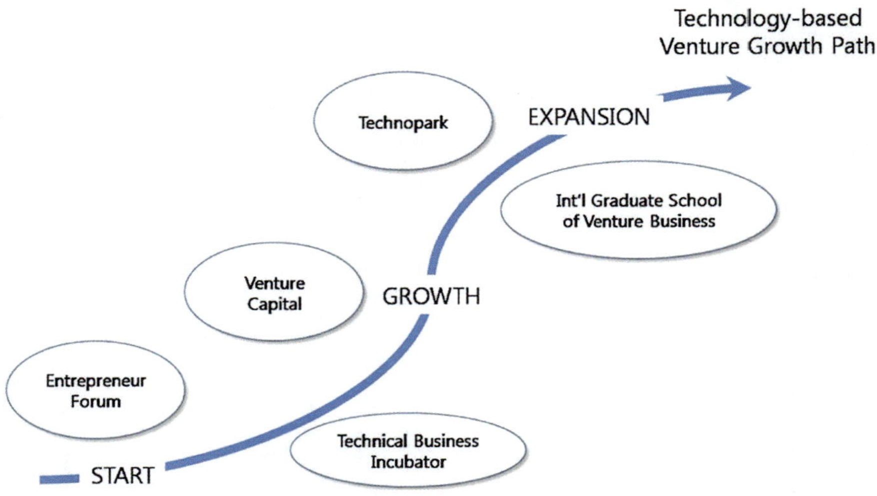


**Fig. 2** Venture infrastructure

Commercialization of innovations and technology transfer are, in effect, an extension of the evaluation and patent activities, although in some cases the university may intercept and improve innovations that have already been deemed marketable and, perhaps, even patented. The emphasis is on: a) identification of potential uses of a product or process; b) preliminary determination of a depth of demand; and c) identification of an economically efficient production process. Again, university faculty and staff involved in these activities will likely have backgrounds in science, medicine, and engineering. Operations of this sort may deal with new and relatively untested innovations or, potentially, they could even focus on new industrial applications for established technologies (Gibson LJ 1992) (Fig. 2).

# 3 Model of Business Incubation in a Science Park

## *3.1 Business Incubation Model*

Three models of technology transfer or commercialization have been most prevalent in the business incubation procedure (Gibson D 1991). The three models include "Appropriability Model", "Dissemination Model", and "Knowledge Utilization Model". The "Appropriability Model" emphasized the importance of the quality of research and competitive market pressures in achieving technology transfer. The "Dissemination Model" emphasizes the diffusion of innovation. The "Knowledge Utilization Model" emphasizes the importance of interpersonal communication between technology researchers and users, and between organizational barriers and facilitators of technology transfer.

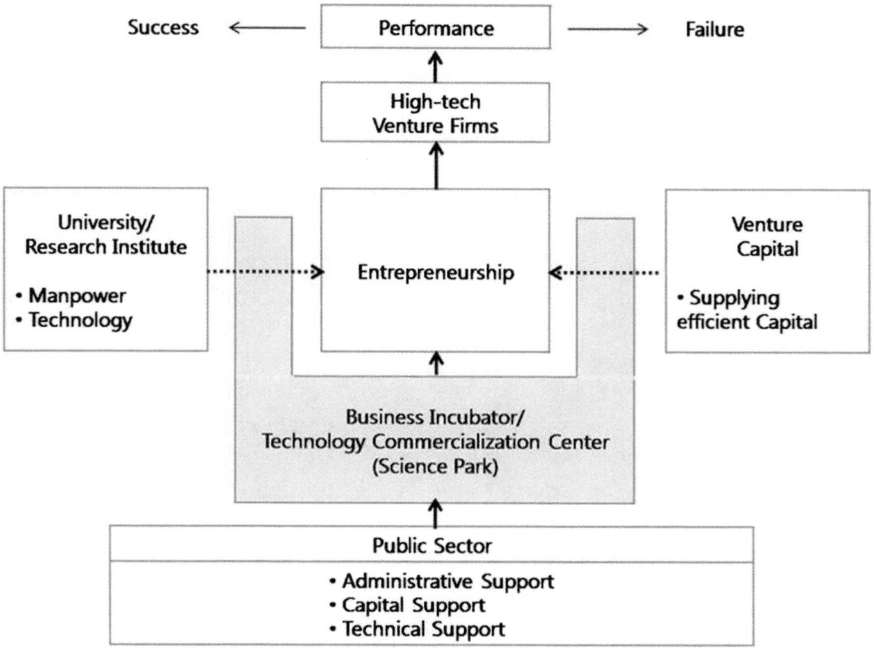


**Fig. 3** Model of venture start-up in a Science park

Figure 3 shows the Model of Business Incubation in a science park. It is necessary to gain a more complete understanding of the relationship between a venture business and a science park and to suggest the policy implications for nurturing venture business. We assumed the following basic hypothesis on the postulated model: venture capital and other infrastructure can help explain the encouragement of entrepreneurship and hatching venture business in the science town. We also assumed that part of venture business through stimulation of entrepreneurship could be explained by activities of infrastructure, such as BI technology commercialization center, venture capital, and so on.

The model explains the hatching process of venture business in a science park or town. It includes the formation of stimulating activities for entrepreneurship in venture start-up, and the success of venture business.

It is important for high-tech venture firms to take sufficient support from the infrastructure to survive as successful firms in the market. Business Incubator should provide financial assistance to small businesses. It takes the form of seed capital, which can be either debt or equity capital, and varies in amount depending on the condition of the small business and the risk involved. There are two kinds of sponsorship for the BI. City-sponsored incubator should provide a business finance coordinator in the building to work with businesses to seek outside seed capital. The revolving fund administrator could act as a finance coordinator as well as a direct provider of funds. Such financial assistance is crucial to the success of a start-up because of the variety of costs faced by the venture firms prior to the stage when sales revenues meet expenses.

Furthermore, incubators should link with the technology commercialization center or one-stop business service center because poor management is the leading killer of small business. BIs should have on-site experts who can identify funding sources in both the public and private sectors and assist in the application process, as well as network experts for general management assistance, technology assistance, and self-employment training.

In order to accelerate technology transfer from a science park to the marketplace for high technology-based venture business a Technology Commercialization Center should be established by the university or the city government. This office is charged specifically to provide businesses with direct access to the research institute in the science town, the universities' technological institute, and their graduate business school. This center as a linking mechanism is very essential to a science park because both research institutes and universities do not know what kinds of specific research industry wants, and industry does not recognize what kinds of research is done.

With the technical-assistance mechanism of the technology commercialization center in place, venture capitals set up by universities, research institutes, city government, and the private sector, should establish a seed-capital and lending fund to encourage entrepreneurship. The venture capital corporation can make both loans and equity investments.

## *3.2 Service Factors for the Successful Technology Business Incubator*

**Service factors**

We can identify the factors of success and services incubators need. Even if the needs of incubators are different from place to place, the areas of support to SME ranked in European countries are modeled as follows (Snyder et al. 1992),

- Technology development
- Facilitation of start-up enterprise
- Financing and exports
- Training
- Information and counseling.

In addition to these factors, access to needed inputs such as finance, raw materials, technology, skills and an ever-opening market is required. The Korean environment for SME creation and growth also highly depends on public policies and business infrastructure within the framework of supply and demand. In Fig. 4, the timely support of incubator is illustrated to deal with problems encountered during incubation.

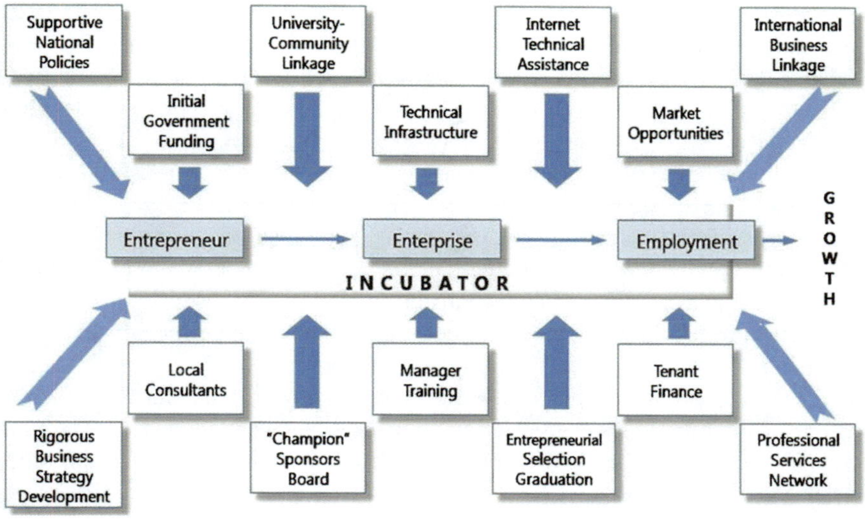


**Fig. 4** Service factors of Business Incubator to success

## Networking Building and Other Infrastructure

Knowledge transfer, adoption, accumulation, and diffusion are key to sustainable economic prosperity in the emerging global economy of the twenty-first century. As stated by Abramovitz and David in a 1996 OECD report, "The expansion of the knowledge base [has] progressed to the stage of fundamentally altering the form and structure of economic growth." Rapid advances in information and communication technologies and declining costs of producing, processing, and diffusing knowledge are transforming social economic activities worldwide (The World Bank 1998). There are four kinds of network for establishing linkage between Firms and Market (Gibson D 1997).

- Networking for Markets: Identification of markets and successful execution of marketing strategies is a determining factor in the success and sustainability of small and medium technology-based firms. Access to adequate financing is one of the most critical factors for the success of technology-based firms. Broad public/private/NGO partnership could be established to offer integrated access to services such as financial planning, support for obtaining grants, opportunities for access to venture, development, and seed capital.
- Networking for Inter-Firm Linkages: A networked approach is ideal for maximizing the impact of programs and projects, such as partnerships, alliances, and linkages to outside suppliers. Most clusters in developing countries tend to rely heavily on the local supplier base, which may become insufficient for their rapidly growing needs. Careful coordination is required to ensure that local suppliers are able to match increases in demand so that jobs may be retained and created and that other substitute supply streams can be brought online as required.

- Networking to Expand Access to Technology Transfer Opportunities and Networking for Technological Support: The use of electronic networks for technology transfer is already being established in several places to stimulate investment in S&T, R&D, technology transfer, development of commercial potential of R&D, and spin-offs. Networks are necessary tools for facilitating access to technology transfer opportunities worldwide. Electronic networks are extremely useful tools to diffuse the benefits of technological support, providing services such as technology assessment and forecasting, technology gateway (assistance on technological choices and on marketing assessment of innovative projects), and access to outside technical information. Theses services could also be concentrated in one or a few centers and could be provided by public agencies, private consultants, and business associations.
- Networking for Talent and Know-How: SMEs often do not have and cannot afford the entire range of technical and business talents and know-how required for success in local and global markets. The process of identifying and hiring such talent and know-how on a short-term as-needed basis is also difficult for smaller enterprises. Networks of talent and know-how would be a great asset that would allow SMEs access to the experts at affordable rates and opportune moments.

Established SME firms will be the initial focus of the network-building program. However, as a longer term strategy there is a critical need to develop the infrastructure and resources of the region in order to promote accelerated development of knowledge-based firms from the bottom-up. The basic need is to improve the process of knowledge transfer, acquisition, absorption, and diffusion. Issues involved include basic and higher education, physical infrastructure construction, and improved policy environments. These are potential areas for action (Quandt 1998):

- Creating and Strengthening Local Technopolis Management Structures: The first step is creation of an organizational and functional structure for the local cluster, preferably by leveraging existing groups and associations. This would involve both private and public sector participants. Establishment of linkages with other technopolis managers will enable a better understanding of stakeholder needs and markets and will improve organizational methods. The creation of a permanent, dedicated business and technology information network would make communications more continuous and interactive, rather than sporadic exchanges that normally occur only at periodic meetings. In particular, visits between key personnel of the regional network would greatly facilitate an exchange of knowledge, technology, and know-how.
- Determining Educational Needs and Offering Training: Based on regional descriptive profiles and targeted interviews with local stakeholders, educational requirements for the LIPs and targeted companies can be ascertained. Courses could then be offered through local workshops as well as via the Internet to help improve the skills of local trainers. It is also necessary to build local skills and

training IT specialists. In order to incubate a local cluster that will depend heavily on virtual linkages, a comprehensive adaptation to the IT paradigm is crucial. Development of new types of specialists will be needed, including technology brokers, research experts, information and technology guides, and animators. Optimizing and Sharing Facilities: For each region, the required facilities for a viable technopolis could be kept to a minimum as long as they are integrated into a shared system. The operational support infrastructure could be optimized and many facilities could be shared over the network, including incubators, prototype centers, pilot plants, online libraries, test laboratories, and online conferencing facilities.

In addition, electronic commerce is quickly becoming an essential business tool in an increasingly integrated world economy. E-commerce capabilities are another kind of function that is much more feasible when implemented on a larger scale.

# 4 Case Study: Application of Business Incubation Model in Daedeok Science Town, Korea

## *4.1 Current Status*

**Overview**

Daedeok Science Town (hereafter DST) was originally designed and constructed as a science park model to achieve scientific developments and to redistribute the concentrated population of Seoul to Daejeon when it was planned. The 20-year project was finished in 1992 and accomplished as a strategic point of national science and technology by the Ministry of Science and Technology (MOST).

In the early stage, only government-run research institutes were located at the west park, and at the second area, research institutes of the major Korean industries settled at the east park until 1995. Since the mid 1990s, efforts are being made to enable high-tech venture firms to be established in DST with mechanisms to support the commercialization of R&D results achieved in DST. The main components of this mechanism are TBI, Post-TBI, and venture park where start-up firms can direct their efforts in commercializing their new technology. In this consideration, Daejeon Metropolitan City Administration mapped out a scheme to develop a high-tech industrial district, Daedeok Techno Valley (DTV). DTV links DST to several other critical areas in Daejeon Metropolitan City. Due to the positive effects of DTV's development on the regional economy and industry, Daejeon City devised a citywide industry development structural plan in 2001 (Fig. 5).

In 2002, central government had designated DST and its vicinity as Daedeok Valley (DV) for the further development of technology commercialization based

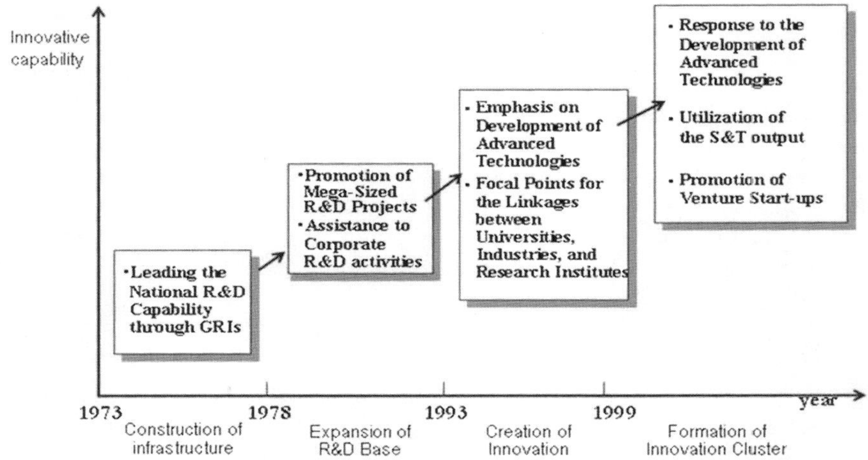


**Fig. 5** Development of Daedeok Science Town in terms of innovation capacity

on R&D. This variety of high technology-based development activities enabled the DV to be redesignated as the first National R&D Special Zone[3]: all accommodated within the Daejeon Metropolitan City—Daedeok Science Town, Daedeok Techno Valley (DTV) local, Industrial Complex, to carry out the pilot project of regional innovation and technology-led economic development with strong support from central government. First of all, forming the largest science and technology research park in Korea could provide an important hub for leading the regional industry through the efficient development of advanced science and technologies.

The statistics of R&D institutes (2006) showed 63 R&D institutes in 2006; 20 government-sponsored research institutes including ETRI, 10 government-invested institutes, and 33 private R&D labs. And also, the number of venture business and SMEs is 824 integrated high-tech companies in 2004 that were established by from HEI's and research institutes.[4]

There are six universities integrated into higher educational institutes with 5,806 Ph.D.s, 6,625 M.A.s, and a majority of Ph.D.s in Daedeok Valley. The total number of employees was about 15,000 in 2006. Highly qualified research manpower in public and private research institutes is highly concentrated in DST. This means that there is a high possibility of the commercialization of research results or production of high-tech commodities from high technology located in DST.

---

[3] National R&D Special Zone: all accommodating within the Daejeon Metropolitan City—Daedeok Science Town (DST), Daedeok Techno Valley (DTV), Industrial Complex, Yuseong Tourist zone.

[4] The members of a venture link called the 'Daedeok twenty-first Century Club' are steadily increasing but after KAIST started its technology innovation center and business incubator (TIC/TBI) in 1992, and ETRI in 1996 the number of the small companies drastically increased to 285 companies in 2000.

According to the report of Daedeok Valley Management Office (DVMO) in 2006, within 10 years DST will have obtained immense growth as well as the number of our occupant companies and R&D institutes and the number of international patent registrations will rise. Especially, the companies in DST are estimated to be 3,000 units in 2015 and the number of foreign R&D institutes is expected to be 20 in 2015, up from 2 in 2004. In addition, DVMO is predicting the international patent registration will be increased about 10 times more than the number in 2004 and sales from technology commercialization are expected to reach 31,296 million in 2015.

**R&D Activities**

There are typically two levels of R&D activities in DST. One is corporate R&D labs, which are engaged in basic research; the other is division-level labs, which are used for research directly related to the business and manufacturing needs of their particular division-level activities and also focus on commercialization of their products for their business. The characteristics of R&D activities in DV have changed according to the changes in the main body of R&D activities. For example, research institutes of private firms in Daedeok are varied in their sectoral pattern of activities, while many government institutes are biased toward basic research and development rather than production and commercialization (Fig. 6).

At an initial stage, DST focused mainly on pure basic research, because most institutes in DST were public organizations. But, in terms of technology-led economic development, this policy predicted development isolated from the regional economy. It is because long-range investigation of a basic research could have little economic impact on the region. In 1988, about 42 % of all government research institutes were located in DST and this percentage had increased until

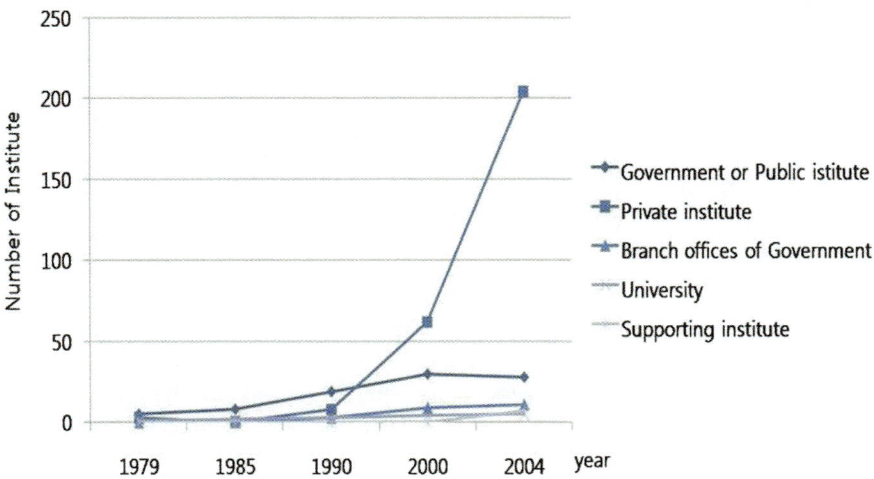


**Fig. 6** Number of institute in DST

**Table 1** Sectoral comparison of R&D centers; number (employee)

| Description | Public sector | Private sector | Total |
|---|---|---|---|
| Comprehensive R&D | 4 (1,230) | 1 (83) | 5 (1,313) |
| Biotechnology | 4 (482) | 5 (455) | 9 (937) |
| Information technology | 3 (2,655) | 5 (848) | 8 (3,503) |
| Precision chemical | 1 (403) | 9 (1,672) | 10 (2,075) |
| New material (including high molecule) | | 5 (549) | 5 (549) |
| Mechatronics (including marine science) | 2 (571) | 2 (305) | 4 (876) |
| Resource, energy development | 5 (2,173) | 2 (76) | 7 (2,249) |
| Astronomy, aerospace, astronautics | 3 (806) | 1 (109) | 4 (915) |
| The others | 6 (2,483) | 2 (112) | 8 (2,595) |
| Total | 28 (10,803) | 32 (4,209) | 60 (15,012) |

*Source* DAEDEOK INNOPOLIS Management Office, 2004 (http://ddinnopolis.or.kr/)

2004. From the late 1980s, private sector research institutes began to relocate from the Capital Region into DST, which resulted from the changed objective of government that promoted applied research by private R&D center. Private sector substantially has been higher than public sector in the R&D centers since 1992. Table 1 shows the increasing development of R&D centers and their employees.

Since private institutes carried out R&D for their mother enterprise, they had more interest in the technological innovation and profits of the mother enterprise than in public benefit. These attributes of private institutes made R&D activities in DST more subdivided and varied. This variety of activities is important, because it suggests the potential for future development in DST as well Daejeon, as it can be attractive to a wide range of research-focused technologically advanced and innovative organizations. For example, the R&D fields in private R&D labs in 2005 were more subdivided than those of 1992. There are 32 private labs in 12 fields related with particular industrial sectors like IT, BT, NT etc. As R&D activities of private institutes became subdivided and varied, the potentiality of DST for technology commercialization has increased.

## 4.2 Spin-Off Activities and Growth of High-Tech Venture Firms

Recently, active spin-offs are increasing in the research institutes and universities in DST. Although several previous assessments indicated that the mere concentration of public and private research organizations in a space could not generate the inertia for high technology spin-offs, DST is a case where this is not so. Significant sources of technology-oriented new ventures are being generated. Actually, 7.1 % of total venture firms in Korea is located in Daejeon City and Chungnam Province (6 % in Daejeon, 1.1 % in Chungnam Province). It is the highest concentration outside the Capital Region (Seoul Metropolitan Area).

**Table 2** Major research institutes in DST and their spin-offs

| Name of research institute | Number of spin-off firms |
|---|---|
| Electric and Telecommunication Research Institute (ETRI) | 57 |
| Korea Standard Science Research Institute | 19 |
| Korea Advanced Institute of Science and Technology (KAIST) | 117 |
| Total of major RIS' spin-offs (88.1 % of total R&D spin-offs) | 193 |

*Source* Oh (2005); Kim (2000)

The start-up activities in DST began to be observed from the time when institutions in DST launched their TBIs. As research institutes and universities including ETRI, KAIST, and CNU participated in TBI business, the number of venture firms in DV suddenly increased, and this made the venture start-up in DST to be accepted as an essential aspect and activity area rather than be understood as a mere part of R&D activity. When venture firms from the major 10 R&D centers in DST were surveyed prior to 2004, there were 2 firms in DST in 1990, 20 firms in 1996, 130 firms in 2002, and an increase to 219 firms in 2004. If job creation is analyzed for the same years, 35 persons were employed in 1990, 187 employees in 1996, 2,212 employees in 2002, and 3,237 persons were employed in 2004.

The total of 219 venture firms created from 10 public research institutes in DST by 2004. ETRI, KAIST, and Korea Standard Science Research Institute are the main bodies, which actively contributed to technology commercialization. And, the venture start-ups extended to all areas of Daejeon since late 1990's. The venture inauguration originating from DST began a full-scale relationship with regional territory not only to Daejeon but also to its vicinity (Table 2).

The experience of KAIST is very interesting, because it shows us the potential of business incubation driven by a leading university in DST.

Since 1994, 127 companies were selected and incubated in KAIST incubators. Since then, 13 companies left KAIST for various reasons, 5 companies canceled the projects, 4 projects failed, but 5 companies graduated in 2000, such as Mari-Telecom, Darim Vision, Intersys, Setri, and Intelligent Telecommunications (IT), which were good examples of successful venture firms in DV.

## 4.3 Support Mechanism by TBI

Prerequisites for creating high-tech start-ups should cover a favorable economic climate, controlled inflation, and currency exchange rate as well as commercial framework. Therefore, it is important that the special influence of DST and complemental structures be established to promote creation and growth of venture. In addition, the software structure of incubation is comprised of open markets, cost effective system for delivering credit and technical support, and protection for both intellectual properties and environment. Also it is expected that the government's role is to promote small/medium enterprises, to prime the pump, and to develop the infrastructure and self-operation for business people during the time of social change.

**Table 3** Incubators in Daejeon metropolitan city area (2004)

| Incubator | Number of rooms | Number of occupancy | Remarks |
|---|---|---|---|
| KAIST TIC/TBI | 31 | 31 | 1992 founded |
| KAIST HTC | 83 | 83 | 1998 supported by MOST |
| ETRI | 70 | 57 | 1996 supported by MIT |
| KAERI | 11 | 11 | 1997 supported by SMBA |
| Taejon SMEC | 27 | 27 | 1998 by Taejon |
| Taejon S/W Center | 11 | 11 | 1998 by Taejon |
| Chungnam University | 21 | 21 | 1997 by MOE |
| Hannan University | 15 | 15 | 1997 supported by SMBA |
| Baeje University | 16 | 16 | 1997 by university |
| Korea Electric Company | 4 | 4 | 1999/2000 |
| Chungnam incubator | 15 | 10 | 1999 by Chungnam Province |
| Total | 304 | 286 | |

★Two private incubators are under the preparation of setting up TBI strategies in 2004

In Table 3, the foundation of incubators in Daejeon area was also noted as being supported by each ministry. The incubator helps in creation and growth of technology-based firms. The incubator also helps the firm to succeed by overcoming bureaucratic obstacles and providing affordable space and shared facilities. To reduce gestation time and startup cost, it provides advice, training, information service, management and marketing support, linkages to research faculty and facilities, and access to capitals.

Reflecting the Korean bureaucracy, however, the policy-makers for venture are distributed among several ministries, i.e., the Ministry of Information and Telecommunication (MIT), the Ministry of Science and Technology (MOST), and the Agency for Small and Medium-sized Business (SMBA). Therefore, a hero is required to promote the regulatory and economic shifts, which induce the active participation of small business people and communities. The leading role of the MOST on DV quickly decreased as the supporting Board of Trustees reorganized into three sectors, such as basic science, industrial, and public technology in 2000.

## 4.4 Demands for Business Incubation

There have been several demands and requests from highly qualified manpower to challenge the opportunity for start-ups from their original R&D centers or universities. Surveys to clarify these have been made twice, in 1997 and 2002.

The most difficult problems when highly qualified manpower, like professors or staff of R&D centers, tried to start their businesses were as follows; funding capital (79 %), lack of demand for products and marketing, shortage of managerial know-how, complicated administrative procedures for opening businesses. However,

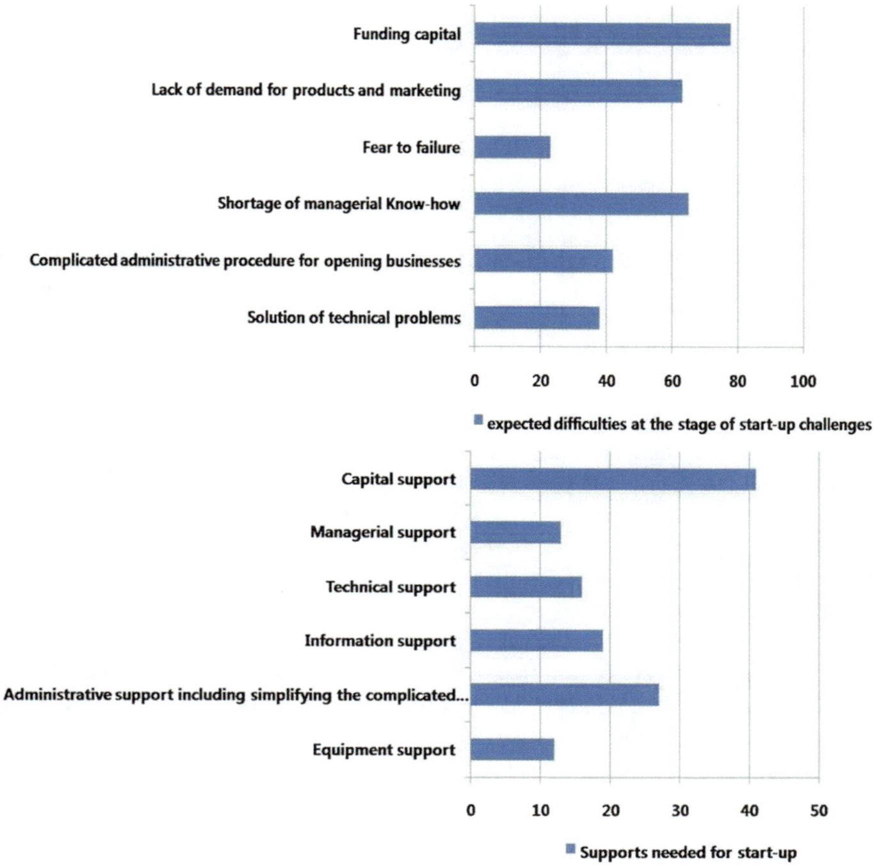


**Fig. 7** The expected difficulties of start-up and supports needed

solution of technical problems and fear of failure were not considered to be major difficulties in starting business. Figure 7 shows that start-up challenges had difficulties similar to those of the supports; funding capital, shortage of managerial know-how, lack of demand for products and marketing, and complicated administrative procedures when considering opening businesses. Both solution of technical problems and fear of failure were relatively minor ones.

When highly qualified manpower in DST tried to establish their venture businesses, the most needed supports from the community were for capital, administrative (including simplifying the complicated procedure for start-ups), information, managerial, and equipment support. The last one needed was technical support. For graduate students, results showed little difference when compared with professors. The survey results of high-tech venture firms from DST show us a clear picture of real demands for technology commercialization. They need venture capital and administrative support. Other factors like managerial and technical support are minor ones.

# 5 Development Strategy: Complex Model According to the Development Stages

## 5.1 Concept

According to development goals and functions of the park, the technopolis development concept can be divided into three kinds (Oh and Kang 1997); incubation center (innovation and incubation center at the universities), a research park (R&D-oriented), and high-tech industrial park (productive-oriented). Although functions of the three parks are different, they must be intimately related to upgrade and strengthen the industrial structure. In the initial stage of technopolis, a research park can be located within a precinct to attract research technicians and technological companies. After the development of the park reaches a certain phase, an industrial park needs to be constructed to produce the results of the R&D-oriented park. In the same way, the incubation center needs to be established next to those parks for the purpose of supporting the business environment for the R&D and industrial activities. Through the combination of those three parks, the high-tech industries can be developed.

This kind of developmental relationship can be seen clearly in the development strategy of DST in Daejeon: As DST permits basic research & prototype manufacturing, there should be another industrial park, which has no limit on the amount of manufacturing performed so long as it is related to the R&D activity. In that area, the main activities of their organizations are for research application and production, rather than for research. By attaching a high-tech industrial park close to the DST, R&D functions can be tied directly to the industrial activities in the region. At the same time, an incubation center needs to be established for the purpose of supporting the spin-off activities from R&D centers of universities in DST. In particular start-up firms, after having been successfully matured in the incubation center can then enter the productive-oriented industrial park. Thus, the development process of technopolis is completed. The combination of these three sectors (research park, high-tech industrial park, and incubation center) can be a kind of complex model for technopolis: research, production, and business incubation.

## 5.2 Policies for Business Incubation

The higher portion of R&D Centers in DST shows the better possibility of developing concept (complex of R&D and industry) than any other regions in Korea. Especially, the development concept of business incubation for supporting spin-off activities must be emphasized in the area of Daejeon, which is the mother city of DST. But, the resident labor force in Daejeon is disproportionately concentrated in commercial and service occupants. They are employed mostly in

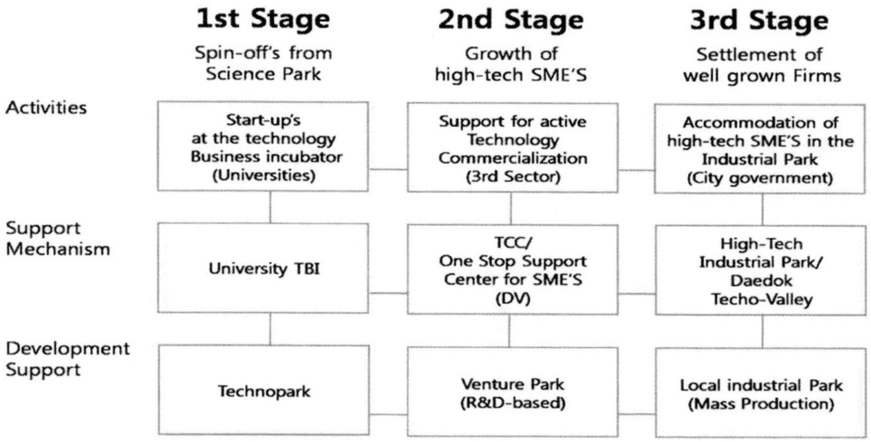


**Fig. 8** Development stages of high-tech venture firms: from spin-off's to settlement

semiskilled and lowskilled occupations. Considering this situation, there needs to be innovative strategies to attract scientific entrepreneurs from universities and R&D centers in DST and to mature on the point of development as viable enterprise. As mentioned above, incubation centers can play the role of these innovative strategies. The development strategies of business incubation and support mechanism can be identified as follows (see Fig. 8).

**First Stage: Spin-Off's from Science Park**

A model type of incubation center was established in DST, which provided an interface between research and production in the field of high technology. Its main aim is to support the spin-off activities from the HEI's or R&D centers and to promote expansion of those start-up firms. It is funded mainly through public subsidies. At present, these kinds of incubation centers are now operating at several universities in DST. They provide units in TBI that can be leased to high technology start-up firms that focus on new materials, computer science, and telecommunication. Business in these centers has a more advantageous position than firms in other industrial estates because of the reasonable technology and business supports. For example, incubators provide inexpensive space to high-tech start-up companies, which also provide common business services. They receive technical assistance through the Technology Commercialization Center and can receive financing through the seed-capital corporation. Four kinds of supportive activities, low-cost space, shared services, technical assistance, and financing, are the major advantages that start-up business requires to begin operating. The seed-capital fund and the technical-assistance operation also draw business to the incubator and all reinforce one another. This support is offered in the first 3–5 years of a new firm's existence, the most crucial period in its ultimate survival.

**Second Stage: Growth of High-Tech SME's**

At the second stage, it is important to establish the mechanism for technological support, collaborative research, and technology commercialization. In DST, there is the one-stop service center that connects industries with government-supported institutes to develop domestic industrial technologies and promote international competitiveness.[5]

Technology Commercialization Center which is an office charged specifically for providing businesses with direct access to the research institutes in the science park, also should have the capacity to pay experts of universities and research institutes to consult for SME's which need specific help in developing their technology-oriented businesses. There are three main strategies of R&D commercialization in DST; (1) strengthens commercialization process, (2) technology evaluation and trade expansion support, and (3) promote commercialization process. More importantly, venture parks of specialized industries like IT, BT, and NT can act as a means of expansion for scientists seeking to commercialize their research results for the market. These parks have lower service level than that of the first incubation center, but they have more rooms and sites for the developing firms. There is also more flexibility of units, leases, and other service activities. Venture parks as a linking mechanism is very crucial to the growth of venture firms in DST because both research institutes and universities do not know what kinds of specific research industry wants, and industry does not recognize what kinds of research are done. In addition to providing sites and facilities, it contains seed-capital and other venture funds to encourage entrepreneurship and new job creation with a particular emphasis on high technology start-ups in DST. The investment corporation can make both loans and equity investments. The DST's mission statement has four aspects: (1) spurring growth of leading edge firms, (2) support trade through technological financing, (3) nurture high-quality human resources, (4) marketing and administration services. The one-stop support center also has this kind of investment corporation. Only through connection with the Technology Consulting Center (TCC), can we cover the need of support mechanism for high-tech SME'S, in particular the spin-offs' from DST.

**Third Stage: The Settlement of Well Grown High-Tech Firms**

By the incentive effect of university-based incubation centers at the beginning stage, venture parks in high-tech industrial parks shall be built at the development stage of incubation policy. Daedeok Techno Valley is the good place for it. These developments are envisaged to promote the third sector planning concept to help business spin out from either the host organizations or other large corporations within the area of Daejeon. Because R&D, production activities of venture firms,

---

[5] Its major activities are as follows:

- efficient utilization of R&D resources, such as manpower, equipment, technology and domestic foreign technological information
- to support high-tech small and medium-sized firms by providing information on technological know-how, funds, etc.

and other enterprises are inseparably related each other, industrial sites for light production are designed in DTV next to Daedeok Science Town. In addition, public organizations give support to the businesses that have been successfully developed through the first and second stages. This area provides them with a site that can accommodate businesses requiring self-containing, prestigious, and stand alone building of their critical mass. They can also construct their production facilities in any other industrial estates. When the start-up firms succeed in standing alone at one of those industrial parks, it is truly the final stage for the incubation of spin-off activities. This is one of the important functions of the high-tech industrial park, which will have a significant influence on the economic development in Daejeon Metropolitan City.

## 6 Conclusion

In this chapter, the infrastructure and support mechanisms for technology commercialization in a science park has been evaluated with particular reference to business incubation of high-tech venture firms.

The critical question is what can be learned from the experiences of successful science park development as to how science parks should develop in order to support the technology commercialization and business promotion, and benefit the regional economy as well as the local industrial structure. There are several important findings from science park development experience that are crucial to the future development of science parks in developing countries.

- First: A popular tool for creating high-tech venture firms is a science park, as it constitutes a channel by which scientific research is linked to commerce. There is a need to create BIs and technological and financial support mechanisms in order to maximize opportunities for technology transfer between academic and research facilities and private firms.
- Second: It is necessary to stimulate researchers' entrepreneurship to create high technology venture businesses by means of a favorable infrastructure that combines each resource of research institute, universities, and a community. Science parks should include four major thrusts: a small BI, a technology commercialization center, a venture capital, and a cooperative linking mechanism for active R&D venture development. The four major thrusts will provide a focal point for the creation of new venture businesses and may serve as magnets to attract professionals in high technology fields and venture capital into the region.
- Third: A science park should develop a strong linking mechanism by network building which is largely composed of action programs that combine each resource of research institutes, universities, and a community. Four kinds of network to make linkage between high-tech firms and market (Networking for markets; Networking for interfirm linkages; Expand access to technology transfer opportunities and networking for technological support; and Networking for talent and know-how) are crucial to the growth of science parks.

- Fourth: There is the need for local initiatives to reinforce technology–led economic development through sustainable development of high-tech start-ups. Their efforts should focus on attracting high-tech industries and establishing promotional organizations. In particular, system approach of supporting entrepreneurship of high-tech venture firms is very crucial in terms of sustainable business incubation strategy. Start-up firms, after having successfully matured in the incubation center can then enter the productive-oriented industrial park. The combination of these three sectors (research park, high-tech industrial park, and incubation center) can be a kind of complex model of sustainable business incubation for high-tech venture firms in a science park: research, production, and business incubation. The establishment of venture parks, high-tech industrial parks like DTV in Korea where R& D activities can be promoted and commercialized, makes sense in this respect.

## References

Allen, D.N. & Rahman, S. (1985). Small Business Incubators: A Positive Environment for Entrepreneurship. *Journal of Small Business Management*, *23*(3), 12–22

Fache, Dominque. (1992). *Cultural and enterprenuerial success: Innovation in science park, the technopolis phenomena*, Maryland, Rowman Littlefield Publishers, Inc., p.193

Feeser, H.R. (1987). *Incubators, entrepreneurs, strategy, and performance: A comparison of high and low growth high tech firms*. PhD Dissertation, Purdue University

Gibson, D. (1991). Key Variables in Technology Transfer: A Field Study Based Empirical Analysis (with R. Smilor). *Journal of Engineering and Technology Management*, Vol. *8*, pp. 287–312

Gibson, D. (1997). Inter-Organizational Technology Transfer: From Standard Technology Packages to Spin-Offs. In J. Sedaitis (Ed.), *Commercializing High Technology East and West*, New York: Rowmon & Littlefield Publications, pp. 77–94

Gibson, L.J. (1992). Science cities or technopolis, science parks and technology parks: alternative models for university participation, Proceedings of the Development Strategies for Science Town (1992) 25–47

Kim (2000). Tendency and prospect of venture business, necessity and development of Venture 627 Town, Daejeon Metropolitan City, Republic of Korea

Larsen, K. & M, Rogers. (1988). *Silicon Valley: The rise and falling off of entrepreneurial fever, creating the Technopolis*. Massachusetts: Ballinger

MacDonald, S. (1987). *British science parks: reflections of the politics of high technology*, R&D Management *17*(1), 25–37

Massey, D. Quintas, P. & Y. wield, D. (1992). *High tech fantasies: Science Parks in Society, Science and Space*. London: Routledge

Oh, D. S. (2002). Technology-based regional development policy: Case study Daedeok Science Town, Daejeon Metropolitan City, Korea. *Habitat International 26*, 213–228.

Oh, D. S., & Kang, B. S. (1997). Technology commercialization and business promotion in Science Park: Case study of Daedeok Science Town. *Journal of the Korean Regional Science Association*, *13* (2), 119–141 (Korean Regional Science Association).

Quandt, C. (1998). Developing innovation networks for technology-based clusters: The role of information and communication technologies. Paper prepared for the workshop on 'Tech-regioes: Ciencia, tecnologia edesenvolvimento—passado, presente e futuro, Rio de Janeiro, December 6

UKSPA (2006). *The Planning, Development and Operation of Science Parks, Second Edition*, UKSPA Publications, The United Kingdoom Science Park Association
World Bank (1998). World development report 1998: Knowledge for development. Washington, D.C.: The World Bank

## Report and Website

Daedeok Techno Valley. (2002). Masterplan of DTV, Daejeon Metropolitan City.
DAEDEOK INNOPOLIS Management Office http://ddinnopolis.or.kr/
Korea Land Corporation http://www.iklc.co.kr
KOSEF (Korea Science and Engineering Foundation). (1992). Development strategies for the future: Taedok Science Town, Daejeon, Korea (in Korean).
Ministry of Construction & Transportation, Republic of Korea http://www.moct.go.kr/
MOST (Ministry of Science parkand Technology). (1989). The Basic Plan of Technobelt in Korea, MOST, Seoul (in Korea).
TSO (Ministry of Science and Technology). (1996). Daedeok Science Town, Information Paper, Daejeon, Korea.
TSO (Ministry of Science and Technology). (2002). Daedeok Science Town, Masterplan, Daejeon, Korea.

# Toward a Sustainable Technopolis

**Fred Phillips**

**Abstract** The chapter offers observations on what makes a technopolis sustainable, and how a technopolis contributes to the sustainability of society outside the technopolis' boundaries. Each technopolis project must attend to sustainability in the scientific/engineering arena, in the social arena, and in the arena of environment and the triple bottom line. The chapter offers criticism of common concepts of sustainability, suggesting that technopolis designers and scientists are well positioned to sharpen our views and practices regarding sustainability.

## 1 Introduction: Sustainability and Technopolis

It has been my privilege, over the course of a long consulting career, to visit many of the world's technopolis projects, including, here in East Asia, Hsin-Chu, Daeduk, Kansai Science City, Tsukuba, and Oita Prefecture. In 2006, I wrote down much of what I learned in these travels, in a book called *Social Culture and High-Tech Economic Development: The Technopolis Columns* (Phillips 2006), a book which emphasized the vital role of social capital in technopolis development. In this chapter, I would like to share, in keeping with the theme of this volume, observations on what makes a technopolis sustainable.

I take a broad view of sustainability, as indeed each technopolis project will have to attend to sustainability in the scientific/engineering arena, in the social arena, and in the arena of environment and the triple bottom line. I will go further, criticizing the foundations of our concepts of sustainability. I feel confident that

F. Phillips (✉)
Department of Technology & Society, College of Engineering and Applied Sciences, Stony Brook University—State University of New York, c/o SUNY Korea,187 Songdo-dong, 406-840 Yeonsu-gu, Incheon, Korea (Rep.)
e-mail: fred.phillips@stonybrook.edu

D.-S. Oh and F. Phillips (eds.), *Technopolis*,
DOI: 10.1007/978-1-4471-5508-9_9, 

**Table 1** Technopolis success factors

| |
|---|
| Embracing change |
| Social capital, especially with cross-sectoral links |
| Cluster strategies that target specific company groups for collocation |
| Visionary and persistent leadership |
| The will to action |
| Action |
| Constant selling |
| Self-investment in infrastructure |
| Outreach and networking |

*Source* Phillips (2006)

this audience includes the thinkers who can fill the gaps in the sustainability concept, and I present these gaps to you as a challenge.

Table 1 lists the success factors shared by thriving technopoleis. We will touch on many of these in this chapter.

It was also my privilege to attend and to publish (Phillips and Eto 1998) the 1997 Tsukuba Advance Research Alliance (TARA) Symposium on returning university research results to society. Japan's Technopolis Act was to be terminated in 1998 (Suzuki 2004), and Japan prepared to look to universities for a sustainable stream of commercializeable research. This is a first lesson in sustainable technopolis: Government technopolis initiatives may not last; universities endure. Private companies may not survive; but universities endure. Technopoleis hoping for a long run must leverage the presence of universities, and partner with them.

## 2 Scientific and Engineering Sustainability

### *2.1 Plan for Massive Tech Shift: Fujitsu and the 2025 Quantum Limit*

The TARA symposium's highlights were the remarks of Mr. Takuma Yamamoto, who was soon to retire as Chairman of Fujitsu Ltd. Mr. Yamamoto repeatedly referred to the year 2025 as a kind of watershed. In 2025, he implied, everything would change for Fujitsu.

Curious, I talked my way past Mr. Yamamoto's bodyguards and asked the chairman, "What is going to happen in 2025?" Mr. Yamamoto replied that he expected chip design to hit the quantum limit to Moore's Law in 2025. Every Fujitsu factory involved in circuits, he said, would have to be redesigned, and that would be in the *best* possible scenario!

Progress has been faster than even Fujitsu expected, and current forecasts of the quantum limit range from 2014 to 2020. The limit seems less like a wall and more like a permeable membrane: In certain niches, slight progress has been made in overcoming the barriers to Moore's Law. These include neural chips and quantum computing.

Nonetheless, new technopolis projects would be foolish to build expensive facilities that may be obsolete in four years. Instead, find science and engineering niches that will allow a long payback period on your infrastructure investment.

## *2.2 Plan for Cessation of Government Funding*

An island in the Mediterranean procured a European Union grant for its technopolis project. An architectural competition was announced, and I was asked to be a judge. The island's ancient and wealthy aristocracy was bitterly opposed to the development. Thus, there were no funds for continuing the project after the EU grant was gone. Moreover, Europe's most prominent architects, who did not get paid, developed a bad attitude about the island and its prospects.

We all know that culture change takes a very long time. The EU did not take this knowledge into account when creating its technopolis grant program. The island's government did not take it into account when launching their project—or if it did, it used the knowledge for short-term political advantage rather than for a sustainable technopolis.

In Japan (Suzuki 2004), the national Technopolis Act terminated in 1998, and the government moved to a policy focused on promoting venture business and links between industry and universities. At the prefectural level in Oita, however, local funds and initiatives continued to build tech and non-tech clusters, Governor Hiramatsu's ties with MITI continuing to serve the region well.

A member of the Oregon Council for Knowledge and Economic Development complained that the financially strapped State slashed funds for nurturing the knowledge economy, but at the same time increased funds for roads and bridges. This complaint is naïve, both from a political perspective and a business perspective. Legislators do not know what "knowledge economy" means. Marketing such an abstract concept to the government, demands cleverness and perseverance.

Roads and bridges get built and maintained in all parts of the state, creating blue-collar jobs everywhere. What could be more appealing to a state legislator? In contrast, technology and knowledge jobs cluster in the districts of the state that are already more privileged. These are few, compared to the number of rural and old-economy districts.

History shows the necessity of keeping the roads and bridges open. After the crashes of 2001 and 2008, however, legislators suspect that the knowledge economy is a flash in the pan or another fraud perpetrated by dotcommers and Wallstreeters.

Smart and sustained effort must be aimed at selling the knowledge economy to political leaders. With luck, you can find high tech companies whose interests are aligned with the technopolis', and recruit their seasoned lobbyists to sell the technopolis project to the government. Even then, do not count on the government being your friend.

# 3 Social Sustainability

## *3.1 Godfathers*

The government is not your friend, but *individuals within the government may be your best friends*. It is beneficial for the technopolis if these friends are re-elected reliably. Then too, an influential individual who is *not* in government may be a technopolis' most effectively advocate and energy source.

The next table lists the "godfathers" of the world's best-known technopolis regions. The asterisks show those who served from government positions. In this sample, the ratio of government godfathers is just more than half.

The godfathers are super-networkers, connecting entrepreneurs, financiers, researchers, legislators, and other technopoleis, harnessing them to the cause of the local technopolis. The godfathers exert a social force, causing others to want to be close to them, and to show a public spirit, in support of the technopolis, perhaps in excess of that to which they would be naturally inclined (Table 2).

Unless your region has an extraordinary amount of research/technology push (like Boston, for example), an extraordinary amount of entrepreneurial pull (like Israel), or both (like Cambridge), or an extraordinary amount of money (like Washington, DC), you should hope that a godfather emerges and jumps to the front of the parade in your locale.

## *3.2 Social Capital*

History matters. Your region may *leverage* a history of science and entrepreneurship. Or, your region may have to *overcome* a history of insufficient education and hostility toward private business.

There is a patent for cloud seeding (rainmaking), filed in the name of the king of Thailand. The Korean alphabet is attributed to a king of Korea. Though we might view these as instances of politically adept inventors allowing their kings to claim

**Table 2** Godfathers of established techno-regions

| | |
|---|---|
| Austin, Texas | George Kozmetsky |
| Curitiba, Brazil | Jaime Lerner[a] |
| Hyderabad, India | Chandrababu Naiu[a] |
| Oita Prefecture, Japan | Morihiko Hiramatsu[a] |
| Silicon Valley, California | Frederick Terman |
| Singapore | Lee Kwan Yew[a] |
| Sophia Antipolis, France | Pierre Lafitte[a] |
| Taiwan | Morris Chang |

[a] Holders of political office

credit, there is more to the story. The king's name, associated with a technological advance that benefits the entire population, does much to influence a culture of innovation for social betterment. In contrast to Thailand and Korea, all technological advance in dynastic China was solely for the amusement of the emperor. It will be difficult for China to overcome this and to establish a view, within the government and among the people, that technology can benefit everyone.

Social culture, says Harvard Professor Howard Stevenson, determines:

- A region's propensity to reinvest the rewards of business success in still more local businesses—rather than in real estate or offshore bank accounts—and in the social welfare of locals.
- A region's attitudes toward the success of others. Does the society ostracize entrepreneurs, or celebrate their successes?
- A region's willingness to embrace change.

The sociologist James Coleman (1988) defined social capital as "the ability of people to work together for common purposes in groups and organizations." Where there is social capital, Fukuyama (1995) claims, there is wealth. Voluntary civil and civic organizations, each of a scope that is wider than family-level yet not organized by state or national governments, show a technopolis region's confidence that it can shape its own future.

External as well as internal networking is an important aspect of social capital. Visits to other technopoleis can build valuable, lasting networks. When they are just a way to find out what other people are doing, the trips are of limited value. A region with historical disadvantages will have to do something original and different from what has gone before.

The will to action—and actually taking action—are the final aspects of social capital that I will emphasize today. As Peter Drucker said in the context of the 1980s manufacturing crisis, "What we have to learn from the Japanese is not what to do, but *to do it.*" All the knowledge, money, and infrastructure in the world, Drucker implied, cannot substitute for will, attitude, and follow-through.

## 3.3 Grassroots Versus Government-Directed Technopolis Alliances

Japanese research shows that designated technopoleis in that country grow faster than other regions, despite having no financial assistance from the central government. One must beware of circular reasoning here; The technopolis regions were identified by the government *because* of their superior potential. Yet the conclusion that it is the characteristics of the region, rather than government help that makes success, is strong (Kyaw 2001).

In his research into obstacles to new technology parks and clusters in Japan, published in *Technological Forecasting and Social Change,* Hajime Eto (2005) concluded that "Cultural factors such as value gaps between the two worlds (government bureaucracy and engineering culture) are… responsible for the unsuccessful outcome of S&T policies."

Peter Hall (1998) analyzed historic creative cities, showing their heydays lasted, on average, a few dozens of years. North (1990) gives wealth-creating economies a few centuries. It is reasonable to think that today's pace of technological change will narrow these windows. Each region must choose a technological pony to ride toward economic development. Today those ponies get winded sooner. The region's external network is also the means for finding a fresh pony. Every region that builds wealth on an industry cluster must, in a few years, stake its next wealth-building strategy on a new or redefined industry.

Should an aspiring technopolis region depend on government support, or drive progress locally? In this section, I have emphasized that local characteristics—including the propensity to network and drive progress locally—are key success factors; that the gap between government and technical cultures is a wide one; and that continual self-renewal is the name of the game, which implies a flexibility that government may not be able to deliver. In an earlier section, I stressed the political risk in extending government assistance to projects that might be seen as helping the few rather than helping the many. Furthermore, regions may feel a responsibility to absorb available government funds, rather than to produce projects of real merit.

All these things speak to the desirability of local, grassroots technopolis initiatives rather than dependence on government initiatives. It cannot be denied, though, that government measures have had terrifically beneficial results. Accomplishments like Hsinchu and Oita are absolutely spectacular and much to be praised (notwithstanding that the Hsinchu park is just now experiencing a turbulent time due to reallocation of central government funding[1]). I conclude that *even with central government support, vigorous local and private initiatives are necessary.* The efforts of Stan Shih and Alvin Tong at Hsinchu are good evidence for this conclusion.

Austin's success was very much grass-roots. There was never a massive government allocation for an "Austin Technopolis," though there were smaller grants for individual projects supporting the technopolis. We always asked local governments for small amounts of money, just to make them feel part of the effort. When a major initiative or success was rolled out, we invited government officials to the press conference. If an impression was created that the credit for the success should go to these officials, so much the better.

[1] *TIME* magazine March 23, 2009: "At Taiwan's Hsinchu Science Based Industrial Park, home to many of the island's flagship tech firms, most workers are taking unpaid leave at least one day a week. Ryan Wu, chief operating officer of the job-search website 1111 Job Bank, say conditions at Hsinchu have never been so dire. 'There's extreme panic right now', Wu says".

## *3.4 Sustainable Technopolis Initiatives*

In all but the most blessed regions, sustained initiatives are a necessary condition for ultimate regional success. By the late 1990s, several local initiatives had shown excellent results in Austin, Texas. Start-up and established companies in software, equipment for oil exploration and semiconductor manufacturing, and computers were flourishing. A number of these initiatives' movers and shakers had done well financially; they were tempted to say, "I've got mine, buddy, good luck with yours." There was, in fact, less public sentiment in favor of continued civic initiatives for entrepreneurship. Good had proven to be the enemy of better; people were content. The present author was not inclined to fight an uphill battle to continue the initiatives. Another civic leader, however, argued against resting on our laurels, insisting the public–private cooperative initiatives were needed "now more than ever". Of course, the Internet bubble burst shortly after that, proving him right. Another impressive Austin leader now says, "Every morning I ask myself, what I can do today to make Austin the best place in the world to live."[2] Through the efforts of these gentlemen and others, Austin launched new initiatives for clean energy and for the computer gaming industry. The gaming initiative is flourishing today, and several energy companies have been launched.

There are no overnight successes. The technological renaissance of Austin, Texas, took 25 years—and by some yardsticks, much longer. The transformation of a region by means of technology entrepreneurship is a long-term process, and it is not unusual for a *crisis* to catalyze a region's entrepreneurial economy. Thus a long-lived regional initiative may be necessary, if only to ensure that the initiative organization is there to catch the crisis (whenever it may happen) and bend it to constructive ends.

Mukherjee (Mukherjee 2005) shows how the same pattern of government investment, slow infrastructure, and entrepreneurial development, followed by a crisis-generated opportunity (the Y2K problem) spurred Bangalore's growth in the twenty-first century. Mukherjee adds, "The 'sudden' buzz in Bangalore is actually just a new chapter in a 100-year-old saga. No amount of planning could have telescoped the process into 10 years."

A crisis in food production, accompanied by sharp increase in population, caused the launch of Greece's trade with Sicily, and thence the "rocketing" Greek economy of the sixth through fourth century (Hall 1998, p. 49).

Crises are not part of conventional cluster theory, which instead emphasizes a critical mass of suppliers and competitors. Critical mass was important for Austin and Bangalore, but it was not the whole story.

Three anecdotes, of course, do not prove that crisis is a needed ingredient. However, Portland, Oregon and Palma de Mallorca are instances of 'no crisis, no entrepreneurial transformation.' Even Washington, D.C., with its amazing collection of technology companies already attracted to the US federal money faucet,

---

[2] These two leaders were, respectively, Pike Powers and Jim Ronay .

**Table 3** Top reasons for failure of cluster initiatives

| |
|---|
| Absence of an explicitly formulated vision for the CI and quantified targets |
| Initiative framework not adapted to the cluster's own strengths |
| No office or an insufficient budget for significant projects |
| Limiting the membership scope |
| Isolated firms and lack of competition |
| Lack of advanced suppliers |
| Basic human capital |
| Lack of trust and networks |
| Few supporting institutions |
| Weak frameworks |
| Facilitator not having a strong network |
| No involvement of influential local decision makers |
| Lack of consensus or difficulties in achieving consensus |
| No brand-building objective |

*Source* Sölvell et al. (2003)

experienced a new burst of activity in security-related technologies subsequent to the 9/11 tragedy. We can be comfortable agreeing with Linstone and Mitroff(1994), who declare, "Crisis may be the best, if not the only, teacher of how to create an economy that is better matched to the needs of today's world."

These concepts allow us to pinpoint the serious risks in using cluster theory or cluster consultants to kick-start a new technopolis:

1. Cluster theory is mechanical ("put this firm here and that company there...") and ahistoric. The mechanical metaphor of cluster theory imposes a mechanical "solution" that ignores the constraints and opportunities implied by the region's history.
2. The question of how the community responds to crisis is not dealt with.
3. A short-term cluster consultant is not an ongoing network facilitator/partner for the client region, and cannot help with the long-term culture change issues that the region almost certainly must face.

Better to call your technopolis initiative a "regional initiative for technology and entrepreneurship" or RITE, than a cluster initiative (CI). Many of Sölvell et al.'s (2003) reasons why CIs fail, however, also apply to RITEs (Table 3).

## 4 Environmental Sustainability and the Triple Bottom Line

Sustainability, I've never seen a precise definition of it. I probably wouldn't recognize it if I saw it. It sounds like a good thing, though.

The eminent dean of a leading school of environmental science snorted when he was asked about sustainability. The whole notion, he replied, seems to run counter to the second law of thermodynamics. At another extreme are advocates of minimizing the impact of people on the planet. Their subtext implies minimizing the *number* of people on the planet. At still another extreme are economists like Solow (1991), who believes everything is sustainable because the price mechanism moderates input substitutions.

We can suspect the dean of taking too literal a view, and suspect the earth advocate of hating people. The economic theorists continue to ignore the externalities that create the environmental problems in the first place. The City Club of Portland (2000) cited one source that advised, "Decouple economic development and population growth from environmental impacts." This is a physical impossibility. Is there a constructive middle ground that is scientifically feasible?

Nothing we do will be sustainable for the very long run. We depend on solar energy (as the dean was no doubt thinking), and the sun will eventually die. Meanwhile, every social process degrades energy, in the aggregate.

- Sustainable cannot simply mean "static;" that would mean the end of innovation and the start of excessive regimentation in all spheres of life. Climate change (that portion that is not anthropogenic) would proceed in any case, and society and the ecology would have to change and adapt.
- Can "sustainable" mean "capable of evolving in a steady, manageable way?" No. There are always Black Swans. Global warming is only one example, and it is one that is more predictable than most.
- People, profit, and the planet? We can all get along sweetly and live lightly on the planet, and still get hammered by a rogue asteroid. Only through technology can we hope to reduce the probability or consequences of an asteroid collision, and better technology will do it better. A static society, creating no new technology, is no answer.
- Notwithstanding that many subsistence economies have lasted for hundreds of years, and have been portrayed by historical writers as noble and fulfilling, we cannot equate sustainability with subsistence regimes. Without surpluses and redundancy, such economies are vulnerable to environmental change.

On what time scale is it realistic to speak of sustainability? How wide are the limits of change, within which we're still willing to say a system has been "sustained?" An impressive number of sources agree the time span is a few generations. The City Club of Portland, the President's Council on Sustainable Development, Sustainable Seattle and others say "for generations to come;" "present and future generations;" "our children and grandchildren." This view sensibly leaves scope for changing the plan when conditions and technologies change.[3]

Thinkers and activists urge companies to attend not only to profits but also to their impacts on the environment and on society. People, profit, and the planet ($P^3$)—can this view be sustainable? Whether the dialog is cast in terms of $P^3$ or $E^3$

[3] Section 4 up to this footnote marker is verbatim from Phillips (2008a, b).

(Environment, Economy, and Equity), technopolis thinkers can help advance it by attending to life-cycle perspectives on products and technologies, and to biomimetic technologies and market mechanisms for recycling/reselling "waste." $P^3$ also implies taking care of (providing new habitat for) the poor who will be left homeless by rising sea levels. Is renewed mining in warming north polar regions immoral? Of course not. If done carefully, it's part of $P^3$. Naturally, forensic investigations will reveal who some of the culprits are in anthropogenic climate change. But the excoriated NASA official who said the climate will change with us or without us, and our major challenge is adaptation, was correct. Let's do less arguing about whose fault it is, and learn to do better as we invent preventive and adaptive strategies.

While the British Stern Review (Norris 2007) concludes, "It would be considerably cheaper to stop current climate trends than to try to adapt to a changed world," Norris goes on to note, "It is really hard to get [heat-trapping] gases out of the atmosphere once we put them there." It is a challenge to those in this audience, and to the scientists who work for you, to determine the facts and the best balance of prevention and adaptation.

I don't think adaptive strategies can rely totally on measuring and pricing externalities. (The carbon rights exchanges have been distressingly scandal-prone, according to *The Economist* magazine.) Does the idea of privatization of your city water supply scare you? If so, consider this more extreme example: You exhale $CO_2$ when you breathe…Will bourses for carbon rights lead to "pay for breath?" We will have to be very clever to design sustainable regimes that are not totally price-based. But we will have to design them.

All in all, sustainability is most workable as a concept when it is defined loosely. In any case, we will not be able to forecast with certainty the impact of a managerial act on people or on the planet any more than we can foresee the impact of a corporate activity on the financial bottom line. (Think about making a sales call on a new prospect, who may or may not purchase your product.) Why not? First, we don't know enough of the applicable science. Second, the complexity of environmental, medical, and psychological phenomena makes prediction error-prone. Third, the impacts on people, on profits, and on the planet will interact with one another!

There are many definitions of sustainability. Perhaps the most familiar is the Brundtland report's: "Sustainable development is development that meets the needs of the present without compromising the ability of future generations to meet their own needs" (World Commission 1987).

An admirable sentiment, but a suspicious one. Suppose I borrow money to buy a house that I will bequeath to my children. I may die tomorrow, and the market may go down, leaving the kids unable to sell the house for the amount of the debt. I have compromised their ability to meet their own needs. Suppose I borrow money to build a green business. Green firms face the same business risks as other companies; my children might inherit a profitable business, or they might inherit nothing, my assets wiped out by the bankruptcy of my enterprise. With no funds to complete their education, they'll have to take out loans, and... well, you get the idea.

Should I then avoid all borrowing?

I cannot guarantee that I won't compromise the future. Should that be an excuse for doing nothing—taking no risk—now?

Obviously, no. A green future depends on innovation. Innovation needs to be financed via debt or equity. And that implies risk. In fact, abjuring risk compromises the future. We dither about the "value" of space exploration, for instance, when resources plentiful unto the *n*th generation are no farther away than the asteroid belt.

The US now has a nine trillion dollar debt. If that doesn't compromise future generations, I can't guess what does. Even fiscal conservatives admit that the large part of the nine trillion spent on bank bailouts was necessary, to keep the economy from death-spiraling.

Yet we continue to talk about sustainability. So I think we don't really believe a definition that emphasizes not compromising future generations. We do care about future generations, but we also have confidence that they'll be smart and compassionate enough to do what we do: Try to fix the broken things we've inherited; innovate boldly; and do what we can to create a better future.

Past generations have done this for us, thus proving that there *is* such a thing as a free lunch. Using a telephone is free, for example. (If you deal with certain mobile providers, you'll argue with that, but bear with me.) Your sixteenth-century ancestors had no phones. You did not personally have to invent the telephone. Yet you get to use a phone—solely by virtue of your luck in being born in the twentieth century.[4] Phone calls substitute for polluting transportation, so they're green. Free, green lunch.

We want to give our descendants more free green lunches. But that part about not compromising their ability to meet their needs—where did it come from?

It echoes injunctions of the Abrahamic religions:

- "Do not charge your brother interest, whether on money or food or anything else that may earn interest" (Deuteronomy 23:19).
- Jesus driving moneylenders from the temple (Matthew 21:12).
- The Koranic prohibition on the charging of interest, which it characterizes as oppressive and exploitative (Qur'an 2:279) (Institute of Islamic Banking and Insurance 2009).

One wonders whether the Brundtland report's slant on sustainability was underpinned, perhaps unconsciously, by the religious feelings of its writers. And, with our current economic mess having been driven by mortgage loans, one cannot call such an underpinning totally wrong-headed (Incidentally, the three Abrahamic traditions do not prohibit equity investment—equity is in fact the central principle

4 Apologies to any 8-year-olds who may read this chapter.

of modern Islamic banking—notwithstanding that ownership can be exploitative as well).

Compared to the world of the Prophets, though, ours is more crowded, interdependent, and specialized. The few weeks we spent this year with no credit market to speak of were not the kind of weeks we would wish on our descendants. It is necessary and unavoidable that we lend and borrow when it seems wise to risk the consequences. The consequences may compromise future generations' assets.

I offer, with apologies to Dr. Brundtland, a modification of her principle: *Let us always do our best for the next few generations, while not compromising (too much) our enjoyment of the present, nor our capacity for taking bold risk.*

# 5 Cases

Additional data about Bangalore and Hsinchu round out the information given above.

## *5.1 Bangalore*

The historic development of Bangalore's technoeconomy is shown in Table 4

## *5.2 Hsinchu*

The Hsinchu Science-Based Industrial Park, opened in 1980, is administered under the National Science Council of the Executive Yuan. The park has 300 employees and is one of twelve science parks in Taiwan.

HSBIP now hosts 432 companies, and 17 more are approved for tenancy. Of these companies, 376 are domestic and 54 are foreign. Collectively they employ 130,000 people (average age 30), generate 1.2 trillion New Taiwan Dollars in revenue, and have given rise to 4,400 patents. HSBIP maintains relations with 24 "sister science parks" in 12 countries.

Only 1.3 % of HSBIP tenant company employees have Ph.D.s. This surprising fact brings us back to the park's name (science-based industrial park): Though it is commonly called a science park, it is really about manufacturing.

HSBIP officials attribute the park's success to the interaction of these factors:

- Environment and services
- Incentives and availability of venture capital
- Quality of human resources
- Industry clusters and industry-academic links for R&D.

**Table 4** Bangalore timeline

| | |
|---|---|
| 1911 | India's British rulers invited Nobel-laureate chemist William Ramsay to help select a site for a science school. Ramsay chose Bangalore |
| 1950s and 1960s | Independent India's first Prime Minister Jawaharlal Nehru set up state-owned engineering companies near Bangalore to fulfill his vision of rapid industrialization. He selected Bangalore because of the talent available at the Indian Institute of Science, the school set up by Ramsay. Non-state companies like Motor Industries Co., a subsidiary of Germany's Robert Bosch Gmbh, moved to Bangalore to supply parts |
| 1977 | A socialist Indian government asked International Business Machines Corp. to leave the country after it refused to dilute its stake to 40 %. IBM's departure became an opportunity for entrepreneurs like Azim Premji, who was then running a Bangalore-based vegetable-oil business started by his father. Premji hired engineers and built his first minicomputer |
| 1981 | N.R. Narayana Murthy, an engineer who wanted to become a communist politician, changed his mind and set up Infosys Technologies Ltd. with $250 in Pune in western India. He moved the company to his hometown Bangalore in 1983 after Motor Industries gave him his first order |
| 1996 | Global companies panicked that the year 2000 date change would crash computers. Premji's Wipro Ltd. and Murthy's Infosys rewrote millions of lines of code for customers worldwide. Bangalore's software industry, which employed only 947 people in 1991, expanded rapidly |
| 2005 | Companies like IBM and Accenture Ltd. are hiring in Bangalore to cut costs. Meanwhile, Bangalore's homegrown software makers are competing for consulting contracts outside India that were once the domain of US and European technology companies |

*Source* Mukherjee (2005), used with permission

The environment includes high quality of life at affordable cost, and business services including 24 h automated customs for tenant companies' exports.

Tax incentives/exemptions, government investment, low-interest loans, and "R&D encouragement grants" round out the picture of incentives. However, the success of HSBIP has led the Taiwan government to plan to discontinue tax incentives for high-tech, and transfer them to solar and green industries.

Tsinghua University (science) and Chiao-Tung National University (engineering) are located on the periphery of the park, enhancing industry-university cooperation. ITRI (the Industrial Technology Research Institute) also has two campuses adjoining the HSBIP, and runs incubators for new firms there.

## 6 Synthesis for the Future

Table 5 brings together the major themes of this chapter.

Table 6 summarizes additional views of a sustainable technopolis based on today's state of the art. It is a completely new way of doing business. New technopoleis cannot hope for sustainability if they follow the practices of the Table's left-hand column.

**Table 5** Key factors for a sustainable technopolis

| |
|---|
| Social capital is an absolutely critical success factor |
| Leverage the presence of universities |
| Balance government support with local initiative. Diversify funding sources |
| History matters. So do crises. Do not be fooled by cluster theory |
| Cultivate godfathers |
| Develop political savvy. Emphasize benefits to the many, not to the few |
| Try things that have never been tried before, but that leverage your region's strengths |
| Act |
| Never give up |
| Re-examine the logical and scientific foundations of "sustainability" |

**Table 6** Shared prosperity initiatives compared with conventional aid efforts

| Old Political Economy Approaches | Newer Shared Prosperity Approaches |
|---|---|
| Paternalistic; one-way initiatives and flows | Multilateral initiatives |
| Strategies imposed from top down | Multiple networked initiatives |
| Driven by a single issue or problem | Multidimensional, attacking related problems |
| Rigid | Flexible |
| Expensive | Inexpensive |
| Large-firm orientation | Entrepreneurial orientation, with the participation of large firms |
| Single industry/agency | Multiple sectors, diversified funding |
| Initiating entity and receiving entity seen as separate and independent | Presumes present or near-future interdependence of participating regions |
| "Developed and undeveloped economies" view. Strict accord with international product life cycle theory | Acknowledges tech leader and tech follower regions, but understands that useful innovation can come from anywhere |
| Program-based | Relationship-based; better able to respect and leverage cultural differences |
| Large-agency programs are prone to bureaucratic inertia and resistance to change | Network initiatives attract innovators and influencers in each region |
| Fixed or inappropriate metrics for success; often discipline-bound | "Fuzzy goals"; interdisciplinary, multiple-perspective, or transdisciplinary |
| Money-focused | Financially responsible, but recognizes that knowledge<br>and sense of empowerment are as important as money to emerging regions |

*Source* Phillips (2005, 2006)

## References

City Club of Portland. (2000). Building a Sustainable Future for Portland. City Club of Portland Information Report, November, 2000.

Coleman, J. S. (1988). Social capital in the creation of human capital. *American Journal of Sociology Supplement, 94*, S95–S120.

Phillips. F., & Eto, M. (Eds.). (1998). Special section revitalizing university research and its contribution to society. *Technological Forecasting & Social Change 57*(3), 205–265.

Eto, H. (2005). Obstacles to emergence of high/new technology parks, ventures and clusters in Japan. *Technological Forecasting and Social Change, 72*(3), 359–373.

Fukuyama, F. (1995). *Trust: The social virtues and the creation of prosperity*. New York: The Free Press.

Hall, P. (1998). *Cities in civilization: culture, technology, and urban order*. London: Weidenfeld & Nicolson. New York, Pantheon Books, 1998.

Institute of Islamic Banking and Insurance. (2009). Retrieved September 2000, from www.islamic-banking.com/prohibition_of_interest.aspx, posted September 2009.

Kyaw, A. (2001). Technopolis and regional development in Japan: A statistical analysis. *Geogr Rep Tokyo Metrop University., 36*, 59–68.

Linstone, H., & Mitroff, I. (1994) *The challenge of the 21st century: managing technology and ourselves in a shrinking world.* New York, Suny Press.

Mukherjee, A. (2005). Can Singapore Become a Fun City? Bloomberg News Service. Retrieved April 21, 2005, from http://www.bloomberg.com/apps/news?pid=10000039&sid=aBfOvonc_Yr4&refer=columnist_mukherjee.

Norris, R. (2007). Global Instability: Climate change is about to come home to roost. *San Diego Union-Tribune,* B1, December 6 2007.

North, D. C. (1990). *Institutions, Institutional Change and Economic Performance*. Cambridge: Cambridge University Press.

Phillips, F. (2005). Toward an intellectual and theoretical foundation for 'shared prosperity'. *Systemic Practice and Action Research, 18*(6), 547–568.

Phillips, F. (2006). *Social culture high tech economic development: The Technopolis columns*. London: Palgrave Macmillan.

Phillips, F. (2008a). Change in socio-technical systems: Researching the Multis, the Biggers, and the More Connecteds. *Technological Forecasting and Social Change, 75*(5), 721–734.

Phillips, F. (2008b). The Godfathers: Characteristics and roles of central individuals in the transformation of techno-regions. *Journal of Centrum Cathedra, 1*(2), 12–27.

Phillips, F. (2009a). Assessing and enhancing the impact of innovation and entrepreneurship. Chapter 15. In F. Phillips (Ed.), *Managing Innovation, Technology, and Entrepreneurship*. Aachen: Meyer & Meyer Media.

Solow, R. (1991). Sustainability: An Economist's Perspective. Eighteenth J. Seward Johnson Lecture, Woods Hole Oceanographic Institution, June 14, 1991.

Sölvell, Ö., Lindqvist, G., & Ketels, C. (2003). *The Cluster initiative greenbook*. Gothenburg: Ivory Tower.

Suzuki, S. (2004). Technopolis: science parks in Japan. *International Journal of Technology Management 28*(3–6) 582–601.

World Commission on Environment and Development. (1987). Our Common Future. (Oxford, Great Britain: Oxford University Press, 1987, p. 8). (Known as the Brundtland report, after Gro Harlem Brundtland, Chair of the Commission).

# French Competiveness Clusters: A Multipurpose Policy Instrument

**Yoslan Nur**

**Abstract** The French government launched, in 2005, a 3-year policy of competitiveness clusters (*pôles de competitivité*), which has since then been extended for the period of 2009 to 2012. It is one of the most important recent innovation policies in the country. Theoretically, a competitiveness cluster brings together large and small firms, research laboratories, and educational establishments with the support of the national and regional government, to work together in a specific region to develop synergies and cooperative efforts. The primary objective of this initiative is to boost the competitiveness of the French economy and to help develop growth and jobs in key markets namely by accelerating innovation efforts, providing support for high-tech and creative activities, and improving international visibility. Active partnerships among the cluster members should contribute to fostering synergies of the sector or technology underlying the cluster. Cluster members are eligible for direct financial aid, tax incentives, and privileges for accessing funding sources. France has now 71 competitiveness clusters and €4.3 billion over 6 years has been dedicated by the public sector to support these initiatives. The competitiveness cluster is a multipurpose policy instrument. It is a convergence of research and development (R&D), industrial, and innovation cluster instruments. This policy not only promotes an outward looking (export-oriented) economic policy for French firms but also constitutes a shift from a top-down to a bottom-up philosophy with competition orientation between initiatives at the regional level. This chapter shall analyze the impact of the public and private partnerships of the competitiveness cluster policy on cooperative research and development (R&D) activities, creation and growth of knowledge-based small and medium enterprises, and the creation of local innovation ecosystem.

Y. Nur (✉)
Division for Science Policy and Capacity Building, Natural Sciences Sector, UNESCO, Paris, France
e-mail: y.nur@unesco.org

D.-S. Oh and F. Phillips (eds.), *Technopolis*,
DOI: 10.1007/978-1-4471-5508-9_10, 

# 1 Introduction

During the past two decades, the concept of clusters has gained enormous popularity to the extent that policymakers, practitioners, and academics increasingly make reference to the term. Notable publications on the concept include: Michael E. Porter's, *The Competitive advantage of Nations* in 1990 and *Clusters and Competition—New Agendas for Companies, Government and Institution* in 1998. Porter (1998) defined that a cluster is a geographically proximate group of interconnected companies and associated institutions in a particular field, linked by commonalities and complementarities. Clusters also often include firms in downstream industries (that is, channels or customers); producers of complementary products; specialized infrastructure providers; government and other institutions providing specialized training, education, information, research, and technical support (such as universities, think tanks, vocational training providers); and standard setting agencies. Government agencies that significantly influence a cluster can be considered a part of it.

The main argument supporting clustering is that individual firms cannot become competitive and stay competitive in the global market on their own. So, to build on and maintain competitiveness, one has to incorporate continuous changes throughout the value chain. The close interplay between firms, their suppliers, and the business environment is why competitiveness theorists and practitioners focus on clusters as the locus of action, as opposed to individual firms or broad sectors (USAID 2003). Cluster initiatives can contribute to comprehensive national competitiveness efforts that include policy reform, trade capacity building, private–public dialogue, regional economic development, workforce development, etc. At times, they become a catalyst, generating broad public understanding and support for the economic reform agenda by working with the media, universities and think tanks, knowledge and technology foundations, industry leaders, government officials, etc. When designed carefully and implemented efficiently, cluster initiatives may well be one of the most effective tools in a broader context of policy reform and other private sector development initiatives (World Bank 2009).

Following to the Report of Christian Blanc (2004)—analyzing the origin of and proposing the solution to the slowing down of the French economic—in 2004, the French government launched the competitiveness clusters (*pôles de compétitivité*) policy. The "competitiveness cluster" is the French equivalent for what Porter calls a "cluster", by definition competitiveness cluster brings together large and small firms, research laboratories, and educational establishments with support of the national and regional government, to work together in a specific region to develop synergies and cooperative efforts. The primary objective of this initiative is to develop the competitiveness of the French economy by boosting local technological and creative industries, attracting business to France (thanks to a higher international profile) and stimulating growth and job creation in key markets. The policy intends to encourage and support projects initiated by the economic and academic agents in a given local area, and to foster public–private local partnerships.

The competitiveness clusters initiative is one of many policy efforts that have been deployed in France in the past 10 years to face innovation challenges. Indeed, with 2.23 % of its GDP devoted to R&D expenditure (GERD/GDP) in 2001, France is not in a very strong position, compared to the 3 % Lisbon objective in 2020, but it occupies 4th place in the EU, after Sweden, Finland, and Germany. According to the latest available data, France's R&D intensity has declined steadily since 2002; it then improved significantly in 2008 and 2009, with domestic R&D spending amounting to 2.21 % of GDP in 2009, virtually back up to its 2001 level. Against this background, France's R&D target for 2020 is ambitious but achievable, given existing and future measures, which will pay off in the years to come, including the investments for the Future programs, the competitiveness clusters initiative, innovation grants, and the development of the digital economy.

The competitiveness cluster policy looks at: (i) facilitating public and private partnership in research and development (R&D), promoting private investment in R&D; (ii) supporting the development of knowledge-based small and medium size enterprises (SMEs); and (iii) reinforcing local competitiveness and attractiveness by facilitating cooperation among regional stakeholders and creating a local innovation ecosystem: (La documentation Française 2008; Bocquet et al. 2009; Bonnafous-Boucher et al. 2010). Thus, this interministerial policy is a convergence among three main policies: innovation, industrial, and regional policies.

The main objective of this chapter is to develop our understanding about the French competitiveness clusters policy, particularly its impact on research and development (R&D) collaboration between the private sector and research institutions on the creation and growth of knowledge-based small and medium enterprises, and on the creation of local innovation systems. This chapter is based on previous studies and evaluations that have been conducted by the French government and relevant scholars.

## 2 Competitiveness Cluster: Basic Principle and Practices

The competitiveness clusters policy was initiated by the Interministerial Committee for Regional Planning and Development (*Comite interministériel d'aménagement et de développement du territoire*—CIADT). The name of the Committee was changed to "Internomisterial Committee for Regional Development and Competitiveness (Comite interministériel d'aménagement et de competitiveité des territoires—CIACT)." The policy was founded on three principal axes: (i) providing financial support to specific cooperative innovation projects; (ii) supporting the governance of the competitiveness clusters; and (iii) coordinating the public stakeholders. The CIACT has an authority over any decisions regarding the national policy of the competitiveness clusters.

During the first phase, from 2004 to 2008, 71 poles were granted the status of competitiveness clusters (66 clusters in 2005, out of 105 applicants, and 6 clusters in 2007, out of 18 applicants). In order to be granted the competitiveness cluster label, a cluster must have: (i) a development strategy that remains consistent with the economic development plan in the clusters local area; (ii) a sufficient international visibility, in terms of industry and/or technology; (iii) a partnership between players, a governance modality, and operational structure ensure to generate synergies in research and development, and (iv) a capacity to create of new wealth with high added value. The evaluation of the first phase was conducted in 2008 by CM International and The Boston Consulting Group (2008) in France (BCG). During the first phase the competitiveness activities are usually dominated by the structuring of the cluster, improving their international visibility and creating R&D collaborative projects (projects fabrication). The evaluation report confirmed that the competitiveness cluster policy has created and accelerated a dynamic to cooperate around an innovation project in all sectors of industrial activities. Regarding the evaluation of the individual performance of each cluster: 39 poles have reached their objectives, 19 poles have only partially reached their objectives and must review their strategy, governance, and management team, and 13 poles need to review their strategy and revise it to be in line with the objective of the cluster policy. Nevertheless, all the poles are renewed; the last 13 poles are provided 1 year by the government to show they have reached the requirements of the policy.

The competitiveness clusters policy 2.0 is the second phase of the competitiveness cluster policy that covers the period 2009–2011. During the second phase, the CIACT granted six new members to the competitiveness clusters. At the same period, the Committee did not extend the competitiveness cluster status of the six clusters which are among the 13 clusters that have had difficulty during the evaluation in 2008. After the first evaluation, the government decided to give a new dimension to the policy by focusing on three priorities: (i) to strengthen the cluster's strategic piloting and management; (ii) to provide clusters with new financing tools (structuring projects); and (iii) to develop support for new dimensions of the innovation ecosystem such as human resources competences, intellectual property rights (IPR), etc. Various measures were applied to support the competitiveness clusters such as supporting the strategic governance of clusters, financing structuring projects, such as innovation platforms; developing other aspects of the cluster ecosystems, such as competence management, international development, IPR management, and an incentive to get more of private funding, specifically to support the growth of SMEs. The evaluation of the second phase was conducted at the first trimester in 2012 by a consortium Bearing Point-Erdyn-Technopolis. The report of the second evaluation was released on 19 June 2012.

The competiveness clusters are classified into three main categories: global clusters (7), global vocation clusters (11), and national clusters (53). The global clusters are Aerospace Valley (Midi-Pyrénées), Lyonbiopole (Rhône-Alpes), Medicen Paris Region (Paris), Minalogic (Rhône-Alpes), Secured Communication Solutions (Provence Alpes Côte d'Azur), System@tic (Paris) and Finance

Innovation (Paris). The classification is based on the economic potential and the R&D capacities of the stakeholders. The calcification has not been modified since 2005, despite the recommendation of the status of the cluster for global vocation cluster. The evaluation report published in 2012 considers that the current cluster classification is obsolete and does not bring any benefit. The report recommended to distinct the clusters into two categories: international competitiveness clusters and competitiveness clusters and innovation. The international cluster must be found on two main criteria: (1) the *strategic character* of the branch, or sector, contributing to France competitiveness at an international level and (2) the *critical size*, including *ex ante* indicators such as the concentration of industrial and scientific resources, and *post ante* such as number and volume of R&D projects (Fig. 1).

Each competitiveness cluster is represented and led by its own legal entity, but this entity is required to give preference to industrial, scientific, and academic stakeholders in its governing body. The local governments concerned with the

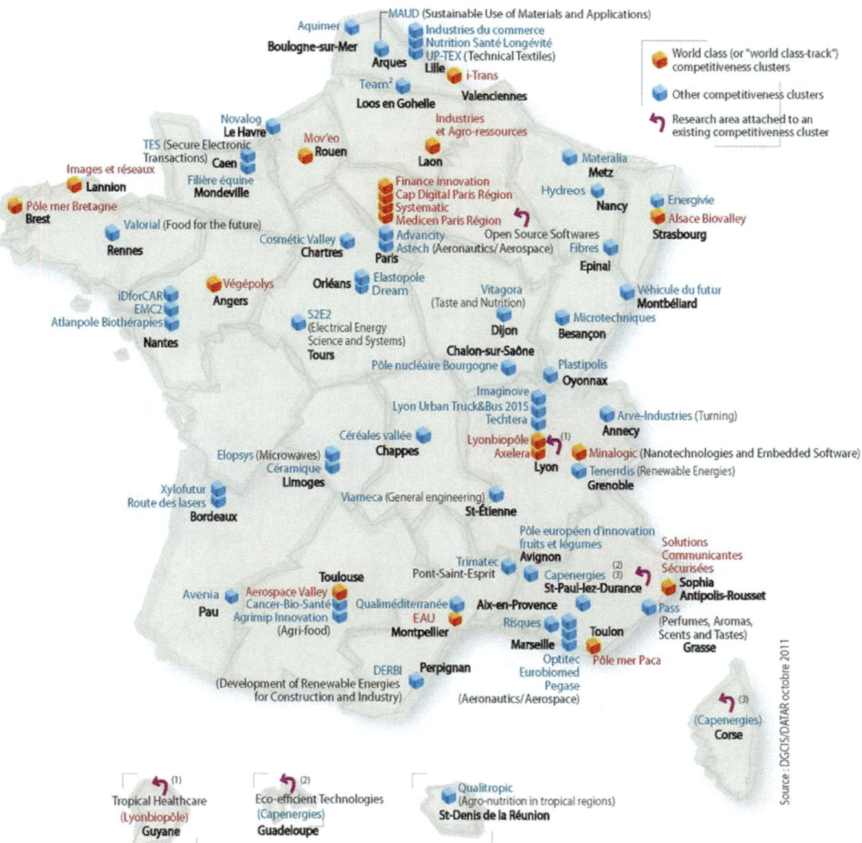


**Fig. 1** Map of Competitiveness clusters (October 2011). *Source* http://competitivite.gouv.fr/la-documentation-sur-les-poles-de-competitivite/les-cartes-des-poles-791.html

project are also duly represented. The projects of clusters are defined by the firms and the governance of the clusters is also defined and run by the firms, according the specificities of each local context. The government is involved through the incentive process it has defined, through a selection process, and though the top-down pressures and organizational role exerted by its local representative when necessary. The governing entity of the clusters is organized by an Association (law 1901), which consists of a general assembly, administrative council, secretariat, president, vice president, and director. The Association has its own legal entities, financed by the State or local governments. The main duties of the Association are to define and implement the overall project of development of the pole, and to foster, evaluate, and select R&D projects submitted for public financing dedicated to the clusters of competitiveness policy. In addition, its main duty is to manage the cluster, the Association deals with external relations including any cooperation with other clusters.

Each competitiveness cluster draws up a 5-year plan, which can be used as a basis to: (i) develop partnerships between the various stakeholders based on their complementary skills; (ii) establish strategic R&D projects that can benefit from public funding, particularly the Interministerial Fund (FUI); (iii) promote a suitable ecosystem of innovation via knowledge-sharing and mutual support among cluster members on topics such as training and human resources, intellectual property, private-sector financing, and international development cooperation.

Therefore, the government strongly supports competitiveness clusters initiatives by providing financial assistance such as: grants for outstanding R&D projects (financed by the Single Interministerial Fund and the Investments for the Future Programme), they partially finance the clusters governance (alongside local governments and firms) and provide financial support for theme-based collective actions initiated by clusters in a wide range of areas, via the various Regional Directorates for Industry, Research and the Environment (DRIRE). In 2006, the Government created the Single Interministerial Fund (FUI), involving six different ministries (the Ministries of Industry, Defence, Infrastructures, Agriculture, Health and Spatial Planning), to finance collaborative R&D projects in the competitiveness clusters. The government also brought additional public institutions on board including: the French National Research Agency (ANR) and OSEO who provide financing for R&D projects, and the *Caisse des Dépôts et Consignations* (CDC) to support innovation platform projects; and other local government agencies who may also provide financial support for cluster projects, both for R&D and innovation platforms. At an international level, the government helps the competitiveness clusters and their member firms find the best international partners and set-up technological partnerships.

From 2005 to 2011, more than €4.3 billion of public funds have been allocated to support 5,750 R&D projects in the competitiveness clusters. This amount is not taking in consideration the contribution made by local governments. During the first phase, from 2005 to 2008, the French Government mobilized funds amounting €1.5 billion to support the competitiveness clusters. This amount consists of

€840 million from the single interministerial fund, €500 million from public agencies that are dealing with innovation, mainly the National Research Agency and OSEO, and €160 million in the form of tax incentives/tax breaks. The R&D projects are the core activity of the competitiveness clusters. More than 1,000 R&D projects, jointly organized by private companies and research institutes, were financed during the period 2005 to 2008.

From 2008 to 2011, on top of the €2.7 billion from the public sector, some €3 billion has been invested from the private sector. Therefore, €5.7 billion has been invested in the R&D projects initiated by the competitiveness clusters. This represents approximately 4.5 % of the national R&D expenses during that period.

## 3 Boosting Innovation by Facilitating Public and Private Partnership on R&D

Facilitating public and private partnerships in R&D is one of the main objectives of the competitiveness cluster. A competitiveness cluster is a platform for the creation of cooperative projects between private enterprises, research centers and academic institutions. The ultimate goal is to innovate together, or to introduce a new or significantly improved product (good or service), or process a new marketing method, or a new organizational method in business practices, workplace organization, or external relations. R&D projects are thus the cluster's core activity and constitute the main factor of their so-called competitiveness. They should involve all the potential stakeholders of the clusters in a process of growing innovative capabilities and competitiveness of the firms, especially the SMEs, which traditionally in France have difficulties accessing R&D resources. The project should also boost research institutes through public–private partnerships. Indeed the projects have to include at least two firms and a research institute of the cluster in order to be labeled as a cluster pole of competitiveness. These projects are the engines for the working of the poles and thus the preconditions of the success of this policy. The subsidies to the poles are not predetermined, they flow from the R&D projects that have gained subsidies, as it will be shown.

In this analysis, the innovation clusters policy is viewed from two different angles. First, we can consider that the competitiveness cluster policy as an integrated series of French policy to promote public and private partnerships in R&D. Second, we can also see the competitiveness clusters policy as an implementation tool of the R&D policies. With the latter, we can identify several policies related to R&D development that have been supporting the implementation of the competitiveness cluster policy.

First, the adoption of the Innovation and Research Act in 1999, the law aims to modernize and reorganize the French innovation system. The main objectives are to encourage technology diffusion and the exploitation of research results and innovation. The Innovation Act has four main areas: (i) encouraging the mobility of researchers toward private enterprises, for instance through the authorization for public researchers to participate in an innovative firm start-up without losing their civil service status for 6 years, or the possibility for public researchers to engage in consulting activities or in private firms' boards of directors; (ii) increasing public–private research collaboration, for instance via incubators in universities and public research organizations, the introduction of industrial and commercial service activities in public research organizations, and simplified administrative procedures; (iii) providing fiscal break measures for innovating firms; and (iv) a providing a legal framework for innovating firms: granting the extension of the simplified status of the joint stock company to all innovative firms (Muller et al. 2009).

Second, the French Innovation Plan of 2003 was produced jointly by the Ministry of Research and the former Ministry of Industry. The main lines of this document concern the diffusion of research results and the facilitation of innovation. Its focus is on publicly funded research and its exploitation for innovation, including providing support for young innovating companies, research funding, industrial strategic research, etc. Research has not only attributed an important role at the national, but also at the international level. Precise measures elaborated in this plan after national consultations, include (i) the status of business angels; (ii) support for projects of *Jeunes Entreprises Innovantes*; (iii) further new instruments to foster innovation, including tax reductions; (iv) simplified access to public funding through attributing a central role to ANVAR (now OSEO); (v) better exploitation of research results for enterprises; (vi) the attribution of a central role to innovation in a national and European perspective; and (vii) the support of industrial strategic research (Muller et al. 2009).

Third, the Research Pact, promulgated in October 2005, pursues the aim of modifying the existing national research and innovation system that had evolved comparatively little in the preceding period. This fact is based on three axes: (i) the balanced development of the three basic components of the research system (basic research, research with societal and economic implications); (ii) the development of interfaces and cooperation between actors of the research landscape; and (iii) the introduction of a global and long-term strategy that aims to increase confidence between research and society (Muller et al. 2009). The Research Pact introduces a rationale of financing for specific research projects instead of regular financial support for the research institutions, promotes cooperation among public research institutions, reinforces the public and private actors (*Pôles de Compétitivité*, Carnot Label).

The collaborative R&D projects are basically the core of the activity of the competitiveness clusters. The call for the collaborative projects, launched twice a year, is the main tool to promote public and private partnerships in R&D. The selection process is operated in three steps: first, a regional evaluation is conducted under the aegis of the regional prefecture; second, an evaluation is performed by an interministerial working group (GTI); and finally an independent audit is carried out by a group of qualified persons (GPQ) in the fields of business, research and higher education. During the period 2005–2011, around €4.3 billion were allocated to support approximately 5,750 R&D projects, which consist of collaborative projects and non-collaborative projects that were supported by the competitiveness clusters. Furthermore, 75 % of the funds were allocated to support collaborative R&D projects. The Single Interministerial Fund (*Fonds Unique Internimisteriel* (FUI)) only finances projects which involve activities with at least two private enterprises and one public research institution, and that are developed within the competitiveness clusters. So far, FUI is a main policy instrument to promote public and private partnership in R&D. During the period 2005–2011, FUI allocated more then €1 billion for R&D collaborative projects. The other agency that supports the collaborative R&D project is ANR. During the period 2008–2011, the Agency spent €748 million to support collaborative projects in the competitiveness clusters.

The collaborative R&D projects have generated 2,500 innovations, with two-thirds of them focusing on product and process innovation. Among the most innovative sectors are IAA, ICT, energy, materials/mechanics, and engineering and service. The projects have produced almost 1 million patents, particularly in the field of ICT, biotechnology, health and energy, almost 6,500 scientific papers published and 93 start-up companies nurtured in the competitiveness clusters.

However, the funds are not equally distributed among the clusters. There is a big discrepancy between the funds distribution and the number of poles concerned; the 6 clusters (four global clusters and two global vocation clusters) out of 71 obtained one-third (33 %) of the total funds. Half of the funds were allocated to 12 clusters. This discrepancy is a result of the bottom-up process of the collaborative R&D projects selection. The selection process starts by a call from the government to the members of the competitiveness cluster to submit their application for collaborative R&D projects. For example, the results of the 12th call, announced on 1 August 2011, 79 collaborative R&D projects, out of 132 applications, were financed in this call, stemming from 55 poles totalling €73 million. This means that 16 clusters (one out of five clusters) have not been able to build at least one project qualifying to receive a funding. Usually, big clusters with better companies and with better research facilities have more chance to win the selection rather than the small clusters. The mechanism should, however, attempt to give more of an opportunity to the members of the small competitiveness clusters to succeed in the selection of the collaborative R&D projects.

## 4 Supporting the Growth of Innovative SMEs

One of the main objectives of the competitiveness clusters is to improve the competitiveness of SMEs. The French government pays special attention to the development of SMEs because they play a very important role in the French economy. They account for more than 2.5 million enterprises which represent 97.28 % of the number of enterprises, they employ more than 9.2 million people or contribute to 47.33 % of the total employment (Saoud 2010). The development of competitiveness clusters benefited from the policy measures that were taken by the French authority related to the promotion of innovation in the SMEs, such as the creation of two main agencies dealing with innovation: ANR and OSEO Innovation in 2005. Other initiatives are the promulgation of the Pact for Research in 2006 and the Law for the Autonomy and Responsibilities of Universities in 2007, the elaboration of the National Strategy for Research and Innovation in 2009, the establishment of the broad outlines of the new French Industrial Policy announced in May 2010. In January 2012, the French Government released the Investments for the Future Program: *Fonds National d'Amorçage (FNA)* which was set up to provide capital to companies developing innovative technologies that generate competitive advantages, jobs, and economic activity.

The ANR is responsible for allocating financial support for basic and applied research. The goal of the ANR is to stimulate the emergence and performance of research projects by participating in competition projects for funding. The Agency has four main programs to support innovation: RIB program (*Réseau Innovation Biotechnologies*—Biotechnology Innovation Network), PRECCOD program (Environmental Technologies and sustainable Development), Biotecs program (Biotechnologies); and RNTL program (*Réseau National des Technologies Logicielles*—National network for software technology).

OSEO was founded by the fusion of the *Agence Nationale de Valorisation de la Recherche* (ANVAR) with the SME Development Bank. OSEO Innovation is a private agency but is part of a public body under the authority of the Ministry of Economy, Finance, and Employment, and the Ministry for Higher Education and Research. Being the major actor for SMEs innovation in France, OSEO pursues three main areas of activity with one aim in common, helping entrepreneurs to take risks through: (i) innovation support and funding for innovative technology-based projects with real marketing prospects; (ii) funding investments and the operating cycle alongside the banks and financial institutions; and (iii) guaranteeing funding granted by banks and equity capital investors. One of the famous programs of OSEO is the Strategic Industrial Innovation (ISI) program. The program supports collaborative industry-oriented innovations, implemented by academic institutions together with medium-sized companies (less than 5,000 employees) and SMEs (less than 250).[1]

Besides the above-mentioned programs, the French government has also implemented several measures to promote and sustain the creation of growth of innovative SMEs, such as the National Contest of Support to the Creation of Innovative Technological Entreprises (*Concours National d'Aide à la Création d'Entreprises de Technologies Innovantes*), Young Innovative Enterprises (*Jeune Entreprise Innovante*—JEI), Young University Enterprises (*Jeune Entreprise*

---

[1] At the European level, OSEO innovation participates in several European Programmes and Networks: FP7 (OSEO is NCP SMEs for France and active member of the French NCP Energy/ Environment consortium), EUREKA and the new transnational initiative Eurostars, ERA-Nets, INNO-Nets (OSEO coordinates the Innet project with 18 partners), aiming at strengthening the SMEs in technology clusters in Europe and OSEO is WP leader for the Competitiveness and Innovation Programme 2007 to2013. KIS-PIMS project (Knowledge Intensive Services in the field of Renewable Energy).

*Universitaire*—JEU) and tax reliefs for innovative SMEs. The government also launched a specific measure to promote the culture of innovation, such as the school of innovation. The measure led to the awareness of research opportunities, the understanding of research partnerships, the identification and targeting of contacts to save time and scientific expertise of project ideas.

The majority of the above-mentioned programs and measures are open to all SMEs and applicable in the competitiveness cluster. They cover the entire value chain steps of research, development and innovation projects, from basic research, design, and development to research exploitation and commercialization of innovation. The French Government also encourages the emergence of new projects for creating firms and/or developing start-ups and rapidly expanding SMEs. In competitiveness clusters, this infrastructure spans technology business incubators and specific frameworks designed to boost growth in SMEs. According to second evaluation report of the competitiveness cluster, in 2011, on average each cluster has 187 members and 72 % of the members are private enterprises. SMEs represent almost 60 % of the members. Between 2008 to 2011, €984 million (€517 million from OSEO, €375 million from FUI, and €87 million from ANR), which represents 36 % of the public funds, allocated to the competitiveness clusters were used to support collaborative R&D projects involving SMEs.

The majority of the funds to support SMEs in the competitiveness clusters come from OSEO (52.54 %). The approved project from the clusters can access two forms of financial support from OSEO: Innovation Assistance (funds) or the Strategic Industrial Innovation program. From 2007 to 2010, between 10 and 13 % of the OSEO Innovation Assistance (funds) were used to support projects in the competitiveness clusters. The Innovation Assistance (funds) provides financial assistance for the realization and development of innovation before its industrial and commercial launching. It concerns the innovative SMEs with less then 2,000 employees. The report of the second evaluation of the French competitiveness clusters states that, in general, the innovation assistance of the OSEO were given to individual enterprises (9 out of 10), not to collaborative R&D projects. This practice is not in line with the policy which is promoted by the competitiveness clusters.

Since 2008, OSEO also allocated financial assistance to the SMEs in the competitiveness clusters through the Strategic Industrial Innovation program. It concerns the collaborative R&D projects which involve activities with at least two private enterprises and one public research institution. Twenty-five clusters benefited from this program from 2008 to 2011.

Others institutions that provide SMEs with financial support in the competitiveness clusters are FUI and ANR. During the period 2008–2011, FUI allocated €375 million, more then 38 % of its funds, to support collaborative project involving SMEs. The SMEs represent around one-third of the participants of the collaborative projects. 45 %of the FUI sponsored collaborative projects were led by SMEs. The public research institutions are the main recipients of the ANR funds, 76.8 % of the funds allocated during the period 2008–2011 (€574 million). The private enterprises benefited from 23.1 %, half of it was used to support SMEs (€87 million).

## 5 Creating an Innovation Ecosystem

When I started to write this chapter, as an urban and regional development specialist by training, I was wondering whether the competitiveness cluster policy is a continuation of the French regional policy, that was instigated in the book by the famous author Jean-François Gravier, published in 1947, entitled *Paris et le desert français* (Paris and the French Desert). In his book Gravier argued that the excessive growth of Paris was resulting in the "desertification" of the rest of France and urged for the necessity to achieve a better balance between infrastructure, industry, and agriculture of modern France. The French Government took heed of Gravier's warnings and created a Division responsible for regional policy in the Ministry of Reconstruction and Urbanism, in 1950, and established DATAR (*Délégation à l'Aménagement du Territoire et à l'Action Régionale*), a State agency responsible for regional planning and policy in 1963. DATAR was assigned three essential objectives: "devolved" industrialization, the pursuit of large-scale regional development projects, and the promotion of the creation of regional centers (*métropoles d'équilibre*). This regional policy was very influential up to the mid1980s.

When I conducted the analyses on the French competitiveness clusters, I realized that the regional policy has not been applied to the competitiveness clusters for the following reasons. First, half of the 71 competitiveness clusters have been created in three leading regions: Greater Paris (Ile de France), Lyon/Grenoble (Rhône-Alpes) and the French Riviera (Provence Alpes Côte d'Azur). Second, 3 out of 7 global clusters are located in Greater Paris. Third, According to the cluster evaluation report published in 2012, 33 % of the public funds allocated for the project went to 6 clusters (4 global clusters and 2 global vocation clusters) and half of the funds accumulated are only in 12 clusters. Therefore, the competitiveness cluster policy has no specific measures to improve the regional balance of distribution of the knowledge-based economic activities.

I do come back to the basic idea behind the creation French competitiveness clusters. The main objective of the competitiveness clusters initiative is to constitute geographical concentrations of actors from the spheres of business, research, and education, inside a structure which is mostly not-for-profit institutions with specific financing mechanisms for collaborative R&D projects involving at least two private companies and one research institution which are expected to bring an innovation to the market within period of 5 years. The competitiveness clusters policy shares a common framework to that of the Triple Helix model, in which the government puts in place the necessary measures to promote cooperation between enterprises (productive sector) and academia (including other research institutions).

The initiators of the French competitiveness clusters understood very well that, one of the key points that can guarantee the success of this policy is to develop networks and networking opportunities between the clusters and other actors from outside the clusters. The collaborative R&D project, involving at least to private enterprises and one research laboratory, was chosen as a strategy to initiate the networking.

In 2012, Rachel Levy, Caroline Hussler, and Pierre Triboulet conducted a study on the evolution (from 2005 to 2010) of the French collaborative network of innovation, by scanning collaborative projects funded by the FUI in the competitiveness clusters. The study covers 779 projects, involving 5,756 organizations. They tested whether the network progressively gets connected, concentrated, or clustered around some (specific) competitiveness clusters or if, on the contrary, it extends on the French territory, act as building collaborations with intra- and also extra-competitiveness clusters' members and becoming more loosely coupled to one another. They also analyzed the existence, creation, disappearance of innovative cohesive groups within the overall network.

As the result of the study, Levy et al. (2012) published their findings, among of them are as follows:

- The density of the innovation network (which are developed around the project supported by FUI) increases through time. This indicates than a given partner tends to collaborate more frequently on different innovative projects in recent years than in 2005. Encouraging competitiveness clusters to organize themselves in networks (as it is done by the new FUI procedure of funds allocation) thus seems to have an impact on the density of the network of innovative projects.
- The number of isolated projects (or grouping of two or three projects of small isolated groups) has been reduced by more than half, as a corollary the size of the big network increased, showing that the network of collaborative projects becomes more and more connected.
- There is a decisive influence of size on the likelihood for a project to be linked to other projects: size of the projects and also of the project's partners. They also reported that projects run by SMEs are more isolated, probably because small firms do not have enough resources to get involved in several innovative projects concomitantly.
- The projects run by distant partners are less connected to one another: indeed, long distance collaboration might be more complicated, and time-consuming for partners who limit their capacity to be involved in several projects at the same time.
- The borders of innovative communities do not coincide with the territories of competitiveness clusters, suggesting thus, that they do not observe a clustering of the French innovative network around some specific competitiveness clusters: public and private actors, when looking for innovation do not limit their collaborative perimeter to the borders of the competitiveness cluster they are members of. The evolution of the structure of collaborative networks of innovation in France cannot solely be explained by a cluster-policy dynamics. They, according to their observation and despite the governmental intervention in a deliberate clustering policy, seem to continue to obey to a spontaneous self organizing dynamic based on traditional collaborative complementarities.

How solid is the existing collaboration among the stakeholders of the competitiveness clusters? If we refer to the basic concept of the triple helix model (Etzkowitz 2002), we can observe that there is a very strong commitment from the

French government to promote the competitiveness clusters. It is expressed by providing financial assistance and the implementation of policy measures. The success of the cluster also depends on the capacity of the enterprises to collaborate with each other and with research institutes. In order to create a solid ecosystem of innovation in the competitiveness clusters, the French government has to gradually reduce the financial assistance provided and improve their involvement in the development of capacity to run the cluster as well as in facilitating international cooperation in innovation.

## 6 Conclusion

The competitiveness clusters initiative is a policy instrument to promote national competitiveness by supporting R&D activities. The second evaluation report of the competitiveness clusters confirmed that the competitiveness cluster policy has continued to develop and accelerate a dynamic to cooperate around an innovation project in all sectors of industrial activities. The competitiveness clusters are attracting more and more members. This dynamic of cooperation is basically developed around subsidized collaborative R&D projects. During the second phase, from 2008 to 2011, the government allocated €2.7 billion to support almost 900 collaborative projects which have produced more than 2,500 innovation projects that have been commercialized.

The competitiveness clusters initiative is one of the numerous programs and measures to boost cooperation between SMEs and public research organizations. It was reported that almost 70 % of public research institutions which are members of the competitiveness clusters have establish R&D cooperation with SMEs (Bearing Point, Erdyn and Technopolis 2012). However, for SMEs, the development of collaborative relationships with research laboratories requires considerable investment in terms of time and skilled human resources. On the other hand, taking into consideration the difficulties that SMEs face to invest in R&D, research laboratories often prefer to work with big enterprises. The competitiveness clusters needs to find a sustainable model to facilitate cooperation among them, grounded on a common interest and trust between the researchers and SMEs.

The competitiveness clusters initiated the creation of local innovation ecosystems by providing an environment to the clusters' member to work and innovate together on collaborative projects. The density of the innovation increases through time. It seems that "natural" networks based on business opportunities are developing. In order to accelerate the maturity of the innovation networks, it is better to gradually reduce the financial support on the collaborative R&D projects and invest more on building the competitiveness clusters' the capacity of governance, strengthening national and international innovation networking and reinforcing innovation information system.

# References


Bearing Point, Erdyn, & Technopolis. (2012) Etude portant sur l'évaluation des pôles de compétitivité. http://competitivite.gouv.fr/l-evaluation-de-la-2e-phase-de-la-politique-des-poles/le-rapport-complet-de-l-evaluation-888.html.

Blanc, C. (2004). *Pour un écosystème de la croissance : rapport au Premier Ministre*. Paris: La Documentation Française.

Bocquet, R., Mendez, A., Mothe, C. & Bardet, M. (2009). "Pôles de compétitivité constitués de PME : quelle gouvernance pour quelle performance?" *Revue Management & Avenir, 25*, 227–244.

Bonnafous-Boucher, M., de Géry, C., & Lavallette, E.M. (2010). A Model of Entrepreneurial Performance in Clusters: Lessons from the French Competitiveness Clusters, Chair of Entrepreneurship and Innovation, Advancia, ESCP Europe and HEC. http://sbaer.uca.edu/research/icsb/2011/402.pdf.

CM International and The Boston Consulting Group. (2008). *L'évaluation des pôles de compétitivité 2005-2008*, Travaux No. 9, La documentation française and DIACT, Paris.

Etzkowitz, H. (2002). The triple helix of university-industry-government implication for policy and evaluation—Working paper, Institute for studier av utbilding och forsking, Stockholm.

La documentation Française. (2008). la politique des pôles de compétitivité 2005–2008.

Levy, R., Hussler, C., & Triboulet, P. (2012). The evolution of the French collaborative network of innovation: towards clustering?, Paper to be presented at the DRUID, on 19–21 June 2012, CBS, Copenhagen, Denmark.

Muller, E., Zenker, A. & Heraud, J.-A. (2009). France Innovation System and Innovation Policy, Fraunhofer ISI Discussion Papers Innovation Systems and Policy Analysis, No. 18, Karlsruhe, April 2009.

Porter, M. (1998). Clusters and the new economics of competition. *Harvard Business Review, 76*(6), 77–91.

Saoud, H. (2010). Comprehensive analysis of programmes and initiatives in France that assists the collaboration between science and SMEs, Research report conducted under the request of European Commission, under FP7-SME-2009-1, SP4-Capacities.

USAID. (2003). *Promoting competitiveness in practice—an assessment of Cluster-based approaches*. Washington: USAID.

World Bank. (2009). *Clusters fro competitiveness—a practical guide & policy implications for developing Cluster initiatives*. Washington: The World Bank.

# Technology, Market and Company Journeys: How Can We Help Them Succeed?

**Malcolm Parry**

**Abstract** The science and technology movement has been developing for over 60 years. In that time these specialist developments have changed from passive developments that are simply property based initiatives that anticipated through co-location to be able to contribute to innovation and technology commercialisation. The need for acceleration of this process has brought together business and the generators of knowledge on these sites which are consistently supported by government programmes to help with the formation of development of new technology-based firms. The combination of features that have been developed are now being coordinated in regional innovation plans which are being supported by local, regional and national policies that are concerned with establishing both the right business, technology and social environments, and the appropriate connections to drive company development. This chapter, in addition to discussing the appropriate macro conditions necessary to support innovation, also looks at the process of business development.

## 1 Science and Technology Parks: An Evolving Model

The science park movement has evolved over the last 60 years from providing passive property based developments to locations that have become active parts of the infrastructure that contribute to economic development by trying to utilise knowledge outputs through commercialisation Parry (2006).

This transformation has involved developing a number of strategies. These include: improvements to the supply of ideas from the discovery phase in science, technology and engineering in the context of their potential for commercialisation;

M. Parry (✉)
OBE, Director Surrey Research Park, University of Surrey, Surrey, UK
e-mail: m.parry@surrey.ac.uk

D.-S. Oh and F. Phillips (eds.), *Technopolis*,
DOI: 10.1007/978-1-4471-5508-9_11, 

the refinement of the range of business development programmes on offer which includes pre, full and post incubation activities; and the integration of these activities on a larger scale across a wider physical region by influencing the key components in the supply chain that link the ideas to the market.

Although these changes may have occurred without the presence of science and technology parks the value of these physical assets is they create a focal point that, if effectively managed, helps to provide the three critical stakeholders of business, knowledge generators and government with an environment in which they can express their own value propositions in an attempt to create new innovation led businesses.

The value propositions for each of these stakeholders in this process vary significantly but it is the combination of these different interests that give to science parks both their importance as projects and the capacity to continue to evolve and push the frontiers of their influence.

Details of these different value propositions broadly reflect economic development for government, gaining a competitive advantage for tenants, and technology and knowledge transfer for hosts.

To deliver these propositions governments have tended to develop macro business, technology and social environments that support innovation, businesses have looked to find technologies and skills that are beyond technology readiness levels 6 and generators of knowledge have begun to adapt to supply the skills and technology that are demanded by business.

## 2 Innovation

Innovation dominates much of the discussion, literature and policies about economic development. The importance of this process is the direct impact it has on the performance of firms or social institutions OECD (2005).

In this context innovation can range from small changes which are incremental in nature but still lead to a commercial organisation maintaining a competitive advantage, through changes which are disruptive which can be very damaging to some industries while launching others. It is also important to recognise that innovation can have a regional context and can be used to describe the introduction of new ideas to a region or country, which may already be prevalent in others and in some instances, can be used as a foundation for autarky. Although the term innovation is most often applied to commercial success arising from technology it can also be achieved in relation to trades and services and changes in government that can lead to significant wealth creation.

The response to these changes or disruptive influences require those industries under threat to themselves innovate in order to survive. Although the need to innovate is widely accepted as important for the necessary business practices that encourage successful execution of the process needs to be widely adopted.

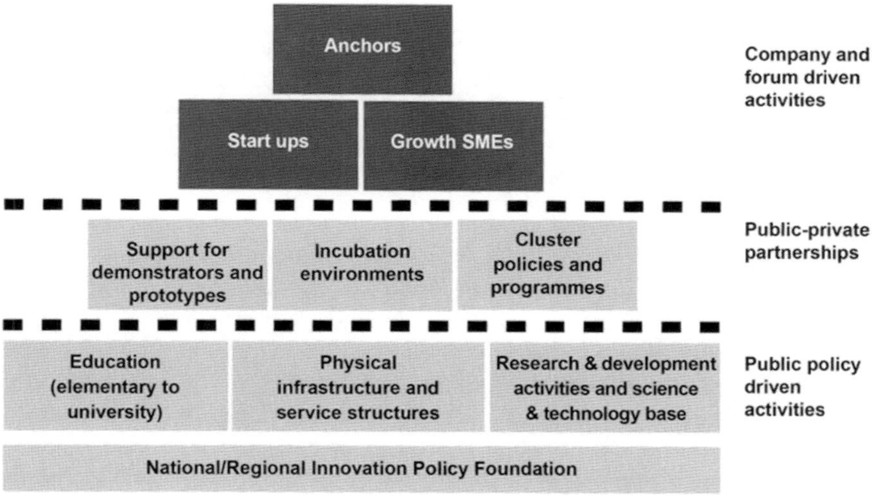


**Fig. 1** Characterisation of the elements and interfaces which are likely to support innovation Launonen and Viitanen (2011)

To understand how to encourage a more widespread adoption of innovation related business practices has led to extensive reviews of the dynamic of the processes. One of the conclusions is that external economic, social, business and technology related conditions are an important influence. The current view is that these conditions can be most effectively created through active public–private partnerships that bring together a number of interconnecting elements which are characterised in Fig. 1.

After Hubconcepts (TM)— Launonen and Viitanen (2011) in Hubconcepts (TM)—the Global Best Practice for Managing Innovation Ecosystem and Hubs published by Hubconcepts Inc (2011).

The current consensus is that drivers that combine to create this environment and push this system require public sector policies and associated investment in education, infrastructure and knowledge generation Research and Development (R&D). The next level of policy and investment needs to encourage the formation of public–private partnerships which share the risks of further development of the outputs from this process. Diminishing risks are eventually perceived as acceptable by the private sector and at that stage the balance of public engagement in the partnership fades significantly; although this element can never disappear completely because they still have to operate in regulated environments.

For the relationships that make up this process to most effect there needs to be structures, mechanisms and instruments that enable the transfer of ideas and the risk associated with their development, between each level. Pre and full business incubators are now commonly adopted as part of the infrastructure that supports the initial handover from being supported by the public sector to entities that are supported by public–private partnerships. Science and technology parks

(Technology Parks) have been develop as one of the types of infrastructure that sit firmly on the boundary between incubation and private sector activities and in some countries taking an active role in shifting the balance of the innovation system from a public R&D driven economy to a private sector based innovation system that acts in concert with the public investment in R&D.

Factors which have proved to be of value in making such a system work include:

- Collaborative relationships which are normally developed through an iterative process that link public and private organisations in complex arrangements which provide infrastructure and business support programmes.
- Creative business practices between knowledge generators and knowledge users that drive technology, company and market journeys.
- Experienced people who can appreciate both the value of technology in a commercial context and have the skill to support the transfer of the technology from the domain of creation to the domain of its use to realise a competitive advantage.
- A funding regime that slowly decreases public involvement in R&D and increases the private sector funding.
- The patience to build the necessary systems over a time horizon of 10–20 years rather than simple often politically driven short-terms horizons: This time horizon is particularly important when benchmarking the impact on the economy of technology parks.

The influence of competition that provides the incentive which gives impetus, through technology push and market pull, to those people and organisations involved in the system to innovate and gain a competitive advantage. The incentives for the public sector include stability that comes with creating employment particularly when this is of high value, tax generation to sustain and grow the process, and the creation of wealth and prosperity.

## 3 Performance of Science and Technology Parks

A study of the performance of companies on science parks undertaken by the UK Science Park Association and the UK Government's then Small Business Service, in 2003 UKSPA (2003) noted that science parks that operate in regional environments that provide all the elements and connections noted in Fig. 2 are more successful than those where the linkages are either missing or not functioning.

The message here is that managers on science parks need to understand where capacity is missing and find ways of filling these gaps.

Knowledge capital includes a wide range of organisations which include universities, public sector research organisations, and private R&D organisations such as corporate research laboratories and contract research organisations. Opportunities exist to develop relationship management programmes to link these

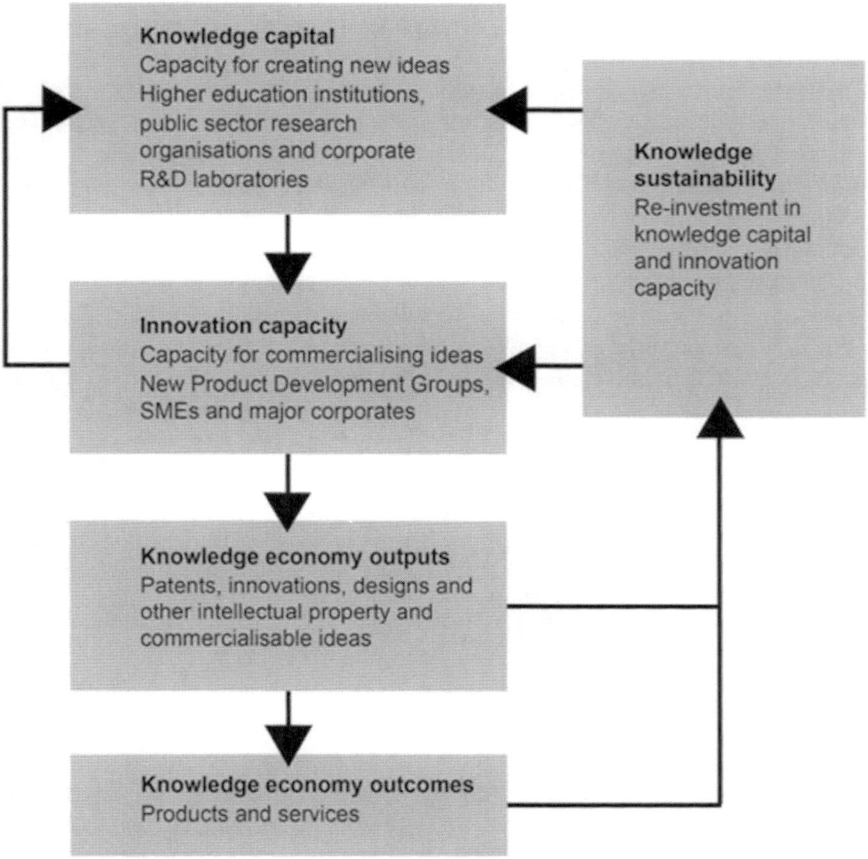


**Fig. 2** Model of the subregional economy (after UKSPA 2003 and SEEDA and Huggins 2001)

organisations into partnership agreements with science parks. If these can be associated with technology transfer funding programmes they become even more attractive links.

The utilisation of the ideas that emerge from these generators requires innovation capacity that is capable of developing and commercialising new products. This capacity derives from a number of organisations that include:

- Small, medium and large enterprises in the region are driven to improve their competitive position in the market place.
- Groups concerned with new product development such as designers, contract research organisations, consultancies and business development teams.
- Organisations which understand the creative process of entrepreneurship helps to wrap ideas in an attractive way that encourages customers or buyers for the outputs.

- Active engagement between customers and suppliers in supply chains helps to drive utilisation of new ideas.

When science parks were first established the networks they created tended not to be supported by any explicit plan; however, the value of these linkages is now recognised and this has prompted the extension of this idea into creating regional innovation systems. These systems use active management, policy frameworks and interventions to try to ensure that the necessary linkages that connect supply and demand are created and each element operates to its full potential. These elements are characterised in Fig. 3.

This system attempts to create the foundations for building capacity for innovation. This connects knowledge creation, encourages the wider use of skills to absorb ideas for commercialisation, helps diffuse these ideas into a wide target

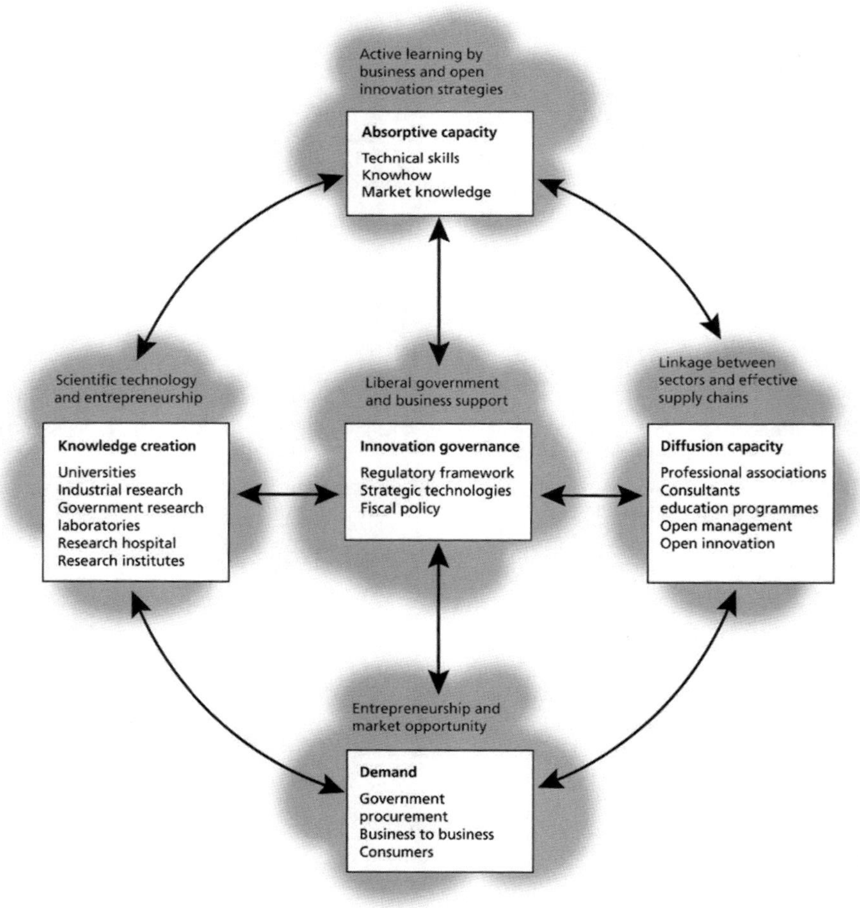


**Fig. 3** Characterisation of an innovation system—after UNECE description UNECE (2009)

group and helps to create new markets or facilitate access to existing or new markets which is described in Fig. 2 as demand.

In this process the production of new ideas and the qualified human resources that can interpret ideas and technologies and translate them into a business model needs to be in place. Without this capacity to absorb ideas some of the elements in any technology, company, or market journey are more difficult to achieve. The process then relies on the ability to diffuse the ideas from these generators into the business community and society and that the ideas that are created find markets that are prepared to utilise the outputs.

Experience has shown that science parks and their host organisations can play a central role in these systems. Factors which help to give this lead to science parks include the brand value, the support they receive because of the broad range of stakeholders which they attract, the presence of on site management that provides the resource to assist in developing the site as a centre for connecting ideas with people through their access to networks and finally their pre and full incubation activities.

Today with shrinking public sector funding available for general business support, government is looking for efficiencies in their spend but still want to realise their value propositions for science parks. This shift in the balance of public expenditure provides science and technology parks with an opportunity to establish and maintain the role of leadership in supporting innovative companies.

To do this the largest parks in a region should try to create connectivity in order to provide a central management support programme for their region. Suggestions to support that include modern video conferencing techniques to link all the parks in a region to provide for example business education and business development programmes.

Innovation governance lies at the heart of innovation systems. This represents the kind of activities which create the right macro-environment that supports business formation and development. Examples of factors which are important in this environment include: creating a good tax environment for investors in companies whether they are active in the management of the company or external investors, making funds available to support the stages of the technology, company and market journeys, establishing a regulatory environment that does not over burden businesses, and creating a regulatory environment which supports enterprise, helps to create markets and does not stifle competition by creating for example state run monopolies or prevent enterprise.

## 4 Functional Links on Science Parks Concerned with Capacity Building and Supporting Innovation

A macro view of the process that is supported by science and technology parks is characterised in Fig. 4.

The elements, activities and linkages characterised in this figure pull science parks away from discovery activities in knowledge generators. The bulk of the

Fig. 4 Characterisation of the role science and technology parks play in innovation systems

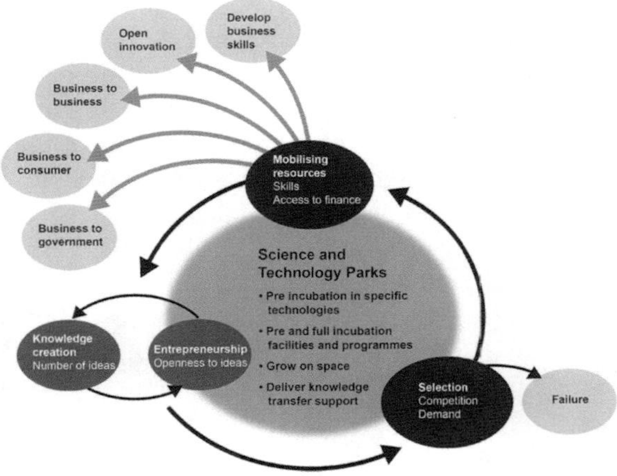


work by science and technology parks is to support entrepreneurs take their ideas through a selection process that selects out those ideas deemed to have low potential and leaves those with some perceived value in place for further development but that requires skills and finance to be mobilised. The stage beyond that is to build businesses that address potential markets or if the idea or skills do not make it, the skills developed in the process should be recycled back into the business community.

Underlying these functional links are a series of technology, market and company journeys Trezona (2008) which if successfully executed lead to an increase in the number, size and efficiency of companies in a region.

## *4.1 Technology Journey*

The technology journey that underpins these links is outside the remit of science parks but becomes interesting when entrepreneurs begin to appreciate the value of these new ideas in satisfying a real or perceived customer requirement. This may result from a response by an entrepreneur to a technical specification which pushes technology towards a market or appreciation of a market by an entrepreneur for a technical specification which helps to pull this towards the market.

This technology journey can be characterised in the following way (Fig. 5).

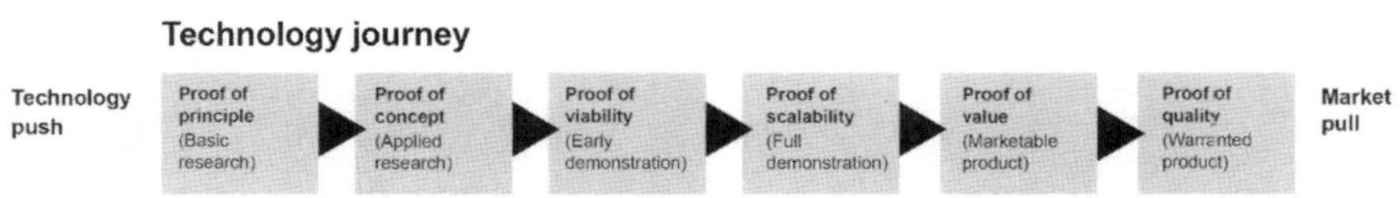


Fig. 5 Characterisation of technology journey in business development

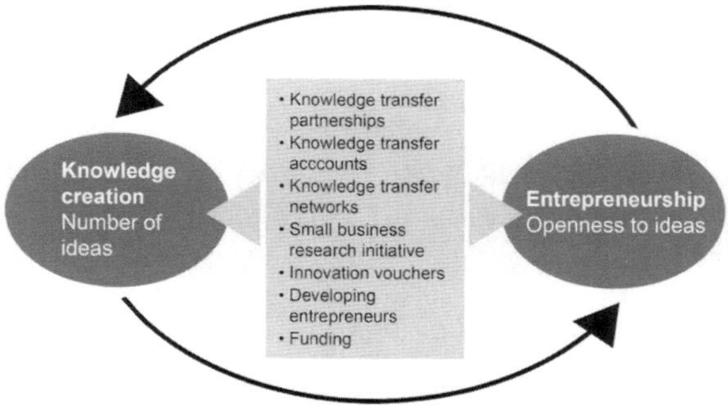


**Fig. 6** Characterisation of technology transfer programmes for linking knowledge generators to the commercial sector

Although some ideas emerge from the business environment through people that are active in a sector, most of the ideas in this process emerge in the first instance from knowledge generators such as universities, government research laboratories, and research hospitals.

Experience suggests that for these ideas to be effective in creating commercial activity there needs to be in place a range of technology transfer programmes. A number of examples of these are described in Fig. 6.

Examples of policies and policy instruments in the UK which support the connection of technology to the market include:

### 4.1.1 Knowledge Transfer Partnerships

These are aimed at helping businesses improve their competitiveness and productivity by using knowledge, technology and skills in the UK knowledge base.

The benefits for academics that are engaged in these programmes include keeping them up to date with commercial needs; helping academics work on real business problems and helping them find new research themes for undergraduate and postgraduate projects. In this, relationship business gains access to skills and expertise for business development.

The kind of projects these support include product design and development, developing manufacturing practices and management processes or working on computing or management information in order to solve problems through innovation.

### 4.1.2 Knowledge Transfer Accounts

These are funded by the UK government with the view to supporting specific skills from a particular university's expertise. Examples of these include accounts in nanotechnology, photonics, communication and signal processing and next generation materials and characterisation. The purpose of these is to help match outputs and capabilities from universities with industrial needs, provide access to funds for pilot and demonstrator programmes, and increase the speed and reduce the costs for access by business to university facilities in order to support innovation.

This programme is also available to support spin out companies that may have been created within the university. They have particular value in supporting demonstrator programme that are concerned with bringing technology closer to the market.

#### 4.1.3 Knowledge Transfer Networks

In the UK there are currently (2011) 16 KTNs which operate in specific fields by linking people who are concerned with driving a specific technology across industrial sectors and encouraging commercially based solutions to be founded by companies.

#### 4.1.4 Small Business Research Initiative

This programme supports the need for the UK governments, like most governments in the world, to reduce the cost to the public purse of R&D. It has been organised to help early stage, high-technology Small and Medium Enterprises (SMEs) gain greater access to R&D opportunities while supporting the future procurement needs of Government Departments by offering competitive R&D contracts that aim to:

- Provide opportunities to those existing small firms whose businesses are based upon providing R&D—by increasing the size of the market;
- Encourage other smaller businesses to increase their R&D capabilities and capacity—to exploit the new market opportunities;
- Create opportunities for starting new technology-based or knowledge-based businesses.

Part of the thinking behind the SBRI is that it enables government departments and other UK public sector organisations to procure new technologies faster and with managed risk through a phased development programme, and it provides paid contracts for the critical stage of product development.

### *4.2 Innovation Vouchers*

These are a mechanism aimed at small and medium-sized businesses which can use the vouchers to buy support from an academic institution to explore potential opportunities for future collaboration.

### 4.2.1 New Frontiers

Ideas which need to be developed include:

- International connections for ideas—building international technology networks that focus on a single sector to try to increase the intensity of commercialisation of ideas by cross licencing arrangement and broadening the exposure of the ideas to the early stage equity market.
- Building networks that connect corporate venturing activities to pre and full incubation programmes.
- Helping governments deliver funding programmes to high growth companies that are active in technologies that have a national priority.

### 4.2.2 Entrepreneurship

This presence in a business community of entrepreneurs is a critical ingredient. If there are high numbers of entrepreneurs with a broad range of skills that are able to both identify business models to develop technology for the market and wrap the right skilled workforce around the technology the greater the chance of success.

Ideas that are now being supported by science and technology parks to help encourage entrepreneurship include the formation of a number of entrepreneur clubs at universities which are concerned with building a culture and ethos of entrepreneurship among students. The experience at Surrey is that an "entrepreneur in residence" is now working in the University's Business School to support the University, its students and the Surrey Research Park.

### 4.2.3 Early Stage Funding

A model for funding technology transfer that is gaining momentum in the UK is based on a contractual relationship between a host organisation and private equity capital group. Around ten universities in the UK have signed such an exclusive arrangement with the IP Group. The business model values the IP at a predetermined value, the IP Group take an agreed proportion of this and then work with the IP to develop its commercial potential.

A similar programme associated with government research laboratories has been established through a privately managed evergreen fund (Rainbow Fund) that provides support to ideas which have commercial potential.

### 4.2.4 Technology Transfer Offices

Today there is increasing emphasis on funding R&D that has a strong orientation towards potential commercial value. This direction when associated with the cultural change engendered by the presence of a science and technology park on a campus helps to increase interest in the commercialisation of new ideas.

Although there are still relatively few companies being spun out of universities the professionalisation of the work of Technology Transfer Offices in the higher education sector through training and provision of funding to support technology commercialisation means it is likely that this will eventually lead to the formation of companies occurring on a larger scale.

The role of science parks in this technology journey is aimed more towards supporting the downstream commercialisation phases rather than in the discovery process. This means that parks have to be more active in mobilising the resources to drive the market and company journeys rather than technology journeys, although this cannot be ignored.

## *4.3 Market Journey*

A full market journey for technology takes the market from complete indifference to widespread adoption. Of course not all technologies start at the beginning of this continuum and not all end in wide spread adoption. Experience has shown that some are taken out of the market by acquisition before they are become fully fledged technologies Fig. 7.

The critical ingredient in this journey are the skills and flair of the entrepreneur. This comes from a combination of an understanding of the market place, recognising how this market would benefit from the technology and then organising the development of the market.

However, there are support programmes which can help with working along the process. These include:

- Providing funds to create demonstrators for field trials of ideas.
- Using networks to link companies still in incubation with larger companies looking to develop by using open innovation strategies.
- Using science park networks to connect companies across wider sector groups.

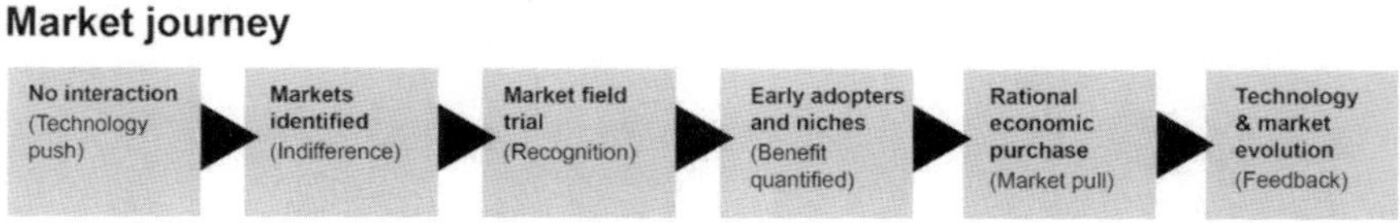


**Fig. 7** Characterisation of the market journey in business development

- Taking an active role in promoting and supporting companies as they link into Knowledge Transfer Networks or their equivalent.
- Creating connections that bring into the network international equity finance helps to widen the exposure of the ideas with commercial potential to an international market.
- Using their connections to promote companies through international trade support channels such as trade missions, commercial attachés, and other science parks with which they have contact.
- Connecting companies and government services that help to provide market intelligence. In the UK a number of programmes are available that can be used to provide this information.

## *4.4 Company Journey*

One of the most common patterns for a company journeys is characterised in Fig. 8. This usually starts with 1 or 2 individuals and numbers of employees are added as new dimensions to the business need to be explored and developed. In the case of spinout companies these individuals are often technology entrepreneurs. As the company develops, the team needs to be supplemented with a wider range of skills. Most commonly the initial increase in skills comes from the investors.

Recruitment of the right people on which to base a team is one of the most critical aspects of building companies. Although this does have element of chance it is possible to reduce the risks by using the networks created by science parks to help to identify potential employees.

In terms of the value of high growth coaching this has particular value when staff levels have reached more than ten in number. Some examples of the kinds of programmes that have been put together in high growth coaching include:

- Finance for Growth—this provides advice on the basics that every business leader should know in managing a growing business effectively and includes details on raising finance.
- Market intelligence—how to increase or better organise company awareness of current and potential markets by developing a marketing intelligence strategy.
- eMarketing and eCommerce—to facilitate the widest possible use of the tools and techniques made available by the e-revolution to open up new sales and marketing opportunities.

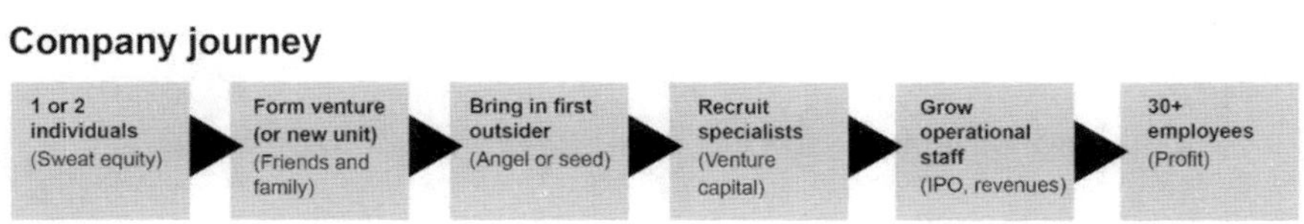


**Fig. 8** Characterisation of the company journey in business development

- Out-thinking the competition—advanced creative techniques to give companies the edge through enhanced decision-making and improved leadership capabilities.
- Building the innovative organisation—to demonstrate how fundamental improvements in organisational behaviour can embed innovation in all aspects of successful business activities and how to protect the results.
- Improving selling skill—to help improve understanding of the sales process and the development of personal selling skills across senior management.
- Leading and managing high growth—training for senior managers who have had little previous formal training and lack the leadership and management expertise to sustain growth.
- Change management—implementing change at middle management level is a major concern in growing companies. This proven change management training helps middle managers deal effectively with change.
- Management information healthcheck—an audit of a company's current management information, followed by a written report on the potential management information needs of the business as it grows.

Providing these programmes through third party contractors that work across regions can help to keep cost low while helping to up-rate a region's innovation capacity.

## *4.5 Regulatory Journey*

Many new ideas have emerged following the liberalisation of markets and deregulation. In addition and conversely a number of new markets have emerged as there is increasing control over some aspects of human endeavour such as development and the need for increased control over the environmental effect of these endeavours.

Many new companies need to be cognisant of the opportunities these changes in the legislative framework can provide or limit Fig. 9.

Examples of deregulation that has driven entrepreneurship include changes in the telecoms sector and in some areas of bio-technology such as allowing stem cell research while in contrast the need to test the environmental impact of development has created an industry that now has to employ ecologists to do the work.

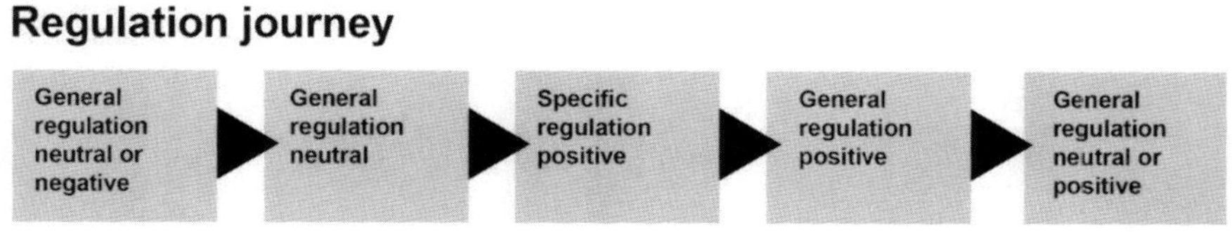


**Fig. 9** Characterisation of the regulation journey associated with business development

Most regulations are promulgated by governments. Today details of the these regulations can be found on such sites as Netregs for the UK regulations on the environment and other business support sites such as Business Links.

One of the roles of science parks is to know how to guide companies in interpreting these regulations.

### 4.5.1 Pre and Full and Post Incubation (Technology Acceleration) and Developing Companies

The process of pre, full and post incubation is an area in which science and technology parks are continuing to evolve and develop new ideas.

Some examples for opportunities for pushing the boundary for science and technology parks include:

- Professionalising this process.
- Concentrating resources in one location and then operating across a network using the professional skills to support the process in a number of locations. There are strategies which can be adopted by science and technology parks in order to create this concentration of activity. These include pricing for accommodation and services, increasing the opportunity to gain access to development finance, provide access to support programmes and the potential to support companies gain access to market related networks.
- Developing specialisms in relation to supporting specific technology sectors.
- Developing working relationships with private sector groups to help develop companies.

To understand better how these kinds of initiatives can be deployed it is of value to review how the processes of pre and full incubation, technology acceleration, investment, growing enterprises and the regulatory frameworks overlay technology, market and company journeys operate.

Although there is no single route through these different parts of the journey for developing companies there are sufficient distinctive elements to enable these to be classified and from that developed to become even more effective Fig. 10.

The science park model adopted in Europe and US does have the capacity to support R&D and some companies on these sites operate corporate development laboratories; however, the majority of the work undertaken on these sites occurs further up the value chain than the discovery stage. However, it is important to have in place the necessary links to help connect ideas to the commercial sector.

In the last 5 years it has become more common for science and technology parks to create and run pre-incubators Fig. 11.

Pre incubation is an activity that has evolved on science parks as a social development that is aimed at reducing the risk to entrepreneurs as they try to find ways of commercialising the output from R&D.

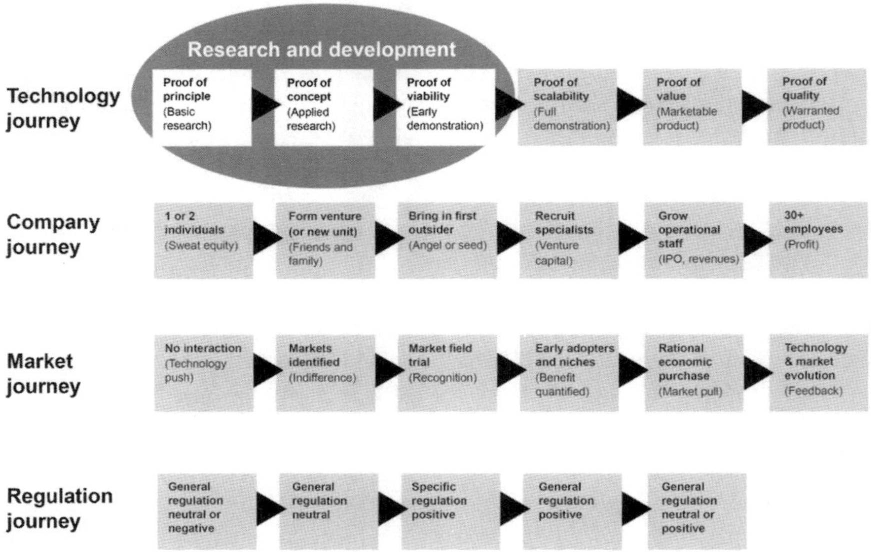


**Fig. 10** Research and development

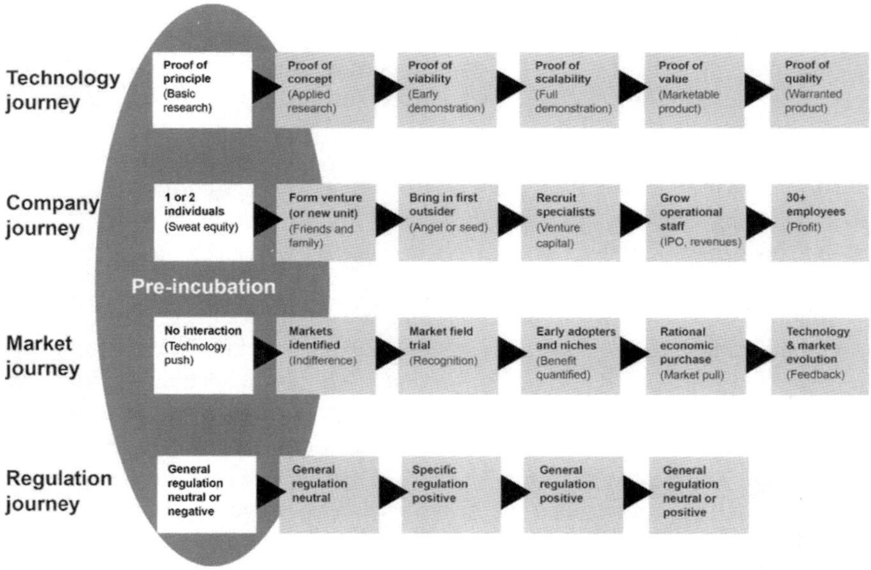


**Fig. 11** Pre incubation

This allows entrepreneurs to have low cost accommodation and free business support to help them with the investigating and defining how they intend to develop their company and embedding this story in a fundable business plan that would normally set out an executive summary, a short description of the business opportunity being pursued, a marketing and sales strategy, the proposed management team, the operational aspects of the business, and the finances of the business.

This process is designed to support entrepreneurs as they work across on developing the foundations to the technology, company, and market journeys for their ideas. In most instances the process is subsidised, but time limited, and company development is measured against milestones which are agreed with the support teams for these social projects.

Once a credible business plan is in place the next and critical stage is to raise finance for the incubation process. In some countries this investment is made available through government financing programmes where as others do not have this benefit and have to look to the private sector to raise finance. In this the process mainly focuses on creating the right team to develop the company and on the early stage of convincing the market that the idea is credible Fig. 12.

If these stages are negotiated successfully the project can then be accelerated by paying particular attention to the development of the market and then on further building the team as the company moves to provide warrantable products that begin to secure market share.

Particular new developments that have been added to the blend of services to support these journeys include:

- Employing an "entrepreneur in residence" in business pre and full incubators.
- Establishing an investment group and trying to increase the capacity and skills base in this by putting in place international links with this group.

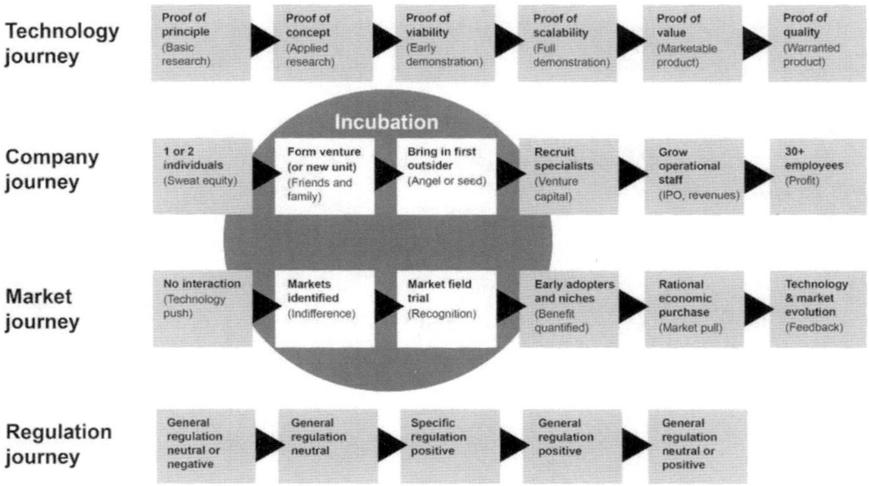


**Fig. 12** Full incubation

- Putting in place pre-incubators through outreach programmes in other local institutions which may have individual entrepreneurs or groups of entrepreneurs that are interested in building a business.
- Working with students on entrepreneurship programmes in the host university's business school.

Once an enterprise has reached the phase of a warrantable product on the market it would be usual to push them into alternative larger but less well serviced accommodation.

At this stage there is some benefit from keeping these companies on the Park as they provide valuable income, they continue to develop and through this development employ and train people which helps to add to the capacity of the skills base in a region and they can attract valuable foreign direct investment Fig. 13.

One of the areas with greatest potential for development relates to raising appropriate finance. The value of appropriate finance is that this by definition implies that those making the investment are well connected to the potential market place for the new business. International connections in this process is important and worthy of further thought and development Fig. 14.

Significant effort has been directed at establishing business incubation programmes which enable companies to take the first steps of taking forward ideas. However, an area of work that has potential for science and technology parks is to develop business acceleration activities. These accelerators utilise formal training for entrepreneurs as part of the process of building a business. An example in the UK is the "New Me" programme that has been developed by The Centre for Micro Business which is based in the UK and offers a business led programme that can be added to a business incubation programme.[1]

These are more focussed on enterprises that have begun to penetrate markets, have a compelling business case but need additional investment in their skills base and the finances of the company.

The transition from the incubator to the accelerator often comes with closer engagement in the company by investors as they expect faster results from the process.

One of the features of business acceleration is access to high growth coaching programmes that assist company executives create internal environments to support growth Fig. 15.

One of the important features of science parks is the ability to support companies above and beyond incubation by providing the space and resources to support them as they grow. To achieve this there is a greater emphasis on the quality and flexibility of business accommodation, good access to skills, technology to support a customer base.

---

[1] http://info@thecentreformicrobusiness.co.uk

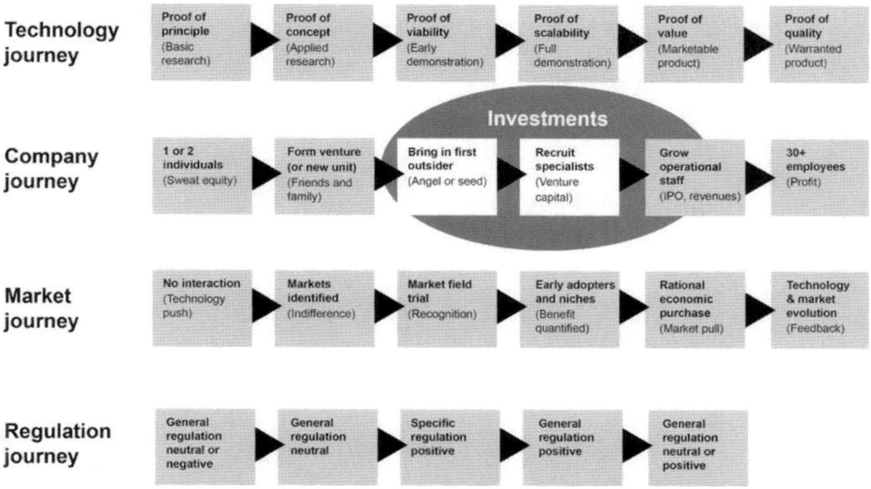


**Fig. 13** Investment

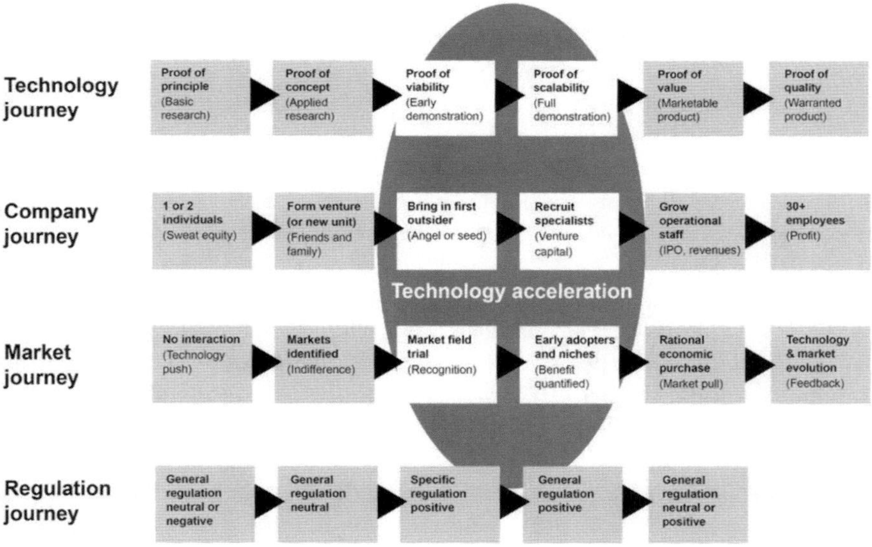


**Fig. 14** Technology acceleration

This capacity to accommodate growth is an important role for science parks as they help companies to emerge from their early technology, market and company journeys.

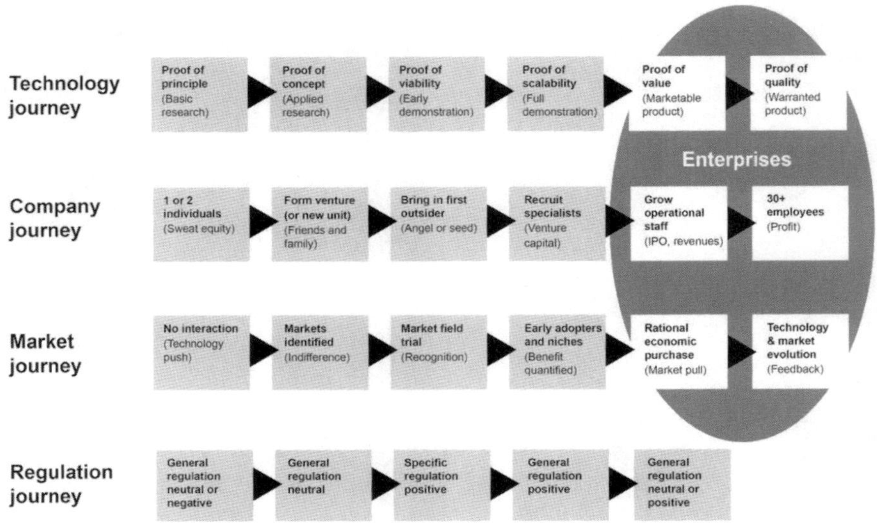


**Fig. 15** Enterprise established

## *4.6 Connecting to the Market*

Critical to the success of businesses in this process is finding markets and in these winning customers. One of the problems for micro-companies (1–9 employees) and small companies (10–50) is finding potential customers and influencing these to the point of converting them into true customers. To support this process a number of regional strategies have been developed that have proved to be a successful formula.

The concept of an outreach programme has been pioneered on the Surrey Research Park which is based on a team of sector specific business mentors that have been set targets for working with 80 micro-companies, 35 small companies with a turnover of between £10 m and £100 m, ten businesses with a turnover of between £100 m and £500 m, and five businesses that have a turnover of in excess of £500 m. The purpose of this process is to link larger companies to relevant small businesses in order to help them develop channels for acquiring technology that has potential in their market place.

To support this a number of large companies are now setting up open innovation teams that act as technology scouts that take an active role to secure new ideas for the future. Beneath this layer of activity by the companies such as ANGLE Technology. Founded in 1994 on the Surrey Research Park, ANGLE now focuses on the commercialisation of technology and the development of technology-based industry by not only creating, developing and advising technology businesses on its own behalf but it also does this for a number of clients.

It is clear from experience of dealing with the commercialisation of technology that the point of entry for either supporting technology as a private investor or adopting a technology for driving a corporate business varies. Business Angels and other private investors can be persuaded at an early stage (technology readiness levels 3 or 4[2]). These investors are often well connected to markets in the relevant sectors and are an important conduit to support market development for micro businesses. In contrast large multinationals that have open innovation programmes have a preference for technologies that are above technology readiness level 6 and have proven defendable provenance and IP and can be brought into play quickly.

An important role for science and technology parks is to continue to develop ways of supporting market connections for the companies they incubate.

## 5 Conclusion

Science and technology parks, although not perfect instruments for facilitating economic development associated with the commercialisation of science and technology, are developments that have been instrumental in helping to shape the business environment to support the process by evolving from passive property developments to active players in business incubation. From this early process and the success of some of these developments a number of innovation systems have been created through policy initiatives which help to create the right environment to support business development.

Parks have pioneered and led on pre and full business incubation and have a proven track record in developing and sustaining innovation systems. This has led to increasing interest in what these sites do that helps the process of creating companies and what can be added to their contribution.

The development of each individual company follows a unique pathway but understanding the broad principles that underlie this process provides insights into the way that support programmes that are promulgated by innovation systems can be implemented to support these individual journeys.

All of these ideas require some public investment but it is clear that these funds are less easily found with growing demands on the public purse so one of the major opportunities is to build links with the private sector to help create funding streams for companies.

## References

After Trezona, R. (2008). Presentation to the The R&D Society, UK.

Launonen, M., & Viitanen, J. (2011). In Hubconcepts (TM): *The global best practice for managing innovation ecosystem and hubs.* Hubconcepts, Inc.

[2] http://www.nasa.gov/topics/aeronautics/features/trl_demystified.html

OECD. (2005). The measurement of scientific and technological activities, oslo manual, guidelines for collecting and interpreting innovation data(3rd edn). San Fransico: OECD and Eurostat.

Parry, M. J. (2006). Setting the scene: *The Planning development and management of science parks*. published by: UKSPA.

SEEDA and Huggins Associates. (2001). SEEDA and Robert Huggins Associates 2001 Global Index of Regional Knowledge Economies: Benchmarking South East England.

UKSPA. (2003). Evaluation of the past and future economic contribution of the UK Science Park Movement. Published by UKSPA in conjunction with the Small Business Service.

United Nations Economic Commission to Europe. (2009). Enhancing the innovative performance of firms. Policy options and practical instruments ISBN 978-92-1-116999-7.

# Accelerating Technology-Based Growth in Developed, Developing, and Emerging Regions Worldwide

**David V. Gibson and Darius Mahdjoubi**

**Abstract** This chapter suggests that the success of selected science parks, innovation clusters, incubators, etc., is due in large part to the regions in which they are embedded. The chapter also discusses processes of Knowledge/Technology Transfer (KTT) to commercial use at the level of the firm and start-ups as well as some ideas on forming and sustaining global networks for sharing knowledge among developed, developing, and emerging regions worldwide.

## 1 Introduction

This chapter was presented at UNESCO-World Technopolis Association (WTA), International Training workshop: Toward Creative Growth of Science Parks and Innovative Clusters, Daejeon Metropolitan City, October 8–11, 2008.[1] Important UNESCO-WTA International Training Workshop Objectives include building the capacities of developing countries in the management of technopolis, with a view toward marshaling science, technology, and innovation for sustainable development to:

- Upgrade the knowledge of managers or future managers of science parks,
- Better understand issues and challenges on creative development strategies of science cities,
- Share information and experiences for the establishment of cooperative regional networks among science cities and for establishing an international network among science cities.

[1] http://www.wta2008.org.

D. V. Gibson (✉) · D. Mahdjoubi
IC[2] Institute, The University Texas at of Austin, Austin, USA
e-mail: davidg@ic2.utexas.edu

D. Mahdjoubi
e-mail: darius@ischool.utexas.edu

D.-S. Oh and F. Phillips (eds.), *Technopolis*,
DOI: 10.1007/978-1-4471-5508-9_12,  

This chapter is based on authors, 25 years working with colleagues on regionally-based technology/knowledge-based growth in developed, developing, and emerging regions worldwide. This work has been a combination of "think and do" often on unstructured problems as advocated by $IC^2$ Institute, The University of Texas at Austin.[2] And these activities have involved innovation clusters, research parks, incubators, consortia, and a range of public and private organizations and partnerships. The main objective of this chapter is to identify key assets and challenges to developing and sustaining regionally-based creative/innovative ecosystems as well as global knowledge sharing networks.

This chapter suggests that the success of selected science parks, innovation clusters, incubators, etc., is due in large part to the regions in which they are embedded. The chapter also discusses processes of Knowledge/Technology Transfer (KTT) to commercial use at the level of the firm and start-ups as well as some ideas on forming and sustaining global networks for sharing knowledge among developed, developing, and emerging regions worldwide. As stated by George Kozmetsky the founding director of $IC^2$ Institute: then the quote by Kozmetsky:

Technology continues to shrink the world. There is no choice other than to participate in the global community. Science and technology is too precious a resource to be restricted from drawing the world together. That is what the twenty-first century is all about.

## 2 Global Perspectives

Worldwide there have been a range of labels for planned regional technology-based growth. Perhaps, the first and most successful regional initiative was Stanford's Research Park which was founded by Professor Frederick Terman and others in the 1950s (1984). One of the first major nationally planned initiatives was in Japan with MITI's Technopolis Law in 1983 (1988). Later national and regional efforts have been labeled Technopoles, Multi-Function Polis, Science and Research Parks, High-Tech Corridors, University and Technology Parks, and Innovation Clusters. Clearly there are different perspectives and orientations underneath these labels but the fundamental goal that links them all is the desire to *accelerate the creation of regionally-based high-value jobs and wealth.* It is also useful to think of these efforts in terms of differences and similarities of theory and practice in developed, developing, and emerging regions including the characteristics of different industry sectors or clusters worldwide.

Two quotes are offered at the global level of analysis, one a vision of the future, the other a warning:

---

[2] "Unstructured Problems" are those that have no clear solution or even methodology and are usually large and complex and require a transdisciplinary approach.

First the vision:

> I cannot tell any society or culture what to say to its children, but I can tell you what I say to my own: The world is being flattened and you can't stop it, except at great cost to human development and to your own future. But we can manage it, for better or worse. If it is for the better we need a generation of strategic optimists, a generation that wakes up each morning and not only imagines that things can be better, but also acts on that imagination.Thomas Freidman, *The World is Flat*: Farrar, Strauss, and Giroux:

Followed by the warning:

> Globalization makes the world smaller. It may also make it—or sections of it—richer. It does not make it more peaceful, or more liberal. Least of all it doesn't make it flat.... Indeed, the coming age could be marked by resource wars, increased religious passions, and serious environmental limits. John Gray, *New York Review of Books,* "Critique: The World is Round," August, 2005.

In short, there are numerous serious challenges to a "Flat World" including national politics, religion, global economies, and regional quality of life issues, especially concerning natural resources and in particular fossil fuels although water, food, and other natural resource challenges are also of central importance and concern. As stated by the US Energy Information Administration, absent mandatory global agreements on capping greenhouse gases, etc. (AAS, A6, June 26, 2008):

- World energy demand will grow 50 % during next 2 decades—oil process will raise to $186/barrel—coal will remain the biggest source of electricity despite its contribution to global warming
  - Emerging economies will see their energy demand grow by 85 %—The steepest increases in energy use will come from China, India, Middle East, and Africa
  - Global demand for oil will grow nearly one-third more than is consumed today and coal nearly two-thirds more
  - Carbon dioxide flowing into the atmosphere will be 51 % higher than now.

However, with a more optimistic view, Cambridge Energy Research Associates conclude that markets respond to higher prices with behavioral change, innovation, and substitution and we have seen the effects of these often beneficial responses worldwide.

Finally within this mix of critical more macro level considerations, Richard Florida suggests—that the world is perhaps more "spiky" than flat.

> Innovative ecosystems matter and there are not many of them—perhaps a few dozen places worldwide really compete at the cutting edge—economic progress requires that the peaks grow stronger and taller. But such growth will exacerbate economic and social disparities that could threaten progress... managing the disparities between the peaks and valleys worldwide—raising the valleys without shearing off the peaks—will be a top challenge in the coming decades. Richard Florida, "The World is Spiky," *Atlantic Monthly,* October 2005.

Again on a more optimistic note, George Kozmetsky, as quoted in *Inc. Magazine*, August, 1996,

> No matter what field you are talking about—electronics, medical, education, the environment, entertainment—the global marketplace opens up more opportunities than I've seen in my lifetime. Very few generations in history, perhaps not since the Renaissance, have been accorded the opportunities this period provides. It is a profoundly different world.

## 3 The Regional Level of Analysis: The Case of Austin, Texas

In the 1980s Austin was nationally, and perhaps internationally, known for being Texas' state capital and a university town with an oil/ranching/cowboy culture. The city was not known for creativity, entrepreneurship, start-ups, venture capital, or High Tech. Jobs were mostly in government and education and the area could not retain most of its educated talent. However, by the late 1990s and following decades Austin has been ranked as the best U.S. city for business and top wealth creator; the best U.S. city for entrepreneurship; and the most creative U.S. city according to Richard Florida (2002). Importantly, Austin has been able to attract creative and innovative talent from such other "spiky" regions including Silicon Valley, CA, Boston, MA and worldwide.

There are certain necessary but not sufficient criteria for the success of Austin and other "technopolis" regions worldwide including at the most fundamental level with being civil societies and following the rule of law to having world-class educational assets and a quality of life that attracts and retains talent. However, *a key reason* for success of the Austin Model is results oriented public–private collaboration of primarily local actors. The University of Texas at Austin, regional industry, local government, and foundations/nonprofits, see in Fig. 1.

This collaboration has been facilitated by visionaries and champions of creative and innovative mechanisms for knowledge transfer (KT) and use; effective action-oriented processes for making things happen; and metrics for success. The central importance of regional collaboration or cooperation is enhanced when one realizes that the Austin region was not and is not the BEST in the U.S. much less than the world on a range of criteria. This is perhaps an encouraging realization for other regions worldwide that might not have, for example, a Stanford University, University of California at Berkeley or Harvard University and MIT, and the other amazing wealth creation assets of Silicon Valley and Boston.

Over time Austin's institutional collaboration/cooperation/synergy has been led by different visionaries and champions in terms of specific regional objectives that involved civic and social entrepreneurship as well as technology entrepreneurship, Fig. 2.[3] Civic volunteers and champions tend to focus their efforts on linking

[3] Collaboration is often associated with working with your enemy; Coordination implies more meaningful sharing of effort/knowledge; Cooperating implies deeper sharing that would include

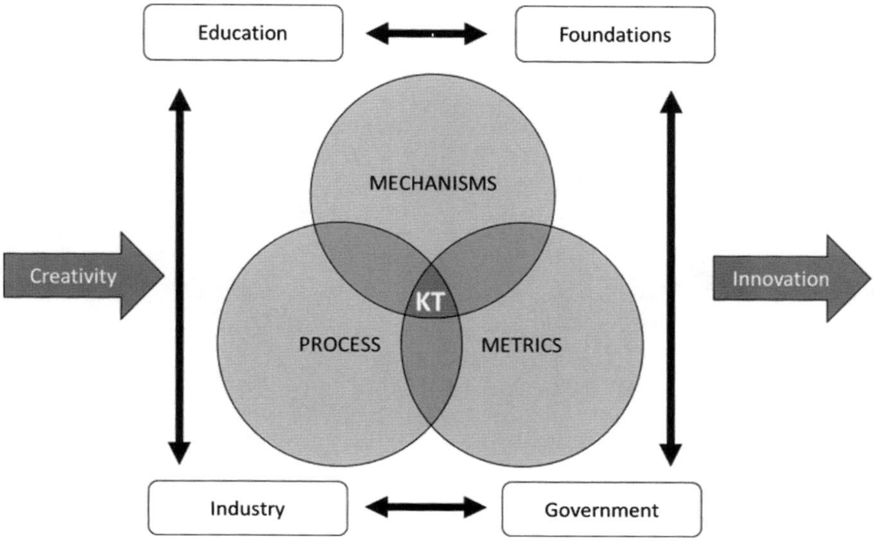


**Fig. 1** Key Sectors of Institutional Cooperation in the Austin Model. *Source* IC$^2$ Institute

academic, business, and local government assets. Social entrepreneurs are more concerned with linking to foundations and non-profits for enhancing regional social inclusion and personal opportunities while technology/business entrepreneurs focused on linking talent, technology, capital, and business know how for accelerated job and wealth creation.

This focus on regional institutional and personal cooperation is a very important conclusion because as IC$^2$ Institute research has shown, while excellence in education, business, government, entrepreneurial sectors is important—the most important factor is effective cooperation across these entities. In short, the quest of institutional or organizational excellence alone, while being important, is not sufficient, Fig. 3.

It is also important to emphasize that the vision and action of Austin's regional visionaries and influencers was not focused on or limited to the construction of world-class science parks, technology parks, incubators, and other physical infrastructure. As witnessed in many regions worldwide, a focus largely limited to the "build it and they will come" strategy has led to largely underwhelming results. For example, a recent seminar in Mexico on technology parks noted with concern that there were 33 technology parks spread throughout the Guadalajara region and surrounding areas with more planned in the coming years. Guadalajara is considered one of the top globally competitive technology regions in Mexico.

(Footnote 3 continued)
Tacit Knowledge; and being synergistic implies that the individual actors or entities work and produce at levels above their individual talent (Gibson and Rogers 1994).

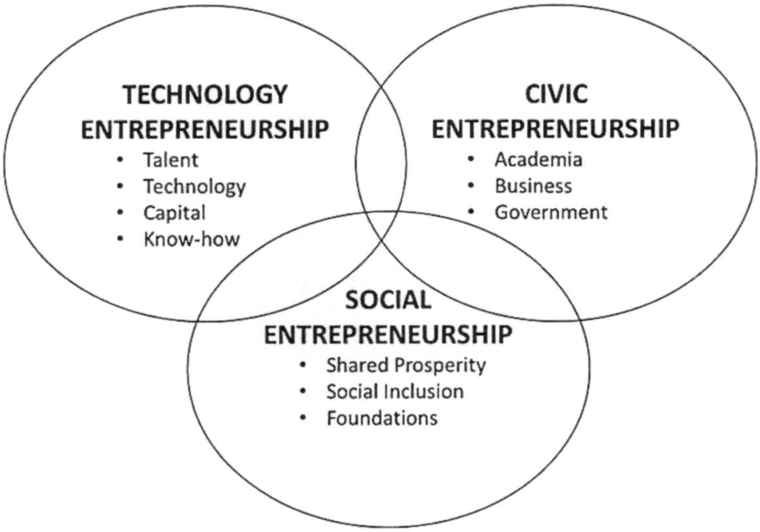


**Fig. 2** Three Types of Entrepreneurship that Promote Regional Development. *Source* Gibson (2002)

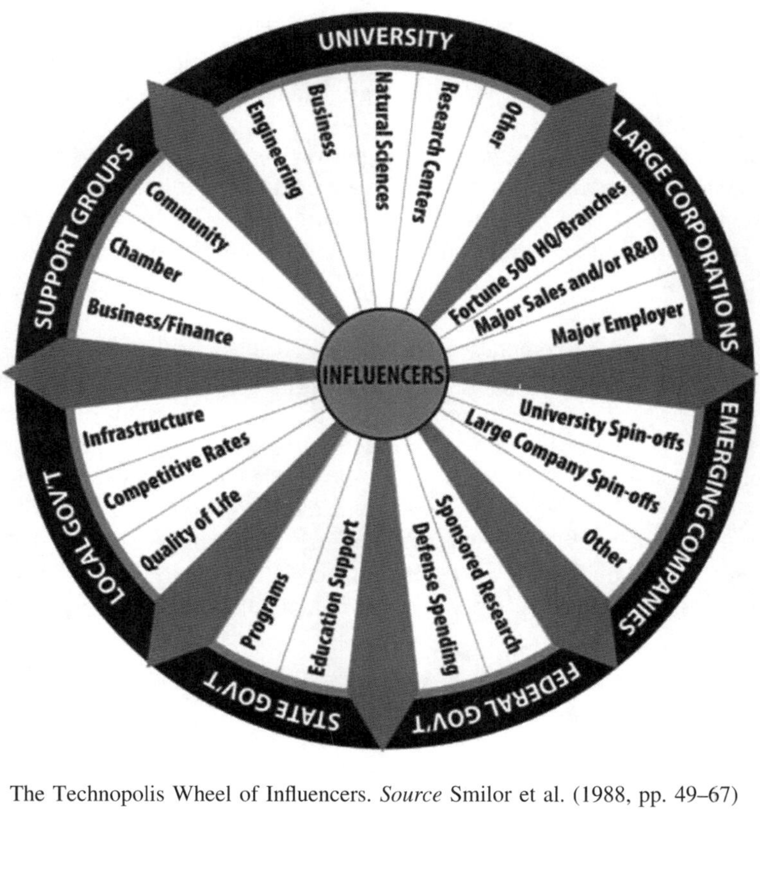


**Fig. 3** The Technopolis Wheel of Influencers. *Source* Smilor et al. (1988, pp. 49–67)

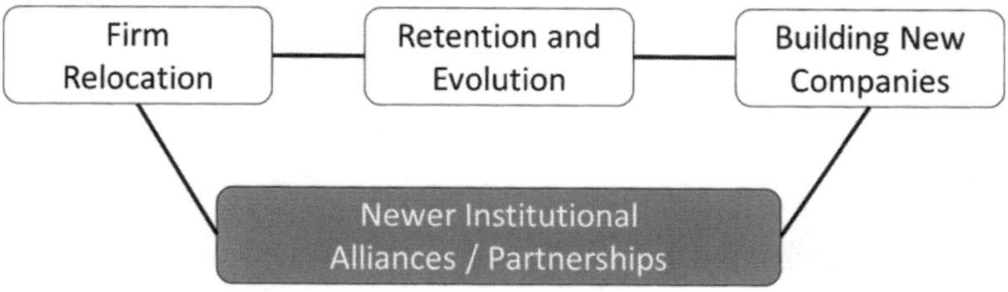


**Fig. 4** Four Main Strategies for Regional Economic Development. *Source* IC$^2$ Institute

The parks contain a large variety of enterprises from automotive, to furniture, to chemicals, to higher tech including electronics and medical devices. Despite these advantages, the report offered these conclusions:

- The technology parks have not been as successful as originally intended in terms of the key objective of job creation
- There is a high turnover of companies in these parks, companies that originally set up to take advantage of economic incentives (e.g., tax abatements)—they used up their Mexican incentives and migrated offshore to India, China, and elsewhere to take advantage of new incentives
- There was no real creation of new industries or companies including spin-offs
- In reality the companies in the parks had very little R&D activity
- Connections between the companies and the local universities had not been formed or were not productive
- There is a general impression among local politicians, universities, and other "powers that be" that the technology parks have not lived up to their potential.

Regional public-cooperation helped Austin to be quite successful in four main strategies for regional economic development: firm relocation, firm retention and evolution, building newer companies, and new institutional alliances and partnerships worldwide, Fig. 4. Firm relocation, while especially important to emerging and developing regions, tends to foster intense competition where there are regional and national winners and losers. Firm retention and evolution, for example, is more likely to concern leveraging regional assets (e.g., the local university and talent) to migrate a firm from focusing on low-cost production to high end R&D as was the case in Austin, Texas when IBM went from building electric typewriters in 1960s to being a major globally competitive R&D center, talent magnet, and patent generator.

However, we consider building new companies and new industry clusters as the key to a region becoming a globally competitive "technopolis." As in the case of Austin, Texas with the launch and growth of global companies with local headquarters as in the case of DELL Corporation, National Instruments, and Whole Foods, all with start-up and growth ties to education and research assets of the University of Texas in Austin. The fourth strategy, Newer Institutional Alliances and Partnerships, regionally and globally, is increasing key to the sustainable success of emerging, developing, and developed regions worldwide. This is a topic of central importance to the UNESCO Workshop and will be discussed in the final section of this chapter.

## 4 The Firm at the Regional Level

While we agree that substantial gains in wealth are to be found in the creation, diffusion of knowledge, we also argue for the key importance of know-how, and needed innovation infrastructure at the regional level. In short, it is important to stress that the accumulation of knowledge leads to the creation of wealth only if the knowledge is effectively used or applied. This UNESCO Workshop is focused on linking creativity (ideas) and innovation (doing) and indeed this is a critical topic for discussion in terms of job and wealth of knowledge leads to the creation of wealth only if the knowledge is effectively used or applied.

In short, the linking of creativity (ideas) and innovation (doing) is a critical topic for discussion in terms of job and wealth creation and competitiveness of Science Parks and Innovative Clusters. As noted by Dr. Charles W. Wessner (2006), Director, Program on Technology, Innovation and Entrepreneurship, U.S. National Academies made the following observations on why innovation is so crucial in today's "Flat World:"

- The U.S. Standard of Living has been built on innovation and competition
- In the current and future global economy, many new talented players are coming on the scene, introducing
  - New competition via low wages and high technical skills
  - Government programs to direct capital to strategic sectors and develop high skill human capital
- Convergence:
  - Competition is getting closer in Skills, Technology, and Capacity.

In *The Innovators Solution* (Harvard Business School Press, 2003) Clayton Christensen and Michael Raynor write of the important differences between sustaining and disruptive innovations as noted in the following quote:

> Although sustaining innovations are critical to the growth of existing businesses, some disruptive technologies offer a higher probability of success in building new-growth businesses. Successful new-growth builders know that disruptive strategies greatly increase the odds of competitive success—disruptive technologies are more likely to beat incumbents—where incumbents are likely to win in the case of sustaining technology improvements.

Figure 5 depicts Type 1 Knowledge/Technology Transfer (KTT) to established firms which is more common with sustaining innovations while disruptive innovations are more often linked to Type 2 KTT leading to spin-outs or start-ups.

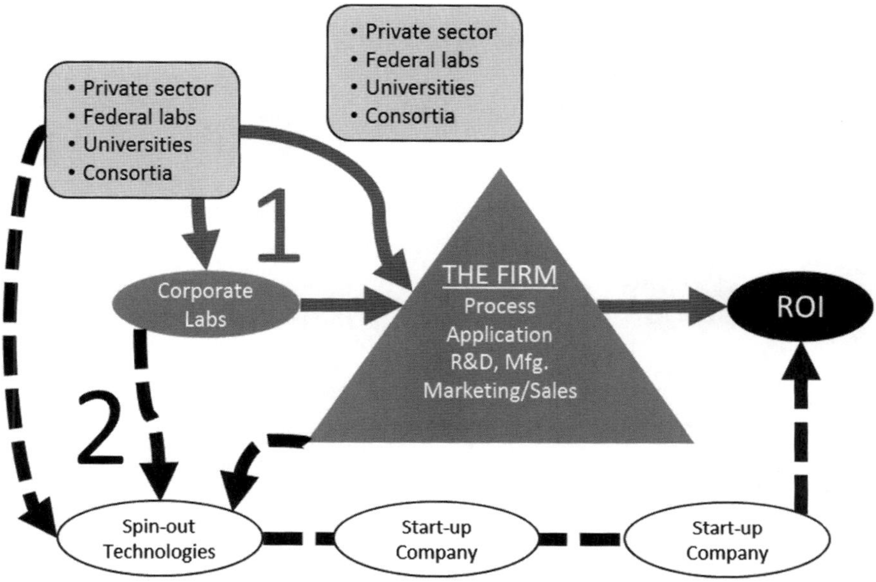


**Fig. 5** Two Key Types of Knowledge/Technology Transfer to Return on Investment (ROI). *Source* $IC^2$ Institute

The $IC^2$ Institute's MS degree program in Technology Commercialization uses the Jolly Model: *From Mind to Market* (1997) to teach about human, institutional, structural, policy, etc. A key focus is on barriers (that cannot be changed at least in the short-term), challenges (can be overcome with creativity and innovation), and facilitators to mobilize energy and resources to get from imagination through incubation, demonstration, promotion, and sustainability within the firm for Type 1 KTT, Fig. 6. It is interesting and thought provoking discussion to assess how key barriers, challenges, and facilitators differ by industry sectors and country or regional locations.

In terms of launching and sustaining start-ups and regional innovation clusters $IC^2$ Institute research emphasizes the importance, at the regional level, of linking entrepreneurial talent, technology, capital (human and financial), and business know-how and market access for established, emerging, and new to the world markets, Fig. 7.

## 5 Facilitating Global Knowledge Sharing Networks

An important objective of this UNESCO Workshop is the establishment of a cooperative regional and global networks and newer institutional alliances among science cities and the sharing of information and experiences with developing

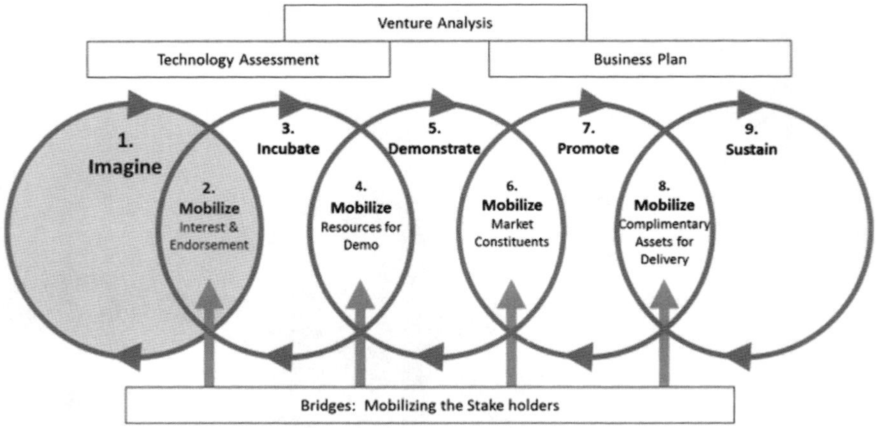


**Fig. 6** A Model of Technology Commercialization: Building the Value of a New Technology. *Source* Vijay Jolly, From Mind to Market, (1997)

regions. This is a very important objective, for as stated in a 1998, *World Bank Development Report : Knowledge for Development:*

> It appears that well-developed capabilities to learn—the abilities to put knowledge to work—are responsible for rapid catch-up…The basic elements [to develop these learning abilities] appear to be skilled people, knowledge institutions, knowledge networks, and information and communications infrastructure

Few emerging and developing regions can hope to match, at least in the near-term, the physical and smart infrastructure of established technopoleis or "spiky regions" where there is a world-class agglomeration of technology, talent, capital, and know-how. However as emphasized by Caircross (1997),[4] the 'death of distance' in the digital era reduces the inherent economies of regionally-based knowledge clusters and opens the field to new entrants. Regional, national, and global digital networks allow for, if not encourage, the development of non-geographically bound or virtual technopoleis. We suggest that the "flat or spiky world" is necessitating enhanced global activities not only for the traditional international activities of major governments, firms, and universities, but also for new entrants such as regional governments, start-up firms, and their local support groups such as entrepreneurial associations and venture financing. These developments may also facilitate access for the developing world to needed information, technical sources, talent, and financial support more commonly available only in more developed regions, Fig. 8.

The importance of global R&D collaboration is also noted, by Professor Keshav Pingali, Professor of Computer Science, as being critically important for more

[4] F. Caircross (1997).

**Fig. 7** Knowledge/Technology Transfer and Application at the Regional Level, $IC^2$ Institute

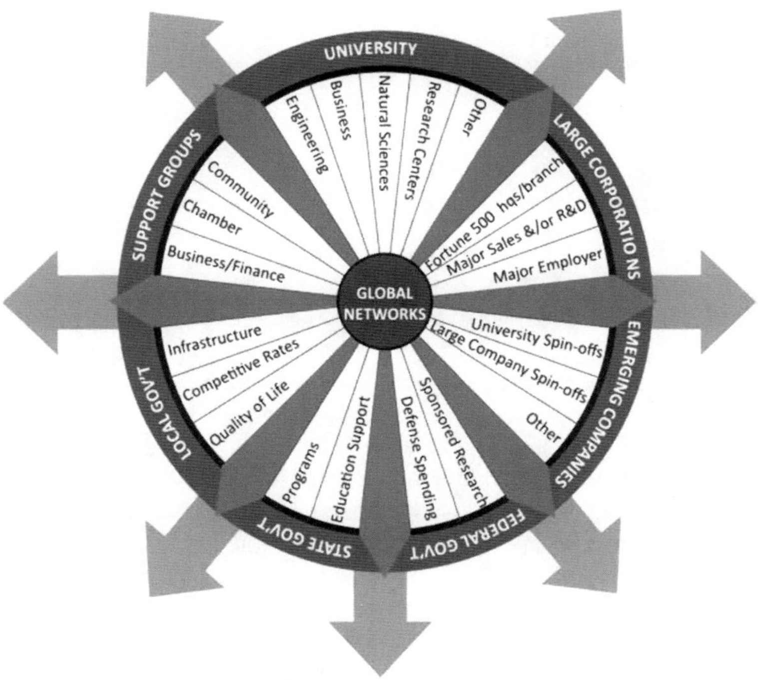


**Fig. 8** Linking the Regionally-Based Technopolis Framework to Global Networks (Gibson and Rogers 1994)

developed regions and universities (*UT-Austin/Portugal CoLab Newsletter* (September, 2008):

> In a recent book titled "The World is Flat", New York Times columnist Thomas Freidman argues that globalization is radically changing business models across the world. These global tsunamis are also changing universities—no country has a monopoly on research talent or resources, so it has become necessary for universities to work together across national boundaries to solve major research challenges.

The objective is networking companies/regions/science parks/innovation clusters for social, economic, scientific, and technological development—through global academic, business, and government collaboration. Gibson and Conceicao (2003), emphasize the Importance of sharing both "software" or codified knowledge and "wetware" or tacit knowledge, an important distinction made by Nelson and Romer in 1996. They also suggest three principles for global knowledge sharing networks while focusing on Learning and Innovation Poles (LIPs).

(1) *Principle No. 1* states that when establishing Learning & Innovation Poles, we must deal with social as well as physical constructs that link participating people and institutions in networks of knowledge production, sharing, adoption, and diffusion while fostering self-reinforcing learning cycles.
(2) *Principle No. 2* centers on fostering networks where interaction leads to increased learning in all network nodes, but in which the rate of learning is higher in the less developed nodes, Fig. 9
(3) *Principle No. 3* centers on fostering regional 'ownership' and sustainability of activities and results for emerging, developing, and developed LIPs.

## 6 The International Collaboratory for Emerging Technologies (CoLab)

The International Collaboratory for Emerging Technologies (CoLab) is an important 5-year project involving the Government of Portugal, numerous Portuguese Universities and public and private institutions, and The University of Texas at Austin (www.utaustinportugal.org). As noted in Fig. 10, the organizing framework for this project includes an Advisory Board, Eternal Review Committee, and three academic programs: Digital Media, Mathematics, and Advanced Computing. The University Technology Enterprise Network (UTEN) is a fourth activity that focuses on the training and internationalization of Portuguese Technology Transfer Officers and entrepreneurs as well as accelerating U.S. and global market access for Portuguese technology-based firms. Each of these programs has a director and co-director located at UT-Austin and at collaborating Portuguese universities. Key CoLab objectives and metrics are as follows:

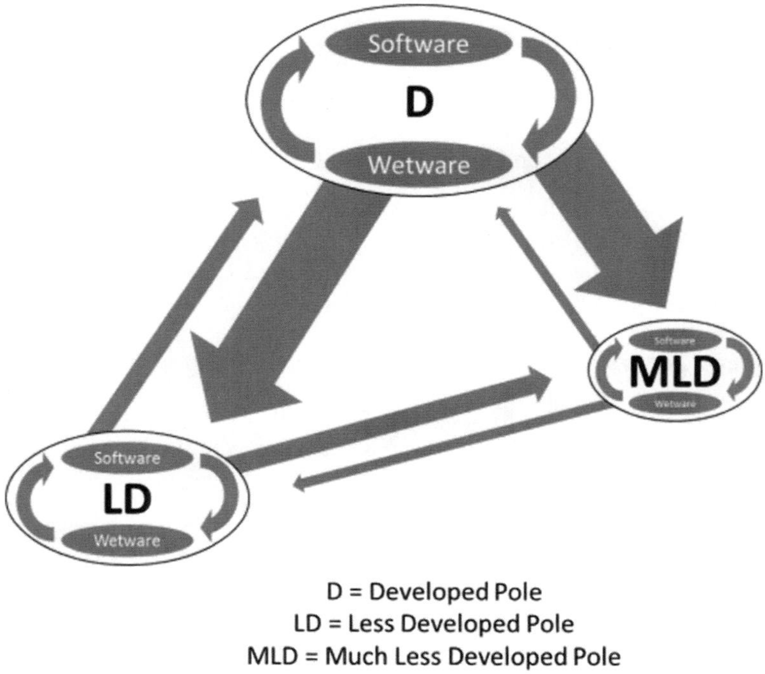


**Fig. 9** Learning Network Based on Proportional Reciprocity. *Source* Gibson and Conceição 2003

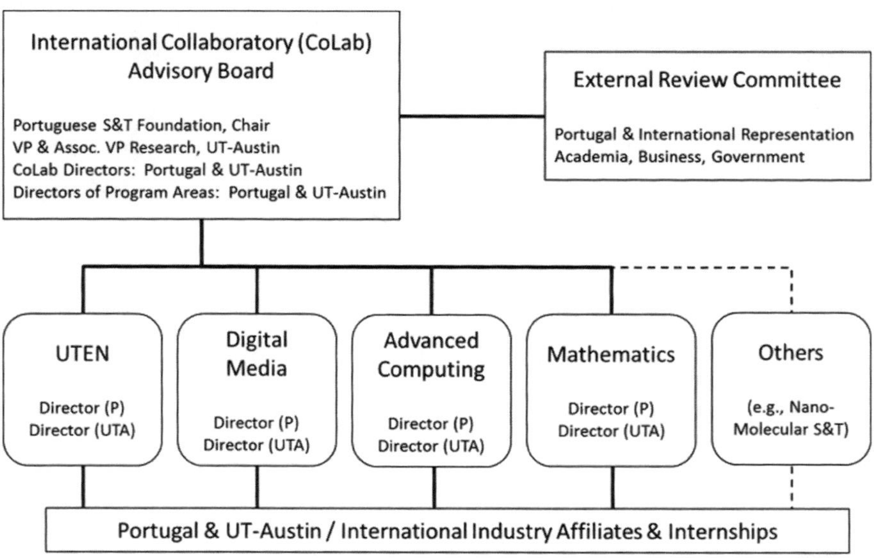


**Fig. 10** UT-Austin | Portugal CoLab (www.utaustinportugal.org). *Source* $IC^2$ Institute

- Academic Programs
  - Relationship Building for Research, Education, and Training
  - Joint and Dual M.A. and Ph.D. Degrees
  - Faculty and Student Research Collaboration
- University Technology Enterprise Network (UTEN)
  - Science and Technology Commercialization Training for University Managers, Staff, and Entrepreneurs
  - Fostering Sustainable Technology Transfer and Commercialization, Entrepreneurship, and Innovation at internationally competitive levels
  - Facilitating Access to US & Global Markets
- Institution Building
  - Build Centers of Research and Education Excellence for Critical Mass
  - Build Sustainable and Globally Competitive Infrastructure for S&T Commercialization
- Industry Affiliates and Internships
  - Enhance Portuguese University-Industry Collaboration

## 7 Key Lessons Learned for Regions, Science Parks, and Innovation Clusters

1. While the numbers of Technology Parks, Incubators, etc., are dramatically increasing worldwide it is an open question how successful and sustainable these efforts will be in the near and longer-term
   - In short, while beautiful infrastructure and buildings are dramatic and visible they are not sufficient
   - Institutional excellence alone is not sufficient

– Regional academic, business, government cooperation is essential to leverage public and private assets to overcome regional challenges and for sustainability
– Well-leveraged action with excellent talent and "know-how" is preferred over average or sub-standard teams and first class facilities
   - Build on regional strengths & not what others have done.

2. It is important to have capable visionaries and influencers who collaborate and cooperate while representing regional business, academic, and government sectors
   - It is a "bottom up" and a "top down" process including both national and regional actors
   - Regional buy-in and action by key facilitators is crucial.

3. The out-migration of regional talent can be reversed and lead to in-migration of talent

   • Regional context is important

– Allow for if not encourage Immigrant Entrepreneurship
– Be a tolerant city/region
– Support social and economic inclusion.

"Spectacular Success" is nice but small wins and regional role models are also important
The right "Branding" sure helps: Be a "cool" place for "hot/creative" jobs.
4. All regions—developed, developing, and emerging—have distinctive challenges and assets.

   And a few questions:

• Are policy makers and managers getting smarter about Knowledge-Based Economic Development?
• Is it increasingly possible to "leapfrog" from being an emerging to being a globally-competitive region?
• Is regional technology-based economic development becoming more virtual and global?
• Given increased R&D and entrepreneurship as well as natural resource constraints worldwide are there limitations for technology diffusion and absorption in global markets?

## References

Caircross, F. (1997). *The Death of Distance*. Boston: Harvard Business School Press.

Christensen, C.M., & Michael, E.R. (2003). *The Innovators Solution: Creating and Sustaining Successful Growth*, Harvard Business School Press.

Gibson, D., & Pablo Rhi-Perez, et al. (2002). *Cameron County/Matamoros: At The Crossroads*, IC2 Institute Report for The University of Texas at Brownsville.

Gibson, D. V., & Rogers, E. (1994). *R&D Collaboration on Trial*. Boston, MA: Harvard Business School Press.

Gibson, D.V., & Conceio, P. (2003) *Incubating and networking technology commercialization centers among emerging, developing, and mature technopolies*.Worldwide International Handbook of Innovation.

Jolly, V. (1997) *From Mind to Market*.

Nelson, R.R., & Romer, P.(1996) *Science, economic growth, and public policy*, in: B.L. Smith & C.E.Barfield (Eds.) Technology, R&D, and The Economy, Brookings, Washington, D.C.

Rogers, E., & Larsen, J. (1984). *Silicon Valley Fever Growth of High-Tech Culture*. New York: Basic Books.

Smilor, R., Gibson, D. V., & Kozmetsky, G., *Creating the Technopolis High-Technology Development in Austin, Texas*, Journal of Business Venturing, 1988, pp. 49–67.

Tatsuno, S., (1988) *Building a Japanese Technostate: MITI's Technopolis program*, Onda, M., *Tsukuba science city complex and Japanese strategy*, in: Creating the Technopolis (Eds.) Smilor, R., Kozmetsky, G., Gibson, D.V., Ballinger Pub. Co.: Cambridge, MA.

# Support System for Entrepreneurial and Small and Medium Ventures in ICT Sector: Case of the Technopreneurship Development Program (TDP) Flagship of Malaysia's Multimedia Super Corridor (MSC) Project

**Avvari V. Mohan**

## 1 Introduction and Conceptual Overview

One of the most important trends in the world economy that has led to a strong interest in clustering has been the globalization of economic and business activity. The notion of clustering or regional development is increasingly being associated with global information resources, creation of new enterprises, and development of advanced skills, continuous innovation, and diffusion of new technologies. These capabilities are considered essential to sustain international competitiveness and build a knowledge-based economy. Numerous cities and countries across the world have started clustering efforts naming themselves Silicon something, e.g., Silicon Island (Taiwan), Silicon Plateau (Bangalore), Silicon Alley (New York), Silicon Hills (Austin Texas), Silicon Fen (Cambridge, U.K.), and attempted to copy the 'Valley's' success story. Malaysia has also initiated the Multimedia Super Corridor project on similar lines to develop the ICT sector. This chapter presents the case Technoprenuer Development Flagship program, developed under the Multimedia Super Corridor (MSC) initiative of the Government of Malaysia, as a support system for the development of entrepreneurial and small and medium enterprises (SMEs) in the ICT sector. It is hoped that the case of the Technopreneur Development program (TDP), will provide some valuable ideas for developing a support for SMEs, to countries or agencies that are planning to develop such systems for entrepreneurship and SMEs.

A. V. Mohan (✉)
University of Nottingham Malaysia Campus, Nottingham University Business School, Jalan Broga 43500 Semenyih, SI, Malaysia
e-mail: mohan.avvari@nottingham.edu.my

D.-S. Oh and F. Phillips (eds.), *Technopolis*,
DOI: 10.1007/978-1-4471-5508-9_13, 

Before going into the discussion about the MSC project and more specifically the Technoprenuer Development programs, a brief overview of the concept of clustering is provided. A cluster has been described as a geographical concentration of mutually dependent companies with vertical as well as horizontal and cooperative as well as competitive relational patterns, and furthermore companies that operate within the same branch or on the basis of the same basic technology (Jacobs and De Man 1996).

While clusters are made up of companies of different sizes and structures, it is the support system to help development of entrepreneurial and/or small and medium enterprises that is of specific interest to this study. Also of interest is the question—what country strategies can be discerned in cluster-based policy, is addressed by authors like Porter (1995) and Quandt (1998) who point to various roles for governments in cluster-based policy. Quandt (1998) identified several generic factors from literature that are associated with the development of a cluster. He states, two factors may be considered as necessary, but not sufficient—they are a capable scientific and technological infrastructure and a critical mass of human resources. This includes universities, technological incubators, science parks, etc. Other generic local ingredients comprise business infrastructure including development agencies, business associations, chambers of commerce. It also comprises financing opportunities through the availability of seed, venture, and investment capital. In addition to grants for training and R&D and government officers providing a wide range of services in the area, Quandt (1998) lists physical infrastructure, favorable business climate, existence of government incentives, and diversified economic base as factors needed to develop a cluster. In addition to the above soft factors like existence of champions, science and business linkages, broad support base, information networks, marketing, and image building are also hard to grasp but important factors for the development of clusters (Quant 1997). Successful clusters depend not only on government policies and infrastructure but also on other aspects like the private and the public sector (usually universities and research institutions), which join efforts to create innovative environments and to build synergies among the agents with complementary capabilities.

In addition to studies on clustering, studies of National Systems of Innovation focuses on flows of knowledge and development in technology sectors is attributed to the linkages and the bonds of interchange among various institutions of a nations system. Policymakers are supposed to be able to find possibilities on ways to embellish innovative exploits in the knowledge-based economies of today by understanding these systems. (OECD 1997) It is in this light that the chapter looks the Technopreneur Development Program as support system for entrepreneurial and small and medium enterprises under the MSC project.

The chapter starts by providing a background to the MSC project followed by a section where the MSC is explained as one of the major regional systems for innovation in Malaysia. Some of the weaknesses of the initiative in terms of supporting entrepreneurial ventures and/or SMEs are identified, providing the background to the development of the MSC Technopreneur Development Program

(TDP). The final sections present the TDP program as a support system to support the development of entrepreneurs and SMEs in the ICT sector.

## 2 Background to the Development of the Multimedia Super Corridor Project

Malaysia's path toward an IT-literate and knowledge society is a part of continuous government policy to chart the country's economic future since its independence in 1957.

The success story of Malaysia becoming a "tiger" nation is largely due to the government's determined effort to transform the society from a predominantly agrarian one where the economy focused on rubber, palm oil, and petroleum to an industrial one with a significant manufacturing sector, focused on electrical, electronic, and other sectors. While earlier policies were formulated to bring Malaysia forward from a commodity-based economy to an industrial one during the 1990 s, the planning for transition to an information society and that to a knowledge-based economy (see Fig. 1) with a focus on the ICT sector was started. The Vision 2020 was introduced for becoming a developed country by the year 2020 and among various other initiatives; the National IT Agenda and the Multimedia Super Corridor were launched toward meeting the objectives of the Vision 2020.

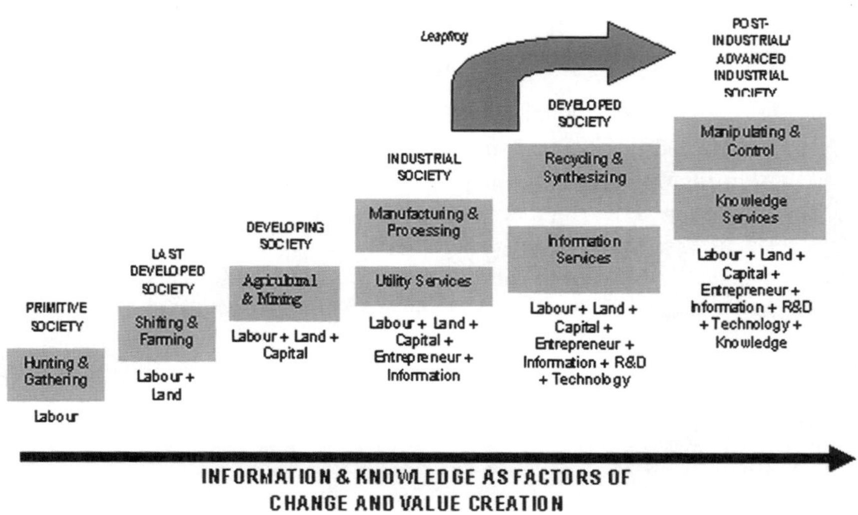


**Fig. 1** Vision 2020- fast tracking the nation's transition source: NITC Malaysia, access, empowerment and governance in the information age, building knowledge societies series, volume 1, NITC Malaysia, NITC Malaysia publication, 2000

The Multimedia Super Corridor (MSC) Project was articulated as a cluster of firms in the ICT (information and communication technology) sector. In August 1995, Dr. Mahathir Mohammed the then Prime Minister of Malaysia, announced the "Multimedia Super Corridor" (MSC) as the centerpiece of the national IT strategy under the Seventh Malaysia Plan (1996–2000). The goal being to enhance the development of the ICT sector in the country and transform the predominately production-based economy into knowledge-based one. The MSC project is a government led development and is not in the second phase of development being supported with various cluster oriented infrastructure and policies.

### *2.1 An Overview of the Multimedia Super Corridor (MSC) Cluster Project*

In this section the Multimedia Super Corridor (MSC) Cluster—a Regional System of Innovation for ICT Companies is described and the lacunae in this project are identified which led to the development of the MSC Technoprenuer Development Program Flagship. The MSC, as it is known, is like a regional system of innovation for firms (including) SMEs in the ICT Sector in Malaysia. The following were some reason why the MSC project was conceived:

- Recognition that Malaysia was losing its comparative advantage in its traditional economic sectors;
- Need to drive the economy toward higher productivity through technology and high value-added economic activities;
- Information Age and converging technologies presented the best opportunities for socioeconomic transformation.

The MSC project was modeled after the Silicon Valley and is a multibillion dollar project started in 1995 and spans an area of more than 750 $Km^2$. A policy-driven cluster-oriented approach, the MSC aimed to develop a friendlier environment for new business development in the ICT sector and also **to spur high technology entrepreneurship** in Malaysia. It uses the 'regional clustering' approach and then intends to roll it out to the rest of the country. The focus was on building capacity in this sector and also to have applications for national development Fig. 2.

## 3 The MSC Regional System of Innovation—Elements of Support for ICT SMEs

While the MSC regional system of innovation was not developed specifically to support SMEs only—many elements of the MSC project have been put in place to support the development of SMEs and building their technological capacities.

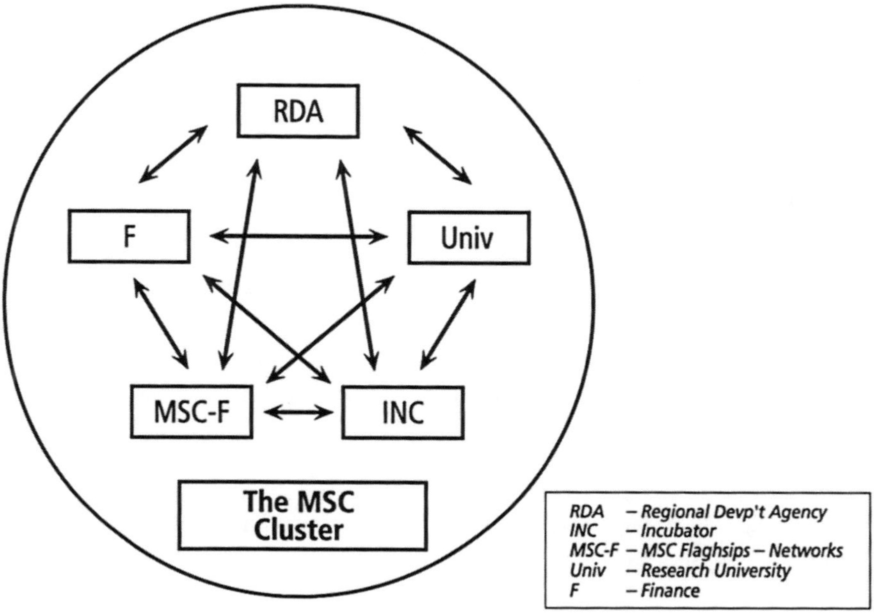


**Fig. 2** Some key elements in the MSC regional system of innovation. *Source*: Mohan et al. (2004)

These include a regional development agency (RDA) called the Multimedia Development Corporation (MDC), innovative finance schemes, the MSC Flagships Networks, A dedicated university and an Incubator, others.

## *3.1 The Regional Development Agency—Multimedia Development Corporation*

The Multimedia Development Coporation (MDeC), the government owned but autonomous organization, plays the role of a champion, facilitator, and partner of companies choosing to operate in the MSC. The MDeC markets the MSC initiative globally. Some of the roles, among many others, set for MDeC to develop the MSC Cluster include the following:

- Foster the development of "web"-based collaboration in the MSC, Malaysia and globally.
- Catalyze and nurture local companies and SMEs to become global players by forging successful smart partnerships between Malaysian and international companies.

- Realize the promise of mutual enrichment by making it easy and cost-effective for companies to do business in the MSC.
- Promote technology and knowledge development in the MSC through incentives for commercial R and D and through the establishment of leading incubation centers.
- Facilitate innovation and entrepreneurship by supporting the development of a financial infrastructure that provides venture capital and public listings for smaller companies.

## *3.2 The Finance Element of the MSC*

In order to build technological capacities among SMEs, it was recognized that the existing traditional financial institutions could not help. In the MSC Cluster more options have made available like venture Capital from the MDeC for creating tech-entrepreneurial culture in addition, to other VCs. The MDeC also provides Special Grants for developing risk taking culture to help in developing an R and D culture. In the recent past a small number of Venture Capital firms have merged in the country that provide some funding to the SMEs—but one issue about VC funding was that it was not easily available to many entrepreneurs/SMEs, especially those who were not doing R&D and more users of technology or technology based (e.g., E-commerce companies). To overcome this lacunae, there some innovative funding products especially from two funding agencies that stand out—viz the Cradle Investment Program (CIP) of the Malaysian Venture Capital Corporation and Malaysin Debt Ventures. Both the organizations have funds designed to the needs of the local market—in this case for Malaysia—and this was done by identifying the areas where there was some lacunae using the framework of the enterprise life cycle.

## *3.3 Research University in MSC Cluster—The Multimedia University*

The availability of technically trained manpower for tech-based SMEs was recognized as a need in Malaysia and also support from a university had been recognized as an important factor for technology development in many clusters. In order to facilitate such support, in addition to identifying universities in the neighborhood, a new university, the Multimedia University was set up within the MSC cluster in order to perform these roles. It not only provides the required manpower—but more importantly serves as a research support base for SMEs which cannot afford to set up expensive infrastructure needed to conduct R and D activities for enhancing their technological capabilities. The university also has institutions like the Center for Commercialization and Technopreneur Development (CCTD) to help in creating startups and also to help SMEs in the cluster.

## *3.4 Infrastructure to Nurture ICT Entrepreneurs and SMEs—MSC Central Incubator*

In order to enhance the numbers of technology-based SMEs in the MSC region—a central incubator was identified to be the nucleus for the National Incubator Network that would link eight other centers which are already in operation. These centers include Technology Park Malaysia, UPM-MTDC Incubator, and Kulim Hi-tech Park. The establishment of this incubator network was considered crucial to help generate the much needed pool of small and medium enterprises (SMEs) to meet the demands of the MSC cluster project when it rolls out nationwide. The incubator is located within the Multimedia University (which is in the MSC Cluster) with 62,500 square feet of space. The incubator is one of the six major ingredients critical to the creation of sufficient technopreneurs. The other five elements being venture capital and financing, research and development, incentives, human resource development, and market access. The MSC Central Incubator is one incubator in the cluster that has been set up to help enhance the capabilities of SMEs by allowing its tenants to have access to the MSC's communication infrastructure, research facilities, networking opportunities and venture capital funding. MSC Central Incubator will also support the tenants by providing seminars, training, and other advance services such as business plan development, accounting, and marketing. The MSC Incubator is now part of a larger nationwide incubator program that is spearheaded by the Multimedia Development Corporation (MDeC).

## *3.5 The MSC Flagships—Institutional and Network Dynamics for Enhancing Technological Capabilities of Firms in the MSC*

Another support element in the MSC Cluster is the "MSC Flagships". The government of Malaysia provided major government ICT projects to consortiums/ network of companies to help kick start the MSC project. This network of firms would also help to develop technological capabilities among the local firms in the networks and the projects themselves were for social development, On one hand this provides 'demand'-oriented incentives to attract the foreign firms to come into the MSC and partner with local firms while facilitate "learning" and some sort of transfer of capabilities to the local companies while providing market to the foreign technology providing companies.

### 3.5.1 Multimedia Development Flagships Clusters

This comprises of firm networks that develop applications to facilitate the development of society and government while they offer concrete business opportunities or in other words create demand for the services of the firms in the MSC. The four identified 'flagship applications' networks of firm within this group are

1. Electronic Government flagship
2. National Multipurpose Card flagship
3. Smart Schools flagship
4. Telemedicine flagship

### 3.5.2 Multimedia Environment Flagships Clusters

In the case of this flagship groups, firms involved carry out their activities within interactive clusters to develop ICT technology, products and applications, designed for enabling the applications in the Multimedia Development Flagship. Thus, the main aim of this flagship is to aid the Multimedia Development Flagship firms. This category consists of firms developing applications and classified under the

1. R&D Flagship Cluster,
2. E-Business Cluster (combining the Worldwide Manufacturing Web and Borderless Marketing Flagships), and most recently the
3. Technopreneur Flagship Cluster—all this with the aim of providing specific support to SME type companies in developing multimedia products and applications.
4. Currently in planning is also a Biotech Cluster.

The two groups of MSC Flagship networks of firms function or operate separately with their own goals. The MSC Flagship co-ordination unit, an institution under the regional development organization MDeC, links the firms in two groups of MSC Flagships to realize the synergies planned between them (Fig. 3).

## 4 MSC Regional System of Innovation (RDI)—Favorable for SMEs?

The MSC project has had mixed reviews, and one issue was that given the number of incentives, not enough number of local start-ups and SMEs were emerging. Some of the elements in the RSI that have been generally helpful are like the existence of a regional development organization—the MDeC as a one-stop shop for all things MSC and also incentives (Financial and Nonfinancial). These have been instrumental in attracting MNCs and larger players to operate in the MSC but not very helpful in attracting and developing startups or entrepreneurial ventures and SMEs very much. The business climate also was perceived to favor larger or MNC players—who are self sufficient in terms of business services. The problems faced by SMEs have been that operating costs in the cluster and related technology

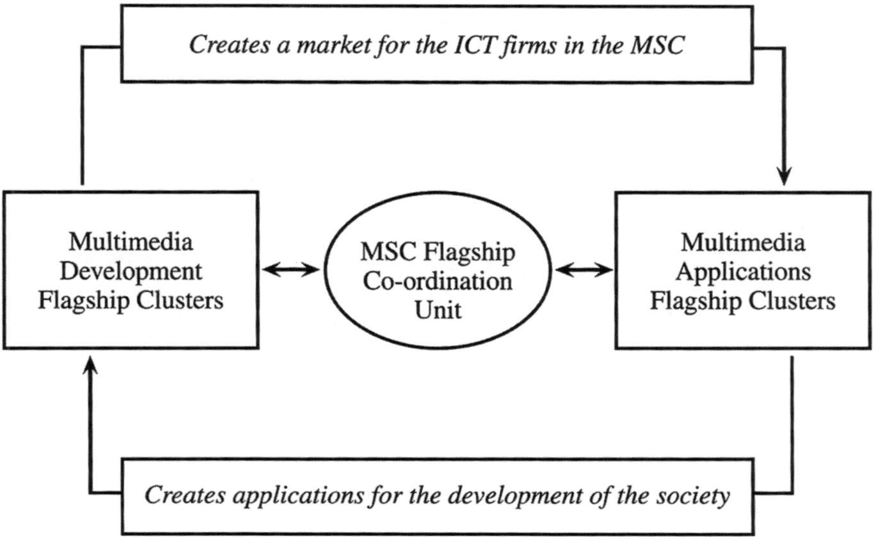


**Fig. 3** The MSC flagships and co-ordination units

parks were very high including many other hidden costs (like infrastructure, distance to market, lack of business services in the region, etc.) and the SMEs appear to lack an overall ecosystem to operate in.

As for the MSC Flagships, initially there was no specific focus on the SME sector. The Flagship Programs with the 'partnering' or networking approach had mixed results—once again this policy-driven system did attract foreign/MNC firms with 'technology' and 'business' competencies—with the lure of a 'market' to partner/network with local firms. Technology Transfer agreements were in place and there was some learning from both the local and foreign players in some flagships (Omar and Avvari 2008), the effect is still sporadic rather than rampant and significant spin offs or spill over effects yet to be seen.

All this led to the revamping of the Flagships Clusters of Firms in the MSC Cluster and more impetus for the development of the MSC Technopreneurship Development Program Flagship, which is described in the next section.

## 5 The MSC Technopreneur Development Program Flagship (TDP)

The MSC Technopreneur Development Flagship (TDP) is a specific flagship cluster designed for the promotion of SMEs in the ICT Sector. In recognizing the need to further enhance the MSC Malaysia's efforts to develop Malaysian SMEs in the ICT and other strategic high technology industries; the Government launched the Technopreneur Development Division (TeDD) in November 2001. The lead agency driving the division is the Ministry of Science, Technology and Innovation with Multimedia Development Corporation (MDeC) acting as the implementing agency.

The core objectives of the Technopreneur Development Division (TeDD) are to:

- To facilitate the development of technopreneurs, start-ups, and existing ICT companies
- To catalyze and nurture a cluster of ICT SMEs
- To assist and facilitate the growth of ICT SMEs into world-class companies

### *5.1 Enterprise Value Chain—An Action Model for the TDP Programs to Catalyze ICT Entrepreneurs and SMEs*

The Technopreneur Development Division uses the Value Chain Model, which it calls the Technoprenuer Value Chain, for its support programs for entrepreneurs and also small and medium enterprises in the ICT sector. Figure 4 below shows the Technoprenuer Value Chan with the associated needs for the enterprise at each stage.

Cognizant that business ventures at different stages of growth require technopreneurs to have different competencies and skills set, the MTD program has designed its development programs to meet the needs of technopreneurs at different phases of the Enterprise Life Cycle (ELC).

The Enterprise Life Cycle (ELC) has the following phases—starting with Discovering and Assessing Opportunities, moving on to Business Planning and Gathering Resources, Managing Growing Business, and finally Harvesting Value—which essentially covers business development from pre-idea stage, idea stage, pre-seed and to the growth and expansion stages of an enterprise. In order to help create and develop entrepreneurs and SMEs the Technoprenuer Development Division (TeDD) has created some program to create a sustainable ecosystem and ensure the continuity across the Enterprise or Technopreneur Value Chain. These programs fall under the following categories:

- Awareness and Development Programs
- Funding Programs
- Incubation Programs
- Enterprise Growth Programs
- Partnership Programs.

In the following section an explanation of the programs under the abovementioned categories' is presented:

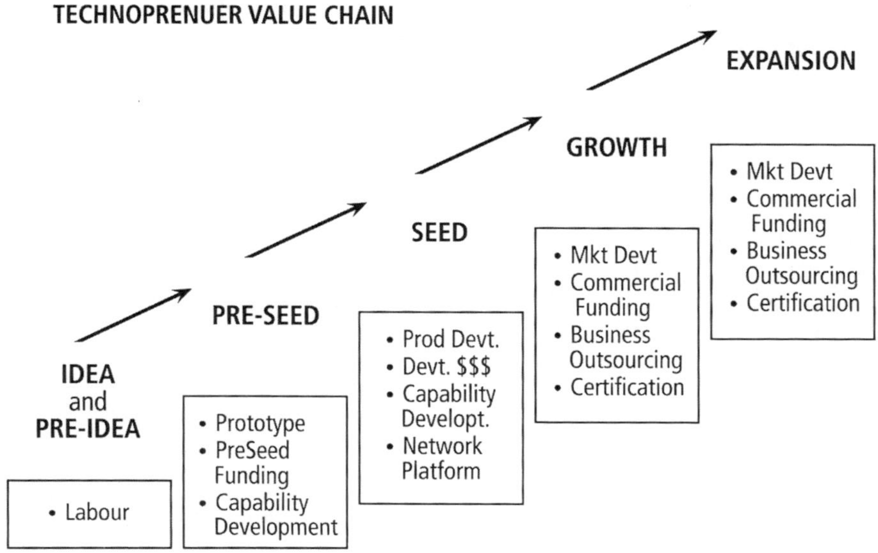


**Fig. 4** Technoprenuer eco system and value chain as envisaged by MDeC-TeDD

### 5.1.1 The Awareness and Development Programs by TeDD

The main objective of this program is to support the preidea and idea phases of the enterprise value chain through activities to promote tech-entrepreneurship in the nation. Some of the programs developed under the umbrella of awareness programs include promotion programs, contests, forum and workshops, and also the creation of an award to encourage entrepreneurship in the country. Recognizing that the students and researchers in educational institutes or institutes of higher learning have ideas and do projects that have potential for business, the different activities under this umbrella are focused heavily on this community. A pivotal program created for this purpose is the 'National Unipreneur Development Program' (NUDP). The aim of the NUDP program is to promote tech-entrepreneurship in the academic community of the country—including students, academicians, and nonacademicians at Institute Of Higher Learning (IHL) on one hand and also to encourage researchers to convert their research into commercializable products.

In order to create awareness of entrepreneurial opportunities for work done in IHLs, there is an annual contest called the **MSC Malaysia—IHL Business Plan Competition** is organized by Multimedia Development Corporation (MDeC—the regional development agency of the MSC). This has been created to encourage IHL community to develop and submit a commercially viable business plans. The winners of this competition in addition to receiving cash rewards can also get induction into TeDD development programs and opportunities to apply for

Pre-Seed Funding. In addition there is a "**Train The Trainers**" program that trains the academicians on how to advice the students and researchers in developing their business plan for the IHL Business Plan Competition.

Some of the development programs as part of the ecosystem include the **University Industry Commercialization Collaboration Forum (UICCF)—**where researchers and students can present their research projects for industry review as well as a platform for the industry to present their industry needs to the researchers. The TeDD also organizes what they call "**Pre-Seed Jump Start**"—a workshop that focuses on guiding the participants on how to submit a solid business plan that addresses all the criteria's needed to qualify for the Pre-Seed Fund Program.

### 5.1.2 Funding Programs by TeDD

In addition to the funding options available in the MSC Cluster, some other funding sources for technology firms include the 500 million ringgit New Technology Investment Fund (NTIF). Its soft loan program "caters to established Malaysian companies or individuals with at least three years of experience in a similar business and a good track record." There is also the Malaysian Venture Capital (MAVCAP), a government VC firm, that has an 800 million ringgit fund and its 100 million ringgit—Cradle Investment Program CIP- fund, In addition, there is the Malaysia Debt Ventures (which was set up with an allocation of 1.6 billion ringgit to spur the growth of ICT and high-growth sectors) and a host of other government funds.

Given all these options, in addition to many other existing options (like banks, etc.) and also new-age funding options (like Angels, etc.)—there was a gap identified in terms of funding for SMEs and entrepreneurs in the ICT sector in the pre-seed part of the Technoprenuer Value Chain. A funding program called the Technopreneur Pre-Seed Funding Program (TPF) has been created. The TPF not only acts as catalyst in the creation of tech-entreprepreneurs and SMEs by addressing the "pre-seed funding gap" in ICT Development Life Cycle but also assists technopreneurs to jump start their business. The Fig. 5 shows the various stages of an SME enterprise and some of the funding options (the shaded ones are some of the innovative funding options created).

The targeted beneficiaries of the TPF program are local tech-entrepreneurs. Only individuals are eligible to apply, not existing companies. The program offers up to RM150 k of funding to develop viable business plans into commercially focused ICT projects with prototype and detailed business plans suitable for venture funding and commercialization. As a developmental program, it is not a pure grant and recipients will also benefit from development, mentoring services, and access to shared lab facilities at MSC Malaysia-Status Incubators provided through MDeC's the Technopreneur Development program (TDP).

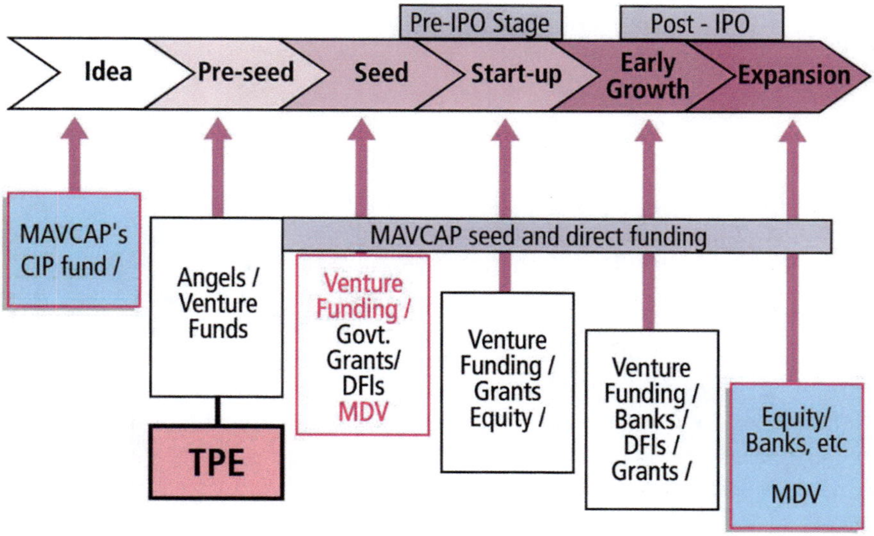


**Fig. 5** Enterprise life cycle phases and funding options

### 5.1.3 Incubation Support Programs from TDP

The National Incubator Network (NIN) is one of the three key components of the Start-up Program under the Technopreneur Development Program(TDP). The NIN role is toward fulfilling one of the main objectives of the TDP vision that is to "capture and cluster" technopreneurs nationwide and to spawn as well as nurture a critical mass of technopreneurs, Small and Medium Enterprises (SMEs) and start-up companies involved in ICT and also biotechnology industries.

The of Model for Incubation program involves leveraging of government agencies, institutes of higher learning (IHLs), finance providers, business chambers, and other private sector resources to help start-ups and SMEs to gain a foot hold in the market. It also involves provision of infrastructure for start-ups like the setting up of a special buildings for startups and SMEs businesses in the regional cluster (at lower costs), including one in the cluster university and also private incubators. The features of the incubator would include low cost rentals, funding access, market access information, basic business services, and in some cases proximity to 'regional development authority the MDC' or access to manpower and technology—from the universities in the proximity.

The Technoprenuer Development Division (TeDD) would be involved in the Management of Incubation Process, Aggregation/QC of All Incubation Services (Space, Shared facilities, Value-Added Business Services) and also in Management Sub-Incubators.

A special feature of the incubation services is the TECHNOLOGY LABS, which include

- Technology Dev Facilities
- Technology Support and Advisory
- Development and Training

These Technology Labs are run with the help of different MNCs that are considered leaders in the ICT sector (a more detailed explanation of the Technology Labs is provided in the Partnership Programs in Sect. 5.1.5). For example, two competing giants Sun Systems and Microsoft have technology support programs run from their offices in the incubator located in the MSC cluster.

In addition to the above, the NIN is to have a unique "community" where incubators throughout the country are linked together for network-sharing and providing opportunities for new ventures, knowledge and expertise on a common platform. To this end, the MDeC spearheads the initiative to create sustainable incubators nationwide that would provide a conductive environment to nurture technopreneurs. Built within the NIN also is the knowledge-based center of expertise which helps in developing best practice incubation models, programs, and talent to mentor and nurture successful start-ups, such as mentoring programs, business advisory, and business plan enrichment services. (some more details of the NIN are provided in Appendix B).

### 5.1.4 Enterprise Growth Programs as Part of the TDP

This program aims at assisting tech-entrepreneurs and SMEs to develop their enterprises for sustainability and growth. The focus is on developing "Go-To-Market" models for business development and access to funding for market development at the growth and expansion phases of the Technoprenuer Value Chain. While funding options like a technology focused stock exchange, etc., was found to be available in the country for this phase it appears that many SMEs and entrepreneurs needed more of advisory type support regarding market and funding access. The TDP includes a **One—Two—One Consulting program and Business Assessment and Advisory** in which TeDD focuses on assessing, enriching and fine tuning Go-To-Market models and also assistance in areas of readiness for market and funding financials and technology. For market access, ventures with the 'MSC' status,[1] get support from a special division of the Multimedia Development Corporation (MDeC). The MDeC identifies some high opportunity markets, their key sectors and the relevant market partner(s) and identifies and match firms with MSC-status to meet these opportunities.

In addition to the support mentioned so far, TeDD in partnership, with a certain NGO for entrepreneurs (called TeAM), provides services for growth businesses like branding and marketing strategies, intellectual property and

[1] "MSC-status" is granted to firms that are qualified to receive special benefits from the MSC project like tax incentives, access to funding etc.

commercialization strategies, legal business set-up and company secretarial services, etc. Overall the aim is to support the ventures to leverage growth options.

### 5.1.5 Partnership Programs as Part of the TDP

The MSC project in general, recognizes MNCs playing a major support role in the development of the ICT sector of Malaysia. The TDP flagship focuses on developing partnerships with MNCs more specifically in terms of enhancing the Technoprenuer Eco-System in the country and also and support the development of new ventures and existing SMEs in the ICT sector. By partnering with MNCs the TDP hopes to benefit the overall eco-system in terms of

- **Leveraging** on the MNCs partners expertise, experience, knowledge, and networks in strengthening the ecosystem
- Adopt the **best practices** in developing and maintaining the best ecosystem
- Improve **service delivery time and cost** to technopreneurs and minimize failure points across the technopreneur value chain
- And develop better **resource capabilities**
- For the entrepreneurs and SMEs, the partnerships with MNCs is expected to provide
- greater **access to markets** like business outsourcing opportunities, etc.,
- wider **networking platform** to build contacts, share knowledge, and share experiences
- greater access and exposure to technical **experts and technology trends**

A more novel partnership program with MNCs is the Technology LABS program to support entrepreneurs and SMEs in the later part of the enterprise value chain or technoprenuer value chain mentioned in the Incubation Support Section. The following are some of the examples of TDP partnerships with MNCs like **Sun Microsystems**, **Basis Bay,** and **Microsoft**, to run the technology labs for prototype development and technology support.

**JTrend (Sun Microsystems)**

JTrend is a collaboration project between Multimedia Development Corporation (MDeC), Sun Microsystems and Multimedia University (MMU). JTrend focuses on supporting and encouraging the growth of the Malaysian ICT industry in the areas of embedded systems, intelligent systems, wireless and mobility, and advanced software development.

**The Microsoft Innovation Center (MIC)**

A joint collaboration as part of the TDB, between the Technopreneur Development Division (TeDD), Microsoft, Hewlett-Packard (HP), and Multimedia University (MMU), the Microsoft Innovation Center (MIC) is a Center for Advanced Software Development providing.NET development related infrastructure, services, and facilities for selected promising technopreneurs. Backed by world leaders in the Information and Communication Technology (ICT) industry, the MIC is a specially designed outfit equipped to offer technopreneurs and ICT companies with potential for rapid prototyping development of best-of-breed.NET-based applications and solutions. Utilizing the complete scale of software and hardware, the MIC will focus on developing companies in the strategic focus areas of embedded systems, intelligent systems, wireless and mobility, and advanced development tools.

**Open Source Lab (Basis Bay)**

Open Source Lab is collaboration between Basis Bay Sdn. Bhd. and MDEC. Basis Bay provides the hardware and consulting while MDEC provides space and networking infrastructure for those entrepreneurs or SMEs interested in developing business using open source software technologies.

### 5.1.6 Summary of TDPs Technoprenuer Support Programs

While the above are the specific programs for the creation of support system for entrepreneurs and SMEs in the ICT sectors—some key services are identified and provided to cover all the stages of the Enterprise Life Cycle, by the Technoprenuer Development Program, like Identifying Industry Demands and Client Needs, E-Services, SYOB, Pre-Seed Funding and Incubation, Business Plan Development services, Market Development Access to Markets, Technology Laps for Prototype Development, and Technoprenuer Competency Development (Workshop and Coaching).

Figure 6 shows how all the above-mentioned services and TDP's five categories of programs fit into the Technoprenuer Value Chain or Enterprise Life Cycle and thus attempt to create a Technoprenuer Eco-System to support the development and growth of small and medium enterprises in the ICT sector in the MSC cluster and Malaysia as a whole.

## *5.2 The TDP Described as Sub-National System of Innovation for SMEs in Malaysia*

In a recent project involving UNDP and UNESCAP, the support systems that exisit in addition to that of the national system of innovation was termed as sub-national system of innovation specific to SMEs. In this section the TDP flagship of the Multimedia Super Corridor project is presented as a sub-system of the larger innovation system or as a sub-national system of innovation for SMEs in Malaysia's ICT sector. The key actors in this "sub-national system of innovation to support the Entrepreneurs and SMEs are identified as:

- Government and Government agencies
- The specific regional development Authority
- Universities or Institutes of Higher Learning

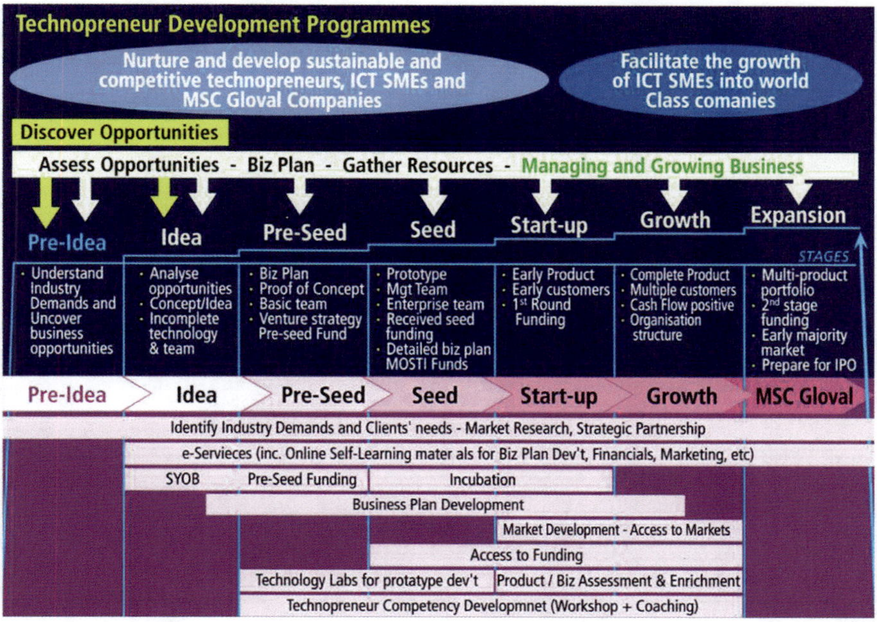


**Fig. 6** Malaysia's MTD programme across the enterprise development life cycle

- Local Firms and MNCs as Partners for Market Access and Technology
- All types of Finance Providers
- Other SMEs in the market.

The Fig. 7 illustrates the linkages in the TDP program between the various actors in the support system for SMEs to enhance their business and technology capabilities.

The role of TDP in development of SMEs in ICT Sector is seen as a facilitator in linking up the actors in the system for building up the business and technological capability of the SMEs who are in various part of the Value Chain (Technoprenuer Value Chain)/An example of its role is when a SME client seeks help, the TeDD personnel assess and then partners these ICT SMEs with companies like IBM, Sun Systems, Maxis (a Local Mobile Telecommunication Giant), all of whom have 'partnership development programs' providing technology support in terms of hardware/software and also training programs for SMEs.

In addition to the SMEs being networked through the MTD flagship program there has been a slow and steady growth of ICT SMEs leveraging on the actors in the sub-national system and creating linkages with firms and other institutions locally and overseas—sometimes on their own initiatives and income cases through interventions from the cluster players and are enhancing their technological capabilities.

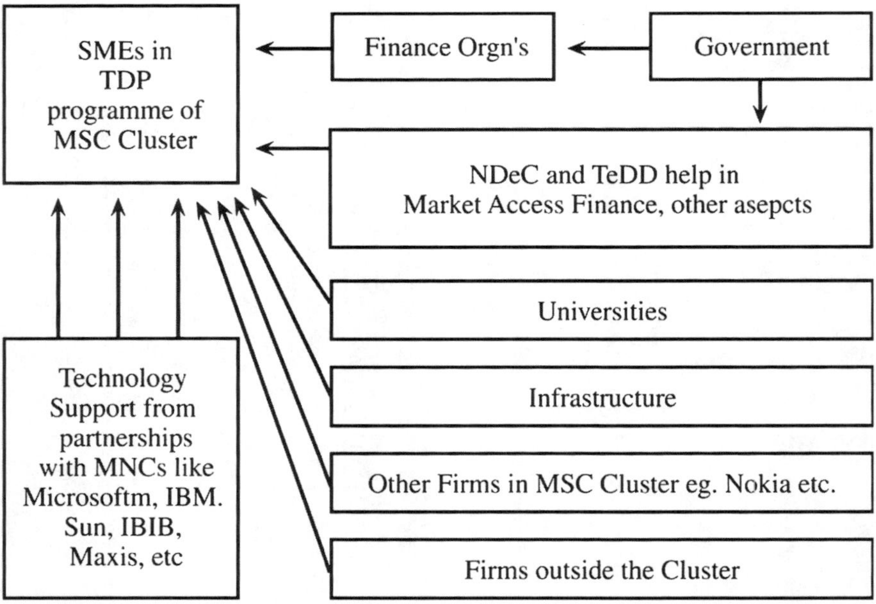


**Fig. 7** TDP flagship catalyzing actors like a support system for entrepreneurs and SMEs in Malaysian ICT sector

## 6 Conclusions and Best Practices from the MSC-TDP Flagship

The Multimedia Development Corporation (MDeC) recently reported in the local press that there are about 569 SME companies with MSC Malaysia-status operating in Cyberjaya—the specially developed technology park type city (Malaysia has a total 2,000 MSC Malaysia-status companies). These SME companies with MSC Malaysia-status posted RM6.16 billion in revenues and contributed about RM1 billion to Malaysian exports last year. The demand for office space in the especially for the purpose built buildings for these SMEs is supposed to increasing especially for start-up ventures. Currently, in the technology city of Cyberjaya, there are about 181 SME companies specially built in the technology buildings and a third center is being built. A third building for the purpose of housing SMEs and Start-ups is under completion and there has also been the second MSC–SME showcase organized this year. Incidentally, most of this growth and development of start-ups and SMEs seems to happening after the setting up of the MSC-TDP flagship indicating some success of such a program.

In addition to the numbers outcomes, interviews with the owners, and managers from some of the start-ups and SMEs reveal, that the support from MDeC/TDP has been forthcoming and crucial for technology development and also in expansion of their business to overseas markets. Again this evidence is also anecdotal but support the view that support systems in terms of infrastructure or programs involving linkages of various actors does help in the growth and development of entrepreneurial and small and medium enterprises in the technology sector.

## 6.1 *The MSC and TDP Flagship—a Sub-National Innovation System for supporting ICT SMEs in Malaysia*

The MSC regional system has been able to attract MNCs that are willing to collaborate with local SMEs for product development. It has also been fairly successful in identifying the regions strengths and embarked on a development strategy of a sector based on that—which is the shared services and outsourcing sector. But more could be done in terms of developing the interrelationships of MSC firms within the MSC itself. This will help to synergize the capabilities of the firms and organizations to make offerings to the market or to come out with innovations. In addition, there is vital need to develop linkages of MSC firms and organizations with outside research institutes, universities, and industries outside to synergies what is already there for the SMEs in the MSC cluster and also to increase the demand conditions for the MSC firms. This is where the TDP flagship program could play a role as a sub-national system of innovation focusing on the development of SMEs.

Overall it appears that Malaysia's MSC cluster, seen as regional systems of innovation, and its TDP flagship, as a sub-system, seem to have helped the development of SMEs in the ICT sector. There is also anecdotal evidence to show the SMEs are able to 'lodge' themselves in global supply chains—giving them market access and also access to technology—needs to be spurred on more. The TDP flagship where the SMEs have got linked with business consulting firms and also with technology providers is gaining popularity. Another element in the system of innovation is the rival technology (SUN, Microsoft, and Basis Bay) providers in the incubator has been helpful and supporting the development of SMEs to learn the 'tools to enhance their technological capabilities in one sphere of activity. While all the elements of a system are in place—there is need to further develop the sub-national system of innovation for SMEs more formally and get a threshold of companies are in place.

The key descriptors for the sub-national system of innovation type MSC-TDP in Malaysia are 'identification of key players/actors in the system', developing linkages among them –the partnerships between public and private organizations and also the partnership between MNCs and local business organizations for enhancing business and technological developments. Other important players in the system include research universities (for manpower and research support) and innovative funding mechanisms—which have been identified as important players in many regional systems of innovation.

In conclusion, the efforts by the government of Malaysia through its policies and institutions has been able to fill the initial lacunae viz support for SMEs, in the Multimedia Super Corridor (MSC)—a policy driven cluster-oriented project for catalyzing the development of the ICT industry. The cluster development organization for the MSC—Multimedia Development Corporation (MDeC) and its Technoprenuer Development Program (TDP) seem to support the arguments that specially developed as sub-national innovation systems (SIS)—be it tech-park style infrastructure or programs for developing interorganizational networks—can complement a more general national or regional innovation systems and help in the development of entrepreneurial and small and medium enterprises in technology sectors.

# Appendix

## Appendix A: Standard Definition of SMEs in Malaysia

**Primary Agriculture—General Definition**:

"A small and medium enterprise in primary agriculture is an enterprise with full-time employees not exceeding 50 or annual sales turnover not exceeding RM5 million."

**Specific Definitions**:

- "A micro enterprise in primary agriculture is an enterprise with full-time employees of less than five or with annual sales turnover of less than RM 200,000."
- "A small enterprise in primary agriculture is an enterprise with full-time employees of between 5 and 19 or with annual sales turnover of between RM 200,000 and less than RM 1million."
- "A medium enterprise in primary agriculture is an enterprise with full-time employees of between 20 and 50 or with annual sales turnover of between RM 1 million and RM 5 million."

Manufacturing (Including Agro-Based) and Manufacturing-Related Services (MRS)

**General Definition**:

"A small and medium enterprise in manufacturing (including agro-based) and MRS is an enterprise with full-time employees not exceeding 150 or with annual sales turnover not exceeding RM25 million."

**Specific Definitions**:

- "A micro enterprise in manufacturing (including agro-based) and MRS is an enterprise with full-time employees of less than 5 or with annual sales turnover of less than RM 250,000."

- "A small enterprise in manufacturing (including agro-based) and MRS is an enterprise with full-time employees of between 5 and 50 or with annual sales turnover of between RM 250,000 and less than RM 10 million."
- "A medium enterprise in manufacturing (including agro-based) and MRS is an enterprise with full-time employees of between 51 and 150 or with annual sales turnover of between RM 10 million and RM 25 million."

**Services Sector (including ICT)**
**General Definition**:

"A small and medium enterprise in services is an enterprise with full-time employees not exceeding 50 or annual sales turnover not exceeding RM 5 million."

**Specific Definitions**:

- "A micro enterprise in services is an enterprise with full-time employees of less than 5 or with annual sales turnover of less than RM 200,000."
- "A small enterprise in services is an enterprise with full-time employees of between 5 and 19 or with annual sales turnover of between RM 200,000 and less than RM 1million."
- "A medium enterprise in services is an enterprise with full-time employees of between 20 and 50 or with annual sales turnover of between RM 1 million and RM5 million."

Source: Bank Negara Malaysia—http://www.bnm.gov.my/index.php?ch=8&pg=14&ac=1037.

## APPENDIX B: Additional Information About Incubation Services of TDP

Benefits to Entrepreneur Tenants of the Incubator Services

Incubators that are members of the NIN would present technopreneurs with the ideal environment to nurture their ideas and transform them into successful businesses. Incubator operators will provide strategic guidance, mentoring and assistance to technopreneurs

1. Technopreneurs located in MSC Malaysia Status incubators (within or outside the MSC Malaysia zone) will be able to enjoy some of the incentives under the Bill of Guarantees.
2. Technopreneurs will be clustered and be able to leverage from the community network and Center of Expertise in terms of knowledge-sharing
3. Through joint efforts at the strategic level and project collaboration with world-class ICT companies/leading international incubators, technopreneurs will be able to gain added value in terms of knowledge, resources and new market access.

Furthermore, these networks of incubators are expected to gradually extend beyond Malaysian shores, linking up with foreign incubators and technopreneurs, with the aim of expanding the "community" of technopreneurs.

Below is a list of MSC-status incubators in Malaysia

| Incubation center | Technology focus | Location |
|---|---|---|
| MSC central incubator | ICT/multimedia | 63000 Cyberjaya |
| Kulim technology park corporation Sdn Bhd | ICT/multimedia/biotechnology | Kulim Hi-tech park |
| BT Multimedia (Malaysia) Sdn Bhd | ICT/multimedia | Cyberjaya |
| N2 N venture solutions Sdn Bhd | ICT/multimedia | Kuala Lumpur |
| YTL E-solutions Bhd | ICT/multimedia | Kuala Lumpur |
| ISpring capital Sdn Bhd | ICT/multimedia | Petaling Jaya |
| Usains tech services Sdn Bhd | ICT/multimedia | Universiti Sains Malaysia, Penang |
| Melaka K-economy incubator | ICT/Multimedia,biotechnology/ bioinformatics | Melaka |
| BioEnterprise Asia Sdn Bhd | Biotechnology/life sciences | Kuala Lumpur |
| SIRIM technology incubator | Biotechnology,electrical and electronic, ICT | SEPANG INCUBATOR CENTER |
| SKALI E-ventures Sdn Bhd | ICT/biotechnology/e-business, virtual incubator | UPM, serdang |
| Bureau of innovation and consultancy (BIP) | ICT/advanced manufacturing | Technovation park, universiti teknologi Malaysia |

Benefits to Incubator Operators

Special incentives and privileges will be awarded to incubator operators under the NIN program. These incentives include MSC-Status and special privileges under the Bill of Guarantee, namely:

- World-class telecommunications infrastructure (currently extended to incubators located in MSC designated cybercities only)
- Employment of foreign knowledge workers
- Freedom of ownership
- Income tax exemption
- Intellectual Property protection
- Guaranteed no censorship of the Internet
- Competitive telecommunications tariff
- Tendering for MSC infrastructure projects
- One-stop shop service of MDC
- Freedom to source capital

In order to attain these privileges, operators must qualify and satisfy the following eligibility criteria:

- Technology Focus
  The Incubator must house tenants that are involved in ICT/Multimedia (e.g. software, e-commerce, telecom), biotechnology and/or bioinformatics.
- Tenancy
  The Incubator must house early stage growth companies (i.e. seed-level or start-up companies, see definition below) involved in the above defined technology focus areas.

| Definition of early stage growth companies | |
|---|---|
| Seed-level companies | Entities that have been incorporated within the past two years, and have ideas or concepts that require funds for proof of concept |
| Start-up companies | Entities that may be in the process of setting up or have been in business for no more than two years and require first round funding capital for commercialization. These companies may not be generating profits yet |

- Facilities
  The Incubator must offer at least the following basic facilities:

  1. Internet access at a prescribed minimum bandwidth
  2. Shared facilities (e.g. meeting rooms, reception, audio-visual systems)
  3. Business advisory support (in-house or outsourced)
  4. Some prescribed Minimum floor space

- Funding and Fundraising
  Applicants must demonstrate the ability to provide direct investment capital or establish channels for financing seed or growth capital.
- Mentoring, Coaching and Training
  Applicants must demonstrate availability of resources in providing mentoring, coaching and training through staffing of incubatee with expert in various relevant business acumen and linkage to experienced and well endowed industry players
  Note: Incubators that are awarded MSC-Status automatically become members of the National Incubator Network. All newly formed incubators must register with the Registrar Of Companies within a month of obtaining MSC-Status.

## References

Jacobs, D., & De Man, A. P. (1996). Clusters, Industrial Policy and Firm Strategy: A Menu Approach. *Technology Anaylsis and StrategicManagement, 8*(4), 425–437.

Mohan, Avvari V., Aziz K., & Omar, A. (2004) ICT clusters as a way to materialize a national system of innovation: Malaysia's multimedia super corridor flagships. Electronic *Journal of Information Systems for Developing Countries (EJISDC),* Vol. 16, January, HK.

OECD. (1997). *The knowledge-based economy, in national innovation systems*. Paris: OECD.

Omar A. A., & Mohan, A. V., (2008). Policy-Driven Collaborative Arrangements and Competency Development—A Preliminary Survey of Organisations Participating in

Malaysia's MSC Flagships, *15th Annual Conference on Multi-Organizational Partnerships, Alliances and Networks (MOPAN)*, Boston, USA

Porter, M.E, (1995) The competitive advantage of nations. In C. Freeman (ed.). The national system of innovation in historical perspective. Academic Press Limited, UK.

Quandt, C. (1997) The emerging high-technology cluster of campinas, Brazil, IDRC Technopolis conference, Ottawa.

Quandt, C. (1998). *Developing innovation networks for technology-based clusters: The role of information and communication technologiesScience, Technology and Regional Development, Past, Present and Future Challenges*. Rio De Janeirio: International Workshop on Techno-Regions.

## Other Reference Material Sources

Unstructured Interviews with: Executives from the MSC Technopreneur Development Flagship Unit of the Multimedia Development Corporation (MdeC), Cyberjaya, Malaysia.

## Websites

The Mutlimedia Super Corridor and about the Multimedia Development Corporation http://www.msc.com.my.

The MSC Technopreneur Development Programme (TDP) Flagship http://www.technopreneurdevelopment.net.my.

Cyberview to build new tech centre for SMEs (2008), The star online, Retrieved August 13, 2008 from http://biz.thestar.com.my/news/story.asp?file=/2008/8/13/business/1790706&sec=business.

All accessed in the months of August—September 2008.

# Innovation and Networks in Industrial Clusters: Empirical Results from Three Large Cities

**Young Sub Kwon**

**Abstract** This chapter examines innovation activities and innovative networks in the strategic industries of three metropolitan cities of Korea, and draws policy implications for reinforcing clustering and networking. It provides the results of a survey of 180 businesses in Busan, Gwangju, and Daejeon.

## 1 Introduction

Innovation and knowledge creation are recognized as the engine for national and regional growth and development in the twenty first century. However, with the progress of globalization and industrial restructuring, many regions in industrialized countries have been faced with challenges over the past several years. These challenges have been comparatively tough in large cities of the countries which have been successfully industrialized. In recent years, these regions have been experiencing various kinds of structural readjustment, there is an urgent need to restructure the industries, and develop high value added new products, and new processes in these regions.

Innovation has a crucial influence on regional growth as well as the growth of enterprises. Accordingly, not only enterprises but also regional and central governments have interest in factors that influence innovation. In creating innovation, the innovative behavior of enterprises and institutional framework of regions are important. This is because the innovative results of enterprises depend on the innovative corporate behavior, or R&D activities, and innovative environment of the region that supports them. In regions where the regional innovative system functions in a systematic way, the innovation between enterprises and knowledge suppliers

Y. S. Kwon (✉)
Korea Research Institute for Human Settlements, 254 Simin-daero,
Dongan-gu, Anyang-si 431-712 Gyeonggi-do, Korea
e-mail: yskwon@krihs.re.kr

D.-S. Oh and F. Phillips (eds.), *Technopolis*,
DOI: 10.1007/978-1-4471-5508-9_14,  

can be created interactively, which is supported by policy decision-making agencies, technology transfer organizations, consulting service providers, brokers, etc.

Regional metropolitan cities of the country such as Busan, Gwangju, and Daejeon lack R&D activities of the enterprises and entrepreneurship since they developed through the mass production and export-oriented industrialization at the early stages of industrialization. Besides, since the proportion of branch plants is high in these cities, they are not regarded as highly innovative as in the innovative clusters in advanced countries. Despite that, they are trying to get over the regional crisis that is unfavorable to innovation, and strengthen innovation through the new types of strategic industries of the region.

The purpose of this chapter is to examine innovation activities and innovative networks in the strategic industries of three metropolitan cities of Korea, and to draw policy implications for reinforcing clustering and networking. To this end, the chapter provides the results of a survey of 180 businesses in Busan, Gwangju, and Daejeon. The structure of this chapter is as follows: The next section gives the concept of innovation and cluster in brief. The third section briefly explains the background of these regions and the methodology adopted for the study, and analyzes the survey results. The last section summarizes the key findings of the study and suggests policy implications for the future development of the strategic industries of the three metropolitan cities.

## 2 Innovation and Clusters

Innovation can be defined in various ways. Schumpeter provides five definitions of innovation as follows: (1) introduction of new products or qualitative change in existing products, (2) process innovation new to the industry, (3) opening of a new market, (4) development of new sources of supply for raw materials or other inputs, and (5) change in industrial organization.

Since the days of Schumpeter, innovation has move to the heart of economic policy-making. The European Commissions Directorate XIII, which is responsible for science and technology of Europe, defines innovation as follows: The commercially successful exploration of new technologies, ideas, or methods through the introduction of new products or processes, or through the improvement of existing ones.

Oslo Manual of the Organization for Economic Cooperation and Development (OECD 1992, 1997) provides a precise definition of technological product and process innovation which is useful for a standardized survey (Holbrook et al. 2000). The OECD definition of innovation classifies the concept of 'new' into three levels: new to the world, new to a nation, and new to the enterprise.

Many theoretical approaches have been suggested to explain the sources, characteristics and determinants of innovation. For example, behavioral theorists have highlighted the important influence and effect of uncertainty on decision-making of a firm. Structural theorists suggest that the structure of an entire industrial system has a key role in influencing the innovative behavior of the firm (Westhead and Batstone 1998). The institutional approach examines the relationships between national institutions of finance, education, law, science and technology, corporate activities and government policies, and their influence on the propensity for innovation (Nelson 1993). The relational approach analyzes the nature of business and social relationships in nations, manifested, for example, in the way user-supplier links encourage shared learning (Lundvall 1992).

Innovation is not an activity of a single firm; it increasingly requires an active search involving several firms to tap new sources of knowledge and technology, and apply these to products and production process (Guinet 1999). In other words, innovation is a result of an interactive learning process that involves often several actors from inside and outside the companies (EC DG XIII and XIV 1996; Simmie 2001). The current focus on knowledge has combined with the interactive theory of innovation-led to the analysis of specific factors which determine successful innovations or which influence the absorption of knowledge created outside the firm (Schibany and Schartinger 2001).

The "cluster concept" is one of these factors studied by the OECD focus group. According to the OECD report (1999), clusters are networks of interdependent firms, knowledge-producing institutions (universities, research institutes, technology-providing firms, etc.), bridging institutions (e.g., providers of technical or consultancy services), and customers, linked to a production chain which creates added value. The main idea of a cluster is that it is considered to be better equipped to succeed in the market place than an isolated company. The 'agglomeration externalities and positive feedback,' or that enterprises within industrial clusters have advantage in terms of growth speed or innovation over those that are isolatedly located, have been verified (Swann et al. 1998).

The enterprises within clusters are located within close proximity to each other; for this, the search cost of customers and relevant enterprises can be reduced. Further, face-to-face contacts appear to be very important as sources of technological information and in the exchange of tacit knowledge. Spatial proximity greatly enhances the possibility of such contacts. Geographical proximity is typical of clusters-although it is not absolutely necessary (Rouvinen and Ylae-Anttila 1997). Because the cooperation between actors enhances mutual trust, this industrial agglomeration of producers, customers, and competitors promotes efficiency and increases specialization.

Learning through networking and by interacting is seen as the crucial force pulling firms into clusters, and the essential ingredient for the ongoing success of an innovative cluster (Breschi and Malerba 2001). The ways enterprises learn in clusters are by embracing user-producer relationships, formal and informal collaborations, inter-firm mobility of skilled workers, and the spin-off of new firms from existing firms, universities, and public research centers (Breschi and Malerba 2001). In particular, universities and research institutes, as producers of new knowledge, may play a crucial role.

According to Asheim and Cooke (1999), there are two types of innovation networks; one is the endogenous innovative network, which is based upon a preexisting regionally or locally delineated cluster of small and medium-sized enterprises. They have a lengthy tradition of interacting and learning from one another, successfully competing on the basis of, as needed, cooperative innovation practices. Examples of such endogenous innovative networks are to be found in southern Germany (e.g., Baden Wuerttemberg) and the Third Italy (e.g., Tuscany or Emilia-Romagna).

The other is the exogenous innovative network, which takes the form of technopoles or science parks. They tend to emerge under two kinds of circumstances: (a) when large firms fragment their production structure and relocate R&D activities to functionally specialized zones where synergies are expected to arise from collocation (as in Sophia Antipolis or Lille in France), or (b) by planned innovative milieu established to promote collaboration between universities and SMEs (as in science parks in the UK and USA).

Industrial clusters are often localized, giving rise to networks and specific innovation patterns in regions. Regions differ in their preconditions for innovation such as qualification of the labor force, universities, research institutions, technology-based firms, knowledge externalities, and spillovers. Cooperation in clusters has increasingly become a requirement for success. Without cooperation, firms almost never innovate in isolation (Roelandt and den Hertog 1999). Moreover, cooperation offers a direct way to improve economic performance and reduce costs (Guinet 1999). Many of these embedded factors in regions are immobile, giving some regions advantages over others.

In many countries, clusters of innovative firms are driving growth and employment (Guinet 1999). They are more concentrated in some cities rather than others (Simmie 2001). That is why innovation is dependent on high quality professional and technical labors. High quality labors are one of the stickiest local factors of production. The primary cities in each regional economy therefore tend also to be the major national concentrations of such labor. Scale advantages in large cities, urbanization, and localization economies are seen as the main reasons for the clustering of innovative activities in metropolitan regions.

## 3 A Descriptive Analysis of the Survey Results

### *3.1 Background and Methodologies*

#### 3.1.1 Background

This chapter is focused on the innovation activities and clustering of the mechanical parts and materials industry of Busan, Information and Communication Technology (ICT) industry in Daejeon and photonics industry of Gwangju. The industrial clusters of Korea can be found in the regions which have the national or regional industrial complexes. Currently, there are 585 industrial complexes with 30 national complexes, 213 regional complexes, and 342 rural complexes (Kim et al. 2006). In particular, most of the metropolitan cities of Korea have one or more national or regional industrial complexes. The case study regions of this paper, Busan, Daejeon, and Gwangju also have national and regional industrial complexes within them.

In the Busan region, which has a strong agglomeration of footwear, apparel, and mechanical parts and materials industries, for example, the mechanical parts and materials industry is one of the main industries of the local industrial complexes. While the enterprises in the industry are located within the industrial complexes of the city, many of them are concentrated in Sasang-gu and Saha-gu. While Daejeon has the National Science and Technology Complex for Research Institutions, Gwangju has the National High-tech Industrial Complex for Photonics-related industry. The ICT enterprises of Daejeon are mostly clustered in Yuseung-gu and Daeduck-gu. The photonics enterprises of Gwangju are mostly located in Buk-gu and Gwangsan-gu. The case studies of the paper, the mechanical parts and materials industry of Busan, photonics industry of Gwangju, and ICT industry of Daejeon have been selected from the strategic industries that have formed clusters centering around the principal industrial complexes of each of the three cities.

Busan, which is located at the hub of the marine transportation of Korea, has the population of some 3.67 million, which accounts for 7.4 % of the entire population of the nation. The mechanical parts and materials industry includes part of the new materials, mechatronics, shipbuilding, mechanical equipment, and steel and iron industry. In Busan, there are more than one thousand enterprises in these industries, with some 25 thousand workers employed in the companies. The sales per worker of the companies surveyed are some ₩ 490 million, which is comparatively high; however, the exports per worker are only ₩ 30 million, revealing that they are simply a domestic consumer-oriented industry. The mechanical parts and materials industry of Busan lacks key leading enterprises, which results from the nature of the industry, and is composed mostly of small and medium-sized enterprises. In addition, supporting institutions such as public research institutions or agencies for technical support, which may lead the technical innovation, are insufficient considering the scale of the cluster.

Daejeon city, located in the central part of the nation, is the fifth largest city in the country among Korea's top seven large cities including Seoul, Busan, Incheon, Daegu, Gwangju, and Ulsan. The population is some 1.45 million, and accounts for 3.1 % of the nation's population. Daejeon is the transportation hub of the country, connecting the whole country from east to west, and from north to south, and easily accessible from anywhere of the country. The Daejeon city is the center of knowledge and information, and situated at the second administrative capital of the country. Besides, Daejeon is home to the Daedeok Valley, the cradle of the high-tech industry of the country, where the Daedeok Science Town—the country's R&D park, is located. This is where the country's leading science and technology think-tank and research institutes are situated with top class human resources concentrated in it—16,000 personnel here hold master's or doctorate degrees with 43 % doctorate degree holders. They work for the government or private enterprises as researcher or higher education researchers, or work for venture businesses or supporting institutions. The ICT industry of Daejeon involves the information and telecommunications components, equipment, software and contents, and services. As of 2004, some 12,000 workers are employed in some 580 enterprises. The sales per worker of the companies surveyed are not high, or ₩ 350 million,[1] with the exports per worker amounting to only 30 million, and they have a profit model that is domestic consumer-oriented.

Gwangju, in the southwest part of the country, is one of the seven major cities of the country coming in sixth in size. It has some 1.41 million populations, taking 3 % of the whole population. The National High-tech Industrial Complex of Gwangju is characterized as a cluster of small firms which have been relocated recently from other parts of the nation, the Seoul Metropolitan Area (SMA), in particular, and spun-off firms, R&D institutes and universities such as Advanced Photonics Research Institute, Gwangju Institute of Science and Technology, private enterprises such as LG Innotech, etc. Since the duration of the clustering is shorter than that of Busan or Daejeon, as of 2004, some 5,000 workers are employed in some 230 enterprises. The sales per worker of the companies surveyed are not high, or ₩ 340 million; however, the exports per worker are ₩ 160 million. This makes the city the most export-oriented of the three case study clusters. The Busan and Gwangju regions do not have enough number of public research institutes that can support the R&D activities of the enterprises compared to the research institute-clustered Daejeon.

### 3.1.2 Methodology

The surveys on the companies were conducted between July and August of 2005. The total number of the surveyed companies is 180 from Busan, Daejeon, and Gwangju, and the surveys provided the base for the following analysis. For the purpose of the innovation survey in three regional strategic industries, innovative firms are defined as those which have introduced a technologically new product or process during the previous 3 years. The survey covers not only the production network but also innovation characteristics and network. However, this paper restricts the scope of the current analysis to the innovation characteristics and network in the three regions' strategic industries (Fig. 1).

### 3.1.3 Company Attributes

Table 1 shows the characteristics of the companies surveyed. Overall, 42 % of the samples were established before the 1997 financial crisis, with 37 % between 1998 and 2000, and 21 % after 2001. Among the companies surveyed in Busan, those that were established before the financial crisis took the most part while in

[1] KRW for USD exchange rate in March 2013 (approx. ₩ 1,000: $0.88).

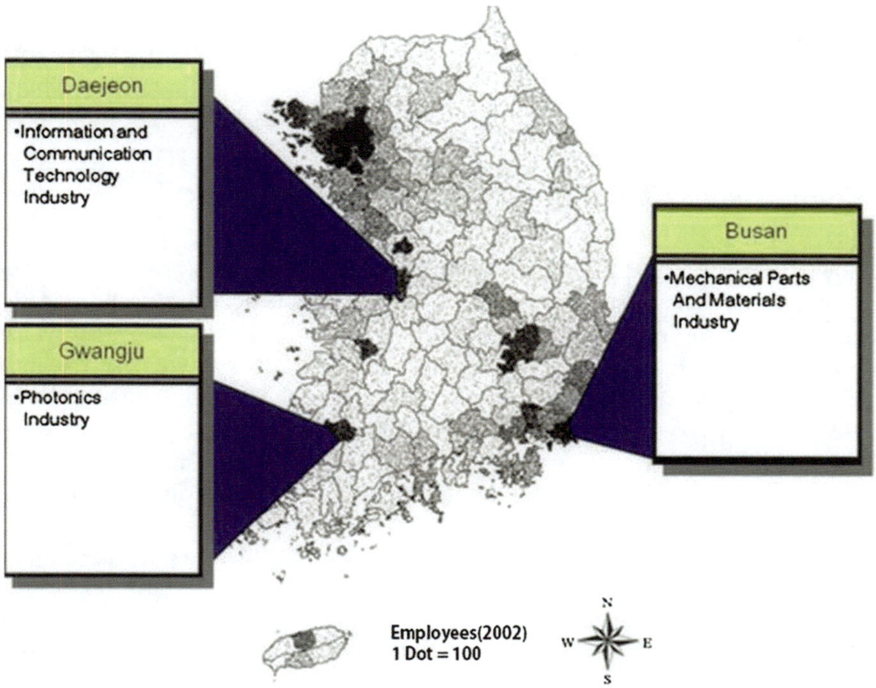


**Fig. 1** Major strategic industries in three large cities. *source* (Kwon et al. 2005)

**Table 1** Characteristics of the sample (percentages)

| Characteristics | Busan | Gwangju | Daejeon | Total sample |
|---|---|---|---|---|
| (a) Year firms started | (N = 60) | (N = 29) | (N = 56) | (N = 145) |
| Before foreign currency crisis | 43(71.7) | 3(10.3) | 15(25.0) | 61(42.1) |
| 1998–2000 | 5(8.3) | 13(44.8) | 35(58.3) | 53(36.5) |
| After 2001 | 12(20.0) | 13(44.8) | 6(10.0) | 31(21.4) |
| (b) Size of firms | (N = 60) | (N = 57) | (N = 59) | (N = 176) |
| Micro < 20 employees | 9(15.0) | 39(68.4) | 30(50.8) | 78(44.3) |
| Small 21–100 | 38(63.4) | 16(28.0) | 27(45.8) | 81(46.0) |
| Medium 101–300 | 8(13.3) | 1(1.8) | 2(3.4) | 11(6.3) |
| Large > 300 employees | 5(8.3) | 1(1.8) | 0(0.0) | 6(3.4) |

*Source*: (Kwon et al. 2005) survey results

Gwangju, companies that were set up after 2001 were surveyed the most. This is reflected on the size of the companies. In Gwangju and Daejeon, the proportion of small-sized enterprises is higher than the average while in Busan those of small and medium-sized, and large enterprises are high.

Despite that, in the mechanical parts and materials industry of Busan, most of the companies are small and medium-sized with large ones taking only 8 %. These companies were established before 1997. The oldest company of those surveyed was founded in 1960. The majority of the small and medium-sized businesses in the mechanical parts and materials of Busan have long history, while companies in the photonics industry of Gwangju and ICT industry of Daejeon are newly founded.

### 3.2 Innovation Characteristics and Innovation Types

Innovation plays a crucial role in the performance of the enterprise. How innovative are the companies in the surveyed regional strategic industries? In accordance with the survey purpose, the innovative enterprise is defined as that which has introduced technologically new products or processes over the past 3 years.

In the study on the 180 companies located in the three metropolitan cities in Korea, the proportion of the innovative enterprises is very high in Gwangju and Daejeon while in Busan, that of innovative firms is only 40 %. Table 2 shows that there are significant differences in the proportion of innovative firms by regional strategic industries.

The innovation types introduced in early 2000s were either product innovations (88 %) or process innovations (12 %). The proportion of product innovation turned out to be high in Gwangju and Daejeon while in Busan that of process innovation appeared high. The mechanical parts and materials industry of Busan is stressing process innovation with some 40 % of the companies introducing innovation. In the mechanical parts and materials industry of Busan, process innovation is considered relatively more important in regions where mature industries are common. In the ICT industry of Daejeon and photonics industry of Gwangju, product innovation is more frequent than process innovation while in the mechanical parts and materials industry of Busan, product innovation is comparatively less frequent.

The OECD definition of innovation has three levels in terms of "new": new to the world, new to a nation, and new to the firm. "New" is necessary but not sufficient for innovation. For a product or process to be innovative, it must have a

**Table 2** Characteristics of innovations (percentages)

| Characteristics | Busan | Gwangju | Daejeon | Total sample |
|---|---|---|---|---|
| Experience of innovation | (N = 60) | (N = 60) | (N = 60) | (N = 180) |
| Yes | 25(41.7) | 53(88.3) | 57(95.0) | 135(75.0) |
| No | 35(58.3) | 7(11.7) | 3(5.0) | 45(25.0) |
| Type of innovation | (N = 25) | (N = 53) | (N = 57) | (N = 135) |
| Product | 15(60.0) | 49(92.4) | 55(96.5) | 119(88.2) |
| Process | 10(40.0) | 4(7.6) | 2(3.5) | 16(11.8) |
| Novelty of innovation | (N = 25) | (N = 53) | (N = 57) | (N = 135) |
| New to firm | 5(20.0) | 8(15.1) | 8(14.0) | 21(15.6) |
| New to the nation | 15(60.0) | 29(54.7) | 41(71.9) | 85(63.0) |
| New to the world | 5(20.0) | 16(30.2) | 8(14.0) | 29(21.5) |

*Source*: (Kwon et al. 2005) survey results

sense of uniqueness to it. This does not mean every innovation must be a world first (Holbrook and Hughes 2000).

In the strategic industry of the three metropolitan cities, only 21 % of the new products were new to the world. At present, in the metropolitan regions, the level of innovation of these enterprises is low and this can be explained by the fact that they are at the early stage of strengthening their strategic industry base. In the Busan mechanical parts and materials industry, 20 % of the enterprises have developed products that are new to the world over the past 3 years. In the Daejeon ICT industry, the proportion decreases to 14 and goes up to 30 % in the Gwangju photonics industry. These innovations are quite interesting since they show the innovation system that produces innovation and the characteristics and competence of the enterprises. This also means that even the novelty of innovation can be differentiated by regional industry.

## *3.3 Innovation Networks*

### 3.3.1 Source of Innovation

In the process of innovation, enterprises utilize various kinds of information sources. In most cases, knowledge cannot be obtained from explicit information alone; therefore, the most efficient means to link diverse activities in the process of innovation is the direct long-term contact among individuals who have the capacity to transfer knowledge. In particular, the long-term personal contact such as cooperative projects or researches is a crucial element for the acquisition and dissemination of tacit knowledge.

As seen in Table 3, the source of innovation originates mostly from cooperative projects or researches. A 45.6 % of the corporations in the Daejeon ICT industry, 54.7 % in the Gwangju photonics industry, and 64 % in the Busan mechanical parts and materials industry consider the experiences from the process of the

**Table 3** Sources of innovations in the innovation process

| Category | Busan | Gwangju | Daejeon | Total |
|---|---|---|---|---|
| Patent information, technology information, paper, Internet, etc. | 1(4.0) | 10(18.9) | 12(21.1) | 23(17.0) |
| Experience from cooperative projects or researches | 16(64.0) | 29(54.7) | 26(45.6) | 71(52.6) |
| Formal consulting | 5(20.0) | 4(7.5) | 10(17.5) | 19(14.1) |
| Formal contacts | 3(12.0) | 10(18.9) | 9(15.8) | 22(16.3) |
| Informal contacts | 0(0.0) | 0(0.0) | 0(0.0) | 0(0.0) |
| Total | 25(100.0) | 53(100.0) | 57(100.0) | 135(100.0) |

*Notes*: considering multiple replies. Percentage of the value added after putting weight value (units: number of firms, %)
*Source*: (Kwon et al. 2005) survey results

cooperative projects or researches to be the most important source of innovation. The result that cooperative projects or researches are the most important source of innovation is not surprising.

The second most important source of information for the Gwangju photonics industry and Daejeon ICT industry, when it comes to innovation, is the patent information, technology information, paper, Internet, and others. A 21.1 % of the surveyed ICT-related enterprises in Daejeon regarded explicit knowledge to be the second most important source. The Gwangju photonics-related enterprises cited explicit knowledge and formal contact as the second most important source. In Busan, formal consulting was cited as the second most important source for innovation. As is the case in the experience of innovation, the Busan mechanical parts and materials industry shows different tendency from the other two industries in Daejeon and Gwangju regarding the source of innovation. Table 3 also shows that informal contacts with ex-colleagues and friends seldom play a crucial role.

### 3.3.2 Innovation Partner

This part deals with the key actors among innovation partners in the strategic industry cluster in the three regions. Enterprises were asked to indicate if they are associated with the following categories of innovation partners: suppliers, clients, supporting agencies, universities, and research institutes. The respondents had to consider only those partners who actually made a contribution to their innovation creation. In general, customers are more important than suppliers who often initiate modifications of products or development of new products (Kaufmann and Tödtling 2000). Sometimes they are even involved in the development process itself (Kaufmann and Tödtling 2000). Besides buyers and suppliers, knowledge providers and service providing firms such as universities, research organizations, and consultants are important innovation partners.

In the strategic industries of the three metropolitan cities, customers are the most important innovation partners. Though the principal value chain of the strategic industry of the three cities are the user-supplier relationship, the significance of universities as innovation partner is very low. In substance, no matter what type of R&D is carried out, in the innovation projects, the public research labs are also more important than universities as innovation partner.

In the Daejeon ICT industry, the contact with research institutes is very strong due to the strong ICT knowledge base such as the ETRI. In the Daejeon region, 28.3 % of the enterprises performed innovation activities in collaboration with public research institutes. In the Busan mechanical parts and materials industry, suppliers are as important, as cooperative partner, as the Gwangju and Daejeon research institutes in the process of innovation. Research institutes are less favored as innovation partner by Busan enterprises than those in Gwangju or Daejeon (Table 4).

A possible explanation for this is the difference in the institutional environment and industrial characteristics. The reason the Busan enterprises less utilize universities and public research institutes as innovation partner than product-

**Table 4** Innovation partner

| Category | Busan | Gwangju | Daejeon | Total Sample |
|---|---|---|---|---|
| Universities | 5.0 | 13.3 | 8.3 | 8.9 |
| Research institutes | 11.7 | 20.0 | 28.3 | 20.0 |
| Customers | 50.0 | 51.7 | 45.0 | 48.9 |
| Suppliers | 21.7 | 0.0 | 1.7 | 7.8 |
| Supporting agencies | 8.3 | 11.7 | 13.3 | 11.1 |
| Others | 3.4 | 3.3 | 3.3 | 3.3 |
| Total | 100.0 | 100.0 | 100.0 | 100.0 |

*Source*: (Kwon et al. 2005) survey results (units: %)

innovating companies in other clusters, can be that the research institutes are not sufficient in the region; that companies here are not capable enough to utilize them; or that they do not recognize the need to cooperate. The relevance between universities and public research institutes as key actors in the innovation network is also determined by the institutionalization of the R&D within the region.

In Daejeon and Gwangju, public policies are dominant which are targeted at the fostering, subsidizing, or supporting the collaboration between public institutions and business organizations. Under the circumstances, it is natural to find that public research institutes are the most frequently used collaboration partners in the innovation process in the Daejeon ICT industry and Gwangju photonics industry.

### 3.3.3 Relationship Between Innovation and Cooperation

Enterprises with a high propensity to collaborate with other partners are more likely to experience innovation. Therefore, it is not surprising to have the result that R&D cooperation plays a crucial role in creating innovation for businesses. This corresponds to the study results that, in Busan, companies with experiences in cooperation successfully created innovation while those with no experiences hardly did (Tables 5, 6).

In the process of innovation of the Busan mechanical parts and materials industry, the R&D cooperation between customers and universities are highly important, in particular. However, with the Daejeon ICT industry and Gwangju photonics industry, the R&D cooperation occurs in a very natural way, and is not a sufficient condition for innovation.

### 3.3.4 Spatial Levels of the R&D Cooperation

What are the relevant spatial levels of the R&D cooperation networks? It can be classified into five types: Local, regional, SMA, national, and worldwide level. For the Daejeon ICT firms, while the R&D cooperation with universities or research institutes is concentrated in the local level, the customers are located in the SMA. For the Gwangju photonics firms, the R&D cooperation of universities and

**Table 5** Relationship between innovation and cooperation with clients

| Category | With clients | Innovator | Non-innovator | Chi square | *P* value |
|---|---|---|---|---|---|
| Busan | Cooperation | 14(63.6) | 8(36.4) | 6.898 | 0.014[a] |
| | Non-cooperation | 11(28.9) | 27(71.1) | | |
| Gwangju | Cooperation | 26(96.3) | 1(3.7) | 3.021 | 0.116 |
| | Non-cooperation | 27(81.8) | 6(18.2) | | |
| Daejeon | Cooperation | 41(93.2) | 3(6.8) | 1.856 | 0.328 |
| | Non-cooperation | 13(81.3) | 3(18.8) | | |

[a] significant at the 5 % level (unit: number of firms, %)
*Source*: (Kwon et al. 2005) survey results

**Table 6** Innovation and cooperation with universities

| Category | With universities | Innovator | Non-innovator | Chi square | P-value |
|---|---|---|---|---|---|
| Busan | Cooperation | 10(66.7) | 5(33.3) | 5.143 | 0.035[a] |
| | Non-cooperation | 15(33.3) | 30(66.7) | | |
| Gwangju | Cooperation | 49(90.7) | 5(9.3) | 3.037 | 0.140 |
| | Non-cooperation | 4(66.7) | 2(33.3) | | |
| Daejeon | Cooperation | 36(92.3) | 3(7.7) | 0.659 | 0.655 |
| | Non-cooperation | 18(85.7) | 3(14.3) | | |

[a] significant at the 5 % level (unit: number, %)
*Source*: (Kwon et al. 2005) survey results

research institutes is focused on the regional level while the customers are situated in the SMA. In the Busan region, all partners for the R&D cooperation are within the region. As a result, while the Busan mechanical parts and materials industry succeeded in achieving the critical mass at the region level, with the Daejeon ICT industry and Gwangju photonics industry, inducing the customers into the region can be a major strategy for the development of the cluster since the customers to cooperate R&D are mostly located in the SMA (Table 7).

### 3.3.5 Source of Tacit Knowledge

Innovation activities involve a great deal of interactions with external sources of knowledge and experiences. Innovation depends on knowledge and assimilation of information through learning and cooperation. Know-how transfer requires personal interactions trough exchanges, training, seminar, cooperative projects, and cooperative work performance. By its very nature, tacit knowledge cannot be written down; therefore it must be acquired by learning and experience, and after that it becomes embodied in a person or organization. This type of knowledge can be achieved by human mobility and personal exchanges through cooperation. These are important instruments for knowledge dissemination.

Table 8 shows 40 % of the companies responded that relevant know-how is transferable to other possible cooperation partners through exchanges among

**Table 7** R&D cooperation

| Category | | Domestic | | | | Overseas | Total |
|---|---|---|---|---|---|---|---|
| | | Local | Regional | Seoul metropolitan area | National | | |
| Busan | Suppliers | 1(4.8) | 11(81.0) | 1(4.8) | 3(14.3) | 2(9.5) | 18(100.0) |
| | Clients | 2(9.1) | 13(59.1) | 5(22.7) | 4(18.2) | 1(4.5) | 25(100.0) |
| | Competitors | 2(18.2) | 5(45.5) | 1(9.1) | 2(18.2) | 1(9.1) | 11(100.0) |
| | Universities | 4(6.7) | 8(53.3) | 2(13.3) | 1(6.7) | 1(6.7) | 16(100.0) |
| | Research institutes | 1(6.7) | 7(46.7) | 2(13.3) | 3(20.0) | 3(20.0) | 16(100.0) |
| Gwangju | Suppliers | 9(31.0) | 5(17.2) | 9(31.0) | 2(6.9) | 4(13.8) | 29(100.0) |
| | Clients | 6(21.4) | 5(17.9) | 14(50.0) | 0(0.0) | 3(10.7) | 28(100.0) |
| | Competitors | 8(30.8) | 6(23.1) | 6(23.1) | 5(19.2) | 1(3.8) | 26(100.0) |
| | Universities | 20(32.3) | 33(53.3) | 6(9.7) | 3(4.8) | 0(0.0) | 62(100.0) |
| | Research institutes | 10(20.8) | 26(54.2) | 6(12.5) | 6(12.5) | 0(0.0) | 48(100.0) |
| Daejeon | Suppliers | 7(15.2) | 14(30.4) | 9(19.6) | 8(17.4) | 8(17.4) | 46(100.0) |
| | Clients | 5(9.1) | 12(21.9) | 22(40.0) | 10(18.2) | 6(10.9) | 55(100.0) |
| | Competitors | 7(17.1) | 20(48.8) | 3(7.3) | 8(19.5) | 3(7.3) | 41(100.0) |
| | Universities | 19(40.4) | 21(44.7) | 6(12.8) | 1(2.1) | 0(0.0) | 47(100.0) |
| | Research institutes | 30(49.2) | 22(36.0) | 6(9.8) | 3(4.9) | 0(0.0) | 61(100.0) |

*Source*: (Kwon et al. 2005) survey results (units: number of firms, %)

**Table 8** Tacit knowledge exchange in the innovation process

| Category | Busan | Gwangju | Daejeon | Total sample |
|---|---|---|---|---|
| Exchanges with the customers | 27.3 | 12.3 | 22.6 | 20.7 |
| Exchanges with the suppliers | 13.4 | 5.6 | 3.3 | 7.4 |
| Exchanges with the competitors | 11.6 | 9.7 | 11.7 | 11.0 |
| Seminars and discussions | 2.3 | 18.5 | 9.7 | 10.2 |
| Joint R&D with universities and research institutes | 13.0 | 19.3 | 23.8 | 18.7 |
| Informal exchanges among researchers | 1.9 | 12.3 | 7.0 | 7.1 |
| Hiring experienced personnels from the same industry | 2.3 | 3.5 | 0.2 | 2.0 |
| Organizational activities for promoting knowledge exchanges | 0.5 | 4.7 | 2.7 | 2.6 |
| Specialized service providers | 0.9 | 0.0 | 0.6 | 0.5 |
| In-house | 25.9 | 14.0 | 18.1 | 19.3 |
| Others | 0.9 | 0.0 | 0.2 | 0.4 |
| Total | 100.0 | 100.0 | 100.0 | 100.0 |

*Notes*: considering multiple replies. Percentage of the value added after putting weight value (unit: %)

*Source*: (Kwon et al. 2005) survey results

firms. In the Busan mechanical parts and materials industry, the most important source of information when exchanging tacit knowledge is the customers. A 27.3 % of the surveyed companies regarded customers as the most important actor. Only a few responded seminars, discussions, and informal exchanges among researchers are the most important source of tacit knowledge in the innovation process.

When exchanging tacit knowledge, the most important source of knowledge is the joint R&D among universities and research institutes for the Daejeon ICT firms. A 23.8 % of the surveyed firms considered the exchanges among universities and research institutes as the most important source. While the exchanges with the customers were cited as the most important relationship for the Busan mechanical parts and materials businesses, they were regarded as the second most important for the Daejeon ICT industry. Table 8 also shows that while the seminars and informal exchanges among researchers are crucial for the Gwangju photonics industry, they do not play any significant role for the Busan mechanical parts and materials industry.

# 4 Summaries and Conclusions

## *4.1 Summaries*

The purpose of this chapter is to compare and analyze the innovation capacity, innovation network, and source of innovation of the regional strategic industries in the three metropolitan cities, and to suggest policy implications for strengthening clustering and networking. In order to examine the status of innovation and networks of enterprises, a firm level survey was conducted among the strategic industries of the three metropolitan cities.

Most of the firms in the Busan mechanical parts and materials industry have a longer history than those in Gwangju and Daejeon with the sales of their primary products two to three times higher. However, the innovation capacity of most of the firms is low compared to those in other regions, and the innovation activities are also highly concentrated in the enhancement of the manufacturing process rather than in the new product introduction.

In contrast, companies in Gwangju and Daejeon were newly founded after the financial crisis with a high proportion of R&D investment. They are more experienced in innovation, and most of the innovation activities are related to the development of new products. For the Busan enterprises, differentiated support should be considered according to each individual company's characteristics since their innovation activities are different—this results from the difference in the companies' employment size and characteristics.

On the other hand, in Gwangju and Daejeon, small and medium-sized firms are common, and the level of cooperation and interactions among them is higher than in Busan. Research institutes play a more crucial role for innovation in Daejeon than in Gwangju. The proportion of new products in the total sales is comparatively high, and a number of firms experience innovation in Daejeon. It can be assumed that this is why the cooperation and knowledge spillover between research bodies such as universities and research institutes, and local businesses play a decisive role in the regional innovation system.

In summary, the strategic industry cluster of the three metropolitan cities has the following in common: First, the premier industry was selected that is the largest in scale and most competitive out of the four strategic industries chosen by region; second, most of the firms are located in the industrial complex or business-agglomerated location; third, the proportion of product innovation is higher than that of process innovation; fourth, the source of innovation in the process of the corporate innovation is the experiences achieved from the joint projects or researches, and informal contacts hardly constitute the innovation source; and lastly, customers as innovation partner play a decisive role in creating innovation for each strategic industry.

On the other hand, the mechanical parts and materials industral cluster of Busan is different from the Daejeon ICT and Gwangju photonics industrial cluster. Firstly, while the businesses of Busan were founded before the financial crisis in 1997, those of Daejeon and Gwangju were set up after the crisis. Secondly, while the proportion of innovation experience of the businesses in Busan was 40 %, that of those in Daejeon and Gwangju was high, or 90 %. Thirdly, for the Busan businesses, open sources such as patent information, technology information, paper and Internet was not important as innovation source; however, it was the second most important factor for the Daejeon and Gwangju businesses. Fourth, for the Busan enterprises, suppliers were the second most important innovation partner while for those in Daejeon and Gwangju, it is the research institutes that are the second most important innovation partner.

Fifth, while cooperation has significant influence on innovation for the Busan businesses, it hardly affects innovation for the companies in Gwangju and Daejeon. Sixth, for the Busan firms, most of the R&D collaboration is carried out within the region no matter who the partner is. For the Gwangju and Daejeon enterprises, R&D collaboration with universities and research institutes is carried out within the region, with customers is conducted in the SMA. This means that the major customers of the Gwangju and Daejeon industries can be found in the SMA; therefore, it is hard to say the clusters have achieved the critical mass. Seventh, while the Busan companies exchange the tacit knowledge with the customers in the process of innovation, those in Gwangju and Daejeon do it in the course of the joint research with universities and research institutes.

Based on these results, it can be said that, the Busan mechanical parts and materials industrial cluster has achieved the critical mass, but that it lacks the leading enterprise which can play a key role for the regional economy, with the venture companies being sluggish in the region. Since the interactions and collaboration are weak among firms, it is proper to say that it is at the early stage of the cluster development at which the base for innovation is provided.

As for Gwangju photonics industrial cluster, the customers exist in the SMA, and the level of critical mass is not satisfactory. Therefore, despite diverse promising bases for innovation, it remains at the early stage of the cluster formation. In terms of the R&D cooperation network, while the research and innovation activities among universities and research institutes occur within Gwangju, it depends on the SMA for the research and innovation activities with the customers.

Compared to the above regions, in Daejeon ICT industrial cluster, there exist abundant resources for innovation such as research institutes, venture companies, universities and supporting institutions. However, since Daejeon does not reach the critical mass which can realize self-reliance, the role of leading enterprises which can provide the backward and forward linkages to local firms, would be critical for future strategies. Concerning practical strategies, Daejeon need to utilize the R&D special zone project.

## 4.2 Policy Conclusions

The efficiency of the innovation system is the key determinant of national and regional competitiveness in the Knowledge-based Economy. Innovation and knowledge generation take place as result of a variety of activities and networks such as user-producer relationships, academy-industry links and spin-offs.

In particular, cooperation is an essential part of the innovation process for most of the innovative firms. Due to the fact that innovation is by far more than a stand-alone activity, policies should be directed toward the systemic aspect of innovation rather than be targeted toward isolated actors. The analysis shows that the presence and quality of the public research infrastructure (universities and public research institutes) and its links to industry are one of the most important assets for supporting innovation.

What can be drawn as policy conclusions from this analysis? It is clear that the basic considerations are also valid for innovation systems in other metropolitan regions. But due to the fact that the structures of innovation-related interactions and institutional environments are different by region and industry, the actual measures should be specific in each region and industry. For the three metropolitan strategic industries, the following measures seem to be necessary to improve the innovative performance of firms.

Firstly, it is necessary to induce clients and key firms into the region and strengthen the network with clients. As for Gwangju and Daejeon, it is necessary to induce customers in the SMA and overseas customers into the region, and expand the size of the cluster, in order to obtain the scale as well. As for Busan, it seems that the critical mass has been already achieved; therefore, the required strategies are to induce public research institutes that are mechanical parts and materials industry-related, and strengthen the network with them.

Secondly, policies concerned with innovation should consider the cooperative aspects of innovation in the Busan mechanical parts and materials industry. For this, it is necessary to add new functions to existing institutions or establish new agencies. It is also required that research institutes and supporting institutions assist the cooperation among small and medium-sized enterprises rather than that among large companies. In order to promote innovation, they need to lead local firms rather than follow them.

Another possible route to encourage academy-industry links is to further develop technopark which is designed to accommodate and support the technology-based firms. Technopark creates the channel by which academic science may be linked to commerce. The technopark model is just one approach to bridging the gap between academic research and commercialization. One of the basic objectives of technopark is to promote cooperation among businesses, and between businesses and universities or research institutes.

Another purpose is to facilitate technology transfer and cooperation for innovation among enterprises. In order to reduce uncertainty and encourage a sense of synergy among enterprises, universities and research institutes, technoparks needs to provide information and organize R&D cooperation as well as formal meeting such as conventions, seminars and conferences.

The third policy conclusions refer to the new roles of innovation supporting institutions such as broker institutions, universities and research institutes. Universities and industrial firms exist for different purposes, and many barriers impede research cooperation among researchers. If there are problem in bridging them, agents are required to mediate between knowledge producers and users. In this respect, public and semi-public technological service institutions are important such as technological innovation centers, regional research centers and technoparks. With the intermediary organizations, communication and cooperation can be further facilitated.

Government-sponsored research institutes and universities are the main performers of generic research and produce a body of basic knowledge for the use and further development by industry. They have many brilliant people making new discoveries but they lack the means or the will to reach out to the market. From the enterprise point of view, firms agree that they could benefit from universities or research institutes. Nevertheless, they may get weary of the cooperation with universities or research institutes, and lack the information concerning the services to be offered. It may be, therefore, necessary for academic institutions to take the lead in establishing linkages through the provision for local businesses of information on the types of linkages available.

# References

Asheim, B., & Cooke, P. (1999). Local learning and interactive innovation networks in a global economy. *Making connections: Technological learning and regional economic change*. Aldershot: Ashgate.

Breschi, S., & Malerba, F. (2001). The geography of innovation and economic clustering: Some introductory notes. *Industrial Corporate Change, 10*(4), 817–833.

European Commission Dgs VIII and XVI. RITTS and RIS Guidebook (1996). *Regional actions for innovation*. Brussels: EC.
Guinet, J. (1999). Introduction. *Boosting innovation: The cluster approach*. Paris: OECD.
Holbrook, A., & Hughes, L. (2000). Operationalizing definitions of innovation at the level of the firm, ISRN Working Paper 2.
Holbrook, A., Padmore, T., & Hughes, L. (2000). Innovation in enterprises in a non-metropolitan area: Quantitative and qualitative perspectives. In H. A. Holbrook & D. A. Wolfe (Eds.), *Innovation, institutions and territory*. Montreal & Kingston: School of Policy Studies.
Kaufman, A., & Tödtling, F. (2000). Systems of innovation in traditional industrial regions. *Regional Studies, 34*(1), 29–40.
Kim, D. J., & Kwon, Y. S., et al. (2001). *Industrial agglomerations and regional clusters in Korea*. Korea Research Institute for Human Settlements.
Kim, W. B., et al. (2006). *Sustainable development on the urban fringe: Korean case study, East Asia and the pacific region*. Anyang: Korea Research Institute for Human Settlements.
Kwon, Y. S., et al. (2005). *A study on the innovation cluster strategy for regional specialization*. Anyang: Korea Research Institute for Human Settlements.
Lundvall, B.-Å. (ed.) (1992). *National Innovation Systems: Towards a Theory of Innovation and Interactive Learning*. London: Pinter
Nelson, R. (ed.) (1993). *National innovation systems*. Oxford: Oxford University Press.
OECD (1992). *Technology and the economy: The key relationships*. Paris: OECD.
OECD (1997). *Proposed guidelines for collecting and interpreting technological innovation data: The oslo manual* (2nd ed.). Paris: OECD.
OECD (1999). *Managing national innovation systems*. Paris: OECD.
Roelandt, T. J. A., & den Hertog, P. (1999). Cluster analysis and cluster-based policy making in OECD countries: An introduction to the theme. *Boosting innovation: The cluster approach*. Paris: OECD.
Rouvinen, P., & Ylae-Anttila, P. (1997). A few notes on finnish cluster studies. *Competitive advantage of Finland*.
Schibany, A., & Schartinger, D. (2001). Interactions between universities and enterprises in Austria. *Innovative networks*. Paris: OECD.
Simmie, J. (ed.) (2001). *Innovative cities*. London: Spon Press.
Swann, G. M. P., Prevezer, M., & Stout, D. (eds) (1998). *The dynamics of industrial clustering: Oxford university press*.
Westhead, P., & Batstone, S. (1998). Independent technology-based firms. *Urban Studies, 35*(12), 2197–2219.

# Fostering Regional Strategic Industries with Reference to Daejeon Metropolitan City

**Byung-Joo Kang**

## 1 Background of the Research

Growth in national and/or regional economies comes with investment, supported by the adoption, and generation of new technology. As productivity of land, labor, and capital increases, the economic competitiveness of a city, region, and nation is enhanced. The key dynamic is the continuing sequence of the application of new technologies in both products and processes.

It is best if the private investors decide when specific industries are to be supported or dropped in order to maximize their profit under the market economy. Theoretically, a firm is started up when its major commodity is composed of sunrise industries but it disappears when its major commodity is composed of sunset industries. Decision making for starting up or closing a firm should be executed by the investors in the market. Government intervention is allowed only when there is a market failure. Market failure occurs when there is an external problem or a problem in providing pure public goods.

We frequently observe that when government intervenes seriously in the market, a problem of distortion in resources distribution or inefficiency in operating the national economy occurs. Market-oriented countries such as the US or Germany do not support any specific industry under the name of national policy. Even Japan had adopted an indirect industry supporting policy after the 1980s, which spread into developed countries after World War II with an active industrial policy by the central government.

However, if local governments select a few industries that seem to have a promising future and are doing well in their regional economy, the possibility of success is higher than it would be if the central government did the same job. Since it is nearly impossible for central government to foster specific industries by giving them general purpose funding under the WTO contract situation, local

B.-J. Kang (✉)
Department of Urban and Regional Planning, Hannam University, Daejeon, Korea
e-mail: Kbj20@hnu.ac.kr

D.-S. Oh and F. Phillips (eds.), *Technopolis*,
DOI: 10.1007/978-1-4471-5508-9_15, 

governments or local industries have become a main body of national competition. Therefore, fostering regional strategic industries is not only important to local governments, but also important to a central government in terms of building a new paradigm of continuous national development.

The aim of this chapter is selecting strategic industries and suggesting some recommendations to foster regional strategic industries in Daejeon Metropolitan City.

# 2 Concept of Industrial Policy, Strategic Industries, and their Goals

## *2.1 Concept of Industrial Policy*

There are many definitions of industrial policy in broad or narrow scope. We could define from the following that an industrial policy is a governmental intention to strongly support a few strategic or sunrise industries in the field of finance or administration in order to enhance national competitiveness.

Adams and Lawrence (1998) define industrial policy as an obvious national policy that enhances potential supply capacity in the national economy through an increase of resources, labor, and capital accumulation for industrial development, and through revision of mobility in production elements and adaptation capability.

Johnson (2005) argues that industrial policy is a governmental intention to adopt or abandon a few various industries in order to attain a superior position in international competition.

Ryutaro (1991) defines industrial policy as a policy that increases the welfare level of a national economy by governmental intervention in industrial restructuring or resource redistribution when there is a certain type of market failure in distributing income and resources.

Lee (1999) asserts that industrial policy is a governmental intervention in production, investment, and transactions of goods in a specific industry or any industry through the backup, regulation, and adjustment of national economy in order to enhance economic growth or national competitive power.

## *2.2 Definition and Goal of Strategic Industries*

### 2.2.1 Definition of Strategic Industries

"Strategic industry" means an industry that is essential for promoting growth of the locality in which the industry is located (Financial Times, LEXICON). Strategic industries are those that have a best location or high growth potential in the future (Park 2005). Best location is evaluated in terms of predominance in a specific industry with a competitive power. It is regarded that an industry that has a best location in the country is the one that has the highest growth potential in a local point of view. This is because an industry can achieve better economic performance when it is located in a best place in terms of natural and innovative resources. A strategic industry could lead other regional industries in this aspect.

#### 2.2.2 Goals of Selecting Strategic Industries

Three goals are frequently cited in selecting strategic industries (Lee 2003). First goal is enhancement of industrial structure and strengthening of national competitive power.

Since the 1990s city and region began to play a more important role than that of a nation in regard to economic activity in the process of internationalization of the economy and localization in political and economic affairs. As a result of this, a city and region appear as a unit of competition, and enhancement of regional competitive power is treated on an equal basis with national competitive power.

Second goal is preparation of an economic base for knowledge-based industries and regional innovation clusters. Since knowledge-based or innovation-oriented industries grow more quickly than others, they should be a driving force in a regional economy in order to gain advantage in industrial structure. Therefore, it is desirable to consider comprehensive developmental elements and location factors when selecting strategic industries for a certain region and to prepare an innovative environment for the chosen strategic industries.

Third goal is formation of industrial structure and organization that lead to balanced and integrated development. Innovation, balance, participation, and networking are being stressed by current central government. Unbalanced development policies, in which large firms, capital region and manufacturing industries are major forces, are being abandoned by the current Korean government.

In this new localization era, large and medium-sized firms, capital region, and local areas are treated equally and balanced and sustainable development strategies are regarded as key factors by the current Korean government. A 5-year development plan for industrial clustering and activation has been initiated and a presidential committee for national balanced development was created to make industrial clustering successful.

## 3 Criteria and Method of Strategic Industry Selection

### *3.1 Key Elements in Selecting Strategic Industries*

There could be many elements that should be considered in selecting strategic industries, but when analyzing this research three core elements were chosen.

### 3.1.1 Degree of Industrial Development

In selecting regional strategic industries, not only current situations and features of the regional economy but also growth potential and desirable structure of the future economy for that region should be analyzed. The key elements to be considered in the selection of strategic industries are degree of industrial development, growth potential, and spread effect. Industries are accumulated and when they enjoy relative superior locations interactions among them happen frequently. The degree of industrial development is an important criterion when selecting a strategic industry from among many existing industries in a certain region (KIET 2004). Industrial clusters are distinguished from others when its economic base has been well structured for a long time and when those industries become representative of the region. Degree of industrial development is analyzed by the degree of industrial accumulation and location quotients.

### 3.1.2 Growth Potential

The focus on degree of industrial development is tuned to current and past situations but growth potential is more future-focused. Since knowledge-based industries are considered a future substitute for existing major industry, an element of growth potential should be a criterion when selecting strategic industries in the knowledge-based economy. Growth potential could be analyzed through the evaluation of growth capability in international and domestic economy, technological innovation capacity and the evaluation of political will from central and local governments (Park 2005).

### 3.1.3 Spread Effect

Spread effect is analyzed by regional input–output analysis. Since regional economic situations are analyzed as a whole or a part through the input–output analysis, it is a useful tool for analyzing regional economic structure. An input–output analysis tool makes macro and microanalysis possible: e.g., such as predicting production, employment, and its importance as consumption, investment, and export are being changed (Choi 1996).

## 3.2 *Measures of Selecting Strategic Industries*

### 3.2.1 Degree of Industrial Concentration

Degree of industrial concentration is measured in two parts separately such as the area of major industrial base and area of knowledge-based industrial base because of limits in statistics. Major industrial base is analyzed in terms of degree of concentration in a certain area, productivity of concentrated region, representative of industrially concentrated region. Knowledge-based industrial base is classified into two groups such as an already highly concentrated region and concentration in an on-going region.

If a few industries are highly specified in the highly concentrated region, those industries are the representative industries of that region. If degree of industrial concentration in the concentrated region is higher than average region and industrial growth rate of this region is very high, this region could step into highly concentrated region within a short period of time.

### 3.2.2 Location Quotient

The location quotient is an indicator that compares the importance of an industry in the region with its importance in the nation.

$$LQ_{ij} = \frac{X_{ij}/X_i}{X_i/X} = \frac{\text{ratio of industry } i \text{ in region } j}{\text{ratio of indusry } i \text{ in the nation}}$$

| | |
|---|---|
| $LQ_{ij} > 1$ | industry $i$ in the region $j$ is specialized compare to the national average |
| $=1$ | industry $i$ in the region $j$ is equally specialized compare to the national average |
| $<1$ | industry $i$ in the region $j$ is not specialized compare to the national average |

In calculating location quotient, employment is most frequently used but output or income data also could be used. This method assumes that tastes and demand patterns are the same in all areas and that the nation is a closed economy.

### 3.2.3 Shift-Share Analysis

Shift-share analysis is a technique used for the analysis of regional industrial statistical data (employment, output, or income). Shift-share analysis measures total change in the performance of the region relative to that of the nation over a given time period (i.e., the nation's actual growth minus its 'expected' growth if it had grown at the same rate as the nation).

G = R + S -----(1)
(Regional Growth) (Regional share) (Shift)

(Shift) divides into:
Differential (Sd)
Proportional (Sp)

Equation (1) could be rearranged such as below

G - R = S

$$E_{ij}(t)-E_{ij}(o)=E_{ij}(o)\times\frac{E(t)-E(o)}{E(o)}+E_{ij}(o)\times\left[\frac{E_i(t)}{E_i(o)}-\frac{E(t)}{E(o)}\right]+E_{ij}(o)\times\left[\frac{E_{ij}(t)}{E_{ij}(o)}-\frac{E_i(t)}{E_j(o)}\right]$$

$E_{ij}$ : Employment in industry i at the region j
$E_i$ : Employment in industry i in the nation
$E$ : Total employment in the nation
$t$ : Terminal period
$o$ : Initial period

This change or shift is then divided into two components. The proportional shift attempts to measure the influence of industrial mix (national industrial composition) on the region. The differential shift, on the other hand, is a measure of the degree to which regional industries grow faster or slower than their national counterparts.

In other words, it reflects the impact of relative location advantages that explain the tendency for industries to be attracted to or grow faster in some regions than others. Each shift component, and the total shift, may be either positive or negative.

The technique may be applied to any one of several economic indicators, but employment is the most common, perhaps because of the accessibility of employment data, perhaps owing to the concern of regional policy-makers with employment growth.

### 3.2.4 Growth Capability of Strategic Industries with a View to Domestic and International Economy

To become a strategic industry among major regional industries, composition of that industry should be big and its impact on the regional economy should be huge. Export, employment and linkage to the new technology are also highly emphasized in strategic industries. Growth rate of an industry in the market is the most important indicator for knowledge-based manufacturing industries.

The second indicator is profitability, which is decided after input and value are decided and is divided into three categories such as rank high, medium, and low. Since export promotion industrial policy in Korea has reached a limit, it is necessary that the portion of manufacturing industry should be decreased and knowledge-based service industry should be fostered. Since knowledge-based service industry has a strong link to information, customer's needs should be surveyed to gain higher value-added in this industry.

### 3.2.5 Capability in Technological Innovation

Existence of RRC (Regional Research Center), TIC (Technology and Information Center), and TP (Technology Park) in a region is important because capability in technological innovation is a critical factor in knowledge-driven economy now and in the future.

### 3.2.6 Government Intention

It is expected that regional competitive power would be decided by superiority among regions in terms of world economy in the future. We are now living in a world where firms choose a county and a region for their profit maximization. Due to this circumstance public–private partnership is needed to gain a better position in the competition.

Government intention is important in selecting strategic industries. Growth potential, synergy effect among industries, and degree of contribution to the regional economy are all critical elements in government intention.

### 3.2.7 Multiplier and Linkage Effect

Multiplier and linkage effect is the most frequently used concept in input–output analysis, and makes economic prediction possible. Industrial linkage effect is divided into two types such as forward linkage and backward linkage. In connection with the process of industrialization in countries undertaking to industrialize, two sequences held promise for generating special pressures toward investment.

First, an existing industrial operation would make for pressures toward the domestic manufacture of this input and eventually toward a domestic capital goods industry. This dynamic is called backward linkage, since the direction of the stimulus toward further investment flows from the finished article back toward the semi-processed or raw materials from which it is made or toward the machines which help make it.

Forward linkage effect is measured by the response coefficient.

$$\alpha_i = \sum_{j=1}^{k} b_{ij} / \frac{1}{n} \sum_{i=1}^{k} \sum_{j=1}^{k} b_{ij}$$

$\alpha_i$: response coefficient
$b_{ij}$: element in the matrix of production induce coefficient
$n$: number of industrial sector

Backward linkage effect is measured by the influence coefficient.

$$\beta_i = \sum_{i=1}^{k} b_{ij} / \frac{1}{n} \sum_{i=1}^{k} \sum_{j=1}^{k} b_{ij}$$

$\beta_j$: Influence coefficient

Another stimulus toward additional investment points in the other direction is called forward linkage. The pressures toward backward linkage investments arise in part from normal entrepreneurial behavior. The pressures toward backward linkage investments come primarily from the efforts of existing producers to increase and diversify the market for their products.

Both forward and backward linkage effect are derived from production induce coefficient, and those two effects are calculated through coefficient and response coefficient.

Selecting industries that have high forward and backward linkage effects is a very important task because these industries are assumed to have a huge influence on the whole regional economy.

## *3.3 Indicators Used in Selecting Strategic Industries: In Case of Daejeon*

Daejeon is an international center for advanced science and technology. The city has a highly intellectual and educated work force, internationally attractive conditions for foreign investors, and various business enterprises.

The "1st 5 Year Regional Innovation Plan" was prepared by Daejeon Metropolitan City in (2004). One of the major jobs in the plan was how to select strategic industries and how to support them under the various domestic and international laws.

Three key elements that were mentioned in Sect. 3 in this chapter were fully considered in the selection of strategic industries, and eight measures were applied in the selection process.

### 3.3.1 Degree of Industrial Development in the Region

*Degree of industrial accumulation* (Table 1).

*Location Quotients* (Table 2).

*Shift-share analysis* (Table 3).

### 3.3.2 Growth Potential

*Growth Capability in international and domestic economies* (Table 4).

*Technology innovation capability* (Table 5).

*Political will from central and local governments* (Table 6).

### 3.3.3 Spread Effect

*Forward linkage effect and the response coefficient*

*Backward linkage effect and the influence coefficient* (Table 7).

**Table 1** Criteria for the selection

| Classification | Highly concentrated area | Potential concentration area | Others |
|---|---|---|---|
| Criteria | ○ | △ | X |

**Table 2** Criteria for the selection

| Location Quotients | Over 1.5 | 1.0–1.5 | 0.5–1.0 | 0.0–0.5 |
|---|---|---|---|---|
| Criteria | ◎ | ○ | △ | x |

**Table 3** Criteria for the selection

| Classification | Industry mix + | Regional share + | Industry mix − | Regional share + | Industry mix + | Regional share − | Industry mix − | Regional share − |
|---|---|---|---|---|---|---|---|---|
| Criteria | ◎ | | ○ | | △ | | X | |

**Table 4** Criteria for the selection

| Classification | Criteria |
|---|---|
| High growth potential industry | ○ |
| Non high growth potential industry | X |

**Table 5** Criteria for the selection

| Classification | Criteria |
|---|---|
| Existence of high growth potential industry | ○ |
| Nonexistence of high growth potential industry | X |

**Table 6** Criteria for the selection

| Classification | Whether or not the industry is chosen as a strategic industry | | | |
|---|---|---|---|---|
| Central government | Chosen | Chosen | Not chosen | Not chosen |
| Daejeon City | Chosen | Not chosen | Chosen | Not chosen |
| Criteria | ◎ | ○ | □ | X |

# 4 Selection of Strategic Industries and the Reasons: In Case of Daejeon Metropolitan City

## *4.1 Selection of Strategic Industries*

It was found that strategic industries in Daejeon Metropolitan City come from the fields of semiconductor, high-tech electronic parts, communication devices, biological industry, fine chemistry/new materials, mechatronics, and information & communication services (Table 8).

## *4.2 Reasons of Selected Strategic Industries*

### 4.2.1 Information and Communication Industry

In the case of information and communication industry, degree of industrial growth is quite high and growth potential is also high. Location quotient reveals that this industry has higher ratio in number of firms and employees than the

**Table 7** Criteria for the selection

| Classification | Response coefficient > 1 Influence coefficient > 1 | Response coefficient < 1 Influence coefficient > 1 | Response coefficient > 1 Influence coefficient < 1 | Response coefficient < 1 Influence coefficient < 1 |
|---|---|---|---|---|
| Criteria | ◎ | ○ | □ | X |

*Legend* ◎: excellent; ○: good; □: mean; X: bad or weak

**Table 8** Chosen strategic industries of Daejeon Metropolitan City

| Ranks from the selection of industries | | Selected strategic industries | |
|---|---|---|---|
| Rank | Industries | Rank | Industries |
| 1st | Communication devices/ information and communication | 1st | Information and communication |
| 2nd and 3rd | Life industry | 2nd | Life industry (bio-industry, fine chemistry) |
| | High-tech electronic parts | 3rd | High-tech parts and materials (next generation battery, new material, Nano) |
| 4th and 5th | Mechatronics | 4th | Mechatronics |
| | Fine chemistry/new material | | |
| 6th | Semiconductor | | |

national average. Since industrial structure and location benefits of this industry are so nice, accumulation benefit is being realized now.

### 4.2.2 Life Industry (Bio and Fine Chemistry)

Level of technology and growth potential is high, and spread effect is huge in this industry. Especially, central and Daejeon City government have strong intention to foster this industry.

### 4.2.3 High-tech Parts and Material

This industry does not have a high degree of industrial growth but is high in growth potential compared to many traditional industries. Many manufacturing factories exist in Daejeon today and the governments have the strong intention to foster this industry which is a high value added one.

### 4.2.4 Mechatronics

Degree of industrial growth, growth potential, and spread effect is generally good in this industry. Number of firms and employees is higher than national average according to the analysis of location quotient. Industrial structure and location factor are very good according to shift-share analysis.

# 5 Strategies to be Used in Fostering Strategic Industries: In the Case of Daejeon Metropolitan City

## 5.1 *Information and Communication Industry*

### 5.1.1 Analysis of Current Status

Information and communication industries have locational advantages because there is plenty of research manpower in Daejeon City and research centers are concentrated at Daedeok Research Town. Since there are many research centers in the Daedeok Area, it is relatively easy to build linkage among research centers and research institutes. It is also easy to build IT accumulation base to develop technologies for mobile communication, telematics, and services in information technology for the sake of IT technology fusion.

This industry has competitive power in the aspect of next generation mobile communication. Base that would lead next generation mobile communication market over the world is already prepared because commercialization of CDMA has already been realized in Daedeok Research Zone. More than 200 patents are acquired annually in this area due to the national investment in 4G technology through the ETRI. There are seven display related research centers including Display New Technology Research Center, Research Center for Semiconductor Processing Technology in the Daedeok Research Zone.

In spite of these advantages, a few disadvantages have also turned up in this area. Lack of expertise in the field of core information and communication technology, and nonexistence of a large manufacturing unit related to mobile communication or display manufacturing are disadvantages in this region.

### 5.1.2 Strategies of Fostering Industries

Three strategies such as construction of exclusive venture town for IT industry, service center for high frequency industry and Cinema Town are recommended to foster information and communication industry.

*Construction of exclusive venture town for IT industry*
Major functions conducted in this venture town would be labs for common use, providing diverse information service, and supporting export and global affairs. Venture town could be built in the Daedeok Techno valley with the 21,487 $m^2$ of site and total construction area of 26,446 $m^2$. Budget for the construction could be \$15 million and \$12 million could be allocated to the central government and \$3 million to Daejeon Metropolitan City.

*Construction of service center for high frequency part industry*
Major functions conducted in this service center would be evaluating design and function of the parts, fostering manpower and re-education for current employees, testing products and issuing certificates and providing support service for technological development and networking. Service center could be built in the Daedeok Techno Valley with the 7,272 $m^2$ of site and total construction area of 3,636 $m^2$. Budget for the construction could be $31.9 million and $23 million could be allocated to the central government, $7.9 million to the Daejeon Metropolitan City and $1 million to the private firms.

*Construction of Cinema Town*
Content of the project would be construction of a broadcasting station for 10 W class with 10 channels, emitting system for three dimensional broadcasting, establishment of digital broadcast receiving facility via inner cable of apartment over 1,000 households or public buildings. Cinema town could be built in the Daedeok Research Zone or at EXPO Science Park or ETRI or Science Foundation. Budget for the construction could be $9 million.

## 5.2 Life Industry

### 5.2.1 Analysis of Current Status

Life industries hold excellent research manpower and facilities in Korea. There are 4 universities and 70 bio-venture firms in the Daedeok Valley. Well-equipped infrastructure for bio-industry and related education, research and supporting services are available in the Daedeok Research Zone. Intensive supporting services from Daejeon Metropolitan City Government are also available and Bio-venture building is under construction by the city government at the Daeduck bio-community site to vitalize whole bio-industries located in the Daedeok valley.

However, there appears to be one disadvantage in this industry. Central government decided to scatter research organizations over the country for the sake of balanced national development.

### 5.2.2 Strategies of Fostering Industry

Three strategies such as construction of an advanced supporting center for bio-industries, supporting Center for Good Drugs Licensing and Bio-Venture Town are recommended to foster life industry.

*Construction of advanced supporting center for bio-industries*
Major functions of the center will promote R&D and advance patent related to life industries located at Daedeok Research Town. Several projects, such as

development of new drugs, clinical research, management of intellectual ownership, and fostering experts for industrialization would also be deployed. Budget for the construction could be $123.5 million and $83.5 million could be allocated to central government, $20 million to city government and $20 million to private firm.

*Construction of supporting Center for Good Drugs Licensing*
Major functions of the center will be supporting foreign market search after establishing "Current Good Manufacturing Practice" for the industrialization of new products and controlling quality of bio-products for the sake of production standardization through the testing of products and licensing with a few research institutes such as Information Center for bio safety, Life Science Research Institute, and Korea Chemistry Research Institute. Budget for the construction could be $50 million and $35 million could be allocated to central government and $15 million to city government.

*Building Bio-Venture Town*
Major functions will be providing space for Post-TBI bio-venture firms, supporting administrative and information service that is needed in managing bio-venture firms, and establishing cooperative systems for machine and materials, production facilities such as pilot system. Bio-Venture Town could be built at Jun-Min dong with 11,570 $m^2$ of site and 9,246 $m^2$ of total construction area. Budget for the construction could be $55.5 million and $37.4 million could be allocated to central government, $15 million to city government and $10 million to private firm.

## *5.3 High-Tech Parts and Material Industry*

### 5.3.1 Analysis of Current Status

High-tech parts and material industries hold excellent research manpower and research centers are concentrated at Daedeok Research Town and linkage among research centers and research institutes could be easily built. Existence of industrial base for new material industries is another advantage for this industry.

Accessibility to integrated research related systems such as excellent research manpower is high and already existing material firms are very good, and forward and backward linkage effect appears quite high in this area. Daedeok Valley holds the highest number of nano experts in Korea and Total Fab-Center is under construction now at Daedeok Valley.

However, as central government recently pronounced a support program for "Incheon SongDo Special Economic Zone", it is no longer easy to induce private companies it locates in Daedeok Research Town.

### 5.3.2 Strategies of Fostering Industries

Three strategies such as supporting R&D, construction of adaptation center, building Nano Fab-Center are recommended to foster high-tech parts and material industries.

*Supporting R&D for next generation battery and technology for new material parts*
Major functions of this strategy is operating technology development team and selecting R&D projects and executing project evaluation. Budget for executing this project could be $40 million.

*Construction of Adaptation Center for Energy and New Materials*
The Center will provide manufacturing space for firms that produce goods in the fields of energy and new materials. The Center will encourage manufacturing process with a step by step approach. Manufacturing process starts with labs and construction of pilot plants followed by manufacturing factories that are built at the end. Site of the Adaptation Center could be 6,611 $m^2$ with total construction area of 3,305 $m^2$. Budget for the construction could be $10 million.

*Nano Total Fab-Center (under construction)*
The Center will be used by industries, universities and research institutes commonly for R&D in nano technology. The Center will attract promising domestic and foreign firms into Daedeok Research Town by the advertisement of facilities and manpower of Nano Fab-Center located at Daedeok Research Zone. Budget for the construction could be $ 30 million and duration of construction would be from 2002 to 2008.

## *5.4 Mechatronics Industry*

### 5.4.1 Analysis of Current Status

There is abundant manpower for robot and mechatronics related research at Daedeok Research Zone and it is easy to link research centers and labs in the Valley. There are many private research institutes and service companies related to mechatronics industry in Daedeok Research Zone. Since three military headquarters such as army, air force, and navy are located in Daejeon Metropolitan City, active commercialization of military technologies related to robotics is expected in the near future.

However, central government does not express strong intentions to foster mechatronics industry, it will take time to foster this industry.

### 5.4.2 Strategies for Fostering Industries

Two strategies such as standardization of intelligent robotics and promotion of common market and privatization of national strategic robotics are recommended to foster mechatronics industry.

*Project for standardization of intelligent robotics and promotion of common market*
The project will provide supporting service for design of intelligent robotics, supplying parts and will provide skills in marketing and problem solving, and create common brand and establish infrastructure for various exhibitions, events and efficient marketing. One-stop service for financial aid, consulting for operation of firms and legal affairs will also be provided. Budget for executing the project could be \$12 million.

*Privatization of national strategic robot*
Major functions of this project are privatization of military and medical robotics, special aimed robots, developing flat-form and intelligent module for personal robots, and supporting for developing anti-terror robotic soldiers and robots for information gathering. Budget for this project could be \$16 million.

## 6 Conclusion

Regional economic development is realized when growth of residential industries is visible on one hand and external factors are well matched with regional economy on the other. Accordingly, to achieve sustainable regional economic growth, it is necessary to foster and attract high-tech and high value added industries to the region and to restructure industrial composition in the region.

To acquire competitive power in technologies and products of local industries is an essential factor in the globalized era. Transformation of main regional industries from traditional ones into high value added ones is one of the major jobs in industrial restructuring at local level. Deregulation, providing financial support, and levying less taxes are important jobs for central government in aspects of external factors.

This research was aimed to review theories of regional economic policy and to find how to foster regional strategic industries. Concept of the industrial policy and definition of strategic industries were identified and the criteria and measures of strategic industry selection were presented. Four industries, such as information and communication industry, life industries, high-tech parts and material industries, and mechatronics industry were chosen as strategic industries for Daejeon Metropolitan City. One of the key questions at this point is under what conditions does a competitive strategic industry arise? The answer is that when industries hold high growth potential, huge spread effect or a high degree of industrial concentration, we may say that it has good conditions to foster strategic industries.

Fostering strategic industries is a more difficult job than selecting an appropriate strategic industry from various industries in a region. Two conditions are suggested as important factors when fostering strategic industries in this research.

The first condition is the presence of high quality R&D manpower and a well organized urban community. There is Daedeok Research Zone in Daejeon Metropolitan City, where 17 higher education industries for R&D manpower and technology development including KAIST, ICU, and CNU exist. Since Daejeon Metropolitan City was selected as a best place to live in by the Gallop Survey Group several times for the last 15 years, the first condition is satisfied.

The second condition is restructuring a regional economy from mass production systems to a high-tech oriented 'just in time' structure. The "5 Year Regional Innovation Plan" was prepared by Daejeon Metropolitan City last year. The major part of the plan is composed of selection of industries for the next generation and strategies of fostering selected industries. Therefore, we expect the second condition would also be satisfied.

## References


Adams, F. G., & Lawrence, R. (1998). *Industrial Policies for Growth and Competitiveness*. Wharton: Econometric Studies.

Audretsch, D. (2001). The role of small firms in U.S. biotechnology clusters. *Small Business Economics, 17*, 3–15.

Boltho, A. (1985). Was Japan's Industrial Policy Successful? *Cambridge Journal of Economics, 9*(2), 187–201.

Choi, J.-S. (1996). *Regional Economics*. Seoul: Beob-moon sa.

Daejeon Metropolitan City, 200, 1st 5Year Regional Innovation Plan.

Daejeon Metropolitan City, 2004, Statistics of Daejeon Metropolitan City.

Dodgson, M. (1992). The strategic management of R&D collaboration. *Technology Analysis and Strategic Management, 4*(3), 227–244.

Etzkowitz, H., et al. (2000). The future of the university and the university of the future: Evolution of ivory tower to entrepreneurial paradigm. *Research Policy, 29*(2), 313–330.

Giesecke, S. (2000). The contrasting role of government in the development of the biotechnology industry in the US and Germany. *Research Policy, 29*(2), 205–223.

Henderson, J. V., Kuncoro, A., & Turner, M. (1995). Industrial Development in Cities. *Journal of Political Economy, 103*(5), 1067–1090.

Hodder, J. E., & Tschoegl, A. (1985). Some aspects of Japanese corporate finance. *Journal of Financial and Quantitative Analysis, 20*(2), 173–191.

Horiuchi, A., & Packer, F. (1986). *The function of financial institutions: What role has the 'main bank' played in Japan?*. Tokyo: University of Tokyo Press.

Johnson, C. (2005). *Industrial policy debate*. Oakland: Japan Policy Research Institute.

KIET (2004). Vision and development strategies of Korean industries for the 21st century.

Kim, D.-H. (2003). *Policy for regional industries in the era of localization in Japan*. Seoul: KIET.

Knight, R. V. (1995). Knowledge-based development: Policy and planning implications for cities. *Urban Studies, 32*(2), 225–260.

Lagendijk, A. (1997). From new industrial space to regional innovation systems and beyond: how and from whom should industrial geography learn? EUNIT Discussion Paper 10. Newcastle upon Tyne: CURDS (University of Newcastle).

Lee, K.-T. (1999). *Theories and realities in industrial policy, Development of high tech-technology industry and industrial policy*. Seoul: KIET.

Okimoto, D. I. (1989). *Between MITI and the market: Japan's industrial policy for high technology*. Stanford: Stanford University Press.

Park, J.-C. (2005). Regional industrial policy through the selection of sunrise industry. *Journal of the Korean Association for Policy Science*, 4(1).

Porter, M. E. (1990). The competitive advantage of nations, Free Press, 2000, "Location, competition, and economic development: local clusters in a global economy", *Economic Development Quarterly,* 14(1).

Rip, A., & van der Meulen, B. (1996). The post modern research system. *Science and Public Policy, 23*(6), 343–352.

Rybczynski, T. M. (1984). Industrial finance system in Europe, U.S. and Japan. *Journal of Economic Behavior and Organization, 5*, 275–286.

Ryutaro, K. (1991). *Analysis of economic and industrial policy*. Tokyo: University of Tokyo Press.

Sheard, P. (1985). Main banks and structural adjustment in Japan. Research Paper, Australia-Japan Research Center, No. 129.

Storper, M. (1992). The limits to globalization: technology districts and international trade. *Economic Geography, 68*(1), 60–93.

# Role of Universities in Contributing Towards Science and Technology Park Development: A Framework of Critical Success Factors

**Quinton C. Kanhukamwe and Maxwell Chanakira**

**Abstract** Universities can play a critical role in the growth dynamics of a Science and Technology Park (STP) as well as economic development of a nation. The research prowess of most universities has afforded them the opportunity to generate research and development outputs which have resulted in new high-tech enterprises being established. Some companies have been enabled to take up new technologies, new ideas and new processes whilst new products and services continue to be churned out. This is indeed a paradigm shift that most universities in Zimbabwe have had to contend with not only for purposes of economic development but for their own survival too, hence the innovation and entrepreneurial-driven thrust these institutions are beginning to take. Whilst there is need to enhance capacity in most of the universities in the developing countries, the critical ingredients for them to adequately perform and contribute to a Technology Park development need to be provided. The innovation systems in these countries also require continuous strengthening and upgrading. The study identifies critical success factors in the establishment of successful STPs, location advantages of the university, the level of skill of research staff, expenditure on R&D and successful Public private Partnerships (PPPs). With an enabling environment, new investments, new companies, new jobs can be created as evidence of universities impact and role in STP growth and development.

Q. C. Kanhukamwe (✉) · M. Chanakira
Harare institute of Technology, BE 277Belvedere, Harare, Zimbabwe
e-mail: qckanhukamwe@hit.ac.zw; qckanhukamwe@yahoo.com
URL: http://www.hit.ac.zw

M. Chanakira
e-mail: mchanakira@hit.ac.zw; pmchanakira@gmail.com

D.-S. Oh and F. Phillips (eds.), *Technopolis*,
DOI: 10.1007/978-1-4471-5508-9_16, 

# 1 Introduction

Link and Scott (2003) observe that there is broad recognition that Science and Technology Parks (STPs) are effective vehicles for promoting new technology-oriented firms, facilitating the commercialisation of scientific research and revitalising regional economies—critical objectives in the pursuit of the sustainable development of economies, especially for developing countries which have to play catch up with advanced economies. Hence, STPs are the perfect habitat for businesses and institutions of the global knowledge economy (IASP 2011).

In the past two decades or so, there has been increasing interest in many countries in promoting the formation and development of national STPs. This follows the establishment of the first STP in the U.S.A. in 1951, followed by the first European STP in 1972. STPs in the Asia Pacific region started to emerge since the early 1980s (Zhang 2002). Inevitably, the increasing number of STPS has appeared under different names and physical manifestations.

The IASP reports that, worldwide, the rate at which STPS have been established rose 11 % between 1980 and 1984, increased by 23 % from 1985 to 1989 and between 2000 and 2006, rose by 26 %. Clearly, this shows an upward trajectory in the establishment of STPS. China, which has been at the forefront of the establishment of STPs in recent years, established its first STP in Beijing Zhongguancun in 1988. The STPs had ballooned to 54 national by 2007 (Zhang and Sonobe 2011).

Zimbabwe is emerging from a debilitating economic crisis and a brain drain that has created problems for the country's science and technology development. Despite these challenges, the country has adopted a national economic development strategy that adopts STPs as vehicles of economic development. For example, the country has plans to develop information and communications technology (ICT) STPs with a bias supporting software development and large-scale assembly of electronic products by 2012 (Zimbabwe MTP 2011).

The adoption of STPs in Zimbabwean universities is still at the formative and experimental stage. What can the universities do to ensure that they establish successful STPs? In other words, what are the critical success factors for STPs in Zimbabwean universities? Focusing on universities in Zimbabwe, the purpose of this study is to develop from extant literature the role of universities and critical success factors for the establishment of successful STPs. The findings and analysis of this study are expected to accelerate the establishment and operations of STPs in the country.

This chapter is organised into four sections. Section 1 provides an introduction to the study. Section 2 briefly explores key theories and empirical findings on STPs. Section 3 proffers a conceptual framework of critical success factors for the establishment of successful STPs that are sustainable. Section 4 concludes the study.

## 2 Theoretical Perspectives on STPs

Although STPs have appeared under different names and physical manifestations across the world, there should be some broad definition that describes them. What are STPs? A Science Park is an organisation managed by specialised professionals whose main goal is to increase the wealth of its community by promoting a culture of innovation and competitiveness of enterprises and knowledge-generating institutions established in the park or associated with it. Science parks include schemes under various similar names such as "Research Parks", "Technology Parks", "Science and Technology Industrial Parks", "High Technology Development", "Innovation Centre" and "Technology Incubator" (Zhang 2002).

### *2.1 Purpose of STPs*

An important dimension in the discourse of STPs is the purpose for which the STP is established to serve. According to the IASP (2011), STPs promote the economic development and competitiveness of regions and cities by:

- Creating new business opportunities and adding value to mature companies.
- Fostering entrepreneurship and incubating new innovative companies.
- Generating knowledge-based jobs.
- Building attractive spaces for the emerging knowledge workers.
- Enhancing the synergy between universities and companies.

In China, for a firm to gain entry into the STPs, it is required to be classified as a high-tech firm and must satisfy three key criteria for qualifying as a high-tech firm (Zhang and Sonobe 2011). First, a high-tech firm is required to develop or use technology in the new and high-tech products or services listed in *the country's Catalog for High and New Technology Products* published by the Ministry of Science and Technology, which includes electronics and information technology, aerospace technology and biotechnology. Second, a high-tech firm is required to spend at least 3 % of its annual gross revenue on Research and Development (R&D) to develop products or services. Third, of the high-tech firm's employees, 30 % or more must have at least a college degree and at least 10 % must be engaged in R&D. Should an enterprise fail to meet these conditions, then it is excluded from enjoying various policy incentives given to high-tech firms. Most high-tech firms are in the manufacturing sector.

The research by Money (1970) in North America on the critical success factors for university related STPs established the following important points:

1. There are principal university services which STP tenants consider significant in their decisions to locate in a university. Such services include the technical

services of a computer and the convenience services of a restaurant, motel, post office and bank.
2. The physical layout of the university where the STP is to be established should enable the expansion of the STP tenants' facilities.
3. Access control and restrictions should be established and enforced to retain STP aesthetic quality.
4. Cost of site should be competitive with alternate sites available.
5. The geographic and environmental factors which tenants consider important in their decisions to locate in an STP in a university are:

- a suburban residential area within commuting distance;
- good public schools;
- established graduate school with significant research programme;
- university programmes which provide library facilities, continuing education;
- a jet airport located in area.

Cox (1985) identified four factors essential for STPs success from the U.S.A. experience about science park development and management. These are a desirable living environment, a major technological university, major institutional research facilities and a skilled labour force. Highly trained scientists, engineers and technical workers are required to work in STPs in universities to raise the profile of the university. Any university community without one of these ingredients will have great difficulty in developing sustainable STPs.

## *2.2 Ingredients of a Knowledge Economy*

The knowledge-based economy should compliment efforts to improve productivity through enhancement in the total factor production as it will add value to existing activities and will be accompanied by improvement in technology, greater innovative capability and the input of higher skilled workforce. An assessment using input–output analysis confirms that knowledge-intensive industries have a higher value-added multiplier and higher productivity compared with non-knowledge-intensive industries. Universities, colleges and research organisations, etc., play a critical role in the development of a knowledge-based economy.

Clearly, the introduction of high-tech and knowledge-intensive and intelligent production systems will offer new investment opportunities within existing industries. Knowledge enabling industries, particularly in the area of information and communications technology (ICTs), will also generate green areas of investment in the development of hardware and software as well as infrastructure. The knowledge economy through its contribution to productivity enhancement and the generation of new ideas of investment will therefore, increase the long-term growth potential and provide the basis for continued sustainable rapid growth of the Zimbabwean economy.

Indeed, ICTs are becoming key to the competitiveness of any nation. Investors are increasingly seeking first mover advantages, new products and services in response to customers and sources of information. The rapid development of ICTs and the Internet are exposing inefficiencies in the functioning of markets, firms and institutions, putting downward pressure on prices and accelerating the need to restructure and adapt to changing conditions (Edigheji 2009).

## 2.3 *Mandate of a University*

Universities are centres for generating knowledge as well as being intellectual and cultural centres. Universities that are effective and continue to influence technopolis growth have been pivotal in several ways (Smilor et al. 1987).

1. Fostering research and development activities;
2. Contributing to perceptions of the region as a technopolis;
3. Attracting key scholars and talented graduate students;
4. Fostering the spinoffs of new companies;
5. Attracting major technology based firms;
6. Nurturing a large talent pool of students and faculty from a variety of disciplines;
7. Acting as a magnet for government and private sector funding, and
8. Providing a source of ideas, employees and consultants for high technology as well as infrastructure companies, large and small in the area.

Clearly, without the universities to offer the above attributes, the basis for development of robust small and large technology-based companies would not be readily available. However, this assumes that the universities in question have attained an acceptable level of overall excellence in their mission. Such universities possess R&D strengths that relate to the local economic base. They are active in the transfer of existing knowhow, i.e. technology to business, governmental agencies, etc., capital investment in the built form and in the equity of private businesses as well as taking leadership in addressing critical social problems (Luger 1997). Universities can certainly serve as a good laboratory for students and faculty to test the applicability of theory to real problems in the various fields including engineering, the environment, economic development and other emerging fields.

## 2.4 *The Necessary Ingredients for Technology Parks Growth*

A Technology Park is a new settlement with many high-tech production firms but relatively few basic research institutes, whereas a science city is dominated by basic research institutes with relatively few high-tech production firms. In order to spur economic development and promote technology diversification, thereby bringing about technopolis growth, three factors have been identified as particularly important (Smilor et al. 1987).

1. The achievement of scientific pre-eminence. This entails amongst other factors quality of R&D outputs including contracts and grants, membership of faculty and researchers in eminent organisations of a scientific nature, etc.
2. The development and maintenance of new technologies for emerging industries. Thus, providing the basis for competitive companies in a global economy thereby laying the basis for economic growth. The industries may be in areas of biotechnology, artificial intelligence, new materials, ICTs, etc. Key in this factor is the commercialisation of university IPs which can then be measured.
3. The attraction of major technology companies and the creation of home-grown technology companies. This factor calls for economic development and technological diversification within the technopolis.

Dr. Fred Phillips (ex Austin, Texas) of the Maastricht School of Management ((Netherlands) (Phillips 2005), an acknowledged authority and experienced expert on the Technopolis phenomenon in the US and Europe, has postulated that Technopolis regions grow by:

- Attracting new companies.
- Nurturing existing indigenous firms.
- Encouraging entrepreneurial start-ups.
- Providing a supportive educational, social, tax, quality-of-life and cultural context for research, technology entrepreneurship and business.
- Networking with other Technopolis worldwide.

Dr. Phillips has also listed the following Technopolis Success Factors:

- Embracing Change.
- Social Capital, especially with cross-sectoral links.
- Cluster strategies that target specific company groups for collaboration.
- Visionary and persistent leadership.
- The will to action.
- Action.
- Constant selling.
- Self-investment in infrastructure.
- Outreach and networking.

From these observations and experiences, we can conclude that in addition to Dr. Fred Phillips' recently identified nine Technopolis success factors, William Gartner's 44 perennial ten ingredients that are pre-requisite for Technopolis creation still hold true: (1) suitable financing, (2) availability of a competent workforce, (3) accessibility to helpful suppliers, (4) government support or absence of obstacles, (5) proximity of universities to assist in research, (6) availability of land or facilities, (7) access to transportation, (8) support of the local population, (9) available support services, and (10) low entry barriers.

Each of the above factors is important. Some of them may be more critical to one Technopolis venture than another, but all of them have a considerable role in ensuring the Technopolis success. Proximity to universities is important in two respects: First, in high-tech start-ups where new inventions or technologies play a dominant role, these institutions can make significant contributions to successful start-ups through research, problem solutions and engineering support.

Another benefit of the university is its business school and the availability of consulting services in terms of marketing, production systems, MIS, accounting and finance advice, etc. Kirzner (1984) has two additional factors to Gartner's list that affect entrepreneurial success:

1. The existence of an entrepreneurial subculture. The tremendous success of Silicon Valley, Boston, Austin and San Diego very much support the notion that entrepreneurs feed off each other in a synergistic fashion and create their own dynamic environment, and
2. The availability of incubator organisations, many of which are initiated by local universities and governments as enterprise centres. The majority of the success stories in Technopolis growth and development have been attributed to the combined support of government leadership in education and industry and in the case of India an actively involved diaspora that invests and develops intellectual capital back and forth.

Zimbabwe cannot therefore run away from the Technopolis phenomenon which is continuously characterised by increasing globalisation, heightened interdependency and the emergence of a new paradigm of regional, institutional and technological clusters, which facilitate innovation and its commercialisation.

## *2.5 The Innovation System in Zimbabwe*

Zimbabwe has 13 universities of which 9 are state and the remaining 4 are private. It also has over 15 Polytechnics and Technical colleges excluding a number of Teacher training colleges. Alongside these institutions are Research institutes which are varied in terms of mandate and these cover areas such as agriculture, veterinary, medical, engineering, fisheries, etc.

A number of Tertiary institutions have adopted an entrepreneurial approach to their existence. The level of support to most of these institutions is insufficient, forcing most of them to look beyond state support. This has seen a number of income generating activities being embarked on by these institutions. Regrettably these activities are in a number of cases not well funded and in some cases organised, to the extent that teaching and research in some instances has suffered.

It is in this regard that some universities and research bodies have begun exploring the need to develop business incubators or as others have done science parks.

Three institutions that have boldly taken on that route are the Scientific and Industrial Research and Development Centre (SIRDC), the National University of Science and Technology (NUST) and the Harare Institute of Technology (HIT). The SIRDC is an industrial and Research Development centre which has several institutes focusing on biotechnology, building technology, electronics and communications, energy technology, environmental sciences, food and biomedical, geoinformation and remote sensing, metrology, production engineering, research, etc. Its mission is to provide technological solutions for sustainable development. It has separated its research and development mandate from that of commercialisation of its Research and Development outputs. This has seen SIRDC create a business innovation park where they have established brands arising from their research into standalone companies (Phillips 2005). All these companies of the SIRDC are run under an investment vehicle called SIRTECH investments (Pvt) Ltd. Private Players, i.e. other companies and financial institutions have been allowed to buy equity in SIRTECH.

The NUST has established a Technopark. It is a unit of the university. The Technopark is in its infancy with a few electronic manufacturing companies having been established. Other potential projects in the offing require financial support of the magnitude that some of these institutions are unable to rise.

Indeed large scientific and economic projects that are generated within these institutions require cooperation from financial institutions and outside. Establishment of partnerships whenever and wherever they are appropriate should be encouraged and this indeed should become policy in these institutions. With the current new inclusive government dispensation prevailing in the country, It is hoped that the efforts by these universities and research entities to mobilise resources in support of their graduates and researchers will result in start-up companies and spin-offs that will contribute to the nation's quest for rapid industrialisation and technopolis growth. This will result in immense economic benefits for the nation at large benefiting including boosting employment levels.

## *2.6 Harare Institute of Technology*

The HIT is one university that has evolved from being a National Vocational Training and Development Centre to a fully fledged university status beginning September 2004. The vision of HIT is to be the leading institution of the development, incubation, transfer and commercialisation of technology and manpower development for greater national industrialisation.

The HIT Mission Statement—to cultivate commitment towards technopreneurial leadership while commercializing technology through professionalism rooted in intergrity.

*Vision: To be the Stimulant of Scholarship in Innovation*

It is clear from the statement that HIT is moulding itself to be the nation's most energetic and responsive institute offering unparalleled educational opportunities for those seeking highest quality undergraduate, graduate, and continuing personal or professional enrichment as well as being at the forefront of growing the nation's industrial base and natural resources beneficiation.

To achieve the mandate strategic units have been set up which are:

1. *Technopreneurship Development Centre*: this unit imparts and creates appropriate business entrepreneurial skills, attitudes and motivation among the university's students.
2. *Technology Centre*: the centre promotes the (development of) production and manufacturing activities from the various academic units within the HIT. It also offers short-term technical training to industry.
3. *Science Park*: the University is developing a science park which brings together ideas and people with knowledge and financial strength. For the achievement of the above HIT has a holding company called Insti-Tech Holdings which has the following companies, Instifoods, Institools, Instisoft, Instiherbs and Institronics.

*Institools:* commercialises research from the department of Industrial and Manufacturing Engineering and produces tools required by industry.

*Instifoods:* produces food products using modern technology and is linked to the Food Processing Technology Department.

*Instisoft:* produces and modifies software and is linked to the School of Information Sciences and Technology.

*Institronics:* designs and produces electronic gadgets and is linked to the Electronic Engineering Department.

*Instiherbals:* produces herbal medicines and is linked to the Pharmaceutical Technology Department.

This is a strategy that the HIT University has embarked upon in its endeavour to promote the development of home-grown technologies and their commercialisation as well as play a catalyst role in the creation of spin-off companies.

## 2.7 Government Role Towards Sustainable Science and Technology Park Development

Generally, there are common attributes that are expected for governmental interventions to succeed. Governments can play pertinent roles by fostering strategic collaborations with other countries to derive synergies for best practices of the world into the country. The enacting of enabling policies that create a conducive environment will certainly stimulate and enthuse would be tenants as well as demand for space in parks by budding entrepreneurs. Attraction of multinational investors who may want to set up in these parks is crucial. International investors will be attracted on the back of a number of more basic structural issues which allow them to flourish.

The parks should also accommodate start-up companies with a ready to go concept as well as graduates who may wish to start a company on the basis of their specialisation or commercialisable research outputs. A well-trained Science, engineering and technology and knowledge human resources base to support business incubation is critical. This should be a manpower that can synthesise and create knowledge. The need to develop an entrepreneurial culture should be vigorously pursued in the country. Universities and colleges in Zimbabwe are in one way or the other embedding to some extent within the curriculum some aspects of entrepreneurship. This hopefully should provoke creativity as well as catapulting some of our graduates to venturing into high-tech enterprise development.

## 3 A Conceptual Framework of Critical Success Factors

The USA, Europe and lately Asia have experienced some success with STPs. Zimbabwe has embarked on the critical path of establishing STPs. Although, beset by major challenges, there is a strong willingness to establish STPs to ensure the sustainable economic development of the country. What are the critical success factors for the successful establishment of STPs in the country's universities?

In this section, the theoretical framework of the study is briefly presented. Based on extant literature, the study proposes the following model:

STPs cannot be studied in isolation from its antecedents. This study argues that the performance of STPs depends on a fit of four key determinants—the strategic objectives of the STP, the university location advantages, the university expenditure on R & D and the successful conclusion of Public private Partnerships (PPPs), which support STPs with the private sector.

### *3.1 STP Objectives*

The objective for which the university wants active roles in the development of STPs is critical for its establishment and success. In the diagram, [Fig. 1, (1)] depicts this factor. Possible objectives for universities establishing STPs include creating new jobs, fostering technopreneurship or fostering synergy with the private sector. Whatever, the objective, it must be clear and unambiguous.

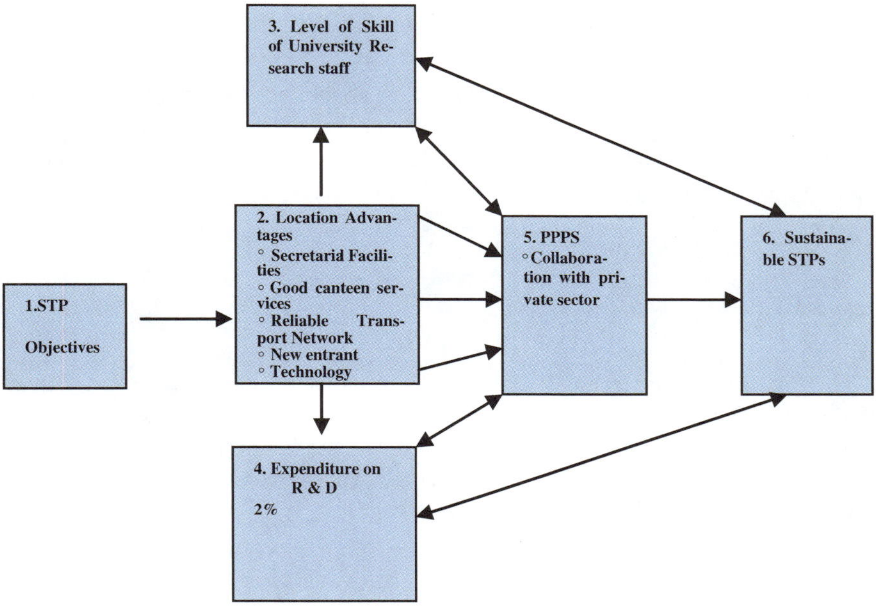


**Fig. 1** Conceptual framework of critical success factors for universities establishing STPs

## 3.2 *Location Advantages*

The location of the university [Fig. 1, (2)] may or may not be attractive for the locations of an STP. While some factors may not fall within the control of the university, nevertheless, there is something that the university can do about its location. The university must be able to offer excellent canteen services, critical secretarial facilities such as photocopying, duplication, binding and typing services to STP tenants. The university must be able to lobby council for reliable transport to and from the university to raise the institution's attractiveness.

## 3.3 *Level of Skill of University Research staff*

The level of skill of university research staff [Fig. 1, (3)] is critical to the research outputs from a university. Ideally, university research staff engaged in STPs should be scientists, engineers or technicians of international repute. Their level of high skills will act as magnets for attracting the private sector to universities. Adopting the Malaysian Agricultural Research Institute (MARDI) model, preferably 25 % of the researchers should have a Ph.D. degree, 50 % a Master's level degree and the remaining 25 % a Bachelor's degree to ensure research output of the highest quality.

### 3.4 Expenditure on Research and Development

The total expenditure of universities on R & D should be at least 2 % [Fig. 1, (4)] of its total budget to get any meaningful research output. In view, of the challenges of funding that Zimbabwean universities are facing, this proposition will not be easy to meet, hence the need for PPPs with the private sector.

### 3.5 PPPs

PPPs [Fig. 1, (5)] are critical in any STP established at a university. Private companies may be tenants at universities, financiers of STP activities or working together with universities in STPs in one way or another. Whatever the situation, there is need for a co-operation framework with the private sector in order to see the successful establishment of STPs at universities.

### 3.6 Sustainable STPs

At this stage [Fig. 1, (6)], the university has managed to establish successful STPs on its campus. A number of key factors have played role to get to this stage—the objectives of the STP, the location advantages of the university, the level of skill of the university, its funding levels of R&D and successful PPPs with the private sector.

## 4 Conclusion

It has been amply demonstrated in this paper that universities do indeed play a critical role in the development of technology Parks. Technology Parks are sprouting up all over the world and are creating a profound impact on global economic landscapes. As knowledge generators which are at the centre of innovation, more investment directed at universities and research institutions is required so as to capacitate them. The role of academia–industry–government collaboration in the creation and success of Science Parks has been highlighted. It has been emphasised that Technology Parks success involves sustained, collaborative efforts by academics, industry representatives, Entrepreneur Support Organisations, Economic Development Organisations, engineers, entrepreneurs, investors, and other practitioners to develop initiatives, plans, methodologies, infrastructure, and action items. This will not only improve our national rankings in terms of knowledge, innovation and ICT application, but have a huge impact on rapid technology Park development.

STPs were the vehicle for the development of the USA, Europe and lately Asian economies. Although Zimbabwe faces challenges across a number of fronts, the country has embarked on the critical path of establishing STPs. The countries universities can play a major a major developmental role through the establishment of sustainable STPs. The above model can prove useful in the country's quest for development through STPs.

## References

Cox, R. N. (1985). Lessons from 30 years of science parks in the U.S.A. In J. M. Gibb (Ed.), *Science parks and innovation centers: Their economic and social impact. Proceedings of the conference held in Berlin* (pp. 17–24). Amsterdam: Elsevier.

Edigheji, O. (2009). *Constructing democratic developmental states in Africa.* Development and its Implications for Higher Education in Southern Africa. SARUA Leadership Dialogue Series (Vol. 1, No 3, pp. 63–74). South Africa.

IASP. (2011). *International association of science parks.* http://www.iasp.ws/publico/index.jsp?enl=10

Kanhukamwe, Q. C. (2007). *Technology and business incubation in Zimbabwe.* Unesco- WTA Workshop, Korea.

Kirzner, I. (1984). The entrepreneurial process. In C. A. Kent (Ed.), *The environment for entrepreneurship* (p. 41). Lexington: Lexington Books.

Luger, M. (1997). Universities and technopolis development. In *Proceedings of WTA conference, Taejon, Korea* (pp. 271–296).

Money, M. L. (1970). *A model for the establishment of a university related research park within a framework of selected management principles.* Ph.D. dissertation, Division of Business Administration, University of Utah.

Link, A. N., & Scott, J. T. (2003). U.S. science parks: The diffusion of an innovation and its effects on the academic missions of universities. *International Journal of Industrial Organization, 21*(9), 1323–1356. http://ideas.repec.org/a/eee/indorg/v21y2003i9p1323-1356.html

Phillips, F. (2005). Sustainability of regional initiatives for technology entrepreneurship. In *EFMD 35th EISB Conference – Sustaining the Entrepreneurial Spirit over Time – Barcelona (Spain)*, September 12–14.

Smilor, W. R., Kozmetsky, G., & Gibson, V. D. (1987). *The dynamics of a developing technopolis.* The Austin/San Antonio Corridor.

Zhang, H., & Sonobe, T. (2011). The development of science and technology parks in china, 1988–2008. *Economics: The Open-Access, Open-Assessment E-Journal, 5,* 2011–2016. doi:10.5018/economicsejournal. ja.2011-6. http://dx.doi.org/10.5018/economics-ejournal.ja.2011-6

Zhang, Y. (2002). *Critical factors for science park management: The North American and European experience.* Department of Marketing & International Management, Hamilton, New Zealand.

Zimbabwe MTP. (2011). *Zimbabwe medium term plan 2011–2015.* Republic of Zimbabwe.

# Part III
# Application and Practice

# The Surrey Research Park: A Case Study of Strategic Planning for Economic Development

**Malcolm Parry**

**Abstract** The development of the Surrey Research Park by the University of Surrey is an addition to a number of existing strategies to collaborate with industry that it has developed over its 120-year history. The potential to undertake this development was based on owning a substantial land holding that the University acquired when the Borough Council for the town of Guildford invited the University to relocate from Battersea in London to its new location in 1966. Initial plans for the Park in 1979 were accelerated in 1981 in response to plans by the government to reduce funding for Higher Education in the UK. Beyond a broad master plan for the site that was based on topography and access to the site the plans that were developed were based on a survey of 100 companies that were deemed to be in the target market for the site and a review of the other seven science parks that were being developed in the UK in 1981. The findings from this proved to be important in developing the master plan for the site. Another important influence on the project were the objectives that were defined for the three stakeholders in the project. Those for the University included commercial potential, knowledge transfer and image and reputation; those for the town primarily related to economic development and the plan was to help tenants gain a competitive advantage by locating on the site. In addition a number of success indicators were defined for the project against which to measure performance and have remained as a useful set of parameters on which to base the assessment of the performance of the site. The chapter sets details about the history of the park and covers the success indicators and factors and reviews these in the context of the original objectives for the site.

M. Parry (✉)
University of Surrey, Guildford GU7 1QB, UK
e-mail: m.parry@surrey.ac.uk

D.-S. Oh and F. Phillips (eds.), *Technopolis*,
DOI: 10.1007/978-1-4471-5508-9_17, 

# 1 Introduction

An important theme that runs through the 120-year history of the University of Surrey is the institution's strong links with industry and business. This long tradition has been important in steering the plans for the development of the Surrey Research Park because the University has been able to use its cultural heritage as a foundation for this project.

Battersea Polytechnic Institute was a purpose built college in the South London district of Battersea. It was founded in 1891 and opened in February 1894. The Institute took on a more scientific and technical leaning from 1920, leading it to be renamed as the Battersea College of Technology in 1957. In 1966 it became the University of Surrey and moved out of london to premises in Guildford in 1968.

The original purpose for the Polytechnic was to provide higher education for the poor and offered courses primarily focussed on engineering, building trades and physical sciences as well as some arts subjects. During the years 1927–1939 the Polytechnic consolidated with a growing emphasis on science, engineering and metallurgy.

During the mid to late 1800 a number of advanced technology companies of the day begun to move to the area of Battersea situated along the south bank of the River Thames in West London. These companies, which included the brewing company Guinness, Garton's Sugars, the Unilever company Prices and the heavy industrial company Morgan Crucible, all needed engineers to develop, manage and operate their respective complex plants. The links between Battersea College and these industrial complexes established an early foundation to the strong links that have endured between the University of Surrey and industrial companies which eventually led to the creation of the Surrey Research Park in order to extend these linkages in the modern era.

In 1966 nearly three quarters of a century after the foundation of Battersea College of Technology was transformed into the University of Surrey through a Royal Charter. which let to a move to Guildford which is the County Town of Surrey that lies just 50 km southwest of Battersea.

This transformation was in response to the growing numbers of young people who were part of the post 1939–1945 World War peak in birth rate and the national need for an increase in the number of in full time higher education (The Robbins Report 1963) and the need to find a larger site on which to develop.

Guildford also has a long history as it has been a settlement of over 1,000 years during much of which it relied on its location as a cross roads between significant traffic travelling from London to the south coast naval port of Portsmouth and more minor traffic following an east west route across south of England as part of the Pilgrims Way. Although primarily a market town in the early twentieth century Guildford attracted some automotive manufacturing and established the first purpose-built motor vehicle factory in Britain however, in the middle of the late 1950s and early 1960s both the automobile manufacturing and the agricultural market activities begun to decline and the Town Council saw the opportunity for a further period of development by attracting a university to the town and laying the foundation for its future as a knowledge-based economy.

In 1966 in collaboration with Guildford's Town Council the University was able to purchase 300 ac (125 ha) of land which was the minimum area required by the UK government if it was to agree to fund the relocation of Battersea Polytechnic to Guildford and establish itself as a university town by hosting the University of Surrey. This minimum land allocation was required to ensure that the University would have sufficient land on which to develop in the future and prevent the problem faced by its pre-cursor organisation which had become land locked on a small site in the south west of London.

In 1966 the new University received its Royal Charter and following a land use planning enquiry which was supported by the town of Guildford and the County of Surrey, planning permission was granted for the University to be established its 300 acre site on the western edge of the town.

The legacy of the origins of the University of Surrey, of supporting an industrial base in London, came with it to Guildford.

Features of this legacy include: significant research links with industry; having in place a department that is dedicated to manage industrial research contracts, technology licencing, and managing a the University's business development outreach unit; and offering each student at the University the opportunity for a 1 year accredited industrial or professional placement as part of the any degree course. This tradition continues, although today the placement year is now no longer a compulsory part of all degree courses; however, its value is that it continues to connect the University to industrial companies and professional organisations across the UK and further abroad and helps the University maintain its leading position for graduate employment in the UK. The tradition of working with the business and industry has helped to build Surrey's reputation of a business focussed research led University.

## 2 Science Parks: An Emerging Trend in the UK

As early as 1964 the British Government urged UK higher education institutions to expand their contact with industry with the objective of increasing the rate of technology transfer to the market place in order to increase the payback from investment in basic research (Edgerton 1996). One impact was that the Mott Committee,[1] in its report published in 1969, recommended an expansion of 'science-based industry' close to Cambridge to take maximum advantage of the

[1] Mott Committee, a special Cambridge University Committee set up under the Chairmanship of Sir Nevill Francis Mott (then Cavendish Professor of Experimental Physics) to consider an appropriate response from Cambridge to an initiative of the Labour government following its election in 1964 published its findings in 1969 in the Mott Report.

concentration of scientific expertise, equipment and libraries and to increase feedback from industry into the Cambridge scientific community.

This change opened the way for creating the Trinity College backed Cambridge Science Park in 1979. The impact of the Cambridge Science Park was described in the Cambridge Phenomenon (Segal et al. 1985) which was characterised by a surge in the formation and growth of high-tech firms and associated services which today employ over 23,000 people[2] and now forms the backbone of the sub-region's successful knowledge-based economy.

In 1979 the idea of extending the University of Surrey's links with industry by creating the Surrey Research Park was promoted by the University of Surrey with the planning authority of Guildford in order to secure an allocation in the County Plan (Structure Plan) for the development of a 70 ac (28.5 ha) site.[3] Their support gave the University the necessary approval in principle to begin to develop the idea for the Surrey Research Park.

In 1981 the UK government reduced its funding for the UK higher education system. This led to those universities which were worst affected by this reduction of state funding to look at other ways to protect their future using their own resources. The University of Surrey used this challenge to accelerate its plans for the development of the Surrey Research Park.

The University formed an internal committee which was chaired by an eminent industrialist and was tasked to oversee the development of the Park. The committee recruited a young academic [4]to take on the position of the Director of the Park to drive the development of the site. The local land use planning authority in Guildford granted permission in 1983 for the initial phase of development of the Surrey Research Park.

In 1984 a wave of seven science parks founded in the UK at that time set up the UK Science Park Association (UKSPA). The Surrey Research Park was one of these parks but is now 1 of 67 operational parks in the UK that are members of the UK Science Park Association. The membership numbers of the UK Science Park Association are noted in Table 1 and Fig. 1.

The location and names of the UKSPA member parks are noted in Fig. 2 and Table 1.

The original eight science parks in the UK were all hosted by universities. However, over the last 30 years other kinds of hosts in the UK have taken the science park model and adapted to suit their own objectives. These hosts now include government defence laboratories such as Porton Down chemical and biological defence laboratories which host the Tetricus Park, and the Malvern Hills Science Park which has as its host another UK defence agency complex. A further important trend that is the location of science parks on corporate research facilities such as the Unilever R&D facility in Colworth Bedfordshire and the Motor

---

[2] http://www.cambridgefutures.org/futures2/report1.htm

[3] Guildford Borough Local Plan, Policies and Proposals 1983, Policy E17.

[4] D. Malolm Party

**Table 1** List of members of the UKSPA 2012

| | UKSPA member parks | | |
|---|---|---|---|
| 1 | Aberdeen Energy and Innovation Parks | 42 | Malvern Hills Science Park |
| 2 | Babraham Research Campus | 43 | Manchester Science Parks |
| 3 | Begbroke Science Park | 44 | Milton Park Innovation Centre |
| 4 | BioCity Nottingham | 45 | MIRA Technology Park |
| 5 | BioCity Scotland | 46 | NETPark—The North East Technology Park |
| 6 | BioPark | 47 | Newark Beacon Innovation Centre |
| 7 | Birmingham Research Park | 48 | No.1 Nottingham Science Park |
| 8 | Birmingham Science Park Aston | 49 | Northern Ireland Science Park |
| 9 | Bristol and Bath Science Park | 50 | Norwich Research Park |
| 10 | Brunel Science Park | 51 | Nottingham Science & Technology Park |
| 11 | Cambridge Biomedical Campus | 52 | The Nucleus Business and Innovation Centre |
| 12 | Cambridge Science Park | 53 | Ocean Village Innovation Centre |
| 13 | Cardiff Business Technology Park | 54 | The Oxford Science Park |
| 14 | CEME Innovation Centre | 55 | Pentlands Science Park |
| 15 | Chesterford Research Park | 56 | Portsmouth Technopole |
| 16 | Colworth Science Park | 57 | Queen Mary BioEnterprises Innovation Centre |
| 17 | Coventry University Technology Park | 58 | Roslin BioCentre |
| 18 | Cranfield Technology Park | 59 | Sand Hutton Applied Innovation Campus |
| 19 | Culham Innovation Centre | 60 | Scottish Enterprise Technology Park |
| 20 | Daresbury Science and Innovation Campus | 61 | St John's Innovation Centre |
| 21 | The Digital Media Centre | 62 | Stevenage Bioscience Catalyst |
| 22 | Edinburgh BioQuarter | 63 | Stirling University Innovation Park |
| 23 | Elvingston Science Centre | 64 | Sunderland Science Park |
| 24 | European Marine Science Park | 65 | The Surrey Research Park |
| 25 | Exeter Science Park | 66 | Tamar Science Park |
| 26 | Granta Park | 67 | Technium |
| 27 | Harwell Innovation Centre | 68 | Tetricus Science Park |
| 28 | Harwell Oxford | 69 | The Bridge, Dartford |
| 29 | Haverhill Research Park | 70 | University of Cambridge West Cambridge Site |
| 30 | Heriot-Watt University Research Park | 71 | University of Essex Knowledge Gateway |
| 31 | Hethel Engineering Centre | 72 | The University of Nottingham Innovation Park (UNIP) |
| 32 | The Imperial Incubator | 73 | University of Reading Science and Technology and Enterprise Centres |
| 33 | Institute of Life Science | 74 | University of Sheffield |
| 34 | Keele University Science and Business Park | 75 | The University of Southampton Science Park |
| 35 | Kent Science Park | 76 | University of Warwick Science Park |
| 36 | Lancaster Science Park | 77 | Wellingborough Innovation Centre |
| 37 | Langstone Technology Park | 78 | West of Scotland Science Park |
| 38 | Leeds Innovation Centre | 79 | Westlakes Science and Technology Park |

(continued)

**Table 1** (continued)

| | UKSPA member parks | | |
|---|---|---|---|
| 39 | Liverpool Innovation Park | 80 | The Wilton Centre |
| 40 | Longbridge Technology Park | 81 | Wolverhampton Science Park |
| 41 | Loughborough University Science and Enterprise Park | | |

*Source* Data from UKSPA.org.uk 2012

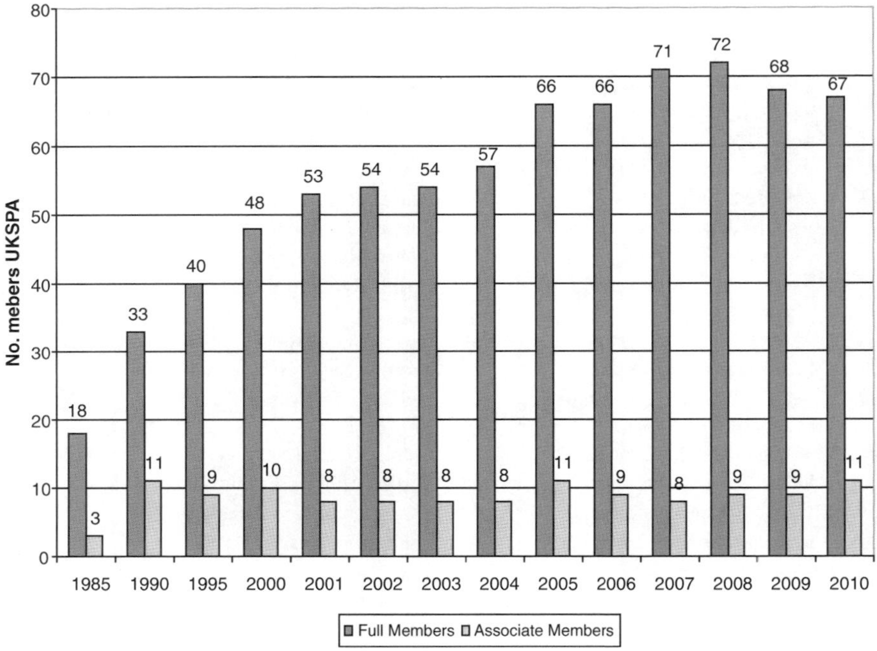


**Fig. 1** Numbers of parks which are members of the UKSPA, Note: full members are parks with tenant companies on site, associate members are those parks that are at the planning stage of their development. *Source* UKSPA statistics. http://www.ukspa.org.uk

Industries Research Association's research facilities in Nuneaton. This diversification of hosts demonstrates the versatility of the European science park model all of which are local initiatives that focus on building technology clusters around their local knowledge infrastructure.

The success of this strategy is noted by the number of companies located on UKSPA member's parks (Fig. 3).

The number of employees in these companies are shown in Fig. 4.

**Fig. 2** The location of the UK science parks that are members of the UK science park association. *Source* Perse Comm-UKSPA 2012

## 2.1 Objectives

Typically, the objectives for science parks combine the interests of the three usual stakeholders of government, the host organisation and the tenant companies. However, a common theme that binds together these stakeholders in the projects is wealth creation.

The original objectives, or value propositions, for the Surrey Research Park that were set out in its business plan relate to its three stakeholders of: the local

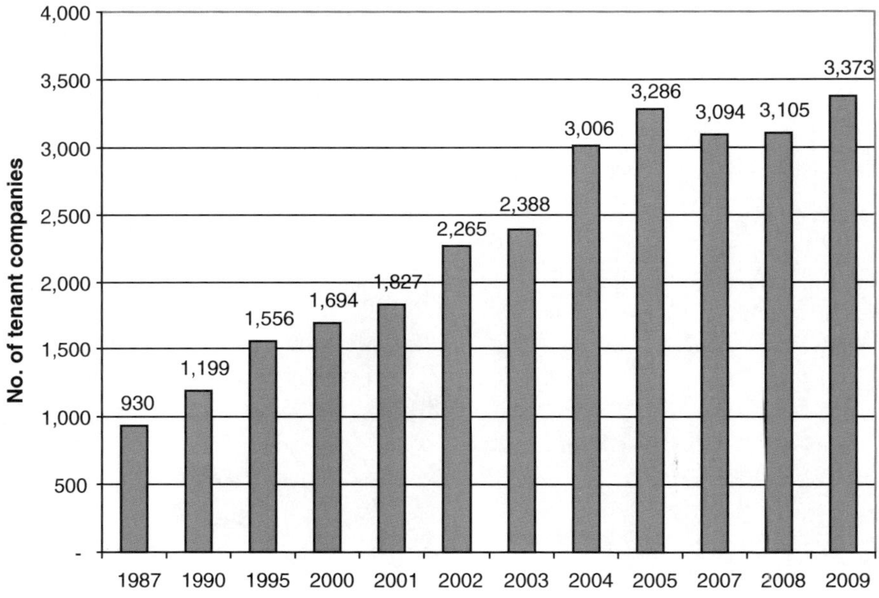


**Fig. 3** The number of technology companies on science parks in UK by year, *Source* UKSPA.org.uk 2012

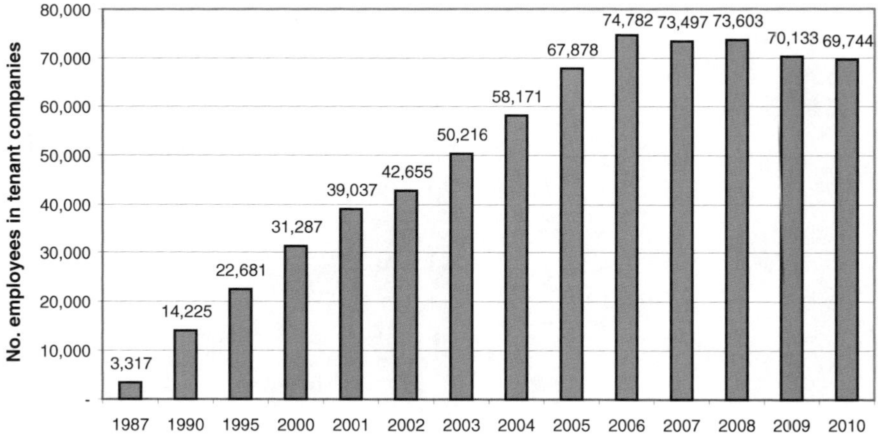


**Fig. 4** The number of employee in UK Science Park tenant companies by year, *Source* UKSPA.org.uk 2012

government, which is the planning authority; the University of Surrey as land owner, investor, developer and manager of the park; and the tenant companies that have a vested interest in the site providing benefits which lead to a competitive advantage for their business.

**Table 2** Strategic objectives for science parks

| Stakeholder | Conceptual objectives for the stakeholder |
|---|---|
| The University of Surrey | To create some independent income for the University of Surrey and create the opportunity for academic staff to secure additional income by working with companies established on the park |
| | To create an opportunity for technology transfer from the University and other sources into the commercial domain |
| | To raise the profile of the University of Surrey as a centre of excellence in technology |
| Guildford Borough Council and Surrey County Council (the planning authorities) | To assist in the process of the economic development of the region and locality |
| Tenant companies | To establish a business in an environment that favours the formation, development and growth in order to gain a competitive advantage through access to skills and technology |

The broad strategic objectives for each of these stakeholders are noted in Table 2.

These conceptual objectives were then translated into a series of development objectives for the site which were based on an analysis in 1982 of the market for suitable accommodation for small technology companies in the region. This comprised a telephone-based survey of 100 technology companies and research laboratories within 25 miles of the proposed park, and an assessment of the other seven science parks that were being established in the UK in that period.

The findings from this study included:

- Access to the commercial property market at that time in the UK was a very significant barrier to entrepreneurs wishing to establish a small technology based company particularly where they had no trading record. All commercial property was offered on long leases of around 25 years and occupation required substantial rental deposits. In addition few if any small high quality units were available for this emerging market of micro and small technology companies. This was recognised as a significant barrier of technology based SMEs establishing an office base from which to develop their activities. The Science Park movement pioneered short occupational leases and licences for high risk technology companies.
- The emerging personal computer revolution which significantly reduced the cost of computing created an opportunity for small science, engineering and technology companies to establish a business without the need for access to substantial capital resources. Access to low cost versatile computing not only supported business in existing markets but created a number of technology enabled markets for software which has for many years and continues to drive the ICT sector.

- Revising the received wisdom that the commercialisation of most R&D required very sophisticated laboratories. Most of the companies that were interviewed were looking for accommodation that would take technology beyond the discovery phase towards the market and it was not straight laboratory space that was required but rather a mixture of laboratories and office space for commercial activities or pure office space.
- New markets were emerging through the de-regulation of such industrial sectors as ICT and the financial sector; there was increasing regulation concerning the environment, the automotive industry and the energy sector; and the change in attitude by government to the release of intellectual property (IP) from its defence laboratories and the management of IP in universities were producing a raft of new technology enabled markets which were attractive to young technology entrepreneurs.

The study led to the conclusion that the facilities that should be built:

- Should be able to provide units for a number of sizes in order to allow companies to move and match accommodation with need.
- The accommodation should be planned to enable its use as wet (chemistry/biotech), dry (engineering laboratories) or for office accommodation.
- The nature of occupancy contracts should be able to offer for lengths which suit high growth companies that need to be able to grow rapidly, or if they sell some of their business, to reduce in size.
- The Park should not restrict tenants to technologies in which the University was a leader because this would limit the capacity of entrepreneurs that were seeking to merger new technologies to develop new markets; the decision was made to make the park technologically promiscuous.

## 2.2 Physical Development and Master Plan

Part of the success for the Surrey Research Park derived from the interpretation and translation of the conceptual objectives into a physical plan that would meet the needs for a new type of small technology company that could be established using the new business tools provided by access to lower cost computing.

The development objectives that were a response to the perceived market were used to establish a Master Plan. This Plan, which still works well after 30 years of developing the site, is based on creating three distinct zones on the park. These were planned to accommodate:

Small units for small start up companies or specialist parts of large companies; this offers units ranging in size from 25 $m^2$ (250 $ft^2$) to 300 $m^2$ (3,000 $ft^2$).

Medium sized units: these are for companies with an annual turnover of between £15 and £40 million or for national research facilities of multinational companies. Large buildings to accommodate headquarters for technology

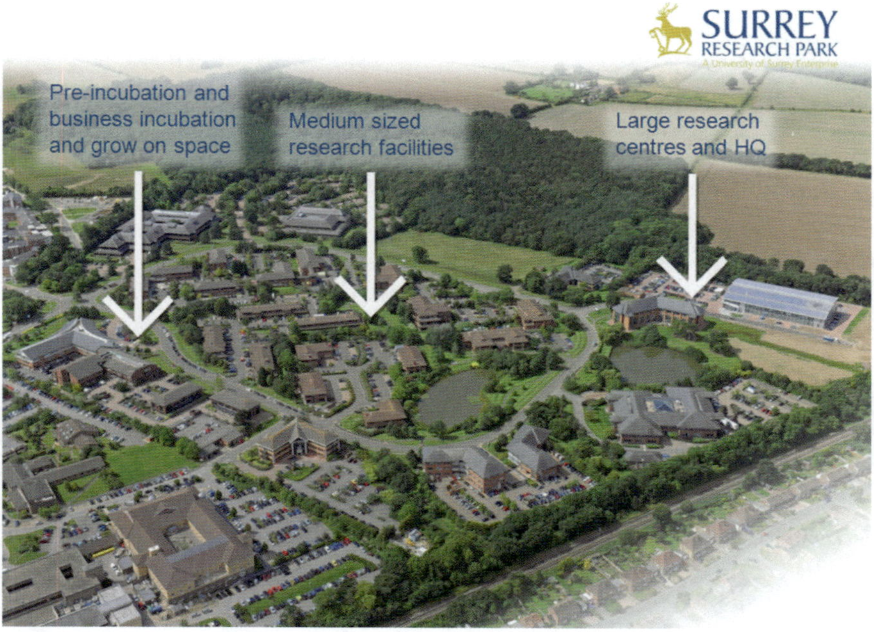


**Fig. 5** Aerial photograph of the Surrey Research Park 2012, This shows the three zones of development for small, medium and large companies on which the master plan for the site is based

businesses, research centres for large technology based companies and facilities for high value manufacturing activities such as building satellites (Fig. 5).

In addition to this broad zoning plan which was proposed by the University, a number of other planning criteria were imposed by the town's planning department as part of the agreed Master Plan. This covered such matters as the width of access roads, parking ratios of 1 car space to 23 $m^2$ of gross space, building lines, architectural style, building materials, building heights, the density of development (25 %), building footprints (15 %) the principle of screening cars behind buildings and a landscape plan.

The total permitted area under the 1984 planning permission allowed 71,250 $m^2$ of gross external space. The original intention was to develop the whole site by 1991 but the University, by undertaking the development itself, has been cautious and has built at a lower rate. In 2012 the development had reach 90 % completion. Between 1983 and 1994 the University developed the site by constructing a series of speculative projects for SMEs. The policy since 1994 has been to build to order rather than as speculative developments. Part of the logic for this shift in strategy was that by 1994 the areas of the site allocated to small start up and grown on space had been fully developed.

Details of the rate of build of space on the Surrey Research Park are noted in Fig. 6.

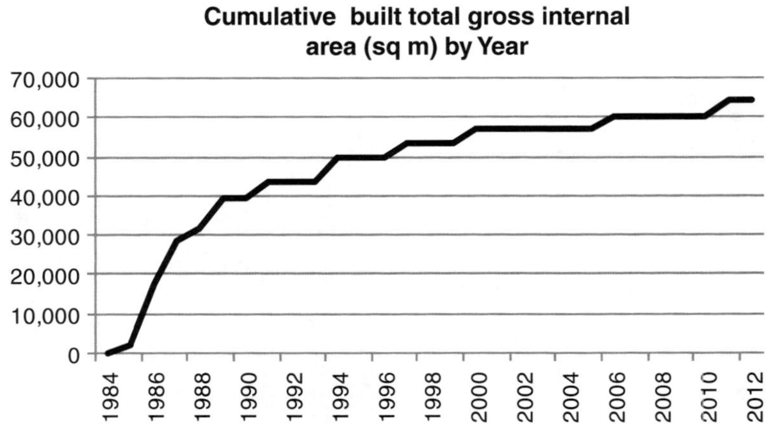


**Fig. 6** Cumulative gross built space on the Surrey Research Park since its inception

The agreed plan has proved to be not only versatile but also gives comfort to potential occupiers as they can see that the Park provides accommodation for future growth of their companies.

## *2.3 Permitted Uses*

One of the features of science and technology park brand of development is the restriction on the permitted uses on these sites. The importance of this is that it maintains a standard of occupier and also prevents mission drift towards accepting tenant companies that may not be appropriate for the site. The permitted use agreed with the planning authority for the Surrey Research Park allows research, development and design in any science including the social sciences and is complementary to the activities of the University of Surrey. This use clause reflects the difference between a science park and a research park. The former also allows some limited high value low volume manufacturing on the site.

In 2011 a 40,000 $m^2$ technical and production facility was completed for a spin out company (Surrey Satellite Technology Ltd—SSTL) from the University. It is now clear that the original planning use for the site was too prescriptive as this failed to provide the flexibility to allow the complete cycle of laboratory to production in one location. This issue has now been resolved with a wider use being permitted on the Park.

The experience in Korea (Oh and Yeom 2012) is that for a long-term economically sustainable activity to develop from a science and technology park there is a need to provide a land allocation to enable manufacturing to be developed from the activities on the Park as a development matures.

There is also evidence that when science parks are located next to a research hospital there is potential for establishing successful translational research clinics on these sites. Experience has shown that the location of the Surrey Research Park next to a National Health Service regional research hospital has attracted a number of hospital consultants to set up specialist clinics on the Park. The work in these clinics does not conflict with the permitted use because the kind of work done by medical consultants develops can lead to advances medical practice which can be justified as an acceptable use. The experience at Surrey is that a number of patents have been granted to clinicians working in these centres.

## 3 Internal Management Structure

In 1983 the University of Surrey concluded that to undertake the development of the Surrey Research Park itself would be the best way to control the development and meet is stated objectives. The management structure that was selected is noted in Fig. 7; however, the University also considered a second option which it may have had to adopt if it could not raise the funds itself for the development. This second option was not adopted in Fig. 8.

In selecting the right management structure the University reviewed the risks associated with the different ownership and management structures. The observations are noted in Table 3.

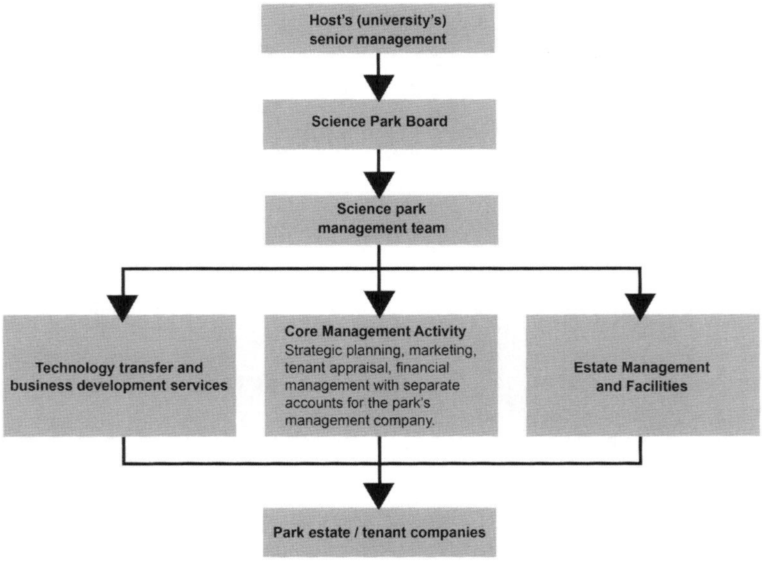


**Fig. 7** Governance structure for single ownership science and technology parks—this structure was selected for the Surrey Research Park

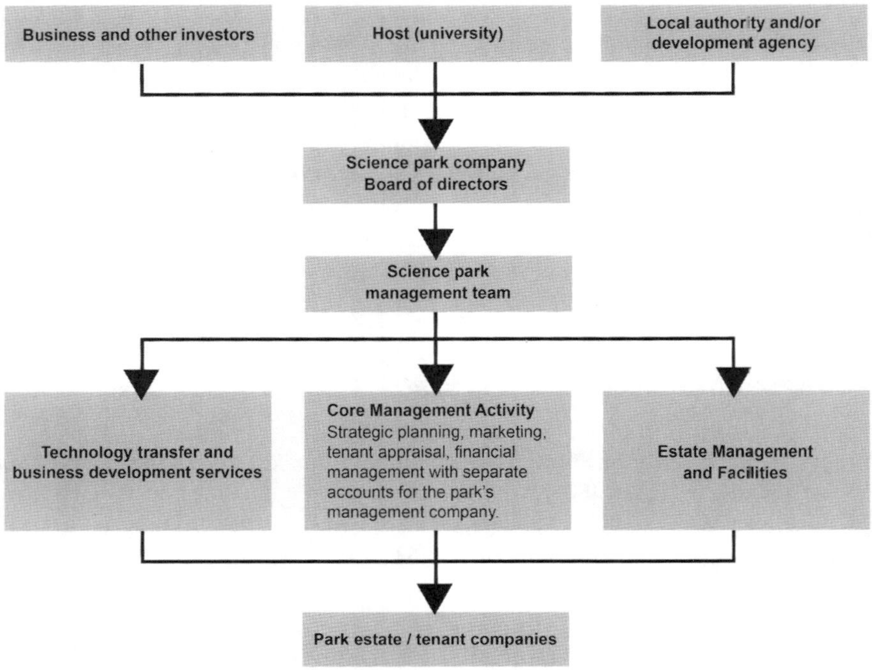


**Fig. 8** Indicative management structure for Surrey Research Park based on a joint venture structure

**Table 3** Details of responsibilities for costs, management control and involvement in tenant selection for single and joint venture ownership arrangements

| Development strategy | Financial cost to university | Strength and nature of control over science park policy | Involvement in tenant selection |
|---|---|---|---|
| Single ownership—university alone | High—needs funds for infrastructure, initial buildings and needs to guarantee any loan finance | High—university has complete control over management policy | High—sole arbiter |
| Joint venture (JV) company | Medium—shares costs with partners in JV arrangement | Potentially high—negotiable between partners but potentially strong where University takes a lead | Usually high—significant or leading influence |

The Management Team and its Board operated under delegated powers from the University's Finance Committee and was able to decide on investments in the Park and to authorise taking loans for development.

The Development Team that eventually changed into the Management Team for the site was established as an Enterprise Group of the University rather than as a separate company. However, this Group operates a separate bank account from the main University account for operational purposes and any surpluses are transferred to the University at the end of each financial year.

### *3.1 Funding*

The Surrey Research Park is one of the few self funded parks in the UK. In the initial planning stage two potential funding plans were explored. The first involved seeking joint venture partners with a commercial developer for each individual phase. The second option was a land sale (on a long leasehold arrangement—125 years) to an anchor tenant which would secure a capital sum to enable the initial phase of the small units to be built.

The Board asked that before seeking a partner with which to create a joint venture an attempt should be made to identify an anchor tenant for the land sale option. A marketing campaign was put in place which was aimed at all UK based companies that were aligned to the University's strengths in chemical and process engineering, toxicology and pharmaceuticals.

This campaign identified the UK company, BOC Ltd (now BOC Linde) which acquired a long leasehold, in 1983, on a 10.25 ac (4.25 ha) for £2.5 m on which to base their UK HQ and R&D management activities. This transaction was sufficient to set out an initial phase of infrastructure and enable the first phase of buildings without the University taking a partner. All subsequent funding of buildings by the University Development Team has been based on loan finance from a number of banks and these loans have been secured against the income stream from let buildings.

### *3.2 Phased Development*

The Park has been developed in phases following the initial sale of land to its anchor tenant BOC. These subsequent phases of development were undertaken by the University on a speculative basis. The plan has been cautious and the University has not over built at any one time which has ensured high occupancy (averaging over 90 %).[5]

This careful approach has been helped by the early development of the Surrey Technology Centre. This business incubator has been instrumental in helping some 500 start-ups companies of which some, if they have not acquired by larger companies, have grown on site using the Research Park's flexible leases to "staircase" their way to quite large and suitable accommodation on the site.

[5] University of Surrey Annual Accounts of the Foundation Fund.

### 3.3 Management Team

The Research Park Management team is located on the Park. It comprises a core management group that is responsible for the overall planning, development and management of the site. The work streams of this group include: dealing with the estate from the perspective of the initial development, finding occupiers and then managing the estate. A second stream of work is concerned with providing the business development service package and links with the University of Surrey's Research and Enterprise Services.

In the UK there has been no formal government support for science and technology parks. To provide the various support packages Park operators have had to rely on building partnerships with those organisations that have from time to time been in place to support general UK businesses. These include the now defunct government funded Business Link programme as well as more specialist government funded group such as the Surrey Enterprise Hub,[6] and the Innovation and Growth team programme for the same area both of which programmes have been cut because of a change in government policy related to providing government support to business. The two latter programmes had their management teams based on the Surrey Research Park and were respectively concerned with helping high growth companies through a coaching and mentoring process and connecting micro, small, medium, large and multinational companies in order to encourage innovation in the region's business community. These programmes proved to be highly effective and their closure was politically motivated following a change of government in May 2010.

In 1997 the UK government formally extended the responsibilities of all universities to extend beyond teaching and research to include business and community development. Grant funding has been made available to all universities to support this initiative. The University's Surrey Research Park Office in collaboration with the University's technology commercialisation office (Research and Enterprise Services) have secured a government grant from the Higher Education Innovation Fund to create a small pre-incubator, known as SET Squared,[7] in the Park's existing Surrey Technology Centre. In 2011 the University secured a further grant with which to develop a space technology incubator known as the International Space Innovation Centre.[8]

---

[6] http://www3.surrey.ac.uk/stc/sehub.html

[7] http://setsquared.co.uk/

[8] http://isic-space.com/

## 4 Measures of Success Against the Original Development Objectives

The performance of science and technology parks is a topic which is of great interest to politicians because many of these projects are publicly funded and compete for public funding and in some instances for the land that is necessary for their development. Although the University of Surrey was not exposed to this political scrutiny there has been concern over time to assess the performance of the Park. In 1996 some broad qualitative and quantitative measures were developed for the measurement of the performance of the Surrey Research Park against its original five objectives of income generation, technology transfer (now better described as knowledge transfer), profile, economic development, and supporting companies on the site in order to help to give them a competitive advantage. This system is still being utilised today to review progress.

Creating some Independent income for the University of Surrey and creating capital value of the assets of the park—success indicators:

*Rate of development*: the original timescale for the development of the Surrey Research Park was 10 years; however, it has taken nearly 30 years to achieve 90 % developed. Only sites for larger building still remain undeveloped after this period. The areas originally designated to micro and small companies was completed within the original planned 10 years. The slower rate of developments for larger companies reflects the fact that today many larger companies are not locating their R&D effort in single large buildings but are looking to deploy their research activities as small specialist groups close to specific host organisations that have defined technology competences.

Evidence of this is shown by the presence on the Surrey Research Park over 19 years of the Mitsubishi Research Centre and for research centres for Kobe Steel, Canon and Hygiena International. The initial Master Plan for the site enabled the Park to develop sites as phases. This flexibility has proved to be important feature of the site.

*Capital value and rental income for the University*: the developed area of the park has been valued as a property asset at £80 million (2006).[9] The rental income generated for the University of Surrey over the period of the development is noted in the Fig. 9.

This rental income has generated surpluses for the University of in excess of £65 millions which have been used to support scholarship and the Park's capital value has enabled the University to use this as collateral for borrowings to support the development of the University.

*Occupancy rates*: the historic data collected by the Park for its annual accounts and financial planning has shown that the park operated at a level of between 90 and 97 %.

---

[9] University of Surrey Foundation Fund Annual Report 2010–2011.

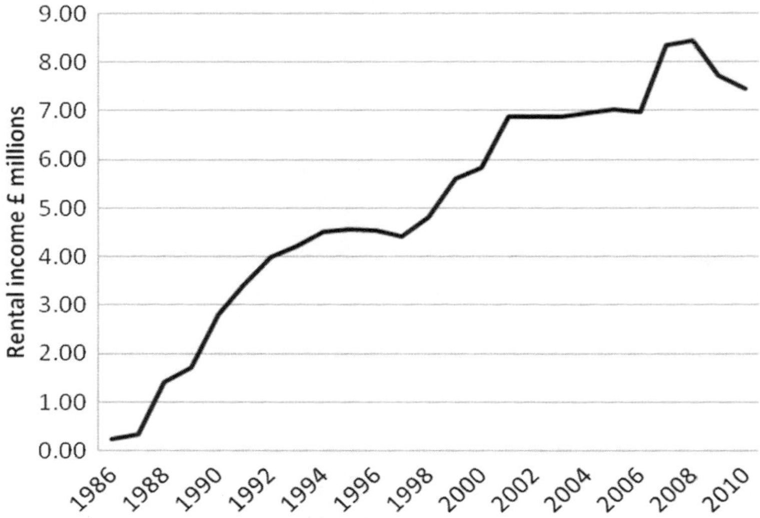


**Fig. 9** Rental income growth (millions) by year

Creating some Independent income for the University of Surrey and creating capital value of the assets of the park—success factors:

*Location and product offered to the market:* the University of Surrey is located in a high property value area where there are few other commercial property projects that offer either the flexibility of short-term rental contracts or the level of support for businesses when compared with those provided by the Surrey Research Park. The development of the Surrey Research Park addressed the issue of market failure in relation to the provision of this kind of accommodation which is favoured by fast growing, venture capital backed technology companies which need, if they are to respond to the market potential of their technology, flexible leases. There is a demand for this kind of space as statistics show that 6 of the 11 districts and boroughs in the County of Surrey are in the top 25 % for numbers of knowledge-based businesses in the UK. In 2007, Surrey had 58 business registrations per 10,000 adult populations compared to a figure of 40 for the wider area of the South East of the UK (Surrey Economic Partnership Ltd. 2010).

*Brand value of science and technology parks*: an informal survey of the companies on the Park in 2009 indicated that the image and reputation of the Surrey Research Park is considered very important for the micro and SME companies that have located on the site. It is clear that science and technology parks have developed as a brand of property development that matches the needs of the types of companies that tend to found on these sites. (UKSPA 2003).

*Presence of business of pre and full incubation facilities:* the statistics for the Surrey Research Park show that 40 % of the occupiers in space on the Park, other than in the Surrey Technology Centre business incubator and its two pre-incubators, have grown out of this Centre. This system has helped the park maintain an occupancy rate of between 90 and 97 % for over 20 years. High occupancy rates translate into rental income and value.

Raising the profile of the University of Surrey—success indicators:

The University of Surrey is a relatively small university with just over 13,500 full time equivalent students[10] which in terms of international comparisons is a relatively small university. However, in addition to the quality of its courses the University has become well known through the success of the Surrey Research Park and its impact in terms of economic development and from the formation and growth of its satellite business, SSTL, which is now located on the Park.

*Origin of tenant companies:* the indicators are that nearly one-third of the tenants on the Park are from overseas.

*International visitors:* the Park also receives on average one international visitor per week that is seeking an understanding of how the project has been developed and its impact on the regional economy.

*Membership of international organisations:* the Surrey Research Park was invited to be a founder member of the World Technopolis Association in 1994 that is based in Taejon Korea and it is also active in both the National UK Science Park Association (UKSPA) as well as the International Association (IASP).

*Business incubation*: the Surrey Technology Centre (STC) has been in operation as a business incubator since it was opened by the University in 1986. Over that time over 500 companies have taken contracts in the business incubator.[11] The success of this building has supported the University's application to government for a grant of in excess of £ 10.5 m to support a pre-incubator. This pre-incubator is now established as Surrey SET Squared and is part of a consortium of the universities of Surrey, Southampton, Bristol, Bath and Exeter.[12] The Surrey SET Squared programme occupies 105 $m^2$ in the STC and provides a range of mentoring, coaching and business development services to entrepreneurs that are trying to build global reach technology companies.

The success of the Surrey Research Park and the SET Squared operation also attracted a grant of £0.96 m to support an International Space Innovation Centre (ISIC) which is a sector specific pre-incubator.[13] This also operates from the Surrey Technology Centre. The objective for the ISIC is to support the development of companies that have a focus on space related technologies which also include remote sensing applications by drawing together academic research partners, international businesses and SMEs into a 'cluster', stimulating an entrepreneurial environment for uniting the upstream and downstream space industry.

Raise the profile of the University of Surrey—success factors:

---

[10] http://portal.surrey.ac.uk/portal/page?_pageid=719,333086&_dad=portal&_schema=PORTAL

[11] Data from the Surrey Research Park tenant occupancy records.

[12] http://www.setsquared.co.uk/home/contact-us/surrey-guildford

[13] http://www.isic-surrey.co.uk/

*Links with business*: the University's long tradition of supporting business and industry through its research activities, its record for graduate employment[14] by creating a pool of well qualified and readily employable people has helped to raise the profile of the university. The University's industry and business facing philosophy has been important as this has helped the Park's development team to promote the Park as a good location for accessing skills and technology.
*Tenant company success*: the success of the tenant companies themselves has helped to raise the profile of the Park. The database of tenants that has been assembled since 1986 has shown over 585[15] tenants that have taken contracts on the Park. The success of many of the Park's tenant companies has helped to raise the profile of the Park and the University. This has been extended because the global reach of products of the majority of the tenant companies and discussions with a number of our major occupiers has revealed that in excess of 50 % of their turnover comes from exports. Examples include the sale of computer games by: Lionhead (subsidiary of Microsoft), IDBS the leading database software company for pharma sector which has won a number of Queens Awards including for export, SSTL which designs, builds, launches and sells satellites to governments around the world, ReNeuron which sells stem cell products to the international market, TMO Renewables which has developed, patented designs and sells bio-fuel fermentation plant into the international market, and Detica which has grown from 30 staff to in excess of 1,400 and has a turnover in excess of £200 m.
*International promotion:* the Park's management is active in promoting the Park and its tenant companies through the UK's extensive national and regional inward investment initiatives which is currently the managed through UK Trade Investment (UKTI).
Technology transfer (knowledge transfer)—success indicators:
One of the most difficult objectives to measure with any success is the level of knowledge transfer to companies on the site. Initial measures were concerned with the transfer of technology from the University to companies on the Park; however, experience has shown that a number of the companies on the site have based their business on technology that has been licenced from other sources such as research hospitals, defence laboratories, corporate research laboratories, and by developing business ideas from their own technology consultancy activities. This broader reach of companies seeking technology and the wider scope of knowledge transfer has increased the potential impact from this objective.
*University spin—out companies*: around 10 % of the companies on the Park are spin-out enterprises from the University. This includes SSTL which is itself a £100 million business and employs in excess of 465 people.[16]

---

[14] http://www.telegraph.co.uk/finance/jobs/8138447/Best-and-worst-universities-for-graduate-jobs.html

[15] Perse Comm. Director of the Surrey Research Park 2012 after analysis of the contracts for occupation.

[16] SSTL 2011 published accounts.

*University linkages*: nearly 70 % of the companies on the Park have links of some form with the University of Surrey. These range from soft links that include attending training courses, using University facilities including the library, and using students for project work, to hard links that include formal research contract, co-purchase of equipment, and running Knowledge Transfer Partnerships.[17] Funded through the University's Research and Enterprise Unit with Higher Innovation Fund (HEIF) money the University runs pre-incubators (SET Squared and ISIC) which have created pathways for entrepreneurs looking to develop their ideas into commercial enterprise.
*Staff and student transfers:* since the park was established there have been a number of members of faculty that have been recruited by companies on the Park and conversely some current members of faculty have been recruited from the companies on the Park. A high proportion of tenant companies on the site employ University of Surrey graduates.

In addition to knowledge transfer from the University to business there are a number of other examples of knowledge transfer from other "discovery" organisations in the region into successful technology companies. Examples of these are set out in the Table 4.
Technology transfer—success factors:
*University of Surrey's research rating*: the quality and reputation of the University of Surrey as a research university has been instrumental in delivering the potential for companies building working relationships with the University. The University has a good overall research activity rating. In 2011 it improved its position from 33 out of 116 institutions in the UK to 28th position. In 2008 the government's Research Assessment Exercise indicated that 88 % of Surrey's research activity was rated either 'world class' or 'internationally recognised' with four areas of research activity in the top ten and a total of nine areas ranked in the top 20. The activities of 425 researchers from across all four faculties were submitted in 14 subject areas. Results show that almost half of Surrey's assessed research staff work in areas that have been ranked in the top ten. Fifty-five percent of Surrey's units of assessment were rated as 'internationally excellent' or 'world class', with a further 44 % ranked as 'recognised either nationally or internationally'.[18]

The government's Research Assessment Exercise has been replaced by the Research Excellence Framework. To improve its level of excellence the University of Surrey established a research strategy in 2011 which runs to 2017.[19] The strategic decision has been made to improve the ranking of the University in the

[17] http://www.ktponline.org.uk/

[18] http://www.surrey.ac.uk/research/rae/

[19] http://www.google.co.uk/url?sa=t&rct=j&q=university%20of%20surrey%20research%20rating%202011&source=web&cd=1&sqi=2&ved=0CFsQFjAA&url=http%3A%2F%2Fportal.surrey.ac.uk%2Fpls%2Fportal%2Furl%2FITEM%2FAB3D9C8D8BBD744AE0440003BA296BDE&ei=l4u3T6DHMaf80QXH-MicCA&usg=AFQjCNGsJ2d791jBdyyTM68evisTzVBXlg

**Table 4** Examples of companies on the Surrey Research Park which have been founded on technology from centres in the region

| Company | Origin | Technology | Funding |
|---|---|---|---|
| ReNeuron | Kings College Hospital, London | Stem Cell technology | Venture Capital (VC) backed |
| Themotech | University of Surrey | Thermodynamic properties of materials | Funded by consultancy and revenue from product |
| Stingray | QinetiQ | Laser technology for oil technology | VC backed company developed warrantable product which was then acquired by Norwegian company that specialises in geoscientific data products fo the oil industry |
| SSTL | University of Surrey | Satellite Engineering and manufacture | Acquired from the University of Surrey by the Dutch company European Aeronautic Defence and Space Company EADS N.V. (EADS) |
| Bullfrog Computer Games | Local start up from business community—technology developed by locally developed skills | Computer games (synthetic environments) | Acquired by Electronic Arts |
| Lionhead Studios | Local start up from business community | Computer games (synthetic environments) | Acquired by Microsoft |
| Omniperception | University of Surrey | Biometrics company | Investor backed |
| Parsortix which is a company developed by ANGLE Technology | Leading Cancer Research Centres across the world | Cancer diagnosis | Backed by ANGLE Technology |
| Criterion Software | Criterion Software Ltd created in 1993 to commercialise 3D graphics rendering technology from within Canon's European Research Lab on the Park | Computer games systems | Initially spun out as a majority Canon-owned start-up but then acquired by EA Ltd which is based in Guildford |

(continued)

**Table 4** (continued)

| Company | Origin | Technology | Funding |
|---|---|---|---|
| TMO Renewables | University of Surrey and University of London | Bio-fuels | VC backed fermentation technology company |
| Actica | Spin out from Detica which is another company on the Park | High level consulting | Funded by consultancy |
| Gold-I | Foreign Exchange Trading Software | Spin out from City of London Banking Sector | Funded by revenue from product |
| IDBS | Data acquisition and storage for pharma sector | Developed idea when undertaking consultancy for the Pharma sector | Funded by consultancy and revenue from product |
| Medpharm | Drug formulation and delivery | A spin-out of King's College London from the Department of Pharmacy | Contract Research Organisation |

Academic Ranking of World Universities[20] (ARWU), which provides an annual global research ranking.

The fit of technical subject in the University of Surrey with the interests of tenant companies is high. This was not a deliberate policy but has been important in driving success.

The University of Surrey has put in place the necessary management structures in its Research and Enterprise Services which support links with tenant companies.

To assist in the economic development of the national, regional and local economy—success indicators:

Commonly used indices of the economic impact of companies on science and technology parks include a change in the number of companies on a park, numbers of employees, the nature of the activities of the employees, salary levels and turnover per employee.

*Employee numbers*: the tenant companies on the Surrey Research Park together employ over 3,500 well qualified staff. A significant proportion of the employment offered on the Park provides new jobs that have been created as a result of company growth. The most significant growth has been by Detica which has increased employee numbers from 30 staff in 1986 to over 1,400 today in 2012, the majority of which are located on the Park.

*Employment:* it is estimated that based on discussions with key companies that at least 35 % of these are routinely engaged in R&D activities.

[20] Commonly known as the Shanghai Jiao Tong Index.

*Wages*: the mean salary in 2012 was $ 85,000[21] this is against the national mean wage of $ 36,200[22] in the UK as a whole. A significant number of employees have been drawn from the local community as companies have expanded. Many of the posts are higher added value compared with the regional average, salaries are also higher than the regional average and the level of qualification of the employees is higher than comparable companies not on the Park. This is consistent with the finding of the UKSPA study on the performance of science parks[23] which was based on a study of matched samples of over 800 companies that clearly showed companies on science parks had better qualified staff compared with their equivalent non park based companies.
*Numbers of companies on the Park:* the current number of companies on the Park is currently 180; however, this includes those in SET Squared and ISIC pre-incubators. The number of those with contracts for occupation of space on the Park is 123 and a proportion of these operate a number of subsidiaries which are developing a range of technologies giving in the region of 140 companies. This includes new companies in the pre-incubation phase. The number of companies that have signed contracts directly with the Research Park Office, which does not include those in pre-incubation, is over 580 since 1986.

The growth of tenant companies on the Park is shown in Fig. 10. This includes some companies that have technology based subsidiaries operating from the site and those in the pre-incubation programme with SET Squared and the ISIC pre-incubators.

Tracking companies over time has proved to be difficult because many companies are acquired by larger companies or competitors in their sector. In this process they lose their identities very quickly.

A number of the original occupiers of new buildings constructed on the Park remain and have renewed their occupancy contracts. This demonstrates that the companies are stable and that they are able to secure the skills that are necessary to continue to develop their business.
*Length of life*: 80 % of the companies on the Surrey Research are more than 5 years old. The survival rate of Surrey companies is 80 % which is much greater than the UK 5-year business survival rate of 44.4 %.[24]
*Capital raised:* including IPOs and acquisitions it is estimated that around £1 bn of investment has been made in the companies that currently operate or have operated from the park in the period since its creation.
*Company turnover*: a study on the turnover (gross sales/number of employees) per employee of the companies on the Surrey Research has shown that the average

---

[21] Data from a sample of accounts of companies on the Surrey Research Park.

[22] http://www.ons.gov.uk/ons/dcp171766_264246.pdf

[23] UKSPA 2002 The evaluation of the past & future economic contribution of the UK Science Park Movement.

[24] http://www.ons.gov.uk/ons/rel/bus-register/business-demography/2010/stb—business-demography-2010.html#tab-Business-survivals

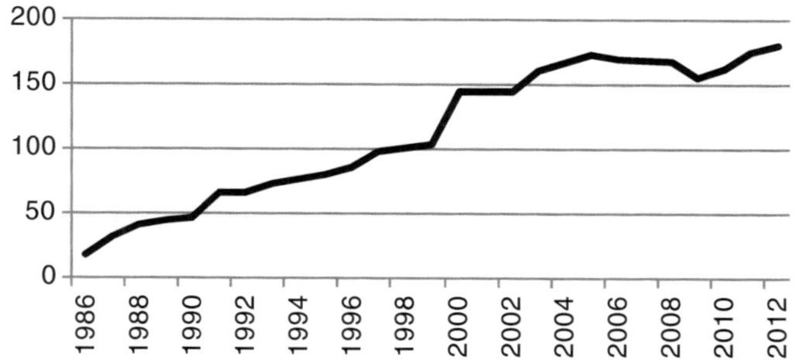


**Fig. 10** Growth of the number of tenants companies on the Surrey Research Park by year

figure is in excess of £150 K with some companies such as SSTL having a turnover per employee of in excess of £225 K. A count of the number of employees working on the Park in June 2012 showed just more than 3,500 staff on site. A calculation based on this figure (turnover per employee x number of companies) indicates the gross sales of the tenant companies of over £475 m. A review of the purchase ledger of two large tenants revealed that 60 % of the company spend was within a radius of 65 km. This figure is influenced by the fact that many of the companies on the Surrey Research Park are still in the development phase of their businesses which is more labour intensive than in companies that are involved in manufacturing. The spend on the purchase ledger of these more labour intensive companies is higher than for manufacturing so there tends to be a greater local impact.

It is estimated that, based on the collective turnover of the companies on Park, their "spend" in the region and the multiplier effect on this spend figures, they contribute in between £350 m and £450 m annually to the region's economy.

To assist in the economic development of the national, regional and local economy—success factors:

The success factors for the Surrey Research Park are based around the location, the offering to tenant companies, the package of business support services offered to occupiers and the attitude to supporting micro and SMEs, of the University that is reflected through the management of the site.

A study by the now closed Regional Development Agency for South East England (SEEDA) noted that the region in which the Surrey Research Park is situated conforms to its designation of an "knowledge heartland economies", in which all the elements and linkages in the sub-regional economy model[25] (Fig. 11) were fully established and working well. Evidence from the UKSPA study[26] showed that Parks in these location were more likely to succeed than those parks in

[25] SEEDA and Huggins Associates (2001): SEEDA and Robert Huggins Associates (2001) Global Index of Regional Knowledge Economies: Benchmarking South East England.

[26] UKSPA (2003) Evaluation of the past and future economic contribution of the UK Science Park Movement. Published by UKSPA in conjunction with the Small Business Service.

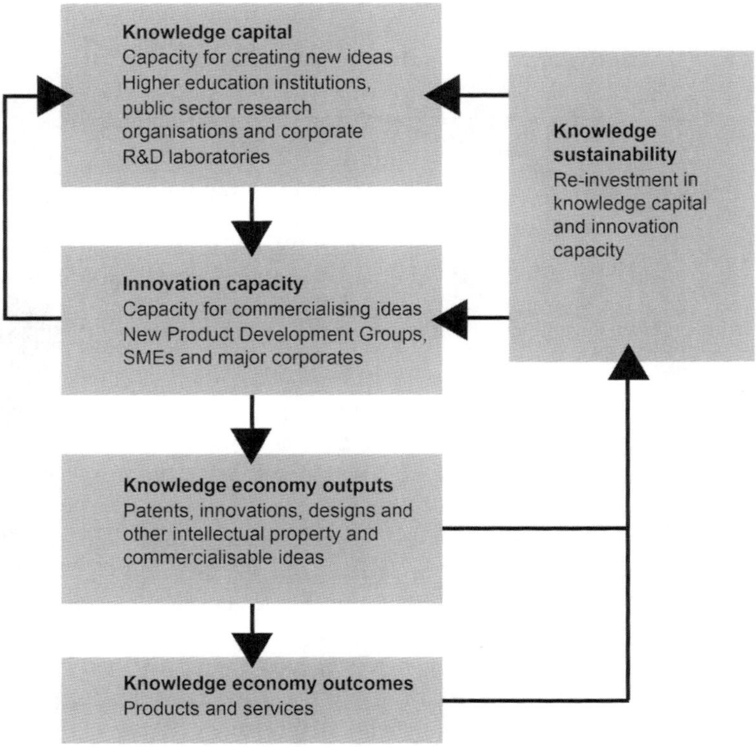


**Fig. 11** Model of the sub-regional economy (after UKSPA 2003 and SEEDA and Huggins 2001)

areas which have "developing knowledge economies" where most elements and linkages are already established; however, some pathways or capacity restrictions hinder the connections, and those areas denoted as "economic development priority areas" where all major elements and/or pathways between the various levels in the model were missing or constrained.

The Surrey Research Park operates in a business environment in which there is substantial "knowledge capital" which provide the capacity to create new ideas. Those organisations that contribute to this include universities, public sector research organisations, and private R&D organisations such as corporate research laboratories and contract research organisations.

The County of Surrey has over 250 major corporate HQs that operate from the County.[27] Data from 2001 showed that the region in which the Park operates has the highest R&D spend which when associated with the level of connections gives some indication of a potential success factor for the site (Table 5).

[27] Perse Comm Surrey County Council Data 2012.

**Table 5** Percentage of R&D expenditure in the UK per region—2001 data[a]

| Regional development agency region UK | % of UK R&D expenditure in each region |
|---|---|
| South East (businesses in south east spent £3.7 billion on R&D in 2001) | 25.7 |
| Eastern | 19.1 |
| London | 10.4 |
| Northwest | 10.2 |
| Southwest | 7.8 |
| Scotland | 6.7 |
| East Midlands | 6.6 |
| West Midlands | 4.9 |
| Yorkshire and Humber | 3.5 |
| Wales | 1.8 |
| North East | 1.5 |
| Northern Ireland | 1.2 |

Perse Comm SEEDA regional development agency

The region also contains substantial innovation capacity if measured by the five pillars that have been used by Lopez-Claros and Mata[28] for analysis of this capacity. This system weights the following in its measure: the institutional environment; human capital, training and social inclusion; regulatory and legal framework; research and development; and the adoption and use of information and communication technologies. The UK also benefits from having a political regime which is classified as a full democracy as opposed to others which include flawed democracies, hybrid regimes and authoritarian regimes within which there is likely to be diminishing innovation performance.

Knowledge-based economic outcomes (Fig. 11) represent the warrantable products and services that are delivered successfully to market. There is no specific data on these outputs for the companies on the Surrey Research Park; however, the rate of growth of companies on the site is an indicator that they are successful. The "gateway policy" for companies coming onto the site, which limits activities to research, development and design in science (including social science), technology and engineering also helps to narrow the likely outcomes to those that contain some proprietary elements.

The final two pathways in the model noted in Fig. 11, which denote feedback loop, if successful, result in re-investment in "knowledge capital" and "innovation capacity" that in turn leads to a "virtuous circle" of success. It is clear from the experience of the Surrey Research Park that many of the entrepreneurs that build and then "exit" from their companies return to the park as serial entrepreneurs and create further new companies. The presence of a science park as a physical asset

[28] Lopez-Claros and Mata (2011), The Innovation for Development Report 2010–2011, Factor, Policies and Institutions Underpinning Country Innovation: Results of the Innovation Capacity Index. EFD Global Consulting Networks.

and focal point for this process helps to build local technology entrepreneurship capacity.

The location of Park on the campus of the University of Surrey in Guildford, which is part of an economic heartland region, has helped attract the right type of tenants which are in high technology businesses with good quality of products and services with strong opportunity for growth and development in global markets. This is further supported by the marketing regime adopted by the Park in relation to the promotion of the services it offers business, its location in terms of the economy and the quality of international air communications which enable companies to link with international markets, and the focus on technology companies that are working on innovation.

Over time the success of the regime of supporting pre and full incubation and grow-on space on the Park has supported a number of companies which have been successful in developing technologies which have then been acquired by large corporations. Examples include the acquisition of Bullfrog by Electronic Arts in 1994 and then the further acquisition of Criterion Software.

The Surrey region is overall an attractive location and appears to have a well qualified workforce. The most recent available National Census data for 2001[29] reveals Surrey's residents as highly qualified. In 2001, 27.2 % of 16–74 year olds (209,100 people) had a degree or diploma (level 4/5) qualification, compared with 21.7 % in the South East and 19.9 % in England. The trend is also the same younger age groups with 10.1 % of Surrey's over 16 year olds are qualified at level 3 (2 or more A levels) compared with 8.3 % nationally and 21.6 % of 16–74 year olds in Surrey are qualified at level 2 (five or more GCSEs at grade A to C) compared with the England average of 19.4 % qualified at this level.

The statistics also showed that there were 28,763 full time students (aged over 18) living in Surrey in 2001. Guildford had 7,004 students (9.6 % of its 16–74 population) and Runnymede had 5,731 students (11.8 %). 4.3 % of England's 16–74 year olds are students, compared with only 3.7 % in Surrey. This suggests that many Surrey 18 year olds go to universities elsewhere in the UK.

To help European based industry maintain its competitive edge through fostering innovation—success indicators:

The performance of companies on the Park is best demonstrated through case studies of some of the high growth companies on the site.

The range of high growth companies that have developed on the site include technology consultancies, companies that have licenced technology from government laboratories, companies that have developed their own intellectual property, companies that have taken advantage of new technology platforms such as low cost and high storage computing power as well as R&D teams working as small specialist parts of large corporations which themselves have spun out companies that have had an economic impact.

---

[29] http://www.surreycc.gov.uk/environment-housing-and-planning/surrey-data-online/2001-census/2001-census-qualifications-and-students

*Smith Associates*: the consulting firm Smith Associates which located on the Park 1986 employed 26 well qualified staff. The company grew very substantially, the founder left after a management buyout and it was renamed to Detica in 2001. Today the company has grown to in excess of 1,400 employees, turnover per employee in 2008 was over £150,000 and the company remains a leader in data and IT security. The company's best known products are anti-fraud software Detica NetReveal, Detica DataRetain, which is software that enables businesses to comply with data retention regulations, and internet security software Stream-Shield. In 2008, Detica was acquired by BAE Systems (Holdings) limited, which is part of BAE Systems plc, and it remains wholly owned by the company. Detica's headquarters remain on the Surrey Research Park in Guildford where they are continuing to consolidate. The majority of staff are located on the Surrey Research Park.

There have been three spin-outs from Detica of which the fastest growing is the privately owned company Actica which has achieved a turnover per employee of three quarters of a million per employee. The ability for the company to find, recruit and retain the best staff because of the location of the site, the flexibility of the University as a landlord that at one time made a loan of £250 K to the company to assist with trading and the quality of its management are all important in this company's progress.

*Stingray:* stingray which started on the site in the Surrey Technology Centre in 2007 was founded by an entrepreneurial team of two technologists who licenced an array of laser technologies from the UK Defense Agency QuinetiQ. This technology was developed for listening to submarines in the north Atlantic during the "cold war" era. The entrepreneurs adapted the technology for use in the oil industry for geophysical data collection. Funded by venture capital the company grew to a team of 11 while in the Surrey Technology Centre. During that time they developed the technology through a proof of principle, proof of concept, and then developed prototypes which were subjected to field trials and were eventually offered as a fully warrantable product. The company's innovative technology prompted its acquisition by the Norwegian geophysical data company. This company is an example of where locally based entrepreneurs used the Surrey Research Park to develop a technology that was transferred from a government funded defence laboratory into a commercial based civilian use.

*Bullfrog:* in 1991 two entrepreneurs started the computer games company Bullfrog and moved to the Surrey Technology Centre. Eventually outgrowing the services of the building the company moved to self contained non managed space on the Park. Growth was driven by the commercial success of both the company's own games of which they established the now well known genre of "god sim games" and by working as games developers for other providers such as Nintendo, Sega and Sony. The commercial success of Bullfrog led to its acquisition by Electronic Arts (EA UK) for an undisclosed sum. EA UK then continued to invest in the company. EA now employs over 300 people in Guildford. One of the founders of the company left EA UK and returned to the Park to establish a new games company Lionhead Studios. The company proved to be highly successful with its Black and White and

Fable range of computer games and was eventually acquired by Microsoft in order to build content for its Xbox games console. The founder left Microsoft in 2012 to found a new computer games company and returned to the Surrey Technology Centre to develop this new business 22 cans which is a games company that is to focus on the new mobile platform. The development of these companies and the games related spin-out, Criterion Software, from Canon Research have created a foundation of a computer games (media cluster) which now has extended to include games testing companies, media related legal services and games related recruitment companies.

There are many other examples of companies that have achieved significant growth. The majority of these have been involved in new commercialising technologies such as stem cells, bio-fuels, internet protocol security, ICT for security systems, building Satellites, as well as some patents for the health care sector.

To help European based industry maintain its competitive edge through fostering innovation—success factors:

*Physical infrastructure and support package:* the factors that have influenced the success of the Park as a centre for economic development not only relate to the physical infrastructure but also to the support service package and management philosophy of the Park.

The market led approach of providing appropriate accommodation and grow-on space for start up companies has provided a valuable opportunity to these companies to become established. The range includes pre and full incubation as well as space in which to grow companies. The approach of using short-term contracts has helped companies that are fast growing to make the necessary moves to adjust their space to their needs.

The presence of the Surrey Technology Centre business incubators since 1986 and more recently since 2004 the presence of the SET Squared pre-incubator have supported the development of a number of technology entrepreneurs which together have helped to raise the levels of competence in technology entrepreneurship. The employment of an entrepreneur in residence and the operation of an Angel Finance Club are both valuable contributors to the growth of more technical companies.

The support of companies in emerging technologies and promoting these has helped to create clusters of companies involved in new innovative fields. An example is the emergence in Guildford of a cluster of computer games companies that are among the world's leaders.

The companies on the Park have been surveyed and there is a high level of awareness of the capabilities of the University of Surrey and the availability of technology services to companies when required.

*Skills:* the quality of the work force on which the companies on the Park can draw that has been established because of the presence of the University of Surrey in Guildford for over 40 years has helped to build a well educated community that are available for recruitment by companies and enable to employ staff that are capable of absorbing and commercialising new ideas.

*Innovation:* the emergence of a number of leading edge technologies from companies that have established on the site, such as satellite engineering, media technology, clean technologies, ICT including bio-informatics, and health care all of which have developed intellectual property have been able to be developed because of the incubation programme offered on site.
*Strong business community:* Guildfords has a strong business community that has been supportive of the companies on the Park as they have developed.
*Good communications:* the Park is well located to gain easy access to international market because of the close proximity to international modes of communication.
*Active management of University links:* the University of Surrey has developed an active programme of support for business through its Office of Research and Enterprise Services that have helped companies connect with the intellectual and technology base in the University. The image and reputation created by the Park as a good place to carry out business. Added to this the management of the Park has had a progressive approach to dealing with emerging technology companies. The University of Surrey has helped to share some of the risk for developing companies by adopting a light touch in relation to its activities as a landlord.

## 5 Conclusions

The Surrey Research Park was first promoted in 1981 in the tradition of the University of Surrey and pre-cursor institutions which had historically worked closely with industry and business since 1896.

Careful definition of the objectives for the Surrey Research Park project provided a valuable framework for developing the master plan, funding, governance and then operational model.

Critical to the Park's success from the University of Surrey's perspective was the successful raising of the initial capital through a land transaction. This enabled the University to control the development of the site and letting policy.

The objectives defined for the three stakeholders which respectively relate to economic development, competitive advantage of tenant companies, knowledge transfer, profile for the University and the capacity to generate income have been critically important in guiding the development; however, their value is diminished without defining success indicators that drive performance and understanding and trying to develop the success factors that assist with achieving a good performance.

The Surrey Research Park defined these parameters in the 1990s as a way of measuring performance of the Park.

The lesson to be learnt from the project is that it has always had a clear set of objectives, supportive management throughout the University and the University of Surrey has undertaken the development of the project from a commercial perspective with a very business like response to the market. It is important to ensure that the academic priorities of a host university are taken into account in a development but do not stifle the enterprise and the market because without the support of the academic infrastructure the foundations for the project cannot be sustained.

The result of adopting a commercial approach to the planning, development and operation of the Park has meant that it has had a significant impact on the economic development of the region in which it is located.

## References

Edgerton, G. (1996). The 'White Heat' revisited: The British government and technology in the 1960 s. *Twentieth Century British History, 7*(1), 53–82.

Lopez-Claros, A., & Mata, Y. N. (2011). Policies and institutions underpinning country innovation: Results from the innovation capacity index. In *The innovation for development report 2010–2011*, Palgrave: Macmillan.

Oh, D. S., & Yeom, I. (2012). Daedeok innopolis in Korea. *World Technolpolis Review, 1*(2), 20–33.

SEEDA and Huggins Associates. (2001). *Global index of regional knowledge economies : Benchmarking South East England.* http://www.seeda.co.uk/_publications/357-Huggins_report_november_2001.pdf

Segal, N., Quince, R., & Wicksteed, B. (1985). *The Cambridge phenomenon: the growth of high technology industry in a university town.* Cambridge: Segal Quince Wicksteed.

Surrey Economic Partnership Ltd. and Surrey County Council. (2010). *Surrey's local economic assessment; executive summary December 2010.* Guildford: Surrey County Council; Surrey Economic Partnership Ltd. http://www.surreycc.gov.uk/environment-housing-and-planning/development-in-surrey/economic-development-insurrey/surrey-local-economic-assessment

The Robbins Report. (1963). *Higher education report of the committee appointed by the prime minister under the chairmanship of Lord Robbins (Cmnd. 2154).* London: Her Majesty's Stationery Office.

UKSPA and Small Business Service. (2003). *Evaluation of the past & future economic contribution of the UK science park movement: Executive summary.* Prepared by ANGLE Technology, October 16, 2003. Cambridge: UKSPA. http://www.ukspa.org.uk/ContentFiles/UKSPAEvaluation-ExecutiveSummary.pdf

# Investing in Nanotechnology in Sri Lanka

**Asela Gunawardena**

## 1 Introduction

The Sri Lanka Institute of Nanotechnology (SLINTEC) was established in 2008 as a public–private partnership between the Government of Sri Lanka and five leading Private sector companies. Since its incorporation, SLINTEC has been able to achieve rapid results in both its directives of realizing success in Nanotechnology research and initiating the development of a Nanotechnology Park.

A team from UNESCO visited Sri Lanka in November 2010 to carry out an assessment of our aspirations and conduct workshops on the development of Science and Technology Parks. The UNESCO team produced a comprehensive report on a Concept Plan for the Development of a Nanotechnology Park in Sri Lanka. Many aspects of that report influenced our thinking and were adopted in the Strategic Plan for SLINTEC.

The Nanotechnology Park will adopt an Open Innovation framework based around a Nanotechnology Center of Excellence (NCE). It aims to attract Multinational Companies (MNCs) as well as Small and Medium Enterprises (SMEs) to the Park, and take advantage of the research, development, and innovation that is bound to flourish in an enterprising ecosystem. This paper describes the journey we have traveled so far, the framework we have adopted, the park we are developing, and the advantages of investing in Nanotechnology in Sri Lanka.

A. Gunawardena (✉)
CEO, Sri Lanka Institute of Nanotechnology, Lot 14, Zone 1 Biyagama EPZ, Walgama, Malwana, Sri Lanka
e-mail: aselag@slintec.lk

D.-S. Oh and F. Phillips (eds.), *Technopolis*,
DOI: 10.1007/978-1-4471-5508-9_18, 

## 2 History

In 2006, Sri Lanka embarked on a new journey in science and technology development when the National Nanotechnology Initiative (NNI) was approved by the Cabinet of Ministers. The Government of Sri Lanka, through the Ministry of Technology and Research and the National Science Foundation (NSF), joined hands with the private sector to form a Public–Private-Partnership (PPP) to realize the vision of the NNI. The private sector partners are:

- Brandix—One of the largest apparel manufacturers and exporters in Sri Lanka.
- Dialog Axiata—The largest mobile telecommunications provider in Sri Lanka.
- Hayleys—A multinational conglomerate with diverse business groups and ventures.
- Loadstar—A joint venture between Camoplast of Canada and Solideal of Belgium.
- MAS—One of the largest intimate apparel and sportswear manufacturers for leading global brands.

To fulfill the objectives of the NNI, two companies were incorporated in 2008, namely: the SLINTEC and Nanco. They were tasked to conduct research on Nanotechnology to make products and industries more competitive, add value to Sri Lanka's mineral resources and develop a Nanotechnology Park.

Research at SLINTEC commenced in August 2009, after the purchase of state-of-the art equipment for Nanotechnology research and the recruitment of top class scientists from Sri Lankan universities and institutes. Within 10 months of initiating research, the team successfully filed five international patents at the United States Patent and Trademark Office (USPTO). This was a great accomplishment for a small team of dedicated scientists and professors.

Nanco continued to explore the feasibility of establishing and developing a Nanotechnology Parke in the 50 acre land allocated by the Government in a suburb of Colombo named Homagama. Feasibility studies were done by a leading consultancy firm at the inception. But the most effective and useful support came from the UNESCO team that visited Sri Lanka in November 2010. The team comprised experts from the World Technopolis Association (WTA), University of Surrey, and UNESCO. They conducted workshops on the development of Science and Technology Parks and understood our vision and our needs. The UNESCO team produced a comprehensive report on a Concept Plan for the Development of a Nanotechnology Park in Sri Lanka, which boosted our confidence and helped us formulate a sound Strategic Plan.

SLINTEC and Nanco were merged in March 2011 to strengthen efforts in fulfilling the objectives of the NNI. The amalgamated company SLINTEC continued its success in research by filing for two more patents in 2011, advancing many projects from research into development and commencing the design and planning of the Nanotechnology Park.

# 3 Open Innovation

Open innovation creates opportunities for all businesses to develop new products and solutions through a collaborative innovation process by taking them from research through development toward commercialization. Technology parks provide this enabling environment for building innovation systems and clusters. SLINTEC aims to provide this environment by developing an open innovation platform.

The SLINTEC innovation platform will have the following five components in our Nanotechnology Park:

1. Research and business development.
2. Technology and business incubation.
3. Capacity building through innovation.
4. Technology commercialization center.
5. Funding mechanisms for business development.

## *3.1 Focus Areas*

In order for SLINTEC to continue to be successful, we must focus on areas that SLINTEC can be most effective and Sri Lanka could have a competitive advantage. Hence we have chosen the five focus areas illustrated in Fig. 1.

In Smart Agriculture, we have chosen to focus on Nanotechnology-based slow release fertilizer. We have plans to expand into sensors, next generation fertilizers, and other advanced technologies in the area of Smart Agriculture.

In Rubber Nano-composites, our focus has thus far been on high-performance tires primarily for solid rubber tires with plans to advance into pneumatic tires.

In Apparel and Textile, we focus on high-end fabric, smart yarn, and other technologies to boost Sri Lanka's robust apparel industry to the next level in global competition.

We have developed a Nanotechnology- based external medical sensor with a view to enabling remote health monitoring under the Consumer Products area. In addition, we conduct research on detergents, cosmetics, and other consumer products.

**Fig. 1** Focus areas of SLINTEC

In the Nanomaterials area our efforts continue to be to add value to Sri Lanka's natural resources such as Ilmenite, Clay, Magnetite, Vein Quartz, and Vein Graphite. We are developing processes to make Titanium Dioxide, Montmorillonite, Nano Magnetite, Nano Silica, and Graphite Nano Platelets.

## 4 Conceptual Framework

In developing our strategy for the Nanotechnology Park, we conceptualized a fivefold framework as illustrated in Fig. 2 and listed below:

1. Center of Research and Innovation (R&I).
2. Cells of Incubation.
3. Corporate Research and Business Development (R&BD).
4. Centralized Services.
5. Community.

The framework is a series of concentric circles developed around a Center of Research and Innovation. This also draws parallels to a Hub and Spoke Model where the hub is the Center of R&I and the spokes being the Cells of Incubation, Corporate R&BD, Centralized Services and the Community.

The Center of R&I is the Nanotechnology Center of Excellence (NCE) that will be developed in the first phase of the Nanotechnology Park. Its primary responsibility is to conduct research on Nanotechnology and facilitate Open Innovation. It will also be responsible for the management and promotion of the Park.

**Fig. 2** Conceptual framework

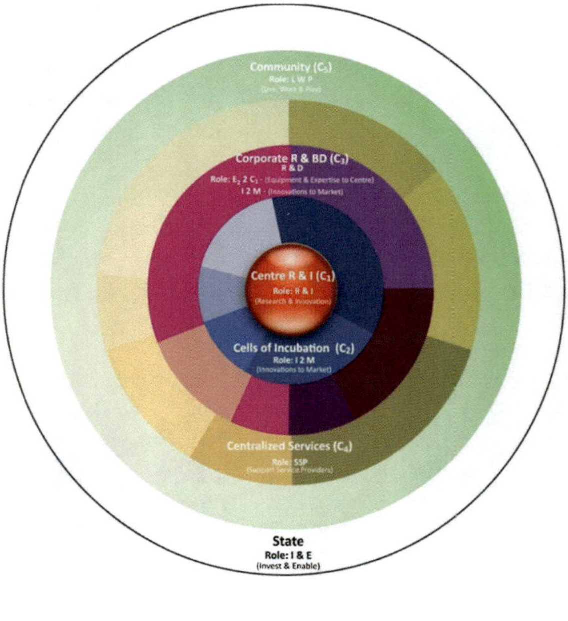

Cells of Incubation are intended to take innovations, from research conducted in the Center, to the marketplace through development and commercialization. Startups that form as a result of research and innovations can begin in the Cells of Incubation taking their journey from pre-incubation through post-incubation to fully fledged companies. This is the facility enabling entrepreneurs to prosper.

Corporate Research and Business Development (R&BD) is meant for Private as well as Public enterprises. It is the aspect of the Park aimed at attracting Multinational Companies (MNCs) as well as Small and Medium Enterprises (SMEs) to establish facilities in the park and take advantage of the research, development, and open innovation.

MNCs as well as SMEs will be provided attractive terms and conditions to invest in setting up their research and development facilities in the Park. They will also be able to invest in Nanotechnology research to make their products more competitive or to develop new products. Corporate R&BD will take innovations to market through commercialization of research. They will also provide market intelligence and product expertise to the Center of R&I. This will be the bidirectional interaction between these two entities.

Centralized Services are those of accounting, marketing, legal, procurement, etc., that the Park would provide to the corporations, enterprises, and start-ups in the park. This is a key enabling factor in the Park ecosystem. The Park would attract and establish satellite offices—for example, a legal and account firm to provide support services to the Nanotechnology Center of Excellence and other tenants of the Park.

The Community will live in a united environment, work in a multidisciplinary setting, socialize, and share knowledge. SLINTEC will provide housing and recreation for the Community and an education facility related to the Nanotechnology Center of Excellence (NCE). The Community will also network with other institutions, universities, companies, etc., worldwide and collaborate on research, development, and commercialization.

Encompassing this framework is the Government of Sri Lanka, which will invest in infrastructure and resources, and enable the creation of a Nanotechnology Park though appropriate policies, laws, and incentives.

## 5 Master Plan

The Master Plan of the Nanotechnology Park is illustrated in Fig. 3.

The Nanotechnology Center of Excellence (2) will occupy about 10% of the land, leaving most of the land for Tenant Development Areas (9) dedicated for Corporate Research and Business Development (R&BD) facilities, Startups, Pilot Plants, etc.

Incubation (7), Administration and Centralized Services (3) will begin at the Center of Excellence and move out into dedicated buildings and spaces with the expansion of the Park. We have dedicated an area in the Master Plan for Advance Agriculture Research and Green Houses to reinforce our focus on Smart Agriculture.

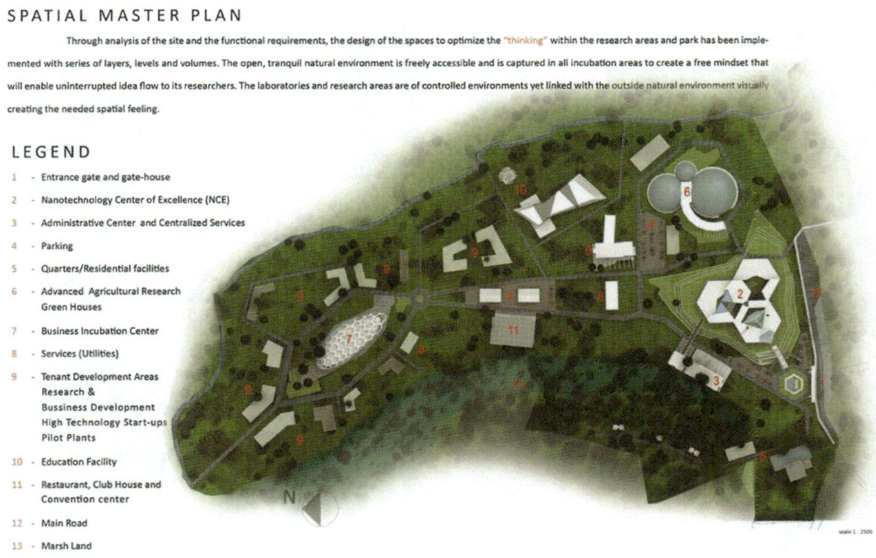


**Fig. 3** The master plan of the nanotechnology

# 6 Phased Development

The Nanotechnology Park will be developed in a phased approach as shown in Fig. 4 through pragmatic investments by the government as well as private investors and companies keen on establishing themselves in the park. A phased approach in

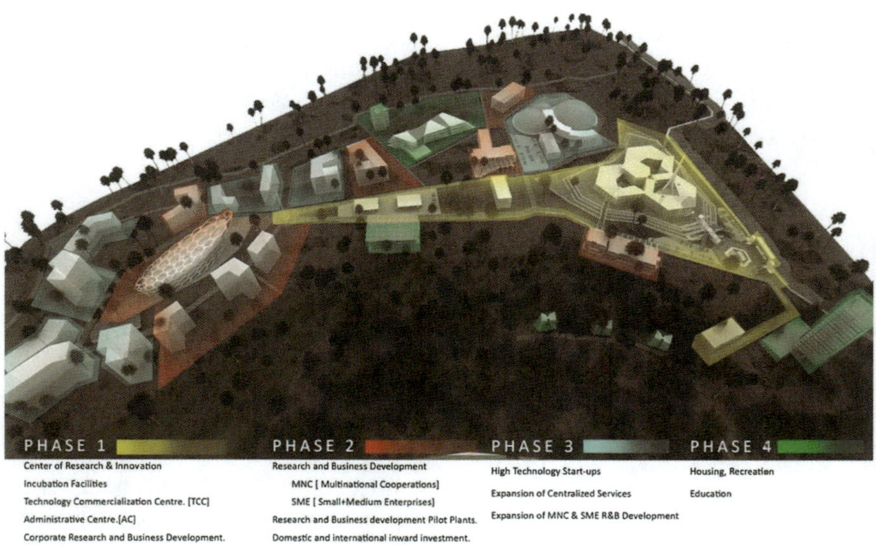


**Fig. 4** Phased development of SLINTEC

developing a Nanotechnology Park is a rational way of planning for the future and expanding the scope based on market demand and research success. SLINTEC is currently focused on developing Phase 1 which will be the Nanotechnology Center of Excellence (NCE). Phase 1 will contain the following elements:

1. Center of Research and Innovation (R&I).
2. Incubation for taking Research into Commercialization.
3. Corporate Research and Business Development (R&BD).
4. Technology Commercialization Center (TCC).
5. Administrative Center and essential Centralized Services.

## 6.1 Phase 1

The first phase is envisaged to be a state-of the art building for the Nanotechnology Center of Excellence (NCE) that will accommodate the following in a sub-phase development approach.

Phase 1a

1. Center of R&I.
2. Cells of Incubation—Pre and Full incubation for start-ups.

Phase 1b

3. Corporate R&BD—Private and Public.
4. Nanotechnology focused departments of other Research Institutes.
5. Technology Commercialization Center (TCC).
6. Administrative Center (AC) and essential Centralized Services.

Phase 1c

7. Library.
8. Auditorium.
9. Accommodation.

While SLINTEC develops Phase 1, we will be promoting and attracting investment in the Tenant Development Areas to MNCs and SMEs to establish their research and development facilities, pilot plants, etc. in the Park.

## 6.2 Phase 2

The second phase will focus on expansion of *Large-scale Corporate Research and Business Development* for both the Private and the Public Sectors. We hope to draw Foreign Direct Investment (FDI) from MNCs as well as SMEs through attractive incentives, terms and conditions.

### 6.3 Phase 3

The third phase will focus on expansion of *High Technology Industries* and expansion of *Centralized Services.*

### 6.4 Phase 4

The fourth phase will provide *Housing and Recreation for the Community and an Education Facility related to the Nanotechnology Center of Excellence (NCE).* The plan is to integrate with the surrounding area and connect with the larger Knowledge Hub master plan for Homagama.

## 7 Investment Advantage

Sri Lanka is enjoying a "Peace Dividend" after the civil war that lasted more than 25 years, which ended in 2009. It has since had an annual GDP growth rate of more than 8 %, with a current (2010) per capita GDP of approximately USD 2,400. The Government has maintained an undivided focus on growth and development and has great ambitions of making Sri Lanka a dynamic global hub in Aviation, Energy, Knowledge, Logistics, and Commerce. The Knowledge Hub initiative is focused on attracting investments in higher education with a view to making Sri Lanka a regional hub for learning and innovation.

Sri Lanka is benefiting from political stability and rapid infrastructure development such as new highways, sea ports, airports, and power plants. The Government is increasing its emphasis on education. Sri Lanka has achieved near universal literacy (91 %); English is spoken and understood widely. These factors make Sri Lanka very attractive to any foreign investor or company seeking a highly educated, qualified, and talented workforce. For many companies and institutes based in North America or Europe, the cost arbitrage of doing research and development work in Sri Lanka is a significant advantage.

The Nanotechnology Park, as illustrated in the Master Plan (Fig. 3), will be subdivided into zones for tenants to invest and develop their research centers, incubation facilities, pilot plants, etc., in an enabling and supportive environment of a Science and Technology Park. These tenants, whether they be MNCs or SMEs will be provided attractive incentives to set up their facilities in the Park. These along with the environment of Open Innovation and the access to the Nanotechnology Center of Excellence will give any company or institute a considerable advantage.

The Government of Sri Lanka is visionary and innovation friendly. It is focused on rapid development through advanced technology. In its Budget for 2012, the

Government has made noteworthy provisions and financial incentives in the furtherance of research and development (R&D) in Sri Lanka. These include a significant reduction in corporate tax on profits derived from R&D-related income, a reduction in personal income tax for individuals engaged in research and technology, and a triple deduction to companies that undertake R&D through research institutes.

The Board of Investment (BOI) in Sri Lanka offers many incentives and tax holidays to companies, both local and foreign, that invest in all aspects of economic development. As such, the BOI will play a major role in the development of the Nanotechnology Park in Homagama, Sri Lanka. We intend to work with the BOI to obtain the maximum tax holiday for all corporations that invest in the Nanotechnology Park. We also plan to secure from the Government other incentives such as tax free import of material and equipment for research, development, and commercialization of Nanotechnology.

Our aim is to provide the maximum tax benefits for individuals, specialists, and experts working in Nanotechnology research and the maximum incentives for companies to invest in the Nanotechnology Park and in Nanotechnology R&D in Sri Lanka.

## 8 Conclusion

The Government of Sri Lanka together with the Private Sector has taken a leap of faith in Nanotechnology. This faith and vision have produced positive results in the formation of SLINTEC and its noteworthy accomplishments in a short period of time.

Guided by a Board of Directors comprising some of the highest ranking officials in technology and research from the public sector and some the most influential CEOs from the private sector, SLINTEC is now poised to advance to its next stage by developing a world-class Nanotechnology Park in Sri Lanka. A technology park ensures maximum sharing of both human and physical resources for multidisciplinary R&D. A technology park is the most effective means to bring research, industries, entrepreneurs, and academia together to achieve a collective greatness.

Sri Lanka has flourishing industries in apparel and textile, rubber, tea, activated carbon, and rich mineral resources ready to move into higher value categories. Sri Lanka has prominent universities in science and technology ready to provide a steady stream of highly skilled talent. The Government has invested in supporting research, provided land for a Nanotechnology Park, and continues to create an enabling environment for local as well as foreign companies and institutes. Now is the time for all to take advantage of this prospect by investing in Nanotechnology in Sri Lanka.

## References

Liyanage, U., Edirisinghe, R., Gunawardena, A., & Team Nanco (2011). *Strategic Plan for Nanco*2011—2015.

Oh, D-S., Parry, M., Silva, R., & Nur, Y. (2011). *Concept Plan for the Development of a Nanotechnology Park in Sri Lanka.*

# Eco-Industrial Park (EIP) Initiatives Toward Green Growth: Lessons from Korean Experience

**Suk-Chan Ko**

**Abstract** The purpose of this article is to provide a more systematic understanding of supporting mechanisms for Green Growth based on the experiences of Eco-Industrial Park (EIP) development in South Korea. Since the enactment of "the Act to Promote Environmentally Friendly Industrial Structure" in 1995, the central government tried to establish a system for cleaner production and environmental management system. In 2005, Korea had also launched an ambitious EIP initiative under the leadership of the Ministry of Commerce, Industry, and Energy. The cases of several pilot projects for EIP in Korea indicate that inappropriate selection of the target industrial parks, conflicts of interests among stakeholders, poor planning, and lack of financial support from the government deteriorated the vitality of the project as a whole. The experience of EIP development in Korea shows that the spontaneous and active participation through training programs for citizens and government officials and the cooperation between the central and local government can guarantee the success of EIP project in the future.

## 1 Introduction

In March 2005, the 5th Ministerial Conference on Environmental and Development in Asia and the Pacific held in Seoul, Korea. Approximately 340 delegates participated and embraced the approach of environmentally sustainable economic growth. The conference endorsed 'Green Growth' as a policy focus and a powerful strategy to promote 'win–win' approaches in reconciling the conflict between the goal of poverty reduction and the goal of environmental sustainability (ESCAP 2006).

The 'Green Growth' approach seeks to harmonize the two imperatives of economic growth and environmental sustainability by promoting "fundamental

S.-C. Ko (✉)
Prof. Urban Planning & Real Estate Studies Dankook University, Yongin-si, South Korea
e-mail: sukchan@dankook.ac.kr

D.-S. Oh and F. Phillips (eds.), *Technopolis*,

DOI: 10.1007/978-1-4471-5508-9_19, 

changes" in the way societies produce and consume. The Green Growth requires the introduction of concept and system changes. Across the world, corporations and small and medium-sized enterprises are becoming agents of change for sustainability. They acknowledge the issues of global warming and green growth and their relevance to businesses. Eco-efficiency and eco-innovation can be good for business and it is becoming widely held view in the business community that solving environmental and social problems is essential for the future growth of firms.

The current economic crisis and negotiations to tackle climate change should be seen as an opportunity to shift to a greener economy. Incremental improvement is not enough. Industry must be restructured and existing and breakthrough technologies must be more innovatively applied to realize green growth.

Industries have traditionally addressed pollution concerns at the point of discharge. Since 'this end-of-pipe' approach is often costly and ineffective, industry has increasingly adopted cleaner production by reducing the amount of energy and material used in the production process. Many firms are paying more attention to the product's lifecycle and are integrating environmental strategies and practices into their own management systems (OECD June 2009 Policy Brief). Some pioneers have been working to establish a closed-loop production system that eliminates final disposal by recovering wastes and turning them into new resources for production. Eco-Industrial Park (EIP) and Eco-innovation help to make possible this kind of evolution in industry practices.

There have been diverse government policy initiatives and programs that promote eco-efficient and eco-innovation. These include both supply-side and demand-side measures. As most countries recognize the need for more collaborative approaches to innovation, many initiatives involve creating networks, platforms, or partnerships that engage different industry and non-industry stakeholders.

The purpose of this chapter is to provide a more systematic understanding of supporting mechanisms for Green Growth such as Eco-Industrial Park (EIP) and Eco-Innovation. First, I would like to introduce the concepts of Eco-Industrial Park and Eco-Innovation, and then I will review the process of EIP development strategies in Korea and their policy implications. Third, I will discuss how these concepts and principles can be applied to the development and management of science park projects in many countries in the future.

## 2 Industrial Ecology and Eco-Industrial Parks

### *2.1 Industrial Ecology and the Role of Government*

Industrial ecology is a strategic approach attempting to reduce environmental impacts by applying the principles of natural ecosystems to the industrial processes (Deutz and Gibbs 2004). Although similar concepts circulated in the 1970s, the concept was systematized by research on 'industrial metabolism' by Ayres (1989).

Industrial ecology studies construction of a ‘closed-loop production system’ which is analogous to natural ecosystems. The closed-loop system assumes the re-input of wastes and by-products into the production system. Existing industrial systems suppose unlimited inputs and outputs of resources by considering resource flows as linear (Korhonen et al., 2004). In industrial ecology, the resource networking between plants is called ‘industrial symbiosis’, as in symbiosis between species in the natural system.

The systematic approach to industrial ecology results in the treatment of individual companies’ economic interests in decreasing input resources and wastes in addition to the social benefit of reducing the load on the environment. The approach is thought to be the realization of the concept of sustainability in terms of considering economic growth and environmental concern simultaneously. The EIP project is to actualize this principle of industrial ecology.

The concept of EIPs was first made known when Indigo Development introduced it to EPA officials in 1993 (Lowe 2001). After that, the President’s Council on Sustainable Development (1997) chose the EIP project as a model project in the Clinton Administration. The EIPs located in Fairfield, MD, Cape Charles, VA, Chattanooga, TN, and Brownsville, TX, are the outcomes of such US government initiatives. The US cases are examples of intentional policy efforts promoted by the government, while the spontaneous appearance of an EIP is found in the industrial park in Kalunborg, Denmark (Ehrenfeld and Gertler 1997).

An EIP or estate is a community of manufacturing and service businesses located together on a common property. Member businesses seek enhanced environmental, economic, and social performance through collaboration in managing environmental and resource issues. By working together, the community of businesses seeks a collective benefit that is greater than the sum of individual benefits each company would realize by only optimizing its individual performance. Lowe (1997) pointed out that “EIP conjoins the principles of mixed use development, recycling business and by-product exchange in coordination with green technology companies that makes eco-friendly products.”

The intention behind the EIP is the formation of a corporate network where pertinent companies cooperate with each other and neighboring communities to accomplish the common goals related to economic interests, the improvement of environmental quality and the fair use of human resources (Cohen-Rosenthal 2003). The EIP provides a participating corporation with the various advantages of curtailing costs of supplying input resources and treating wastes, and publicity of an environmentally friendly corporate image. In addition, there are great social benefits from the construction of environmentally friendly communities that namely reduced energy and resource consumption, and sustainable treatment of wastes, and the reduction of social costs created by conflicts between companies and local communities. The government also enjoys a few advantages, reducing some regulation costs thanks to corporations spontaneously joining in ‘green business’ and accomplishing social integration with small expenses (von Malmborg 2004). As mentioned, many governments of the world have enthusiastically been promoting EIP projects for these reasons.

However, there are some opposing views about the government's involvement in promoting EIP projects. The question is whether or not the public sector should actively lead the EIP project. Some insist on a government-led approach while others advocate a market-led approach. The American and Korean cases are typical examples representing the active role of the public sector, while the case of Kalundborg, Denmark, is a representative example of a spontaneous project promoted by the private sector. Both approaches have strengths and weaknesses. The government-led approach has the advantage of being able to easily initiate the project. Meanwhile, it has the disadvantage of not being able to guarantee the persistence of the project. The market-driven approach's strength is in the EIP's firm persistence once the project begins, while its weakness is in the difficulty encountered for the project to begin. Therefore, it is hard to assess which acknowledges the role of public authorities in the development of EIP projects, while others such as Ehrenfeld and Gertler (1997), Bass (1998) and Waller (1999) emphasize corporate spontaneity and the role of the market mechanism in the project.

## 2.2 Eco-Industrial Park Management and Support Services

As a community of companies, an EIP needs a more sophisticated management and support system than a traditional industrial park and a science/technology park. Management or a third-party supports the exchange of by-products among companies and helps them adapt to changes in the mix of companies through its recruitment responsibilities. Management may maintain links into regional by-product exchanges and a site-wide telecommunications system. The park may include shared support services such as a training center, information center, offices for purchasing common supplies, transportation logistics office, and cafeteria. Companies can add to their savings by sharing the costs of these services.

An EIP encompasses two distinct but overlapping business entities. It is a real estate development property that must be managed to provide a competitive return to its owners. At the same time, an eco-park is a community of companies that must manage itself to gain common benefits for its individual members. The full range of management functions to be performed by the combination of business community and park management systems include the followings:

- Maintain the values, culture, and identity of the eco-industrial park as a community.
- Resolve conflicts between companies, between park management and tenants, and between the needs for future viability and present efficiency.
- Facilitate the self-organizing community development process among tenants.
- Recruit firms to keep the park fully leased and maintain the mix of companies needed to best use by-products as companies change.
- Coordinate recruitment with local and state economic development agencies.

**Table 1** Environmental management continuum for industrial parks

| Standard Industrial Park | Concentrated industrial and business activity within a defined planning boundary with organized infrastructure |
|---|---|
| Eco-labeled Industrial Park | A labeling scheme developed in France to recognize an organizational set of enhanced environmental practices and amenities in industrial parks and zones |
| Environmental Industrial Park | Clusters of manufacturers of environmental products providers of environmental services and developers of environmental technologies |
| Eco-efficient Park | Cluster of companies working to reduce resource intensity, control pollution and minimize collective waste outputs |
| Environmentally balanced industrial cluster | Clusters of industries co-located such that the by-products of one become the inputs or materials for other businesses or industries to minimize waste and dissipation of resources |
| Eco-Industrial Park | Clusters of companies taking into account of ecological limits, using resource-efficient infrastructure, buildings and processes, networking purchases and a balance of producers, scavengers and decomposers |

- Track present trends and emerging challenges and opportunities, including: patterns of inter-company collaboration, technologies, and firms that support by-product exchange, changes in regulations at all levels of government.
- Support continuous evolution of economic and environmental performance for individual companies and the park as a whole by managing a learning center and designing new inter-firm initiative.
- Conduct audits of successes as well as failures in EIP performance to assure learning and improvement.
- Coordinate provision of shared support services.
- Set up a project operations room to support effective work by the park management company and the community self-management system or tenants association (Lowe 2001).

EIP requires an ecological or systematic approach and must be more comprehensive involving more aspects and both management and tenants. EIP can be a part of environmental management continuum for industrial parks as shown in Table 1.

## 3 Eco-Industrial Park Development in Korea

Industrial policies in Korea have been changed drastically since the Ministry of Commerce, Industry, and Energy MOCIE(currently the Ministry of Trade, Industry, and Energy) since February, 2013 enacted 'the Act to Promote Environmental Friendly Industrial Structure' in December 1995. Based on this Act, the ministry established an institutional system for cleaner production (CP) and environmental management system based on ISO 14001 as an implementing tool. The first comprehensive master plan for environment friendly industrial development was made and operated based on this Act. The plan includes: streamlining the supporting system, cleaner production transfer and dissemination, promoting environmental industry, and stimulating environmental management. The cleaner production transfer and dissemination deal with technology transfer, international collaborative projects, supply chain environmental management, and eco-industrial park development (Park et al. 2008).

More recently, Korea has launched an ambitious EIP initiative in 2005, under the leadership of the Korea National Cleaner Production Center (KNCPC) and the Korea Institute of Industrial Technology (KITECH). Six industrial parks or complexes out of 35 large national parks were designated as EIP projects, and some of these projects are actually clusters of several industrial parks. Eventually, this initiative would encourage all 504 industrial parks in Korea to achieve the transition to become EIPs. The six industrial complexes that are pilot projects in the EIP initiative are listed in Table 2.

**Table 2** Six industrial complexes selected as EIP projects in Korea

| Name of Complex | Land Area (hectare) | Number of Companies | Major Industries and Characteristics |
|---|---|---|---|
| Banwol and Siwha | 3,180 | 5,400 | Located in the southern part of Seoul metropolitan area<br>Typical industries include textile, dying, pulp, chemical plants, small manufacturers, and waste incinerators |
| Mipo and Onsan | 5,557 | 700 | Located in Ulsan city, industries are automobile manufacturing, ship building, and one of the world's largest petrochemical complexes; Nonferrous metals, steel, and metal manufacturers are the major industries |
| Yeousu | 3,130 | 149 | Located in the southern part of Korea and is primarily a petrochemical complex and refinery with 149 companies |
| Cheongju | 410 | 200 | Major industries are textile, paper-mill, petrochemical, electronics, nonferrous metals, metal manufacturing and assembly |
| Machun, Chilso, Sangpyeng | 581 | 550 | Three clusters of smaller industrial parks include Machun Industrial Complex at Jinhae, Chilso Industrial Complex in Hanam, and Sangpyeing Industrial Complex at Junju. The three clusters are around 50 km apart. Most of the companies in these locally managed parks are small and medium enterprises. Industries at Jinhae include nonferrous metal, steel, and machinery and at Jinju, food, textile, pulp, and chemical |
| Pohang | 2,010 | 220 | Located in Pohang city, major industries include cement, steel, metal processing, fine chemical, and waste disposal |

The Korean EIP initiative is notable for the potential impact on all industrial parks in Korea, because this initiative is a 20-year, 3-phase long-term project. Korea has a total of 504 industrial parks, with 35 large national parks or complexes on two-thirds of the total industrial land (about 66,635 ha) and the government intends to transform the major industrial complexes into EIPs in the long run.

The first phase (2006–2010) of the developmental plan strives to perform trial projects for two industrial parks in order to shift them to EIPs, with prior understanding of the material and energy flow analysis, input and output of raw materials, products, by-products, and wastes. An energy efficient by-product exchange network would be created using the basic concepts of industrial ecology. Pollution monitoring systems are installed to envisage the existing wastewater and waste treatment systems. Additionally, an integrated environmental management system would emerge together with detailed analysis of the infrastructure. Sustainable education and awareness campaign has been conducted. The development so far has envisioned for further phases that would upgrade the existing manpower resources in conjunction with an organizing group that manages the operation on a timely basis (Park et al. 2008).

The second phase (2011–2015) would provide conceptual ideas and disseminate understanding of the designed concept to twenty other industrial parks. It would also help in spreading the environmental management system and sustain a balance between the different key factors that are likely to influence economic growth. A system of common sharing and practice, common purchase and common transportation system would be organized to establish an enlarged infrastructure that is capable of handling joint ventures.

The third phase (2016–2025) would overview the flaws and constraints envisioned in the earlier phases and strive to rework and reinvent the existing system of practice. The performance indicators would be analyzed and evaluated by an expert committee to redesign any missing components and infrastructure. The ultimate aim would realize and pave the way to provide zero discharge in all process industries within the EIPs.

The Korean EIP Model is characterized as a cluster of inter-networking businesses, which perform individual and collective cleaner production program prior to by-products exchange network within an environmental management system (Chiu 2005). Under such framework, the KNCPC and Korea Industrial Complex Corporation will be the main actors in implementing the different phases and strategically supervising the development and implementation programs.

In November 2004 the KNCPC organized the 2nd International Conference on Industrial Ecology and Eco-Industrial Park in Seoul, with Indigo Development providing presentations on a system view of EIPs. Indigo also led two workshops with delegates from 5 of 6 pilot complexes. On the basis of these experiences, Lowe and Chiu (2005) developed the following critical success factors for the transition from industrial park to EIP in Korea. These critical success factors emerged from Indigo's consulting and research in Korea but are generally applied to the EIP projects in other countries.

These can be highlighted as follows:

- Good cooperation among the national agencies with responsibility for EIPs.
- Each pilot industrial park requires an adequate management structure for coordination and cooperation supporting the transition to an EIP.
- Both public management authorities and business associations require capacity development and education so they can participate effectively in the EIP initiative.
- Businesses in the park need to be involved from the beginning of the planning process. They are the ultimate actors in the system.
- The high level planning process for the transition to EIPs must be supported by a strong bottom up planning process, i.e., a dialog between top down and bottom up.
- An evolving long-term vision of the whole system is required to make effective decisions about the specific strategies used in each phase of the transition.
- An EIP is much more than an exchange of by-products among companies.
- Strong support to the growth of the environmental technology and services cluster will provide Korean industrial parks with many of the solutions they require.
- Green chemistry is an important field for petrochemical EIPs as well as customers using chemicals.
- Resource-based policies.
- Policy in support of the EIP initiative should take an integrated view of all aspects of cleaner production as complementary to eco-industrial strategies.
- National policy should support excellent management of the eco-industrial park initiative and individual industrial parks (Indigo Development 2005).

The Korean EIP initiative is relatively new and participants are still learning the basics of eco-industrial park development. Recently, there have been a few efforts to evaluate the outcomes of EIP demonstration projects. Several researches found that the total of 45 pilot projects were successfully implemented and resulted in $14 million of economic benefits mainly from energy exchange and recycling by-products. Most of pilot EIPs could establish resource circulation networks (Ban 2008).

## 4 The CMS EIP Development Case in Gyeongnam Province, Korea

The Gyeongnam Regional Environmental Technology Development Center suggested that building a 'resource-symbiotic network' unifying a few industrial complexes in the exchange of by-products is relatively easy, as local industrial parks in Gyeongnam province consist of various types of industries. The Center investigated nine industrial parks in the province from the viewpoint of material flows and proposed the construction of the Chilseo-Macheon-Sangpyung (CMS) resource symbiotic network.

From the beginning, however, the project has not been carried out following EIP principles. The three industrial parks are as far as 80 km away from each other. This fact is contrary to the EIP principle of short commuting distance between plants (Lowi 2001; Deutz and Gibbs 2004). Also, the principle of an EIP is to basically build a resource-symbiotic network within an industrial park. Therefore, the network between the industrial parks has some problems in that it has weak economic feasibility and can cause secondary pollution in the transport process of wastes and by-products. The Center's initial plan was to promote the project in Chilseo Industrial Park alone. Then, Macheon and Sangpyung asked to be involved in the project for several reasons. The MCIE accepted their requests and the project in Gyeongnam Province developed this unusual form as a result. It is clear that the unreasonable structure continues to threaten the rationality and justification for the existence of the project (Kim 2007).

Macheon Industrial Park (MIP) is a local industrial complex officially approved by the MCIE in 1993. The main type of business is small foundries. These small businesses moved to the outskirts of Jinhae, Gyeongnam, from the border of Busan City in the late 1980s due to expansion of the residential district and civil complaints. The area of MIP belongs to the Busan-Jinhae Free Economic Zone (BJFEZ), and the BJFEZ authority takes charge of the area's environmental and industrial affairs.

MIP has generated civil complaints concerning the stench created by the combustion process of molds made of sand. Ammonia and phenol gases are the main source of the stench, produced when melted iron combusts sand molds. The stench might be partially treated through pollution prevention facilities, but complete treatment is impossible because a great deal of the work is carried out in the open air. The stench and air pollution are serious in the summer season when the south-southwestern wind tends to concentrate pollutants in the basin area. Residents complained of administrative irresponsibility in placing the residential area beside the industrial complex, which generates a large quantity of pollutants and demanded strict supervision and even the removal of MIP.

In addition to the unfavorable quality of life of the residents, the issue was also related to fiscal matters. Due to the stench problem, the housing development project promoted near MIP did not pass the environmental impact assessment twice, in 2002 and 2005. If the quality of a site does not sufficiently meet the standards of the environmental impact assessment, the inevitable use of the land will be as a factory site. The residents are discontented with this notion because this affects the property value of the residents' houses. The BJFEZ has feared that the MIP problem could ruin the master plan of the entire free economic zone and has come up with two alternatives to resolve the problem.

The first alternative is to convert MIP into an EIP. The managerial board of MIP felt that the EIP approach promoted by MCIE could settle its stench problem. The chief manager strongly requested the BJFEZ and MCIE to involve the industrial park in the project. The BJFEZ sympathized with the view of MIP and requested the involvement of MIP in the pilot project to MCIE. However, this happened because of their conceptual misunderstanding of EIPs (Kim 2007).

In principle, the EIP is a resource-symbiotic network constructing a food-web of resources and wastes, creating a closed loop. However, the stench is not a resource in which the project is interested. A zero-emission approach which is popular in Japan, can be considered in this case, but the governing ministry is the MCIE and not the Ministry of Environment. MCIE is more interested in recycling and energy-efficiency issues than environmental ones. This implies that MIP and BJFEZ began the project with poor assumptions.

The cooperation between businesses and local authority in the EIP project is essential for the successful promotion of the project. Despite this need the attitude of the public officials was not only passive but skeptical due to sectionalism and display administration which are prevalent in the Korean administrative culture. MCIE was in charge of the pilot project to construct the EIP. However, the post of environmental affairs in the local authority manages the project. Due to sectionalism, the EIP project is considered an extra duty by the local environmental post. In this situation, the pertinent post has been rarely interested in the EIP project and cooperation with MCIE has not been effective. Particularly, the City of Jinhae and Gyeongnam Province seems passive in developing the project (Kim 2007).

After the central government launched the pilot EIP project in 2005, the central government created relevant policies and directed local authorities to promote the project substantially. Since then, local governments have developed very few ideas about the project or policy means to promote it. In addition, government authorities, regardless of their level, have not secured the budget to support the project. Local authorities do not concentrate efforts on securing extra funds because the project is considered an extra duty.

Previous studies on EIP in other countries indicate that the EIP project requires more funds than ordinary industrial park management costs. Therefore, the financial support from the public sector is necessary considering the social benefits. However, in the case of Macheon Industrial Park, there is no substantial support plan for funds from central government or the local government. There is no financial support program because the basis of environmental policy is the polluter-pays principle.

The existing support program suggested by the public sector requires some modification to draw more businesses into the EIP project. Businesses want substantial subsidies but the government does not have any specific support program. The reality is that just emphasizing the conceptual superiority of the EIP cannot realize the project without any support in Korea. Structured display administration and the absence of concrete policy measures generate administrative lethargy and risk-averse behavior. The EIP project in the view of local public employees is nothing but a mere fad, and thus given a low priority.

Because the long-distance network between the pertinent industrial complexes makes the original concept of EIPs unclear, the project manager tried to exclude MIP from the entire project. MIP has not been excluded from the project yet, but the fact is the manager tried to launch a separate project targeting MIP. The interim assessing group of MCIE pointed out that the pilot project in Gyeongnam Province is not feasible economically due to the long-distance transport of wastes and by-products between industrial parks. KNCPC knew that the idea of the long-distance network was theoretically and practically unreasonable from the beginning, but they accepted the present structure because of the local politics. The Center could not refuse the local authority's requirement to include MIP in the project, and subsequently the poor initial selection of the target industrial park deteriorated the vitality of the project as a whole.

## 5 Lessons and Policy Implications for Science Parks

This chapter discussed the promotion process of the EIP project in Korea through the case of MIP. The need for the EIP project is mainly found in the pollution and civil complaints in Korea. The conceptual understanding of the EIP in Korea is quite different from the original idea. That is, the EIP primarily aims to construct organic material flows and a resource-symbiotic network within an industrial complex. However, the approach was adopted to deal with environmental complaints from residents caused by the industrial plants in the case of MIP. This might pervert the aim of the project and finally threaten the sustainability of the project itself. Therefore, a re-conceptualization or a training program for ordinary citizens is needed for the successful promotion of the park.

The EIP initiative in Korea is in its early stage and is experiencing trial and error. There are sharp gaps between the expectation of the interested parties and those actually benefitting from it. This problem mainly results from a conceptual misunderstanding of the project. The public sector should play a more active role for the environment and economies in connection with the EIP project. Furthermore, private companies seem too passive in the promotion of the project. They have tendency to wait for governmental action and to simply want to get a free ride on the support. This passive stance of businesses makes the future of the EIP project gloomy. Only the spontaneous and active participation of businesses guarantees the success of the project. Businesses need to realize that the project is an effective solution for the environmental regulation, which is becoming strict. The active participation in and accurate understanding of the project by the local residents is also essential for the success of the project.

The experience from the EIP initiative in Korea can be applied to make our science and technology parks more eco-friendly and greener science parks.

First, it is necessary to gather accurate data for environmental and industrial fields as well as material balance from the activities in science parks. We also need to review the experience of existing and technology parks and identify their institutional and/or organizational structure; and estimate what fraction of the activities might be considered relevant to eco-innovation for sustainable development. From these analyses, the experience garnered in previous science/technology park efforts can be utilized to help establish a new arrangement for green growth.

Second, while a wide range of technologies might be appropriate for green growth efforts, it may be appropriate for the science and technology park to be become more focused on the specific area of eco-innovation. It may be appropriate for the proposed entity to focus on clean energy technologies, eco-environment protection, cleaner production technologies, water efficiency technologies, agriculture and health issues, or other topics of immediate concern to each society. The selection of technologies and the area of eco-innovation can be done effectively by establishing governance system with interested parties.

Third, networks between science-parks and eco-industrial parks should be established to explore potential opportunities for new business. Public sector can facilitate the establishment of these networks by providing incentives to encourage park entities and businesses to participate.

Fourth, when a potential institutional arrangement for the network has been developed, the next step will be to estimate the financial needs necessary to establish the project, and to sustain its future efforts.

A more comprehensive understanding of the interaction between supply and demand for eco-innovation will be a pre-requisite for creating successful eco-innovation policies. Eco-innovation has the potential to lead to significant economic opportunities. But the costs of some innovations may be very high initially, and government will have to share the risk of new technologies with the private sector in some circumstances.

A number of other measures are already being employed by countries to support environment-related R&D. An analysis of the results of the OECD survey on current government innovation policies reveals several areas for improvement among such measures: Supply-side measures include equity support, research and development, pre-commercialization, education and training, networks and partnership, and information services. Demand-side measures include regulations and standards, public procurement and demand support, technology transfer.

With eco-innovation gaining ground within both industry and government as a way to tackle environmental degradation and to foster green growth, both developed and developing countries are intensifying its work in this area. But research on eco-innovation is still in its infancy, particularly concerning systemic eco-innovations, which have greater potential for overall environmental improvements but are also highly complex, involving non-technological changes. In this context, the management entities of science and technology parks need to intensify activities such as

- Develop a toolkit to help businesses benchmark their performance and improve their production processes and products.
- Gather examples of eco-innovations, particularly those of more integrated, systemic approaches, and conduct in-depth analyses of such innovations to deepen understanding and extract lessons for practitioners and policy makers.
- Identify promising policies that encourage eco-innovation by sharing best practices among countries with science and technology parks.

## References

Ayres, R. U. (1989). Industrial Metabolism. In J. H. Ausubel (Ed.), *Technology and Environment*. Washington,DC: National Academy Press.

Ban, Yong-Un. (2008) Korean EIP Development Strategies Including Legal System Improvement Measures (http://www.wbginvestclimate.org/advisory-services/cross-cutting-issues).

Bass, L. (1998). Cleaner production and industrial ecosystems: A Dutch experience. *Journal of Cleaner Production, 6*, 189–197.

Chiu, A. S. F. (2005). *Training package for NCPC on CP/environmental management of industrial estates*. Ver. 3.3. UNEP.inWENT, Paris. http://www.uneptie.org/pc/ind-estates/event

Cohen-Rosenthal, E. (2003) A Case study of the Louisiana Army Ammunition Plant Re-use. In Cohen-Rosenthal, E. & Musnikow J.(Eds.). *Eco-industrial Strategies Unleashing Synergy between Economic Development and the Environment* (pp. 330–340). Sheffield:Greenleaf Publishing Ltd.,

Deutz, P., & Gibbs, D. (2004). Eco-industrial development and economic development: Industrial ecology or place promotion? *Business Strategy and the Environment, 13*, 347–362.

Ehrenfeld, J., & Gertler, N. (1997). Industrial ecology in practice- the evolution of interdependence at Kalunborg. *Journal of Industrial Ecology, 1*, 67–79.

ESCAP (2006). *Green Growth at a Glance: The Way Forward for Asia and the Pacific*. U.N. Economic and Social Commission for Asia and the Pacific.

Indigo Development. (2005). South Korean EIP initiative.

Kim, H. (2007). Building an eco-industrial park as a public project in South Korea. The stakeholders' understanding of and involvement in the Project. *Sustainable Development, 15*, 357–369.

Korhonen, J., von Malmborg, F., Strachan, P.A., & Ehrenfeld Jr. (2004). Management and Policy Aspects of Industrial Ecology: An emerging research agenda. *Business Strategy and the Environment. 13*, 289–305.

Lowe, E.A., (1997). Regional Resource Recovery, and Eco-Industrial Parks: An Integrated rategy. Prepared for the Symposium, Verwertungsnetz Obersteiermark Innovation durch regionale Recycling-Netzwerke at Karl-Franzens-Universität Graz. April 28–29, 1997. Available at:http://www.indigodev.com/Eipresrecov.html

Lowe, E. A. (2001). *Eco-industrial development in Japan*. Emeryville, C.A: Indigo Development.

Lowe, E. A., & Chiu, A. S. F. (2005). *Consultant's Report on Korean EIP Project*. Korea National Cleaner Production Center.

OECD. (2009). Sustainable manufacturing and eco-innovation: Towards a green economy.

Park, H. S., Rene, E. R., Choi, S. M., & Chiu, A. S. F. (2008). Strategies for sustainable development of industrial park in Ulsan, South Korea—from spontaneous evolution to systematic expansion of industrial symbiosis. *Journal of Environmental Management, 87*, 1–13.

Von Malmborg, F. (2004). Networking for knowledge transfer. *Business Strategy and Environment., 13*, 334–346.

Waller, H.P., (1999). Toward sustainable development of industry: Networking complexity and eco-clusters. *Journal of Cleaner Production, 7*(1):49–58.

# Green Development Strategies at Southern Taiwan Science Park

**Shiann-Far Kung**

**Abstract** Taiwan is an island with a total area of 36,000 $km^2$ and densely populated by some 23 million people. Because of the scarcity of land, spatial planning has become instrumental to the supply of land for industrial development. The first planned industrial parks was initiated by the central government in 1960, the first export processing zone in 1965, and then the first science park was announced to be built in Hsinchu in 1976 and started operation under the supervision of the National Science Council in 1980 (Hong et al. 2000; Kung 2000). Since Hsinchu Science Park (HSP) has been widely recognized as successful in attracting high technology industries, more science parks have been established or proposed to be built since the 1990s. Currently there are 11 science parks in operation and two sites in planning stage as of year 2011.

## 1 Introduction

Taiwan is an island with a total area of 36,000 $km^2$ and densely populated by some 23 million people. Because of the scarcity of land, spatial planning has become instrumental to the supply of land for industrial development. The first planned industrial parks was initiated by the central government in 1960, the first export processing zone in 1965, and then the first science park was announced to be built in Hsinchu in 1976 and started operation under the supervision of the National Science Council in 1980 (Hong et al. 2000; Kung 2000). Since Hsinchu Science Park (HSP) has been widely recognized as successful in attracting high technology industries, more science parks have been established or proposed to be built since

S.-F. Kung (✉)
Department of Urban Planning, National Cheng Kung University, Tainan, Taiwan
e-mail: sfkung@mail.ncku.edu.tw

D.-S. Oh and F. Phillips (eds.), *Technopolis*,
DOI: 10.1007/978-1-4471-5508-9_20, 

**Table 1** Science parks in Taiwan

| | Name | Year of establishment | Area(ha) | Target industries |
|---|---|---|---|---|
| | Lungtan S.P. | 2004 | 107 | OE |
| | Hsinchu S.P. | 1980 | 653 | IC, C&P, OE, TC, PM, BT |
| Northern Taiwan | Chubai S.P. | 2008 | 38 | BT |
| (1342 ha) | Chunan S.P. | 2001 | 123 | TC, OE, BT |
| | Tungluo S.P. | 2004 | 350 | TC, OE, AA |
| | Ilan S.P. | 2009 | 71 | TC, OE |
| | Houli S.P. | 2006 | 255 | IC |
| | Taichung S.P. | 2003 | 413 | IC, TC, OE, PM, AA |
| Central Taiwan | Huwei S.P. | 2004 | 97 | OE, LCD |
| (1657 ha) | Erlin B.P. | Planning | 631 | BT, OE, TC |
| | Higher Research S.P. | Planning | 261 | – |
| Southern Taiwan | Tainan S.P. | 1996 | 1,043 | OE, BT, IC |
| (1613 ha) | Kaohsiung S.P. | 2001 | 570 | OE, TC, BT, MD |
| Total | 4,613 | | | |

*Note BT* Biotechnology, *C&P* Computer and peripherals, *LCD* TFT-LCD, *Med* Medicine, *OE* Opto-electronics, *PM* Precision machinery, *TC* Telecommunications, *IC* Integrated circuit, *AA* Aeronautics and astronautics
*Source* HSPA, CTSPA

the 1990s. Currently there are 11 science parks in operation and two sites in planning stage as of year 2011 (Table 1).

Tainan Science Park was established in 1996 as the second science park in Taiwan, originally titled as "Tainan Science-Based Industrial Park" (TSIP). The designated site was a flat and low land of 638 ha (2,565 ac) in the middle of Taiwan's agricultural heartland, located across the border of three rural townships—Hsinshih, Shanhua, Anding of Tainan County; these three nearby major settlements altogether housed around 100,000 residents in the mid-1990s. At a walking distance, there is a renowned international research institution—Asian Vegetable Research and Development Centre. However, it locates some 12 km from Tainan, the nearest major city and National Cheng Kung University (NCKU), the key higher education institution in Southern Taiwan. From the very beginning, there had been doubts of its possibility to succeed. And indeed, it even survived the 1997 Asian Financial Crisis and gained substantial growth amidst the economically pessimistic period of the late 1990s.

Because of the fast industrial growth and the leasing system, by early 1999, only 70 ha of planned industrial area was left. In December 1999, the Ministry of the Interior approved the "Plan of the TSIP Special Zone" which embraces the idea of building a science city of 3,299 ha (8,152 ac) around the original TSIP with another 400 ha (988 ac) adjacent land for future industrial expansion of TSIP. In June 2001, another expansion site of 570 ha (1,409 ac) for TSIP was approved

**Table 2** TSP industrial statistics in 2011

| | IC | OE | BT | TC | PM | C&P | Others | Total |
|---|---|---|---|---|---|---|---|---|
| Number of firms | 9 | 47 | 39 | 11 | 44 | 3 | 5 | 158 |
| Number of employees | 16404 | 36220 | 1939 | 947 | 4684 | 249 | 1901 | 62344 |
| Turnover (NTD billions) | 233.9 | 301.3 | 5.5 | 3.2 | 31.3 | 1.5 | 2.7 | 579.4 |

*Note BT* Biotechnology, *C&P* Computer & peripherals, *IC* Integrated circuit
*OE* Opto-electronics, *PM* Precision machinery, *TC* Telecommunications
*Source* STSPA

at Lujhu, Kaohsiung County. In January 2003, the Southern Taiwan Science Park Administration (STSPA) was officially formed, under the National Science Council, to manage the operation of Tainan Science Park (TSP, the expanded TSIP) and Lujhu Science Park (LSP) and began to provide services to their tenants. Lujhu site was renamed as Kaohsiung Science Park (KSP) in 2004. Because of the expansion and frequent change of titles, in addition, there is a Tainan Technology Park established and managed by the Industrial Development Bureau (IDB) at nearby Tainan City; the science parks in the southern region are very often confused by the people. In this article, the focus is on the Tainan Science Park and its surrounding area TSP Special Zone.

Tainan Science Park has covered an area of 1038 ha since 1999. According to the STSPA statistics of 2011, as shown in Table 2, the total number of tenant companies reached 158, with 105 in TSP and 53 KSP, respectively; the total employees in these companies amounted to a new height of 62,344, with 57,447 in TSP and 4,897 in KSP; the total turnover of these firms was NTD 579.4 billion (USD 19.3 billion) which is nearly 10 % lower than the NTD 605.9 billion of the year 2010, and TSP accounted for NTD 555.4 billion (USD 18.5 billion). From all of the three variables, opto-electronics is still the largest industrial sector within the TSP, followed by the semiconductor industry; precision machinery comes third. Comparing with its predecessor Hsinchu Science Park (HSP), the statistics show that HSP houses 430 companies with 130,577 employees and a total revenue of NTD 1,080 billion (USD 34 billion) in 2008; and at its 12th year after establishment in 1980, HSP registered 148 tenants, 25,148 employees, and the total revenue amounted to NTD 87 billion (USD 2.8 billion). Although the industrial categories are not exactly the same as it was planned in the beginning, nevertheless, the burgeoning TSP tenants are often grouped into four major industrial clusters: opto-electronics, integrated circuits, precision machinery, and biotechnology (Figs. 1, 2, 3).

The opto-electronics industry has become the dominant sector in the TSP since 2001, the total number of employees reached 36,220 in 2011, accounting for 61 % of the total employment in the TSP. It has the most complete production chain among Taiwan's three TFT-LCD industrial clusters (Fig. 4), led by the Chimei Optoelectronics, the second largest producer of TFT-LCD panels in Taiwan and the fourth in the world (Kung and Chen 2008). Because of the increasing importance of photovoltaic technologies in future energy production, this sector is

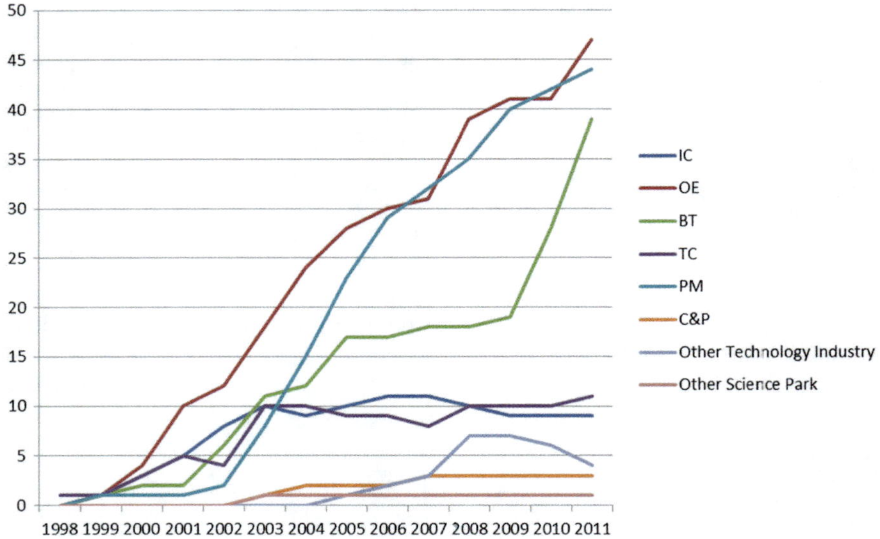


**Fig. 1** Number of firms in STSP. *Source* STSPA

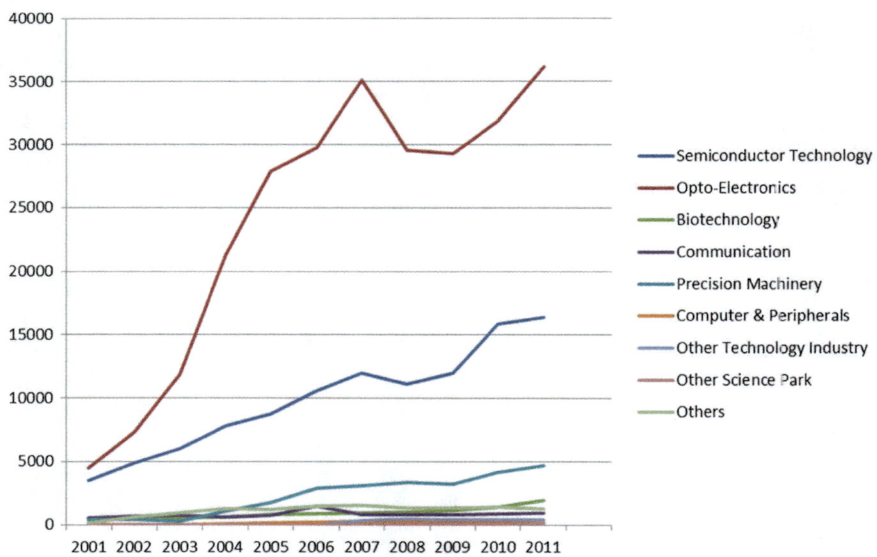


**Fig. 2** Number of employees in STSP. *Source* STSPA

now regarded as having high potential for greener development which will be further discussed in this article.

In general, TSP is now about half the scale of HSP, and has been growing at a faster speed than that of HSP; this matches well with the prime goal of any science

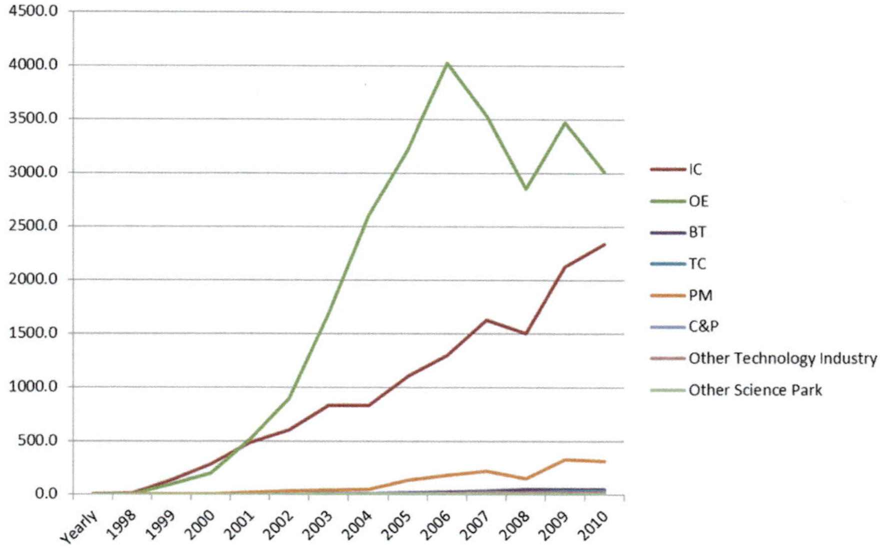


**Fig. 3** Turnover (NTD hundred million) in STSP. *Source* STSPA

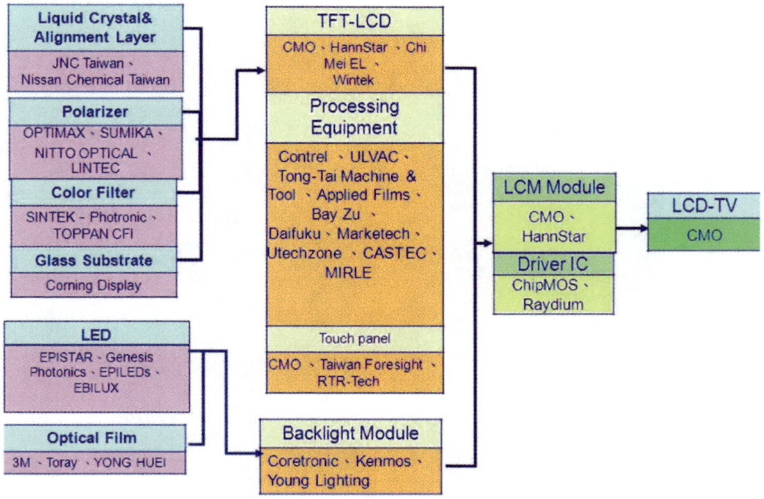


**Fig. 4** TFT-LCD cluster in STSP. *Source* STSPA

park which is the development of high technology industries. However, the more important feature of TSP is that it is leading in the realization of green development which encompasses spatial, industrial, and cultural perspectives.

## 2 Development of Green Urban Planning

The earliest noticeable green development concept at TSP is like that of most of the other science park projects, the notion of combining the "science" or "technology" with the "park" or "garden." Indeed, the HSP had already boasted its much greener environment in comparison with other industrial parks or cities in Taiwan. Following HSP spatial planning experience, the park-like environmental image can be gained through the gateways and the major boulevards; as shown in Map 1 is taken from (National Science Council 1995), there are eight parks in the 1995 TSIP proposal plan. But the distribution of these green spaces along the major north–south transport corridor and especially at its junctions with the east–west major road, the northern gate and the southern gate, clearly indicates that the formation of visual corridor was the major consideration. Thus, green spaces at this planning stage were basically used as a tool to gain good images for visitors.

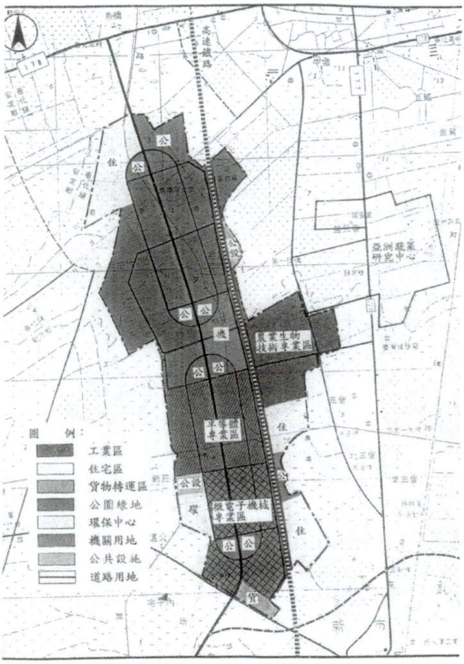

Except for a larger percentage of parks and other green spaces in the TSP than its predecessor, the more significant difference between TSP and HSP evolved through the very necessity of flood prevention. Unlike the HSP which is developed from tea farms on tender hillside topography, the TSP is located on a very flat and low land in the middle stretches of Yenshui River. Moreover, the site is sloping from northeast to southwest, and this area is also part of the watershed of Tachou Drainage, Anshunliao Drainage, and upper reaches of Yenshui River's tributary.

The area had been prone to flood hazards even when it was still cultivated as sugarcane farms before the construction of the TSP. Therefore, when China Engineering Consultants was called to make the first official urban plan for the park, flood risks were taken into account. The proposed solution was to create four detention ponds, mainly located along the western side of the park. While the proposal was to be reviewed by the concerned authorities, however, because the "detention pond" was a new land use category to Taiwan's mandatory urban planning system, the detention pond sites were marked separately from parks and other green spaces, even when some of them are located side by side is taken from (China EngineeringConsultants 1997).

During the process of environmental impacts assessment, considerations of newly found archaeological traces and habitat of endangered birds Glareola Maldivarum led to the designation of a conservation zone of 30 ha at central west end where the site was originally proposed for the waste treatment facilities. Because the estimated average space per Glareola Maldivarum would exceed at least 1 ha, pro-development commentators even criticized that the ecological conservation zone devalued human beings to the extent that even the smaller birds can enjoy much larger space. Regardless of the arguments, the designation of the ecological conservation zone in such a national priority industrial project is also one of the first landmarks in Taiwan's environmental protection history, and it clearly demonstrated that ecological value should be respected in the industrial development process. With these adjustments, in the Detailed Plan approved in May 1997, a central green belt consisting of detention ponds, parks, and conservation zone appeared along the east–west major road; there is also a cluster of ponds and parks in the northwest corner (Map 3). This kind of layout is unlikely to be proposed by traditional urban and industrial planners alone without consultation from other disciplines that have a deeper respect and understanding of ecological values and natural processes.

While the detention pond idea was taking shape on the TSP Detailed Plan, monsoon shower rains in May 1997 were causing floods in southern Taiwan, in particular, the industrial parks in Kaohsiung suffered heavily. Even if the TSP was then in the early stage of construction, they worried about the consequence of land-use change because floods alerted local communities again. Officials also noticed the potential threat from natural forces to this flagship project in the South, and the existing drainage system could not effectively solve the problem. Another heavy rain hit TSP again in July 1, pouring a historical high record of 420 mm rainfall in one single day which caused 2000 ha of flood area and two-thirds of the TSP was included. President T. H. Lee and Premier C. Lien immediately called for the joint effort from Ministry of Economic Affairs (MOEA), NSC and Tainan County Government to prevent TSP from flood threat in July. In May 1998, NSC and Water Resource Agency, MOEA jointly announced a two-year regional drainage improvement project with a total budget of NT dollar 1.8 billion. Yet, before any concrete work was done, the next monsoon on 8 June 1998 seriously flooded the park and, in particular, caused great damage to one of the major local investment projects—Chi Mei TFT-LCD factory, which was under construction. Owing to the industrial importance of the enterprise and the strong political connections of the entrepreneur, Premier Lien visited the TSP again and demanded the acceleration of drainage system improvement project.

By September 1998, six emergency systems of drainage and flood preventions were completed, including: the Tachou drainage passage, the basic work of detention pond D, temporary drainage network, increasing the height of ChiMei waterproof dyke, completion of drainage and flood control measures before the beginning of construction, and establishment of the united typhoon defence organization in the park. In September, NSC submitted The Revision of TSP Detail Plan (Extension of Detention Ponds) report to MOI, and it was approved in June 1999. It is recognized that the original detention pond system had underestimated the drainage need in the southern part and the existing Tachou drainage capacity would not match future catchment area flood volume. Although Tachou drainage flows almost along the eastern and southern borders of the park, hence almost half of its catchment area is outside of the park, NSC decided the project would include the whole catchment area. Except for the rearrangement of drainage passages in the park, a significant change in land-use planning is the enlargement of detention pond in the south-end, increasing its area from the original 2.5 ha to 20 ha Map 4 from (China Engineering Consultants 2002), together with the changed drainage course and other open spaces, forming a southern blue-green corridor. In March 1999, the Tachou drainage remediation engineering was completed. There has been no flood hazard in and around the TSP since 1999.

Based on the development experience of the HSP and its mother city Hsinchu, local communities have been expecting that industrial growth in the TSP would eventually attract migrant population, new business, and benefit the surrounding property market. Local pressure for new urban planning in the surrounding areas has been high since the beginning of the TSP. Tainan County Government started such effort—Plan of Tainan Science-Based Industrial Park Special District which covers a quite large area and surrounds the whole area of TSP—in late 1997. Because of the flood awareness, the project consultant company seriously conducted flood potential simulations and took TSP Plan as key reference. This is the first urban planning project known to have seriously combined hydraulic analyses in the spatial planning process. The Plan of the special district (covering 3,299 ha of land) was approved by the Regional Planning Committee in August 1999. Except for the land for building purposes, the Plan reserved a very high proportion of land for parks and productive green spaces, and the blue and green corridors are well connected with the previously planned TSP corridors (Map 5). As a retrospect, the urban planning efforts between 1996 and 1999, more or less following the principles of landscape ecology, had laid the backbones for the green infrastructure of the TSP and its surrounding special district. They also stand as key landmarks in the evolution of "green" urban planning in Taiwan.

Map 5 Tainan Science-Based Industrial Park Special District Plan – Green and Blue Corridors

## 3 Development of Green Industries

The concept of "green industry" did not exist in the original TSP industrial planning. There are three target industries in the original TSIP proposal: microelectronics and precision machinery, semiconductor, and agricultural biotechnology industries. They were expected to be geographically co-located in three specialized zones (National Science Council 1996). A review of the sub-industries within the target industrial clusters will find that the central idea was to develop a semiconductor industrial cluster and its supportive precision machinery industries, and to utilize the agricultural base of the southern region to develop related biotechnology industries (Tables 3, 4).

Without the policy guidance, the development of "green industries" in the TSP may be regarded as a process and the result that started from the entrepreneurial insight of private investors. This entrepreneurship first came from those who saw the new industrial advantages, not necessarily thought as "green," from the existing industrial clusters, including opto-electronics, integrated circuits, and precision machinery. Motech was established at TSP in 1999 to catch the newly rising solar energy industrial opportunities in the global market, and became the first silicon solar cell producer in Taiwan in the next year. While most of the other TSP opto-electronic tenants were competing in the TFT-LCD display-related R&D and production investment, it gradually expanded into a vertically integrated firm. It produced 272 MV of crystalline silicon solar cells in 2008, ranked as the eighth solar cell maker in the world. Other major players in the industry include Delta Electronics which expanded into the field of power management solutions from its already wide-ranging electronics product line and invested a new branch in the TSP.

From the industrial policy aspect, the term "green industry" was formally used in the 2007 MOEA policy program "2015 Economic Development Prospect." Three industries were: renewable energy, energy saving, and cleaner production.

**Table 3** Industrial clusters featured in TSIP 1996 plan

| Target industrial cluster | Target sub-industry |
|---|---|
| Microelectronics and precision machinery zone | Wireless communication |
| | Precision machinery |
| | Medical instrument and materials |
| | Semiconductor equipment |
| | Computers and peripherals |
| | Micro-electro-mechanical systems (MEMS) industries |
| Semiconductor zone | Microwave communication semiconductor |
| | Power electronics |
| | Special-purpose integrated circuit industries |
| Agricultural biotechnology zone | Flowers and ornamental plants |
| | Biopesticide |
| | Livestock vaccine |
| | Aquaculture industries |

*Source* National Science Council (1996)

**Table 4** Industrial development target of Tainan science-based industrial park

| Target year and value<br>Target industry | Production value (USD million) | | Number of employees (person) | |
|---|---|---|---|---|
| | 2005 | 2010 | 2005 | 2010 |
| Microelectronics and precision machinery | 2,600 | 4,600 | 5,000 | 7,000 |
| Semiconductor | 13,000 | 27,700 | 15,000 | 30,000 |
| Agricultural biotechnology | 370 | 685 | 380-860 | |
| Total | 15,970 | 32,985 | ca. 20,000 | ca. 37,000 |

*Source* Adapted from National Science Council (1996), pp. 35, 47, 53

**Table 5** Green energy firms in Tainan science park

| | Category | Firm | Production date |
|---|---|---|---|
| Solar cells | Crystalline silicon | Motech | 2000 |
| | Thin-film | Kenmos photovoltaic | 2008Q3 |
| | | Auria Solar | 2009Q1 |
| | | Chimei energy | 2009Q2 |
| | CIGS | Axun tek | 2009 |
| Solar module | | Gloria | 2007 |
| System | | King ener tech | End of 2009 |

The MOEA 2009 "The Green Energy Industries Development Program" put new emphasis on "green energy," and identified two major industrial sectors: solar voltaic, LED lighting. According to this latest policy category, the solar voltaic sector has stronger presence in the TSP. There are 13 tenants approved until 2008, six of them are in production, one is under construction; Table 5 shows that Motech alone stands for the first wave of investment in the sector at the TSP, then the second wave of investment rushed in after 2007. Total capital investment reached NTD 17.7 billions (USD 0.55 billions) in 2008 and the total revenue of these firms reached NTD 22.8 billions (USD 0.71 billions); according to STSPA, new investment in the first half of 2009 totalled NTD 7.6 billions. The solar voltaic production chain may be divided into four steps, the TSP tenants in the sector are grouped as follows: (1) silicon wafer: Motech, Gloria, and MPI; (2) solar cell: Motech, Auria, Kenmos Photovoltaic, Chimei Energy, CAxun Tek, Solar Energy; (3) solar voltaic module: Gloria, Lightwave Solar Power; (4) solar voltaic system: Motech, Delta Electronics, King Enertech System, Everphoton. In addition to these, equipment producers include: Ulvac, AMC and Mirle. There are five companies working on LED lighting industry, including the EPISTAR. Boyam Power System, a subsidiary of Canada-based Palcon, is dedicated to fuel cell production. Although this is still a relatively small and evolving cluster, member firms already cover quite a variety of component and product, forming an industrial chain that include from upstream to downstream producers. It is expected that the green energy industrial cluster may contribute both to a low carbon emission world as well as a very competitive new sector to the TSP.

## 4 Development of Green Buildings

To the end of "create a green environmental-friendly subtropical Taiwan," the Architecture and Building Research Institute (ABRI), Ministry of Interior collaborated with Ministry of Economic Affairs in the promotion of green buildings since 1999. Through the establishment of the Taiwan Architecture and Building Center (TABC) in the same year, a green building evaluation system was introduced to the public. The original evaluation system covers seven indicators in 1999 and two other were added in 2002, the building which successfully passed the evaluation would receive a "green building badge," meaning that the building is sound in "ecology, energy saving, waste reduction and health" (EEWH) respects. According to ABRI, this is the fourth green building evaluation system under operation in the world, after United Kingdom, United States and Canada, and the only such system applied to tropical and sub-tropical environment. The Executive Yuan announced the "Green Building Promotion Program" in March 2001 marks the beginning of official action. The action started mainly from public buildings, and all the major new government building projects are required to be evaluated. Private building projects which applied green building badge have been primarily the luxurious commercial, office or residential projects; industrial buildings have been very "quiet" in this arena.

However, industrial property comprised a significant proportion in Taiwan's built-up landscape, if the industrial buildings are still not involved in the movement, the purpose of promoting green building would be greatly handicapped. The major types of industrial property include: science park, industrial park, and industrial estate. While the subject of this article—science park—is under the administration of National Science Council, industrial parks have been promoted vigorously and directly by the Industrial Development Bureau (IDB), MOEA since 1960, and individual industrial estates are basically owned and managed by the private investors. From this respect, the public owned or managed science parks and industrial parks have the potential to contribute in "green development." Kung et al. (2002) proposed that industrial parks should try to find opportunities for using existing industrial clusters as foundation for the construction of ecological industrial park in Taiwan, and within this, green building badge should be also promoted to the industrial buildings within the industrial parks managed by the IDB. Yet, this suggestion did not get immediate response from the government sector. This has to wait for the initiation from private investors.

On 4 October 2006, Delta Electronics held a ceremony not only for the opening of its new office/factory building at the TSP, but also for the building successfully being awarded the green building badge (STSPA 2006). Delta Electronics founded in 1971 and now a world-class supplier of power management system, started its new investment at the TSP in 2004. Owing to the social responsibilities concept and the goal of enhancing its place in energy saving industries, the company

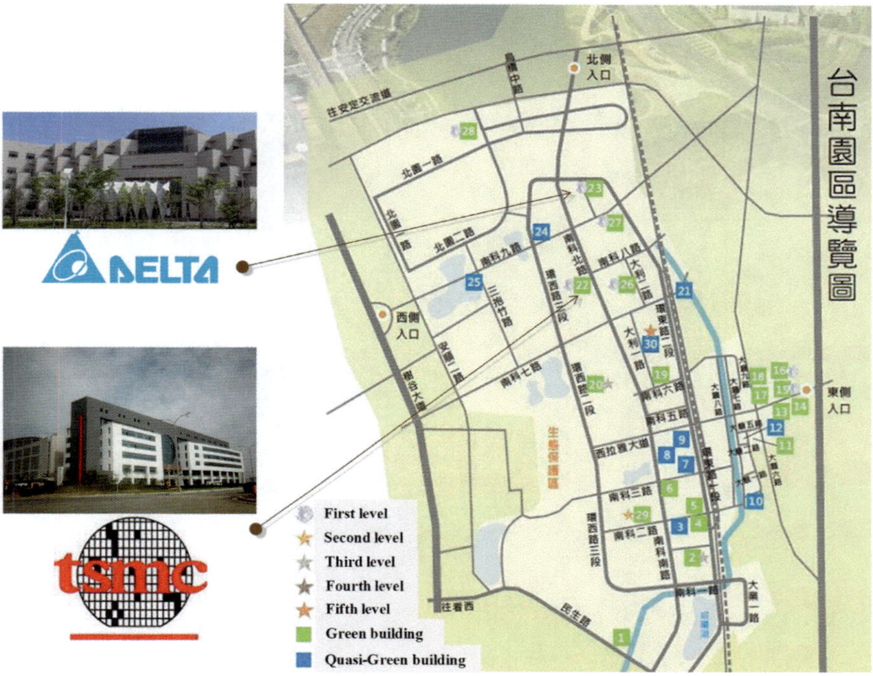


**Fig. 5** Green factory buildings in STSP. *Source* STSPA

decided to take the green building badge. Dr. H. T. Lin, professor of architecture at National Cheng Kung University, who has collaborated with ABRI in the study of green building subject for many years, became the chief consultant of this new project. Although it is the 100th green building in Taiwan, nevertheless, it is the first industrial architecture that passed all the nine green building indicators and certified as the gold-rated EEWH (Fig. 5). The Delta example of realizing enterprise's social responsibility and its success in obtaining green building badge has encouraged other tenants and even the STSPA such as the IC industrial giant Taiwan Semiconductor Manufacturing Company (TSMC). Indeed, the TSMC 14Fab P3 is certified by LEED as the gold-rated factory and by EEWH as a diamond-rated factory not long after (Fig. 5). In the public sector, energy saving photovoltaic devices have been installed (Fig. 6), for example, at Park 17, STSP Commercial centre and the Administration Building which is itself a certified EEWH building. According to STSPA statistics, there were 26 private EEWH green buildings and 25 public EEWH buildings in the park in 2009. Perhaps, TSP might have the highest green building density in Taiwan.

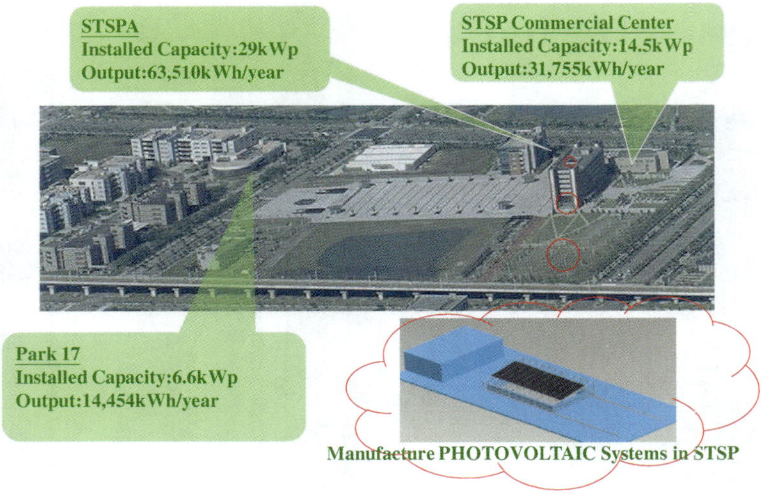


**Fig. 6** Green public buildings in STSP. *Source* STSPA

# 5 A Green Science Park Policy

In sum, the green development of the Tainan Science Park until now is not the result of a far-sighted comprehensive planning. As previously described, it is the archaeological remains, the Glareola Maldivarum birds, and the unexpected strong floods that alerted people of the different risks from industrial development. However, the changing of government's attitude toward these anti-development factors from negative to positive is important. These early interactions that mainly happened before this millennium, have paved the green infrastructure and spatial layout of the TSP. The private sector has also played an important role in two aspects; first, in the investment and formation of the energy saving industrial clusters and second, in the introduction of green buildings. Yet, the STSPA may have also contributed to the evaluation of investment proposals and assisted in building evaluation processes.

However, it is more apparent now that the STSPA is going to play a major role in the future green development. Based on existing blue and green corridors, green buildings, archaeological sites and exhibition centers, and other interesting resources, the STSPA has planned bicycle routes of 58 km to promote LOHAS and environmental education to local communities and TSP employees (Fig. 7).

It is also planning to improve the quality of detention ponds and parks with the introduction of TSP Green Map. Perhaps the most important initiative is the establishment of an internal steering team for the promotion of "sustainable environment and Green Park" in early 2009. With green development as the clear goal, the organization is headed by the TSP Director-General. There are four mission teams, covering green industry, environment protection, ecology and

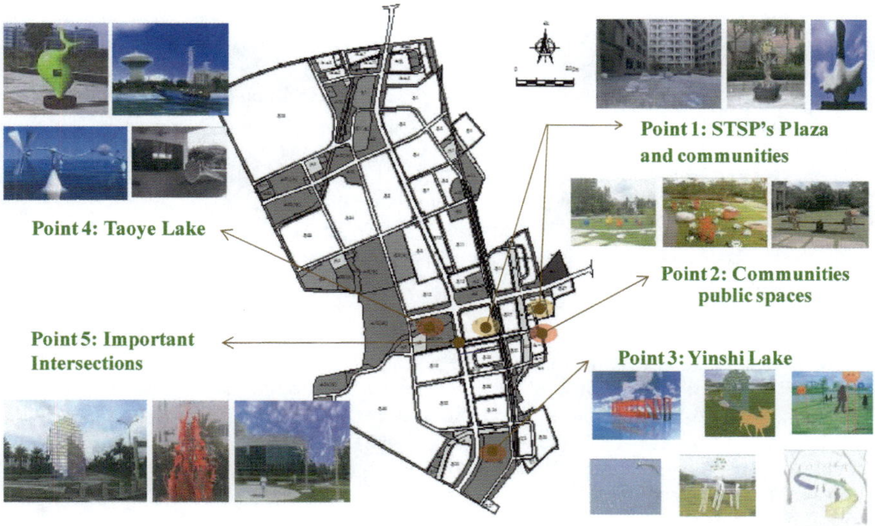


**Fig. 7** Green environment and public art in STSP. *Source* STSPA

landscape, and green building. It is hoped that through this organization the individual green actions within the park will be orchestrated into an integrated green development plan (Fig. 8).

**Fig. 8** Green environment and public art in STSP. *Source* STSPA

# References

China engineering consultants, INC. (1997), *Detail plan of Tainan Science Park,* Science Park Administration (in Chinese).

China engineering consultants, INC. (2002), *Detail plan on the second phase of Tainan Science Park development,* Tainan Science Park provisional office (in Chinese).

Hong, J. K., Kung, C. C., & Kung, S. F. (2000). Export processing zones in Republic of China. In A. K. Kundra (Ed.), *Asia's export processing zones and science parks in global markets* (pp. 12–23). Asian Productivity Organization: Tokyo.

Kung, S. F. (2000). Science parks in Republic of China. In A. K. Kundra (Ed.), *Asia's export processing zones and science parks in global markets* (pp. 85–93). Asian Productivity Organization: Tokyo.

Kung, S. F. et al.(2002). Sustainable management indicators of industrial parks in Taiwan. In *2002 International Conference on Sustainable Mechanisms for Industrial Park* (September 20), Tainan: National Cheng Kung University (in Chinese).

Kung, S. F. & Chen, C. W. (2008). Role of science parks in the formation of high technology industrial clusters—the case of Southern Taiwan Science Park. In *XXV IASP World Conference on Science & Technology Parks* (14–17 September, Johannesburg).

National Science Council. (1995). *The Southern Tainan science park preparation plan.* National Science Council (in Chinese).

National Science Council. (1996). *Environmental impact assessment of Tainan science park development project (Final edition).* National Science Council (in Chinese).

STSPA. 2006. *Southern Taiwan Science Park Newsletter*, no. 94.

# Science and Technology Park as Regional Innovation Platform: A Case of Chungnam Techno Park, Korea

**Hak-Min Kim**

**Abstract** Chungnam Techno Park (CTP), established in 1999, is one of the first 6 technology parks in Korea. It became a hub of regional innovation system (RIS) within 10 years by the collaborative network of enterprises, universities, research institutes, governments and business service agencies in the Chungnam area. The professionals of these network groups work together to develop 4-step business incubation program, a new campus of technology park, 4 regional strategic industries for the Province, and technology park management system. The case of CTP clearly shows how science & technology park (STP) creates RIS and functions as a platform of RIS by the author who was the chief of the master planning group in 2000 and the CEO of CTP in 2007–2010.

## 1 Science and Technology Park as a Regional Innovation System

### *1.1 Concept of Science and Technology Park*

Science and Technology Parks (henceforth STP), defined by the IASP (International Association of Science Park), are organizations managed by specialized professionals, whose main aims are to increase the wealth of their communities by promoting the culture of innovation and the competitiveness of their associated businesses and knowledge-based institutions. To achieve these goals, STPs stimulate and manage the flow of knowledge and technology among universities, R&D institutions, companies, and markets; they facilitate the creation and growth of

H.-M. Kim (✉)
Professor of Public Administration/Dean of Industry-Academy Cooperation,
Soonchunhyang University, Chungnam, Korea
e-mail: hakmin@sch.ac.kr

D.-S. Oh and F. Phillips (eds.), *Technopolis*,
DOI: 10.1007/978-1-4471-5508-9_21, 

**Table 1** Types of science and technology parks

| Types | Physical characteristics | Main functions | Examples |
|---|---|---|---|
| Business incubator (BI) | (R&BD: research and business development) facilities | Creating new businesses | University BIs |
| Research park | Large-scale land with basic R&D facilities | R&D + prototype production | Surrey RP |
| Science park | Large-scale land with applied R&D facilities | R&D + light production | Cambridge SP |
| Technology park | Large-scale land with post BI(Commercialization) facilities | Light production + business development | Chungnam TP |
| Science city | Large-scale town with SPs universities, and research institutes | Basic R&D | Daedeok SC |
| Technopolis | Large-scale town with TPs, R&D and production facilities | High-tech production | Sophia Antipolis |

innovation-based companies through incubation and spin-off processes; and provide other value-added services together with high quality space and facilities (IASP International Board, 6 February 2002).

STP should be the perfect habitat for businesses and institutions of the global knowledge economy. The major role of STP is to promote the economic development and competitiveness of the region. For this role, STP should create new business opportunities and add values to mature companies, foster entrepreneurship and incubate new innovative companies, generate knowledge-based jobs, build attractive spaces for the emerging knowledge workers, and enhance the synergy among universities, R&D Institutions, and companies (see Table 1).

## 1.2 Concept of Regional Innovation System

Regional Innovation System (henceforth RIS) is a network where actors from enterprises, universities, research institutes, governments, and business service agencies work collaboratively in a region to improve technology, production system, and business services. When the various actors learn collaboratively with each other to improve R&D, product quality, public service, and culture, the region, in turn, becomes an innovative region with dynamic economic development.

STP should be a hub of RIS. There are three major areas for STP to provide innovation: science and technology innovation, production system innovation, and business service innovation. These innovations should be collaboratively conducted by a network of enterprises, universities, research institutes, governments and business service agencies (see Fig. 1).

Table 2 shows the components of RIS where system and activities are essential. The innovation system should be created for official as well as non-official positions. The innovation system is carried out by direct as well as indirect activities of innovation activities. Although a region has a formal innovation system and direct

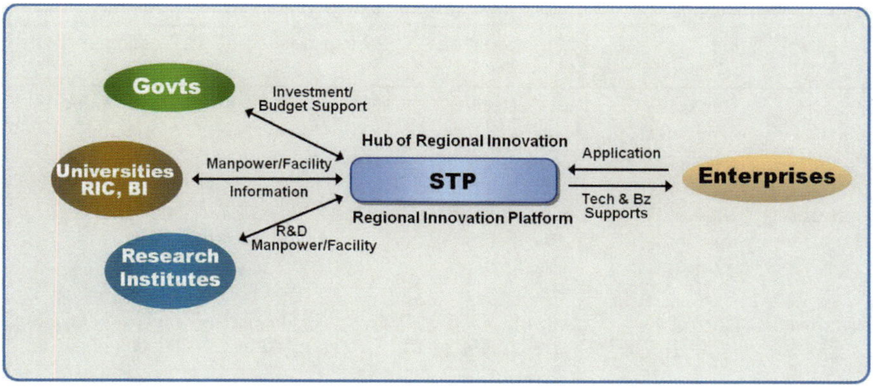


**Fig. 1** Function of STP

**Table 2** Two components of regional innovation system

| Categories | Innovation activities | | |
|---|---|---|---|
| | Direct | | Indirect |
| System condition | Formal | Innovation network<br>– Enterprises<br>– Universities and R&D institutes<br>– Government and agencies<br>– Technology policies | Official system<br>– Education/finance system<br>– Labor market<br>– Legal and tax system<br>– Policy |
| | Informal | Innovation patterns<br>– Trust among clients and suppliers<br>– Collaborations among enterprises<br>– Collaborations among scientists<br>– Familiarity on tech policies | Cultural factors<br>– Value of region<br>– Civic awareness if the region<br>– Education supporting<br>– Financial support environment |

cf: modified from Lundvall (2007)

innovation activities, the region is not classified as a completed RIS. To become a perfect RIS, the region should have innovation patterns and cultural factors, which are informal system and indirect innovation activities.

Figure 2 shows the completion of RIS, where innovation infrastructure, governance and system, and financial incentives are critical. Governance and system represent the government supporting system, actors like facilitator and coordinator, and social capital. Except social capital, the components of governance and system in Fig. 2 are the similar characteristics of a formal system in both direct and indirect innovation activities in Table 2.

There are three components of innovation infrastructure; physical, technical and knowledge infrastructure in Fig. 2. Except knowledge infrastructure, the physical and technical infrastructures are also similar concepts of a formal system in both direct and indirect innovation activities inn Table 2. The financial incentive in

**Table 3** Four steps for business incubation at Chungnam Techno Park

| Categories | Pre-Incubation | Main Incubation | Post Incubation | Production Site |
|---|---|---|---|---|
| Actors or Functions | University Students Researchers/ Professors | R&D Activities Business Development | Proto-type Production Commercialization | Mass-Production Marketing |
| Supporters | University Research Institutions | University Incubators Technology Parks | Technology Parks | Industrial Park |
| Performances | Contest for Start-ups | Over 400 Companies in 17 Universities | Over 130 Companies w/2,000 Employees | Over 60 Companies w/3,000 Employees |

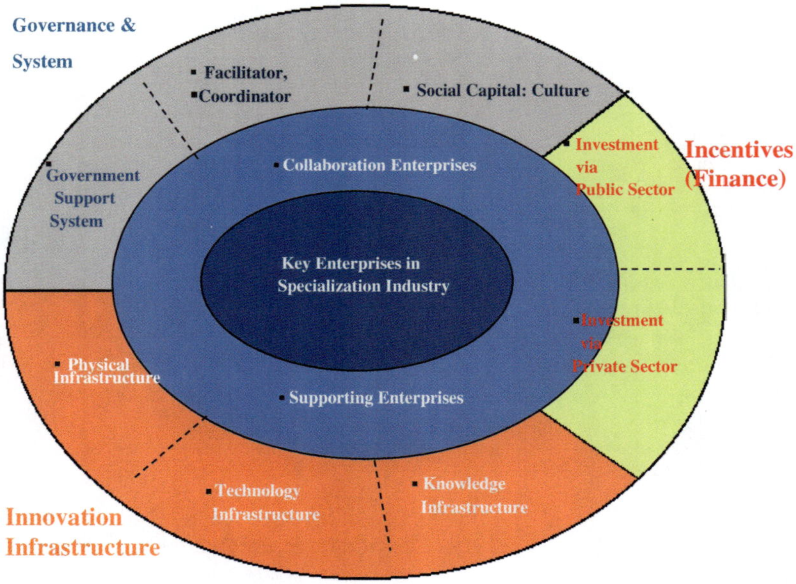


**Fig. 2** Completion of regional innovation system. Cf: modified from Andersson and Karlsson (2004)

Fig. 2 is comparable to the position where formal system and indirect innovation activities are met in Table 2.

Thus, the formal system, whether direct innovation activities or otherwise, are not enough to complete the RIS. The completion of RIS requires innovation patterns where informal system and direct activities are met in Table 2. It is a knowledge infrastructure in Fig. 2. The cultural factors in Table 2 are the similar concept of social capital in Fig. 2.

### 1.3 RIS Versus NIS

When a Regional Innovation System is completed in a region by STP, then the sum of RIS will become National Innovation System (henceforth NIS). If a region has a competent STP, as discussed in the previous pages, the region would have three innovation areas; science and technology innovation, production system innovation, and business service innovation. When these innovations are collaboratively conducted by a network of enterprises, universities, research institutes, governments and business service agencies, the region can be classified as completed RIS (Fig. 3).

In Korea, there are 18 Technology Parks (henceforth TP). Each TP is supposed to build an RIS platform in its metropolitan government area. There are seven metropolitan city governments in Korea such as Seoul, Pusan, Incheon, Daegu, Daejeon, Gwangju, and Ulsan. In addition, there are nine provincial governments, which are also classified as metropolitan governments. Each of 16 metropolitan governments has its own STP. There are two more STPs which are owned by universities. Each of these 18 STPs builds the RIS in its own region. The sum of these 18 RIS becomes NIS.

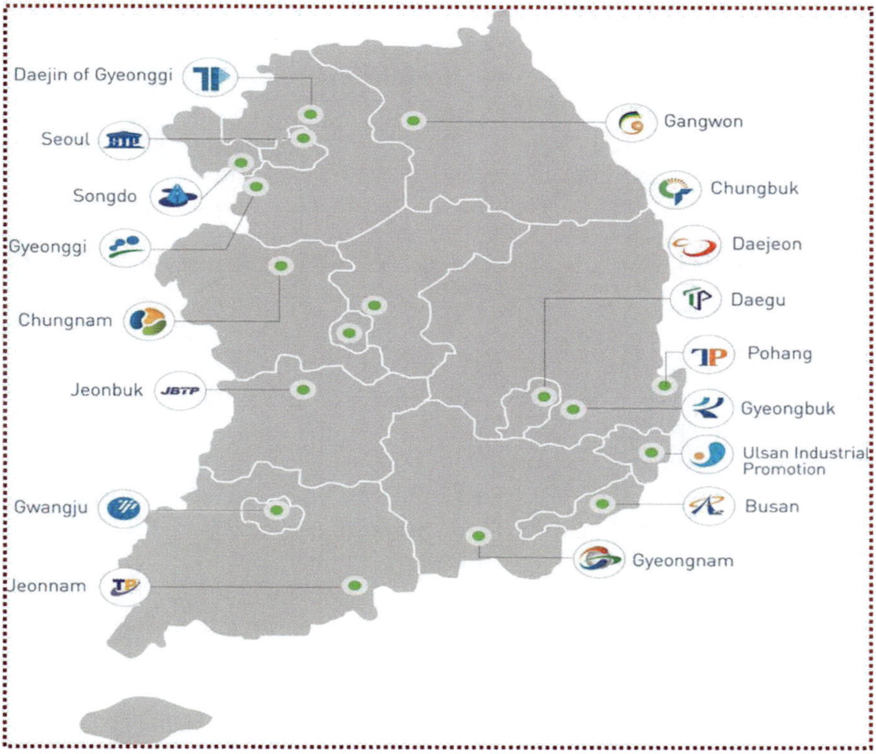


**Fig. 3** National innovation system by STPs in Korea

## 2 Foreseers and Innovators of Chungnam Technopark

### 2.1 Early Foreseers

The province of Chungnam used to be an agricultural area in Korea until 1995. In May 1995, local university professors began to discuss about the creation of a technology park in Chungnam Province. They organized the research group to persuade the local university presidents. The association of 17 university presidents supported the professors' vision to create a technology park in Chungnam, and became a strong lobby group to the provincial governor.

When the local university professors and provincial government officials proposed the plan to create a technology park to the central government, the central government responded that the program was too good to designate only one province. That was the first step to create a technology park in Korea. It was initiated by the local university professors in Chungnam Province. They were foreseers.

In two years, the central government enacted the special bill on "technology-driven industrial complex," and prepared the budget for creating six model technology parks in the nation. The local university professors actively contributed to the implementation of the project for the central government. But, Chungnam was a weak candidate at the moment, compared to the industrialized cities such as Seoul, Busan, and Daejeon. All of 16 provincial level governments applied to the technology park project. Professors and government officials from Chungnam worked hard to complete a proposal for the project. They received a big Christmas gift from the central government on 23rd December 1997. It was surprising news that Chungnam was chosen. Seoul, Busan, and Daejeon failed to get the technopark project at that time.

In 1999, CTP began its business at the old city hall of Asan. The Asan city government donated its old city hall to induce the CTP's headquarter. The city hall became a business incubation center as well as CTP's main office. But CTP needed a bigger space rather than a small office building. The first President of CTP whose previous position was deputy governor found a decent farmland of 45 acres in Cheonan city. CTP obtained the Farmers' Education Institute from the Chungnam provincial government as a form of donation.

The university professors organized a research group to conduct a master plan for CTP on the farmland in 2000. They brought the government officials and staff members to advance countries' science parks for benchmarking. Their benchmarking parks include Surrey Research Park, Cambridge Science Park, Sophia Antipolis, Dortmund Technology Park, and Kumamoto Technopolis. They learned that their technopark should be a campus for researchers, entrepreneurs, business service professionals, and local people. They also learned that their technopark should stimulate and manage the knowledge to commercialization (Chungnam techno park 2009a).

## 2.2 Active Innovators

The main actors at the beginning of CTP were professors, provincial and municipal government officials, and entrepreneurs. There were about 70 professors involved in the early stage of CTP (Kim 2006). Some of them became deans, vice presidents, and presidents of the local universities. One of them became the President of CTP. Most of them are still working as advisory groups of CTP and enterprises. With the spirit of cooperation among 17 universities, there are more than 300 professors in the CTP's working group in 2010. They participate in the four specific industrial advisory committees based on their major fields. The specific areas include semiconductor and display industry, automobile industry, multimedia and contents industry, and agricultural bio industry.

Some of the government officials became national congressmen. Some are still in the government with high ranking positions such as deputy ministers in the central government and deputy mayors of the local governments. They are still strong supporters for CTP in terms of policy and budget. The early residents of CTP were venture businessmen. While many of them disappeared, some of them became the owners of $100 million companies. They organized the CTP Alumni Club in 2009. The graduates are active as business mentors and 'angel investors' to their junior members who are new residents of CTP.

They were dreamers for CTP and became innovators for the region. Due to their dreams, the farmland became a perfect habitat for innovative business. They are still dreaming together for the future of the region. They believe that CTP as a regional innovation platform will make their next dream possible. Their next dream is to make their region a competitive cluster for high-technology industry. The high-technology includes green and clean technology, and contents industry.

Chungnam is now a growth engine of Korea's high-tech industries, where the major manufacturing sites of Samsung Electronics and Hyundai Automobiles are located. With 36 universities 623 research institutes, Chungnam leads the technology development and commercialization of the nation. CTP plays a pivotal role in vitalizing Chungnam economy as a regional innovation platform. The beginning was the research group of local university professors and government officials in May 1995.

# 3 The Main Function of CTP

## 3.1 Theoretical Background of CTP

Since multiple actors are involved in the project, CTP needs to organize the working flow among universities/research institutions, enterprises, and governments. Figure 4 tells how these groups work together. CTP as a regional innovation platform should be a base to link these groups. Universities and research institutions

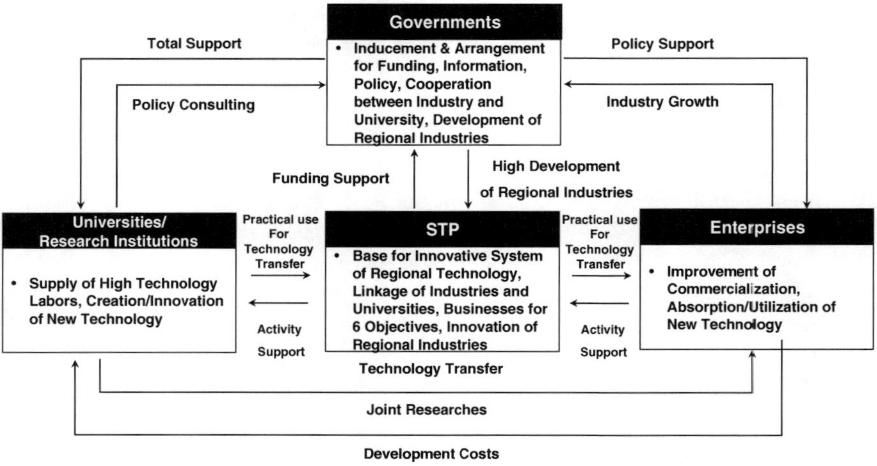


**Fig. 4** Regional innovation platform by CTP (2009)

should supply new sources of high technology. Governments should provide policy and budget for cooperation among university/research institution, and enterprises as well as CTP.

CTP plays as a coordinator or a matchmaker between university/research institution and enterprises. The first function of CTP is technology transfer between university/research institution and enterprises. The second function of CTP is to support R&D program to universities/research institutions when they conduct researches with enterprises. The third function of CTP is to provide investment fund from the government budget or private investors to the resident enterprises. The fourth function of CTP is to provide human capital from universities to enterprises. For this process, CTP provides training programs to match the needs of enterprises.

## *3.2 Four Steps for Business Incubation*

The fifth function of CTP is business incubation. There are four steps of business incubation program at CTP: pre-incubation, main incubation, post-incubation, and graduation. The pre-incubation is done at the labs and public sponsored offices. When entrepreneurs bring their business ideas, CTP arranges them to universities or research institutes to develop their business items. The CTP also provides seed money for their early stage incubation activities.

Once their items are developed as business models, they can enter into one of the 17 business incubators as start-up companies. The 17 business incubators are located in 16 universities and 1 research institute in Chungnam province. CTP provides managerial know-how and a part of operation cost to 17 business incubators. A start-up company normally stays 2–3 years at a business incubator. The managers, professors, and researchers provide their knowledge and skills to the start-up companies at their business incubators.

When the enterprises are successfully incubated at the 17 business incubators, they come to CTP for the post-business incubation program. The CTP's post-business incubation program includes pilot plant space, technology transfer, prototype production facilities, professional business services, investment fund operation, and global marketing. The "Contact Center" of CTP is a problem solver for enterprises. The staff members and retired professional experts are waiting for any kind of needs of enterprises. The enterprises normally stay 3–5 years at CTP's post-business incubation program.

Once the enterprises exceed their annual production amount over $10 million, they are asked to graduate from CTP. The enterprises develop their own mass-production facilities near CTP. Some enterprises allocate their factories in the industrial parks, while some others in private sites. Although they left from CTP, they are still connected with CTP by the "Family Enterprise" program. The CTP's family enterprises are supported by CTP's investment program, R&D fund, human capital development, and global marketing. This is the fourth step of the business incubation program.

Enterprise Support Center of CTP is the key staff for business incubation. The staff members are able to conduct professional business consulting. They provide the management and marketing services as well as technology transfer. CTP also operates Human Resource Development (HRD) Center where enterprises learn from entrepreneurship to three-dimensional computer graphic techniques. HRD Center of CTP plays a key role in networking various industrial fields through the CEOs' education program.

## 3.3 *Creating a Habitat for Innovative Businesses*

The sixth function of CTP is to create a perfect habitat for innovative businesses. All of the four steps of business incubation program can be implemented at CTP. Thus, CTP has its own pre-incubation and main-incubation facilities on the campus. The major spaces are allocated for post-incubation program, however. For the fourth step incubation program, CTP also provides the mass production spaces on the campus. Figure 5 shows the habitat of innovative business by four steps (Kim 2000).

As seen in Fig. 5, the campus of CTP provides all steps of business incubation. The business support building contains the "Contact Center," the "Technology Transfer Center," the "Global Marketing Center," and private business service agencies. The precision-machinery building provides prototype production facilities. There are three industrial support centers at CTP. The multimedia industry center is on the CTP campus. Two are allocated separately within 20 km. The empty spaces with green grasses will be filled with national level labs and multinational firm's R&BD complex (Korea Technopark Association 2004).

**Fig. 5** The bird-eye view of CTP (2010). ① Business service building, ② Post-incubation plant 1, ③ Post-incubation plant 2, ④ Precision-machinery building, ⑤ Post-incubation plant 3, ⑥ Logistics building, ⑦ Business incubator ⑧ Multimedia industry support center, ⑨ Display industry center (separate campus), ⑩ Automobile industry center (separate campus)

## *3.4 Development of Strategic Industries*

The seventh function of CTP is to promote the regional strategic industries of Chungnam Province. The provincial government of Chungnam selected four strategic industries. They are electronic–display industry, automobile industry, multimedia and contents industry, and agricultural bio industry. As a think tanker of Chungnam, the Policy Planning Agency of CTP plays an important role in conducting researches on how to develop the regional industries. The Policy Planning Agency designs the roadmaps for four strategic industries as well as for municipal level traditional industries.

CTP operates Display Industry Center, Automobile Industry Center, Multimedia Industry Center, and Agricultural Bio Industry Center. Each center is equipped with state-of-art facilities and professional staffs. They provide high quality of R & D and technical services. Each center also nurtures start-up companies in its incubator as well. Each center plays an important role in networking its own specific industry as a hub. Thus, each center has a pool of university professors, researchers, entrepreneurs, and business service experts. The pool ranges between 100 and 300 people based on the industrial field.

The Multimedia Industry Center, for example, leads Korea's visual media industry through state-of-art facilities, technical consulting programs, start-up incubation spaces, and enterprise training programs. In 2010, the Multimedia Industry Center created 50 start-up companies in the visual media field, supporting the main and/or post-production works for more than 30 movies, TV dramas, and virtual games. Recently, the center supports the 2D to 3D converting business. One of the resident enterprises in the center began to convert Hollywood movies from 2D to 3D. The center provides the enterprise with a space for 400 people, world class equipment and facilities, and 3D converting education programs. For this enterprise, CTP utilizes 17 universities in recruiting people, conducting researches, and even sharing space and equipment. This is one of the good examples of how CTP works as a hub of regional innovation system.

# 4 The Major Achievements of CTP

## *4.1 Setting Up of Operation System*

One of the most important achievements of CTP is to set up a habitat for innovative business in the region in a relatively short period of time. It took less than ten years to set up the whole system from nothing. It took five years to establish the hardware system. In this period, CTP completed the major buildings by the master plan. Then it took another 5 years to establish the software + hardware system. In this period, CTP completed the roadmaps for four strategic industries. Based on the four strategic industries, CTP built four centers equipped with state-of-art equipments (see Fig. 6).

From 2007, CTP began to focus on system building for enterprise support programs. CTP can concentrate on software programs now. The major functions of CTP nowadays are technology transfer and commercialization, financial investment for enterprises, research and business development projects, human resource development, and global marketing. The enterprises are satisfied with the one-stop service that CTP provides with a network of universities, research institutes, governments, and business service agencies.

The essence of the working system is people. The technology park is not about just technology. It is all about people. Since CTP has several functions from business incubation to provincial industry planning, CTP needs professionals and

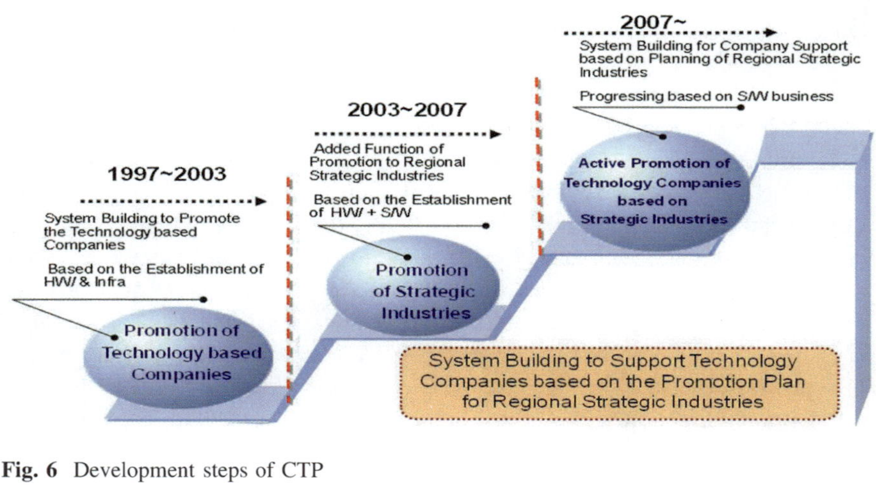


**Fig. 6** Development steps of CTP

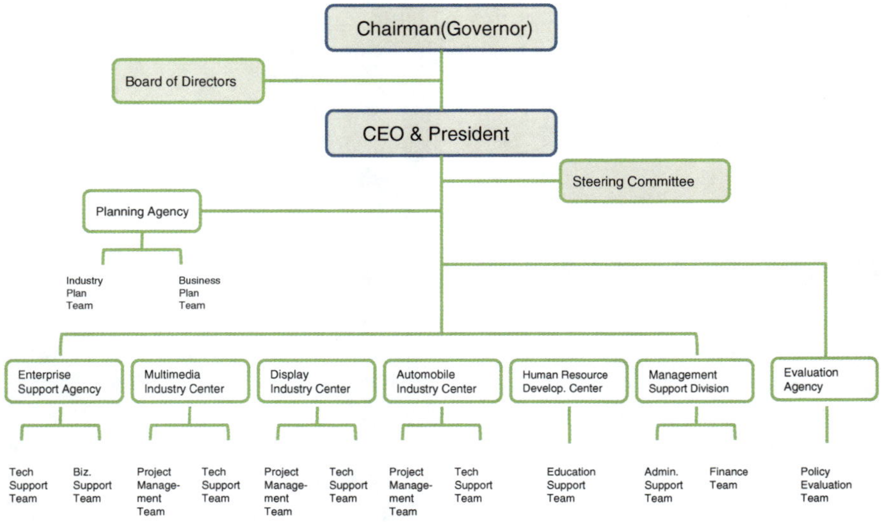


**Fig. 7** Organization chart of CTP

experts on several different fields. For the past 10 years, CTP has tried to recruit the best members in each field. The current staff members of CTP are 120 persons. More than half of them are graduate degrees holders. Most of them hold technical licenses in their fields (see Fig. 7).

The staffs of Planning Agency are researchers on industry planning and enterprise management. The staffs of Enterprise Support Agency are business consulting experts who can deal with technology transfer, financial investment, global marketing, etc. Each industrial center is filled with operating managers or technicians of specific industries such as multimedia and movie business, semi-conductor and display industry, and automobile area. The staffs of Enterprise Training Center are education program experts. The staffs of Management Support Division are accounting and personnel management specialists. Two of them are provincial government officials. The staffs of Evaluation Agency are researchers in the field of four strategic industries.

Next to the people, the management system is another important issue to run a technology park. Figure 8 shows the CTP's management system. The organization's vision of CTP is "promoting local industry to global standards." To accomplish the vision, CTP aims to stimulate the local economy by advancing its industry structure as the organizational mission. There are two categories in the management process: system and network. For the system management process, CTP operates SMI (system management integration), BSC (balanced score card), and CRM (client relationship management). For the network management process, CTP operates service network with various business service agencies, sharing network for information and outcomes and learning network with enterprises by co-education program.

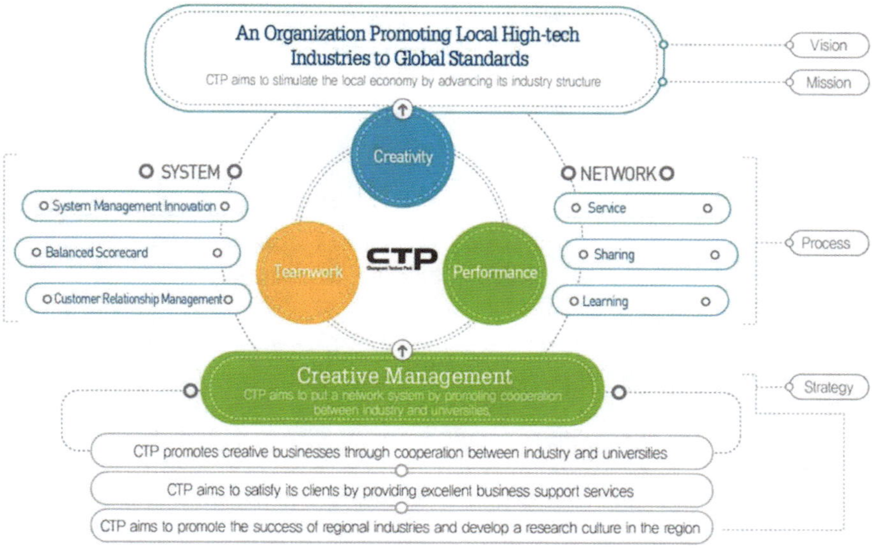


**Fig. 8** Creative management system of CTP

The motto of CTP is C + T = P, meaning that Performance will be obtained by Creativity and Teamwork. CTP is the place where creative people work together for great performance. This motto can be accomplished by a network of industry and universities. This is called CTP's "Creative Management" system. To make this motto come true, CTP has four strategies. First, CTP promotes creative business through cooperation between industry and universities. Second, CTP aims to satisfy its clients by providing excellent business support services. Third, CTP promotes the success of regional industries and develops a research environment in the region as the final strategy.

To visualize its quality of service, CTP obtained an accreditation as the Excellent Service Quality Organization by the Ministry of Knowledge Economy in 2007, and was awarded as the Best Practice Organization in 2008. CTP was also evaluated as the Best Technology Park in Korea by the central government in 2009.

## 4.2 *Performance of Business Incubation and R&D Development*

CTP's primary goal is to nurture enterprises. The performance of enterprises by the business incubation program is a direct output of CTP. The enterprise performance is measured by the number of enterprises, annual sales, and employment (Kim 2003).

When CTP opened its business in 1999, 32 enterprises began their businesses at CTP's business incubation center. The number of start-ups has dramatically increased since 2006: 56 start-ups in 2006 108 in 2007, 113 in 2008, and 132 in 2009. The number of employees has also increased quickly since 2006 as well; 651 persons in 2006, 1,021 in 2007, 1,223 in 2008, and 1,481 in 2009.

The most interesting statistics is the sales amount of the enterprises: $63 million in 2006, $131 million in 2007, $314 million in 2008, and $442 million in 2009 (Chungnam Techno Park 2010). Although there was an economic downfall during this period, the performance by the amount of sales jumped significantly. It was possible due to the fact that the automobile industry center newly opened at the period. Moreover, the enterprises are advanced technology-driven ones (see Fig. 9).

The total numbers of start-up enterprises at CTP campus from 1999 to 2009 are 282. Among them, 56 enterprises successfully graduated from CTP, moving out of CTP to their own independent production sites. Among 56 graduated, 35 are doing their businesses successfully and 21 have been merged with other companies or stop their businesses. There are still 132 enterprises at CTP campus and 94 enterprises failed during the incubation process (see Fig. 10).

The aggregated amounts of 282 enterprises produced are $5,962 million and the aggregated numbers that 282 enterprises employed are 14,884 persons for the last 11 years. In 2009, 35 graduated enterprises produced $285 million and employed 1,311 persons. The graduated enterprises will continue to increase when the business incubation program works properly. It is believed that CTP has rapidly increased its reputation among entrepreneurs. The brand name value of CTP has recently increased as well. Thus, many more graduated enterprises will reside around CTP and in Chungnam province.

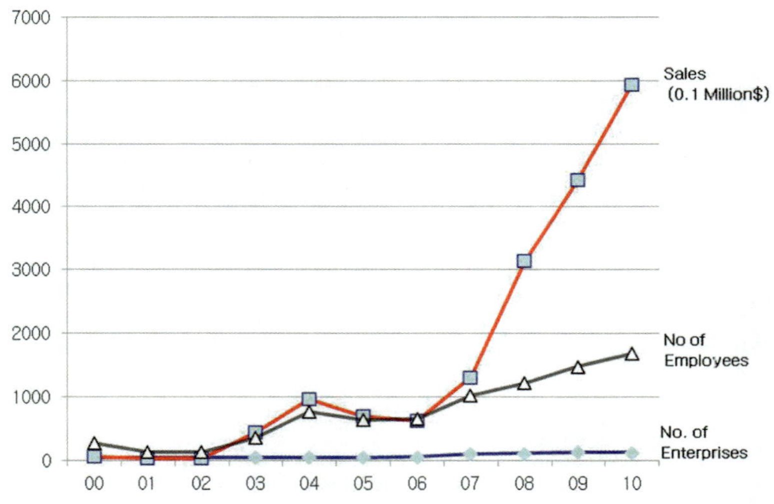


**Fig. 9** Performance of start-up enterprises at CTP (2000–2010). *source*: annual report of Chungnam Techno Park, 2011

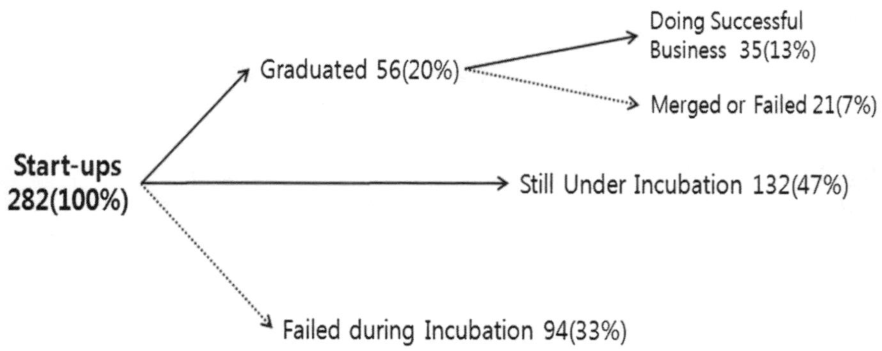


**Fig. 10** Performance of CTP's start-up enterprises (1999–2009)

In addition to the business incubation program, CTP puts its effort in R&BD (research and business development) activities. The R&BD program began in 2003 at CTP. The performance of enterprises by the R&BD program is another indirect output of CTP. The sales amounts of the enterprises that participate in CTP's R&BD program in 2003 were $1,215 million. This number became $ 4,013 million by 250 enterprises in 2009. The total employed persons who were working at the R&BD enterprises were 4,206. This number became 10,191 in 2009.

## *4.3 Five Star Enterprises of CTP*

Among many successes of CTP for the last 10 years, the Star Enterprise Project is the most tangible achievement. Five enterprises have been selected as Star Enterprises so far. One of them is Evertechno Co., Ltd. The enterprise began its business in 2000 with three people at CTP, listed on the Korean Stock Exchange Market in 2007, and reached its production amount to $200 million with 450 workers in 2009. The enterprise has won the Grand Prize in 2009 ASPA(Asian Science Park Association) Awards, which annually selects the best venture enterprise among 13 Asian countries' candidates. The enterprise expects $1 billion in production in 2013. More than 130 enterprises are following the path of Evertechno Co., Ltd at CTP today.

One of CTP's Star-Enterprise, Young Hwa TECH Co., LTD, successfully developed a superior technology in automobile parts, and has reached an agreement providing its products to one of the biggest automobile companies in the world, General Motors. It will provide $30 million worth of junction boxes to GM, which will be assembled for GM's Cadillac Models. Junction box is a core part in the hood to divide electricity and protect circuit, which provide electric energy and information to other components. The company's quality of technology is considered as one of the best in the world. The enterprise has been a first vendor of domestic automobile manufacturers such as GM Daewoo, Ssangyong Motors, and Renault Samsung.

The manufacture of golf distance measurer based on GPS, DECA System Co., Ltd is also a resident enterprise of CTP. The product name 'Golf Buddy' displaying distance and map of each golf course, has been exported $15 million from the end of 2008 to the first quarter of 2009, winning $10 million Export Tower Prize by KITA (Koran International Trade Association) in 2009. This product's main markets are USA, Europe, Canada, and other countries and it shows the course information about 28,000 golf courses around the world. In 2010, the company will service 40,000 golf courses in the world and expect to export $30 million expanding its markets to Australia, China, Japan, and so on.

CTP's TTM Co. LTD., the global provider of the epochal thermal solution, developed its strategic products in just 6 months since it started its business and the sales increased drastically. The creative thermal management solution developed by TTM is used in various fields such as computers, LED products, and other heat related products. The total sales reached up to $5 million in 2009 and were expected to be $20 million in 2010. TTM's core technology that handles the heat of electronics is considered as one of the best in the global market today. It has already received the validation from big global companies such as Intel, HP, and Samsung Electronics. With this great technology and recognition, the company exports its products to USA, Japan, Israel, and other developed countries in IT system.

The functional cosmetic manufacture that is popular on TV home shopping channels, Cotde Co. Ltd., is a company of CTP, too. The ingredients of products are mainly natural preservatives and bio-compatibles. The main product, Zymogen won the gold medal of Alternative Medicine field in 2005 INPEX (Invention and New Product Exposition–USA), one of the biggest invention expo in the field. The enterprise has exported $4 million in 2009, expanded its markets to eight countries including Japan and Singapore. The enterprise expects $10 million of exports in 2010.

CTP published a book named *Seven Stars' Story* (Lee et al. 2009) in 2009. The book contains success stories of seven enterprises supported by CTP's comprehensive service program. Their successful stories show knowledge of management to entrepreneurs and provide vision and dream to those who want to be a successful entrepreneur in the future. This book is often used in college classes in the business administration department.

Upon these achievements, CTP's comprehensive nurturing and supporting system itself has become a good example to the developing countries' economic innovation system. In 2009, over 100 government officials, entrepreneurs from about 30 countries visited CTP in order to learn CTP's innovative system. Moreover, CTP has been consulting its know-how of operation to many developing countries of Northern Africa, Central Asia, Southeast Asia, and Middle East. CTP hosted 20 policymakers from the above continents to CTP's global training workshop for 2 weeks in May 2010.

## 5 The Future of Chungnam Techno Park

CTP is in the middle of its second period of the roadmap. CTP has changed the business environment of the region with its comprehensive business support programs emphasizing the global standards of technology development and commercialization, and global marketing competitiveness for regional enterprises. CTP keeps the path of success as an innovative companion of all enterprises in the region. However, CTP needs one more step forward to make CTP's ultimate goal to make Chungnam as the world class high-tech cluster in a sustainable innovation system (Chungnam Techno Park 2009b) (Fig. 11).

From 1999 to 2007 CTP created innovation infrastructure as the first period of the roadmap. The major issue in the first period was developing Cheonan Valley. The headquarters of CTP was located in Cheonan Valley. Now CTP is in the second period. The major issue in this second period is to develop three more local valleys. Here, the valley means the land space for technology park business. Thus, CTP creates three more technology park business space. CTP already created Asan Valley. Yesan Valley has completed in 2010, and Dangjin Valley is under research. Cheonan Valley, which was completed in the first period, will be expanded as a technology city (technopolis).

The third period of the roadmap concentrates on the sustainable innovation system. Cheonan Valley will be developed as a world class cluster. Three local valleys will become technopolis and national level clusters. By doing so, Chungnam province will be developed as a technology cluster by the end of 2010. CTP's endeavor will assure the sustainable development of the Chungnam and, thus, the quality of life for the people of Chungnam.

In the third period, CTP will expand its global network through collaboration programs with foreign science and technology parks. CTP plans to create several satellite incubators abroad. The first one is in Thailand Science Park, Bangkok, Thailand. The CTP's enterprises will do their business with Thai partner

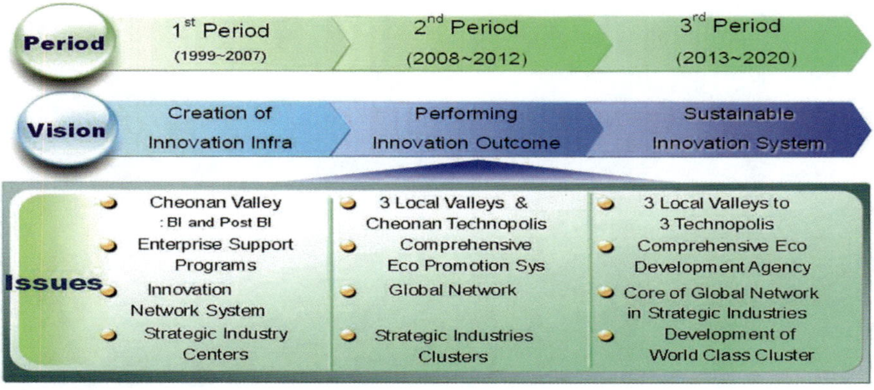


**Fig. 11** Roadmap of CTP

enterprises at the CTP's satellite incubator in Bangkok. In addition, CTP will stimulate the joint researches between CTP enterprises and foreign innovation institutes. The main industry in the third period will be green and clean business as well as contents business.

In 2020, CTP will create 20 star enterprises whose annual sales exceed $100 million and two enterprises whose annual sales exceed $1 billion. CTP will be remembered as a regional innovation platform that makes impossible possible.

## References


Andersson, M., & Karlsson, C. (2004). *Regional innovation systems in small and medium-sizes regions: Critical review and assessment, working paper series in economics and institutions of innovation*. Sweden: Royal Institute of Technology, CESIS.

Chungnam Techno Park. (2009a). *10 years history of Chungnam Techno Park*. Korea: Seongwoo Adcom.

Chungnam Techno Park, (2009b) *Fusion creative management: Business plan for 2009*, Chungnam Techno Park Press, Korea.

Chungnam Techno Park. (2010). *Annual report on business performance 2009*. Korea: Chungnam Techno Park Press.

Kim, H.-M. (2003). *Evaluation methodology for technology park formation*. Korea: Annual Conference of Korea Public Administration.

Kim, H.-M. (2006). *Innovation capacity analysis for regional innovation system*. Korea: Journal of Korea Industry-Academy Cooperation Society.

Kim, H-M., et al., (2000) *Master plan of Chungnam Techno-Park*, Korea.

Korea Technopark Association. (2004). *A report on creation of regional technology innovation hub by techno parks in Korea*. Korea: Korea Technopark Association.

Lee, S.-G, Lee, H.-J and Lee, H.-G. (2009). *7 Star stories: Report on the endurance and passion of the 7 venture businessmen*. Korea: Seongwoo Adcom

Lundvall, B-A. (2007). Innovation system research : Where it came from and where it might go, Globelics working paper, No. 2007–01.

# Building STP's Ecosystem: Kitakyushu Approach

**Sangryong Cha**

**Abstract** In the last three decades since the 1980s the concept of science and technology park (STP) has emerged from many academic and practical debates on regional development. The concept has come to be used frequently, but often with quite different meanings and images. A key reason for the popularity of this concept is its capacity to cover the whole range of institutions and relationships involving in process of creation of science, innovation of technology, improvement of industry, and evolution of market based on its networking. It has given an impression on policymakers who have been eager to find some useful measures for regional economic development since an era of so-called new economy. Therefore, STPs have been built as a core of supporting system for R&D, business incubation, industrial clustering (networking and governance among entrepreneurs), management of sustainable innovation system, and infrastructure development for regional economic prosperity in many regions of the world. In this chapter, we call it STP's ecosystem.

## 1 Introduction

In the last three decades since the 1980s the concept of science and technology park (STP) has emerged from many academic and practical debates on regional development. The concept has come to be used frequently, but often with quite different meanings and images. A key reason for the popularity of this concept is its capacity to cover the whole range of institutions and relationships involving in process of creation of science, innovation of technology, improvement of industry and evolution of market based on its networking. It has given an impression on

S. Cha (✉)
Faculty of Economics and Graduate School of Economics, University of Nagasaki, Nagasaki, Japan
e-mail: cha@sun.ac.jp

D.-S. Oh and F. Phillips (eds.), *Technopolis*,
DOI: 10.1007/978-1-4471-5508-9_22, 

policymakers who have been eager to find some useful measures for regional economic development since an era of so-called new economy. Therefore, STPs have been built as a core of supporting system for R&D, business incubation, industrial clustering (networking and governance among entrepreneurs), management of sustainable innovation system, and infrastructure development for regional economic prosperity in many regions of the world. In this chapter, we call it STP's ecosystem.

Kitakyushu City, one of the oldest industrial cities in Japan, is no exception to it. In this chapter, we will review the Kitakyushu approach on building STP's ecosystem and elucidate its process with clustering of networking sequences from the point of view of regional system of innovation.

Now, let us begin with a brief review of Japan's experience to understand the Kitakyushu approach on building STP's ecosystem.

## 2 A Historical Sketch of Japan's Experience to Understand the Kitakyushu Approach

### *2.1 Integration and Network Linkages: Reminding Chris Freeman and Luc Soete's 'National System of Innovation'*

As you know, Japan experienced amazing economic growth from the 1950s to 1980s such as Ezra Feivel Vogel, an American sociologist, described 'Japan as Number One' in his book published in 1979. In his eyes, the 'Japanese Miracle' seemed to occur based upon basic education, low crime levels, an all-wise-all-powerful bureaucracy, even a well-functioning democracy (Scalise 2001). *Foreign Affairs* gushed "Vobel's book helps explain why Japan is the most dynamic of all modern industrial nations" about it (the Economist, 2009.11.12).

However, there are some researchers or their groups who found the reason for the Japanese success story from other fields: national environment related with science, technology, and innovation. Chris Freeman and Luc Soete are also one of them. They emphasized that the national environment can have a considerable influence in stimulating or facilitating the innovative activities of firms and Japan had it then. In their eyes, especially, the Japanese economy appeared particularly successful in bringing technology to the marketplace and in building up the scientific and the technological capabilities needed to sustain the process (Freeman and Soete 1997).

As empirical evidences and analyses have been accumulated on industrial R&D and innovation in Japan, the United States, and Europe, it became increasingly evident that the success of innovations, their rate of diffusion, and the associated productivity gains depended on a wide variety of other influences as well as formal R&D (Freeman and Soete 1997). Freeman and Soete observed that these kinds of systemic aspects of innovation were increasingly influential in determining both the rate of diffusion and the productivity gains associated with any particular diffusion process in a national economy. To argue it, they used a term which has become known in some literatures as 'national system of innovation' which is a kind of reinterpretation of 'national system of political economy' called by List (1841).

Like List had paid attention to a broad range of national characteristics concerned with leaning about new technology and applying it for accelerating or making possible industrialization and economic growth, Freeman and Soete also gave attention to some national systemic factors related with science, technology, and skills in the growth of nations such as UK, US, and Japan. Particularly, they took note of the systemic interactions and its structures as characteristics of national system of innovation (Tables 1, 2, 3).

In Table 3, we may notice that Freeman and Soete emphasized 'integration' and 'network linkages' using the word 'strong' in the characteristics of the Japanese national system of innovation in the 1970s. They laid stress on the importance of the integration of R&D, production, and technology imports at firm level, and the user-producer and subcontractor network linkages as the strongest features of the Japanese system (Ibid., p. 304). In the Japanese case, moreover, the long-term 'visions' generated by an interactive process involving not only MITI[1] and other government organizations but also industries and universities made the integration and the network linkages more effective for industrial development and economic growth. The Japanese competitive success in the 1970's international economy based on industries as diverse as shipbuilding, automobiles, and color television is eloquent evidence of it. And then, Japanese managers, engineers, and workers grew accustomed to considering the entire production process as a system and to thinking in an integrated way about product design and process design (Ibid., p. 148).

## 2.2 *For Changeover: Science City in Center Area and Technopolis in Periphery*

It can be said that the Japanese national system of innovation above-mentioned was for the widespread and effective agile use of 'reverse engineering' at firm level. In fact, in the 1950s the first Japanese production models, whether in automobiles, TV sets, or machine tools, were often of relatively poor quality (Baba 1985). A determined effort to overcome these defects included a systematic review of all the possible sources of such social innovations as 'quality circles' and to the development of greatly improved techniques of quality control, not simply at the

[1] Ministry of International Trade and Industry (Ministry of Economy, Trade and Industry (METI) at present).

**Table 1** Characteristics of British national system of innovation, eighteenth and nineteenth centuries

| |
|---|
| • Strong links between scientists and entrepreneurs |
| • Science has become a national institution, encouraged by the state and popularized by local clubs |
| • Strong local investment by landlords in transport infrastructure (canals and roads, later railways) |
| • Partnership form of organization enables inventors to raise capital and collaborate with entrepreneurs (e.g., Arkwright/Strutt) |
| • Profits from trade and services available through national and local capital markets to invest in factory production especially in textiles |
| • Economic policy strongly influenced by classical economics and in the interests of industrialization |
| • Strong efforts to protect national technology delay catching up by competitors |
| • British productivity per person about twice as high as European average by 1850 |
| • Reduction or elimination of internal and external barriers to trade |
| • Dissenters' academies and some universities provide science education. Mechanics trained in new industrial towns on part-time basis |

*Source* Freeman and Soete (1997, p. 296)

**Table 2** Characteristics of US national system of innovation, late nineteenth and twentieth centuries

| |
|---|
| • No feudal barriers to trade and investment; slavery abolished 1865; capitalist ideology dominant |
| • Railway infrastructure permits rapid growth of very large national market from the 1860s onwards |
| • Shortage of skilled labor induces development of machine intensive and capital intensive techniques (McCormick, Singer, Ford) |
| • Abundant national resources exploited with heavy investment and big scale economies (steel, copper, oil) |
| • Mass production and flow production as typical US techniques |
| • Strong encouragement of technical education and science at federal and state level from 1776 onwards |
| • US firms in capital intensive industries grow very large (GM, GE, SO, etc.) and start inhouse R&D |
| • US productivity twice as high as Europe by 1914 |
| • Major import of technology and science through immigration from Europe |

*Source* Freeman and Soete (1997, p. 296)

end of the production run but at every stage.[2] It affected especially the characteristic R&D strategy of the major Japanese companies and their staffs. Japanese engineers and managers grew accustomed to the idea of 'using the factory as a laboratory'.[3] The work of the R&D department was very closely related to the

[2] Freeman and Soete, Ibid., pp. 150–151.

[3] Baba, Ibid.

**Table 3** Characteristics of Japanese national system of innovation, 1970s

| |
|---|
| • High GERD/GNP ratio (2.5 %) |
| •. Very low proportion of military/space R&D (<2 % of R&D) |
| • High proportion of total R&D at enterprise level and company financed (approx. two-thirds) |
| • Strong integration of R&D, production, and import of technology at enterprise level |
| • Strong user-producer and subcontractor network linkages |
| • Strong incentives to innovate at enterprise level involving both management and workforce |
| • Intensive experience of competition in international markets |

*Source* Freeman and Soete 1997, p. 303

work of production engineers and process control and was often almost indistinguishable.[4] As a result, the integrative effect of learning by creative reverse engineering conferred a major competitive advantage on many Japanese firms.[5]

With the overwhelming national growth based on the system of innovation aforementioned, Japan had its politico-economic status elevated in the international economy. In 1963, Japan's membership in the OECD was approved and IMF recommended that Japan is granted the status of an Article 8 nation, and GATT recommended granting it the status of an Article 11 nation. This new status promoted the internationalization of Japanese industry, which was now pressed to develop original technologies to compete with foreign companies in the world market. It means that the existing system of innovation suitable for the widespread and effective agile use of reverse engineering had to be made to changeover to a new one. And then, two key strategies guided those efforts: science city development in the metropolitan core of the country, and technopolis development in the periphery (Miyakawa 2006).

In 1963, the Cabinet of Japan decided to erect Tsukuba Science City in the northeastern part of the Tokyo metropolitan area. It would relocate 47 government-owned institutes there and develop original advanced technologies in modern, newly established institutes. As a result, Tsukuba Science City has become one of the most important and indispensable incubators of Japan's latest technologies. It attracts R&D institutes from all over Japan and the world (Fig. 1; Tables 4, 5, 6, 7).

The development of Tsukuba Science City led to a boom in building central research institute by the major Japanese companies there. However, it means that the work of R&D department is separated as more pure form from the work of production department which is not moved in Tsukuba Science City. Although the specialty of R&D activity may be strengthened through it, it is clear that some changes occur in the traditional way of the integration between R&D and production at firm level through it.

In addition, as a counterpart to the development of Tsukuba Science City in the metropolitan core, MITI introduced technopolis projects in peripheral areas from

[4] Freeman and Soete, Ibid., p. 148.

[5] Ibid., p. 149.

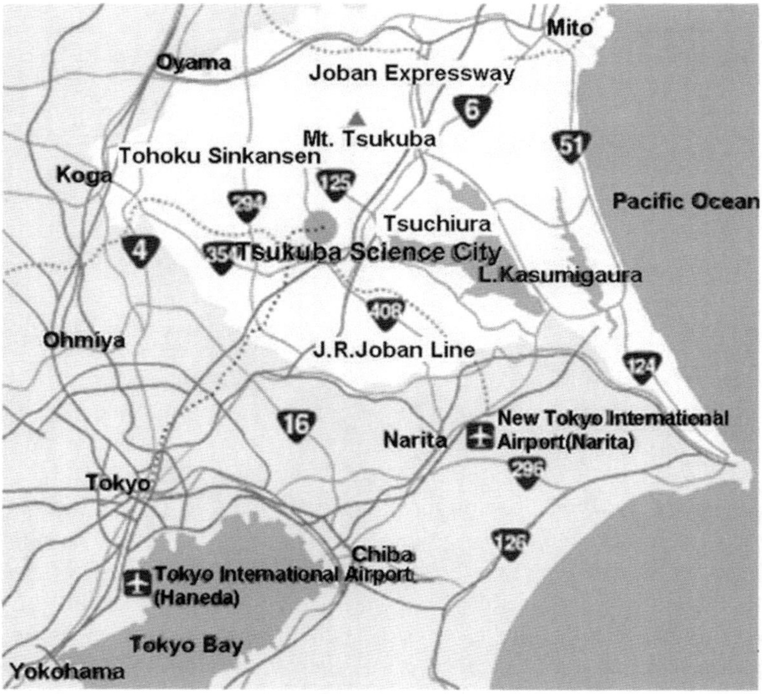


**Fig. 1** Location of Tsukuba Science City. *Source* http://www.mlit.go.jp/crd/daisei/tsukuba/english/city/002.html (2011.9.10)

1983. They were modeled after industrial research parks in the US and science parks in the UK. The idea behind technopolis is to infuse rural serenity in the city, and in turn, based on the Garden City model, inject urban activities into the countryside, so that peripheral communities can expand their employment based on high-technology industries.

The technopolis project is defined as a strategy to achieve two goals; knowledge intensification and heightening of value added of industrial structure, and regional development headed for the twenty-first century by introducing high technology industry into culture, tradition, and nature of regional society through accomplishing town building which is harmonious with industry, academic center, and community facilities (S. Yazawa 1990). The technopolis concept is a synthesis of three streams of thought: MITI's high-tech research strategy, Japan's regional development programs, and Silicon Valley's process of innovation (Tatsuno 1986) (Fig. 2). The first two streams provide the framework for the concept of technopolis, while Silicon Valley's process of innovation (entrepreneurialism) just provides the inspiration and drive. MITI selected 19 technopolis zones involving 20 prefectures among 40 candidate regions (Fig. 3).

**Table 4** Brief history of Tsukuba Science City

| 1961–1969 | 1970–1977 | 1978–1987 |
|---|---|---|
| 1961—Cabinet decided to examine concrete measures for collective transfer of government offices | 1970—A law for construction of Tsukube Science City was promulgated | 1980—All of the 43 research and educational institutions were started operations A construction plan for the Research and development district was decided by the National land agency |
| 1963—Cabinet approved selecting the Tsukuba area as the site of a new science city | 1972—The National Research Institute in Inorganic Materials was transfer red to the city | 1981—A arrangement plan for the surrounding urban district was decided by lbaraki Prefecture |
| 1964—Cabinet decided to start construction from 1965 and to complete it within about 10 year | 1973—The University of Tsukuba was founded | 1982—The construction of Tokodai research park is completed |
| 1968—The construction of large-sized experimental equipment for an earthquake engineering investigations laboratory attached to the National Research Center for Disaster Prevention was started. It was first institution to be built in the city | 1975—Cabinet decided to postpone the target date for completing the transfer of designated institution to the city until the end of 1979 fiscal year | 1985—The Joban Highway directly connected to Tokyo Tsukuba Expo was opened |
| 1969—Cabinet decided to complete the transfer of the institution which were expected to move to the city, within about 10 years from 1968 | 1977—The Tsukuba Science City Associated was started | 1987—Tsukuba city officially formed by merger of existing towns and villages Tsukuba Linked to Tokyo Joban Highway Bus |

*Source* Author based on the information from the website of MLIT (http://www.mlit.go.jp)

## *2.3 Policy Synthesis: Cluster Projects*

The purpose of science city and technopolis development in Japan was to erect major incubators of original technologies under keen international competitions, not only in the dominant metropolitan areas but also in the frontier areas of Japan. However, the development lacked the synchronism, synergism and synthesis of related policy, planning and practice within the same region for making the integration of R&D, production and technology learning at region level, and the user-producer and subcontractor network linkages. This situation, moreover, was not so much changed for a long time until the 1990s in spite of some strategic trials

**Table 5** Facilities for education and research of central government, etc., in Tsukuba

| Classification | Authorities | Facilities |
|---|---|---|
| Education (7) | Ministry of Education, Culture, Sports, Science, and Technology (MEXT) | Tsukuba University |
| | | Tsukuba University of Technology |
| | | High Energy Accelerator Research Organization |
| | | National Center for Teachers' Development |
| | | Tsukuba Research Center of the National Science Museum |
| | Cabinet Office | Tsukuba Annex of the National Archives |
| | The Ministry of Foreign Affairs | Tsukuba International Center of Japan International Cooperation Agency |
| Construction (5) | MEXT | National Research Institute for Earth Science and Disaster Prevention |
| | Ministry of Land, Infrastructure, and Transport (MLIT) | Geographical Survey Institute |
| | | National Institute for Land and Infrastructure Management |
| | | Public Works Research Institute |
| | | Building Research Institute |
| Cooperation (1) | MEXT | Tsukuba Center of Institute |
| Science and Technology (7) | MEXT | National Institute for Material Science |
| | | Tsukuba Space Center of Japan Aerospace Exploration Agency |
| | Ministry of the Environment | National Institute for Environmental Studies |
| | Ministry of Economy, Trade and Industry | National Institute of Advanced Industrial Science and Technology |
| | MLIT | Meteorological Research Institute |
| | | Aerological Observatory of japan Meteorological Agency |
| | | Meteorological Instrument Center |
| Life (12) | MEXT | RIKEN Tsukuba Institute |
| | Ministry of Health, Labour and Welfare | Tsukuba Primate Center for Medical Science of National Institute of Infectious Diseases |
| | | Tsukuba Medicinal Plant Research Station of National Institute of Health Science |
| | The Ministry of Agriculture, Forestry, and Fisheries | National Agriculture and Bio-oriented Research Organization |
| | | National Institute of Agro-biological Science |
| . | . | National Institute for Agro-environmental Science |
| | | National Institute for Rural Engineering |
| | | National Food Research Institute |
| | | Japan International Research Center for Agriculture Sciences |
| | | Forestry and Forest Products Research Institute |
| | | National Center for Seeds and Seedlings |

*Source* Author based on the website of MLIT (http://www.mlit.go.jp)

**Table 6** Facilities for research of public-service corporations in Tsukuba

| Classification | Facilities |
|---|---|
| Public-Service Corporation (12) | ZEN-NOH Central Research Institute for Feed and Livestock |
| | Civil Engineering Research Laboratory |
| | Foundation for Advancement of International Science |
| | Tsukuba Research Institute of Ocean Policy Research Foundation |
| | Technical Laboratory of Public Works Research Center |
| | Laboratory for Fuudo Technology of Public Works Research Center |
| | Nippon Agriculture Research Institute |
| | Japan Automobile Research Institute |
| | Tsukuba Building Test Laboratory of Center for Better Living |
| | Technology Development Center of Japan Pressure Welding Society |
| | Institute of Society for Techno-innovation of Agriculture, Forestry and Fisheries |
| | The Corporation for Production and Research of Laboratory Primates |

*Source* Author based on the website of MLIT (http://www.mlit.go.jp)

**Table 7** Features of five major cities on number of employees in research institutes

| Rank | City | Research institute | | No. of employees per institute (A) | No. of total employees in the City (B) | A/B |
|---|---|---|---|---|---|---|
| | | No. of institutes | No. of employees | | | |
| 1. | Tokyo | 653 | 30,310 | 46 | 7,134,000 | 0.42 |
| 2. | Kawasaki | 61 | 23,603 | 387 | 499,000 | 4.73 |
| 3. | Tsukuba | 140 | 15,912 | 114 | 88,000 | 18.08 |
| 4. | Yokohama | 124 | 11,158 | 90 | 1,347,000 | 0.83 |
| 5. | Atsuki | 24 | 10,596 | 442 | 142,000 | 7.46 |

*Source* Cha 2004, p. 101

such as the Super Techno Zone Concept (1996) by MITI and the Regional Platform Project (1999) by METI (Fig. 4).

Since the 2000s, however, the situation has begun to change with the start of two cluster projects: the Industrial Cluster Project from 2001 by METI and the Knowledge Cluster Initiative from 2002 by MEXT.[6]

The Industrial Cluster Project is a program in which regional SMEs and start-up companies utilize innovative research results or seeds obtained at universities and research institutes to form industrial clusters, in fields such as IT, BT, environment, and manufacturing. The goal of this project is strengthening the competitiveness of the Japanese industry. To strengthen the global competitiveness of the Japanese industry and to invigorate Japan's local economies, this project carries out the missions stated below with the objective of businesses, universities, and other institutions, in regions throughout Japan, forming wide-area networks

[6] Ministry of Education, Culture, Sports, Science and Technology.

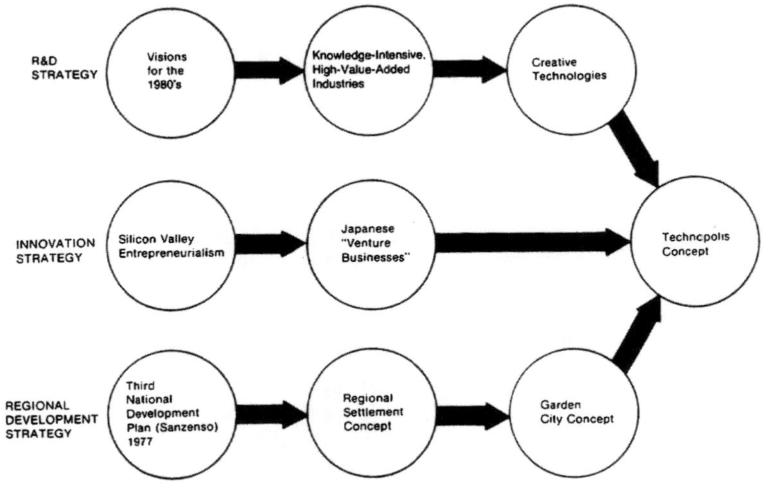


**Fig. 2** Synthesis of three streams of thought in the technopolis concept. *Source* Tatsuno (1986, p. 123)

through partnerships between businesses in the same industry and across industrial sectors. Those networks, through synergetic sharing of participant's intellectual and other resources, will then reach the state in which they generated new industries and businesses, mainly in their regions.

The most salient keyword in this project is innovation. In concrete terms, it means exerting a significant impact on the economy and society by marketing competitive products and commodities based on new technologies and ideas. The basic idea of this project is to prepare regional environments to give rise to a stream of innovations. Three major missions of this project are as follows:

1. Preparing a business environment that promotes innovation.
2. Creating new industries in the fields defined as of strategic importance in the Economic Growth Initiative and other national programs.[7]
3. Fostering the emergence of synergistic effects through tie-ups with the regional promotion policies that local governments and other bodies implement.

In a change from implementing policies uniformly throughout Japan under centralized management, policies are implemented first and foremost by those actually working in their fields under the field workers who know the best principle. The project secures the appropriate tools for constructing the networks that are the cores of clusters and strategically applies the policies and programs of other ministries, departments, bureaus, and organizations in and out of the region to support research and development of other tools, partnerships between businesses,

[7] This project is positioned as a core strategy in Japan's Economic Growth Initiative and the Third Term Science and Technology Basic Plan.

**Fig. 3** Distribution of technopolis zones and their Mother Cities in Japan. *Source* Tatsuno (1986, p. 127)

development of marketing channels, entrepreneurship, human resource development, and other critical factors[8] (Figs. 5, 6).

On the other hand, the Knowledge Cluster Initiative is a program to form knowledge clusters that a system for technological innovation organized by local initiative around universities and other public research institutions with original

[8] Cha (2008, pp. 181–182).

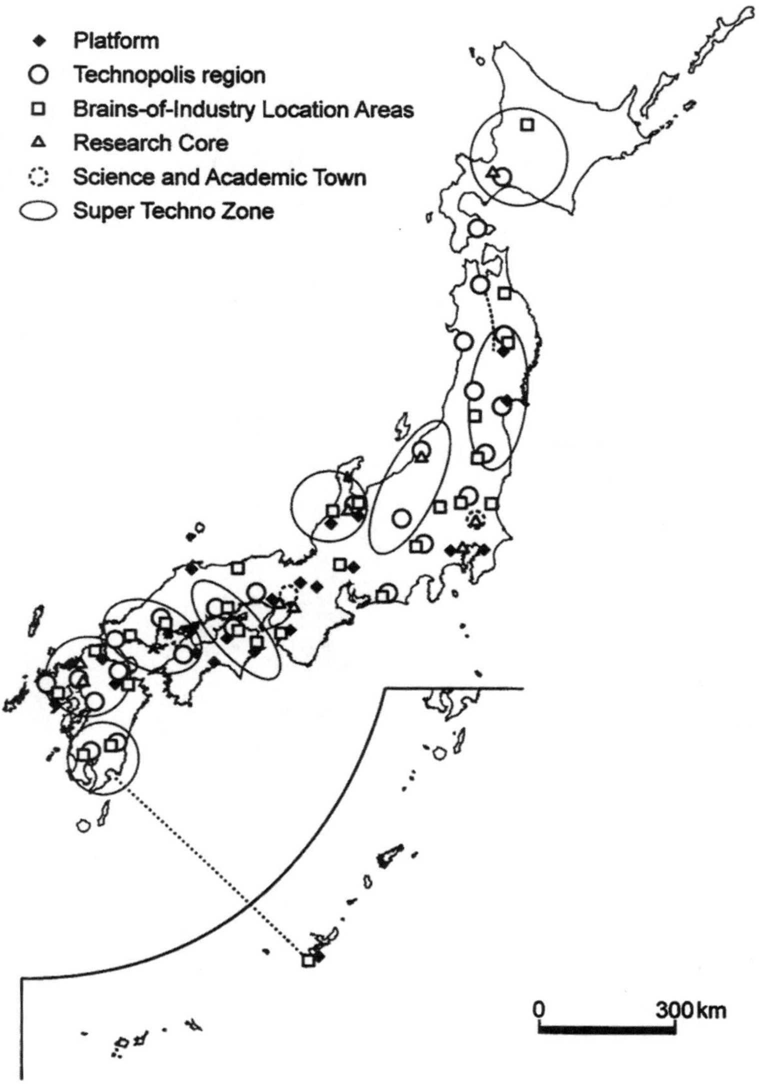


**Fig. 4** Distribution of technopolises and related projects areas in Japan. *Source* Miyakawa (2006, p. 102)

R&D topics and potential in the regional levels. The system, of course, also features participation by companies inside and outside the region. More specifically, the Knowledge Cluster Initiative establishes networks of individuals in academia and the private and government sectors—through project planning, joint-research, and exchanges of ideas. This project creates a system successively fostering technological innovation while stimulating interaction between the

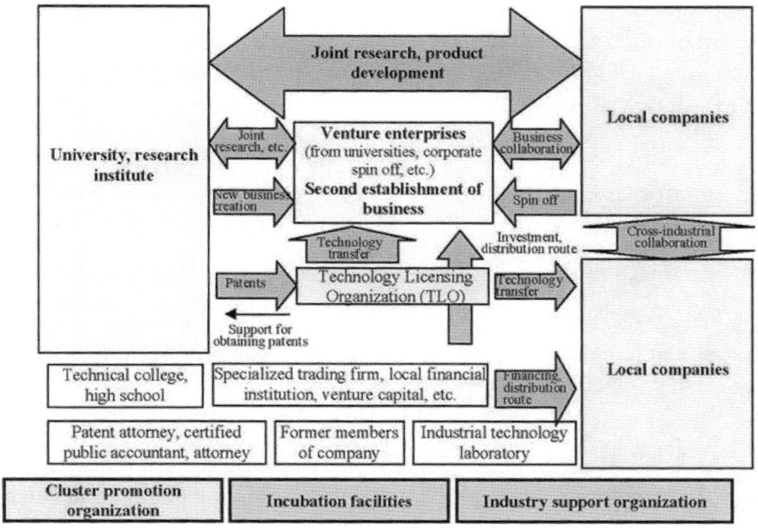


**Fig. 5** Conceptual framework of the industrial cluster project. *Source* METI (2008, p. 4)

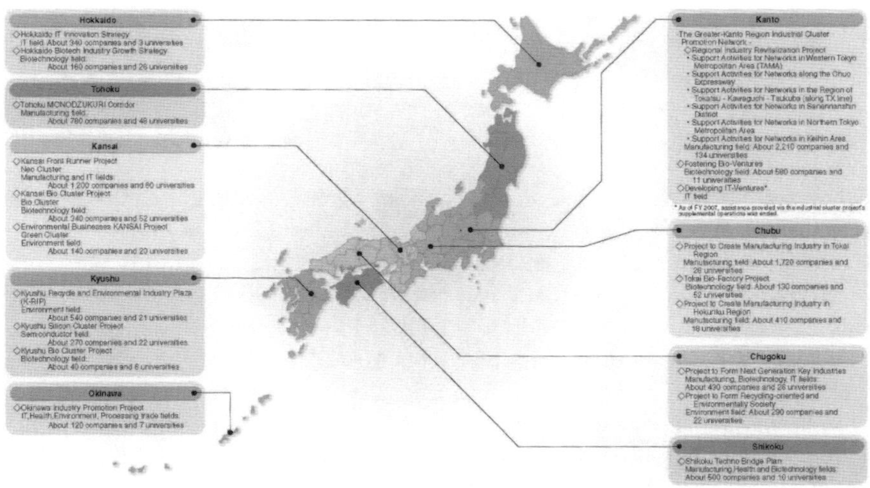


**Fig. 6** Distribution of the industrial cluster project areas (second term). *Source* METI (2009, p. 4–5)

technological seeds of core research institutions and the practical needs of the business world based on close collaboration and daily communication in the region. Points about the Knowledge Cluster Initiative by MEXT are as follows:

1. Strategic implementation based on the cluster concept for the region.

2. Using accumulated knowledge: fostering technological innovation centered on universities and other public research institutions.
3. Technological innovation on a global level: accumulating people, things, and money from other regions.

As part of policies to achieve their cluster vision, local governments, based on a cluster vision for an individual region, are expected to implement the following in concert with other measures (of their own or of other ministries and agencies):

1. Conducting joint research by industry, academia, and government.
   - Conducting joint research by industry-academia-government at university joint research centers or other institutions to produce new technology seeds in light of corporate needs.
   - Patenting research results and conducting R&D relating to incubation.
2. Using projects by local governments, related ministries, etc.
   - Fully implementing projects, from R&D to commercialization, and using R&D systems controlled by related agencies and ministries like METI.
3. Other
   - Setting up a Knowledge Cluster Headquarters in each region as a control center for project implementation (staffed by a President, Project Director, Chief Scientist, and others).
   - Assigning science and technology coordinators (experts), with emphasis on expertise, and using advisors like patent attorneys.
   - Holding forums and other meetings to announce and discuss research results.

These are achieved by research organizations, R&D-oriented companies and other participants, working around a core of universities and public research institutions that are centers of knowledge creation, and the Stress of the project is placed on the autonomy of local governments[9] (Figs. 7, 8).

Collaboration between the Industrial Cluster Project initiated by METI and the Knowledge Cluster Initiative promoted by MEXT has been further promoted (Example: Establishment of Regional Consortium Research and Development Project, Interministerial Collaboration Framework). The Regional Cluster Promotion Council has been established jointly with METI and MEXT, and Joint Achievement Announcement Meetings have been held in each region. Through this, policies of both ministries have been mutually complementary, e.g., the seed generation achievements of MEXT policies have been applied and commercialized by METI, and conversely, feedback of market needs has led to R&D of new seeds[10] (Fig. 9).

---

[9] Ibid, pp. 184–187.

[10] Ibid. pp. 188–189.

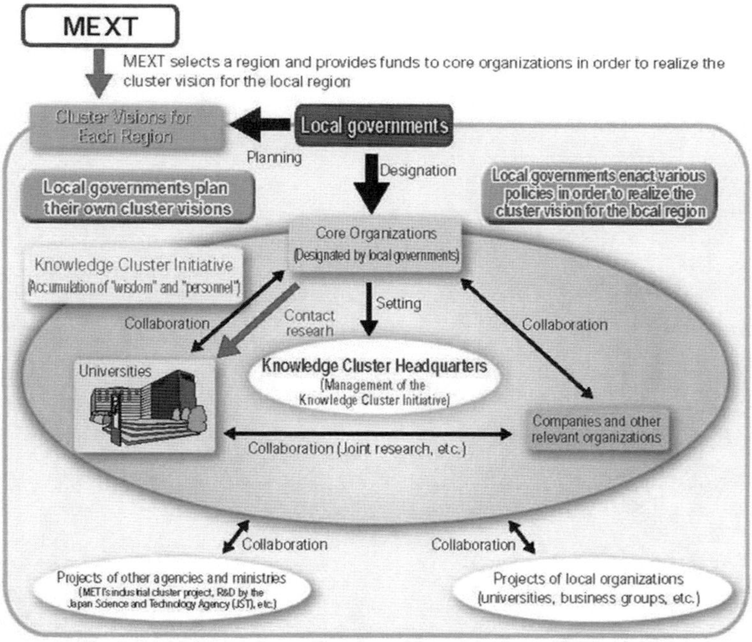


**Fig. 7** Overview of the knowledge cluster initiative. *Source* MEXT (2009, p. 3)

## *2.4 Decentralization of S&T Promotion: Building Regional Innovation Networks and Establishing Prefectural S&T Councils*

In the Research and Development Enhancement Act enacted in 2008, innovation was positioned as an objective of national strategy legally for the first time in Japan. Japanese S&T system, together with it, was needed to reform for creating innovation. Building regional innovation systems for creating regions full of vitality is a critical one of major measures to implement it.

The Regional Innovation Networks Project subsidized by METI is a good example of it. This project was implemented to make a creative basement of regional innovation by subsidizing formation of integrated network of a large region including universities and public research organizations carrying regional innovation through using research resources mutually. When it ended in 2009, nine regional innovation networks in total were formed in each jurisdiction of the regional bureaus of METI: Hokkaido, Tohoku, Kanto, Chubu, Kinki, Chugoku, Shikoku, Kyushu, and Okinawa.

On the other hand, prefectural governments established councils in charge of deliberating S&T policies and made active contributions to the S&T promotion by formulating their own plans and guidelines related to S&T (MEXT 2010).

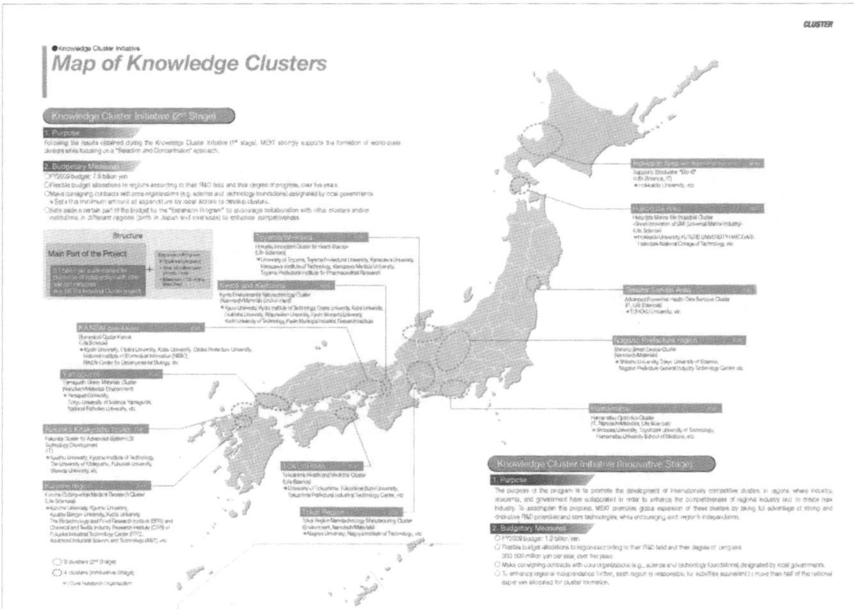


**Fig. 8** Distribution of the knowledge Clusters initiative Areas. *Source* MEXT (2009, p. 4–5)

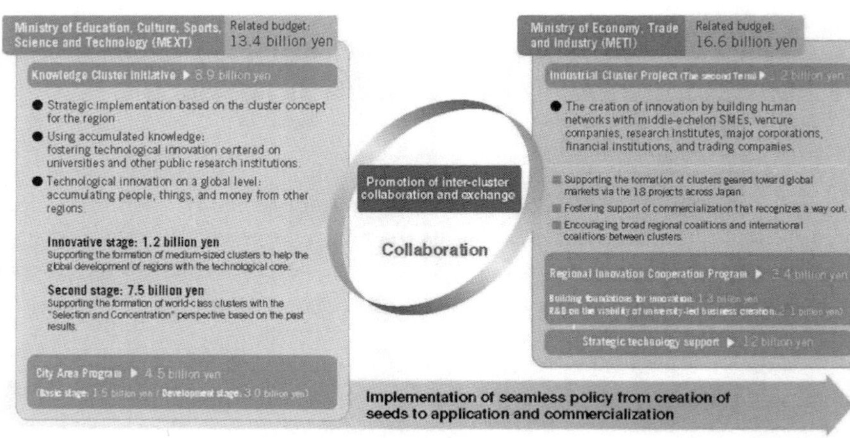


**Fig. 9** Outline of regional cluster-related budget for FY 2009. *Source* MEXT (2009, p. 11)

Regional promotion of S&T by the councils helps to revitalize local industries and enhance local residents' quality of life, which in turn contributes to the advancement and diversification of S&T in Japan as a whole and the enhancement of the competitiveness of the national systems for innovation (Table 8).

**Table 8** S&T councils established at local governments

| Locality | Name of council | Locality | Name of council |
| --- | --- | --- | --- |
| Hokkaido | Hokkaido Science and Technology Council | Hyogo | Hyogo Science and Technology Council |
| Aomori | Aomori Industry, Science and Technology Council → Aomori Research and Development Conference | Nara | Nara Prefecture Committee on Guidelines for Science and Technology Promotion → Nara Prefecture Science and Technology Promotion Conference |
| Akita | Akita Council for Science and Technology | Wakayama | Wakayama Prefecture Science and Technology Strategy Council |
| Iwate | Iwate Science and Technology Promotion Council | Tottori | Tottori Science and Technology Promotion Council |
| Miyagi | Miyagi Conference on Guidelines for Promoting Science and Miyagi Technology | Shimane | Shimane Science and Technology Promotion Council |
| Yamagata | Yamagata Science and Technology Council | Hiroshima | Hiroshima Science and Technology Promotion Conference |
| Fukushima | Fukushima Science and Technology Promotion Council | Yamaguchi | Yamaguchi Science and Technology Promotion Conference |
| Ibaraki | Ibaraki Science and Technology Promotion Council | Kagawa | Kagawa Science and Technology Council |
| Tochigi | Tochigi Science and Technology Promotion Council | Ehime | Ehime Science and Technology Promotion Council |
| Gunma | Gunma Science and Technology Promotion Headquarters | Tokushima | Tokushima Forum for the Promotion of "Vision for Science and Technology" → Tokushima Prefecture Science and Technology Promotion Plan Formulation Committee |
| Saitama | Saitama Science and Technology Council | Kochi | Kochi Science and Technology Academy |
| Chiba | Chiba Science Council | Saga | Saga Science and Technology Council |
| Kanagawa | Kanagawa Science and Technology Council | Nagasaki | Nagasaki Science and Technology Promotion Council |
| Niigata | Niigata Science and Technology Council | Kumamoto | Kumamoto Science and Technology Council |
| Toyama | Toyama Science and Technology Council | Oita | Oita Science and Technology Promotion Committee |
| Ishikawa | Ishikawa Industrial Science and Technology Council → Ishikawa Industrial Innovation Council | Miyazaki | Miyazaki Science and Technology Council |

(continued)

**Table 8** (continued)

| Locality | Name of council | Locality | Name of council |
|---|---|---|---|
| Fukui | Fukui Science and Technology Promotion Council → Council for Fukui Production Planning Strategy | Kagoshima | Kagoshima Science and Technology Promotion Council |
| Yamanashi | Yamanashi Science and Technology Council | Okinawa | Council for Promotion of Science in Okinawa → Okinawa Science and Technology Council |
| Nagano | Nagano Prefecture Investigative Commission on Science and Technology Industry Promotion Initiative | Kawasaki City | Kawasaki City Innovation Promotion Meeting |
| Gifu | Gifu Science and Technology Promotion Council | Yokohama City | Yokohama City Council for Promotion of Cooperation between Industry and Academia |
| Aichi | Aichi Science and Technology Council | Kyoto City | Kyoto City Conference on Projects for Promoting Industry, Science and Technology → Kyoto City Committee on Promotion of Industrial Science and Technology |
| Mie | Mie Science Academy Representative Conference → Science and Technology Exchange Council → Science and Technology Promotion Conference | Osaka City | Osaka Science and Technology Promotion Advisers Council (literal translation) |
| Shiga | Shiga Science and Technology Promotion Council | Hiroshima City | Hiroshima City Science and Technology Advisory Council |
| Kyoto | Kyoto Science and Technology Council | Kitakyushu City | Kitakyushu City Science and Technology Promotion Council |
| Osaka | Osaka Science and Technology Roundtable | Fukuoka City | Fukuoka City Adviser Meeting on Vision for Promotion of Science and Technology |

*Source* MEXT (2010, p. 233)

## *2.5 For Sustainable Growth and Social Development over the Future: The 3.11 Earthquake and Green Innovation*

On August 19, 2011, the 4th S&T Basic Plan (the 4th Plan) was decided by Cabinet. The 4th Plan was to build a new strategy and mechanism leading to various value creations by comprehensive promotion of science, technology, and innovation policy. The 4th Plan positioning S&T as an engine for value creation targets two major innovation fields strategically: Green Innovation and Life Innovation. The former is to realize low carbon society with sustainability and the latter is to realize the high quality of life in an aging society.

Under the influence of the 3.11 Earthquake, especially, the 4th Plan puts stress on Green Innovation to develop renewable energy, low carbon of energy supply and demand, saving energy, and green infrastructure because it is a critical issue to deal with the limits of nuclear energy and the crisis of global warming simultaneously. Green Innovation aims to stimulate technology innovation for environment and energy, diversify energy supply sources, reform social system for the innovation of energy use, build long-term stable energy demand structure, and realize the leading edge low carbon society in the world. The important issues for Green Innovation in the 4th Plan are as follows:

- To realize stable energy supply and low carbonization.
- To achieve high efficient and smart use of energy.
- To greenize social infrastructure.

To stimulate Green Innovation, it is needed to construct a new system that promotes issue-driven innovation including the consideration and reform of regulation and institution. In particular, the 4th Plan puts emphasis on construction of a new social system such as "smart community" by academia-local government-industry collaboration based on local characters (MEXT 2011).

## 3 Building STP's Ecosystem: The Kitakyushu Approach

### *3.1 Backgrounds: Industrial Decline and City Renaissance Plan*

Kitakyushu City is a core of the Kitakyushu Industrial Area that was one of the four major industrial areas in Japan. The area of city at present was gained through the merger of five old cities which had functional identity severally: Wakamatsu (port for coal distribution), Moji (trading port), Yahata (ironworks), Tobata (coastal industrial area), and Kokura (military base and arsenal). These five cities had constituted the consecutive and the industrial development mechanism which embedded in the regional structure like demand creation (Kokura)—resource procurement (Wakamatsu)—production (Yahata and Tobata)—marketing (Moji), and it was considered as a unified industrial area. In other words, it can be said that the space structure of Kitakyushu City was a form which the mechanism of industrial development that brought the improvement of productivity of steel industry appeared as a space structure of the city (Fig. 10).

However, the industrial development mechanism of this area, as a base of iron manufacture and coal distribution which had supported the industrialization and

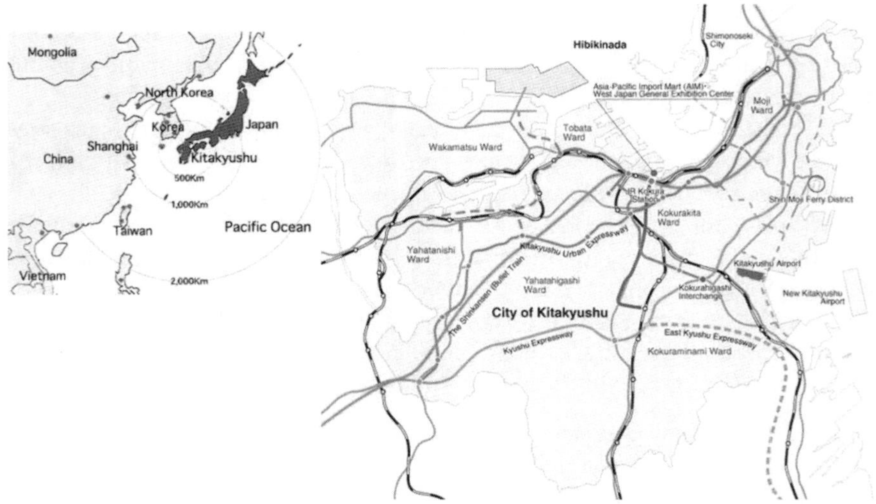


**Fig. 10** Location of Kitakyushu City and its spatial structure. *Source* Author based on the figures from some brochures of the City of Kitakyushu

the military power of Japan, did not operate well from 1955 because of some structural shifts on trades, industries, and energies. As a result, a lot of companies were moved to other industrial areas such as Keihin Industrial Area and Keihanshin Industrial Area, and the number of employees decreased. The industrial decline of this area did not get better in spite of the merger of five old cities in 1963, and the regional economy began to be sluggish (Table 9).

In 1988, the Kitakyushu Renaissance Plan to regenerate the locality was drawn up. This plan pursued the keynote theme "international technology city" as a new industrial city model based on rich natural environment and active exchange supported by high-technology until 2005. And, "Science and Research City" based on STP development was a city image to be aimed at the Plan.

## 3.2 STP Development: Scheme on New University of Kitakyushu and Kitakyushu Science and Research Park

According to the master plan drawn up in 1989, the purpose of the Kitakyushu Science and Research Park (KSRP) development was to build a core of S&T in Asia for technology advancement and new industry creation making the most use of the geographical advantage, local condition as the biggest cluster of industrial technology in Western Japan, and experience of technological cooperation in environmental fields with other Asian countries.

**Table 9** Industrial decline of Kitakyushu area

| Classification | No. of companies | | | | | No. of employees | | | | |
|---|---|---|---|---|---|---|---|---|---|---|
| | 1965 | 1975 | 1985 | 1995 | 2002 | 1865 | 1975 | 1985 | 1995 | 2002 |
| Total | 2,075 | 2,819 | 2,689 | 2,502 | 1,327 | 129,605 | 119,270 | 89,140 | 75,932 | 53,067 |
| Food | 674 | 595 | 470 | 373 | 211 | 10,632 | 8,990 | 6,032 | 6,519 | 5,323 |
| Lumber, Furniture and pulp | 340 | 419 | 369 | 277 | 111 | 8,431 | 6,926 | 4,096 | 3,369 | 1,377 |
| Publication and Presswork | 144 | 241 | 281 | 278 | 116 | 5,954 | 6,667 | 7,346 | 6,538 | 3,171 |
| Chemistry | 51 | 49 | 42 | 51 | 38 | 11,234 | 9,262 | 6,343 | 5,725 | 3,343 |
| Oil and Coal | 14 | 16 | 16 | 18 | 17 | 829 | 1,031 | 538 | 425 | 410 |
| Ceramics and Earth and Rock | 101 | 140 | 113 | 106 | 76 | 9,921 | 8,122 | 5,449 | 7,319 | 6,731 |
| Steel | 67 | 89 | 109 | 97 | 65 | 47,249 | 32,219 | 22,004 | 11,958 | 7,666 |
| Metal | 178 | 367 | 350 | 346 | 206 | 12,698 | 7,932 | 7,048 | 8,475 | 6,288 |
| Machinery | 319 | 540 | 516 | 512 | 283 | 18,922 | 31,972 | 24,103 | 19,444 | 11,110 |
| Etc. | 187 | 363 | 423 | 444 | 204 | 3,735 | 6,149, | 6,181 | 6,160 | 7,557 |

*Source* The City of Kitakyushu (2003)

It goes without saying that the collaboration between academia and industry is very important for high technology development. Unfortunately, however, it is hard to say that Kitakyushu City as a typical industrial city already had enough foundation for collaboration although there are many good high education institutes (HEIs) such as the Kyushu Institute of Technology in the city. In the master plan, therefore, it was emphasized that the KSRP aims to create new industries and further development of technology by promoting collaboration between academia and industry there.

The conceptual framework of academia-industry collaboration in the KSRP is shown on the Scheme on New University of Kitakyushu. It is to make an active and advanced environment for research and education, gather HEIs which are national, municipal, and private ones with science or engineering faculties into one campus, and encourage competitive collaboration between academia and industry within and outside the campus through various joint researches. It was the first attempt ever made in Japan to develop a high-level education and research environment by gathering several HEIs, which have very different academic traditions from one another, into one campus (Fig. 11).

To realize the scheme, the KSRP has been developing with the idea of multi-perspective town development taking advantage of the surrounding nature and urban environment, which provides a favorable residential environment and the accumulation of education and research institutions related to leading-edge science and technology. The KSRP, which is developed on the site of about 335 ha in the northwestern part of Yahatanishi Ward, has a population of 12,000 and 4,000 households in the plan. The site was developed by the land readjustment project,

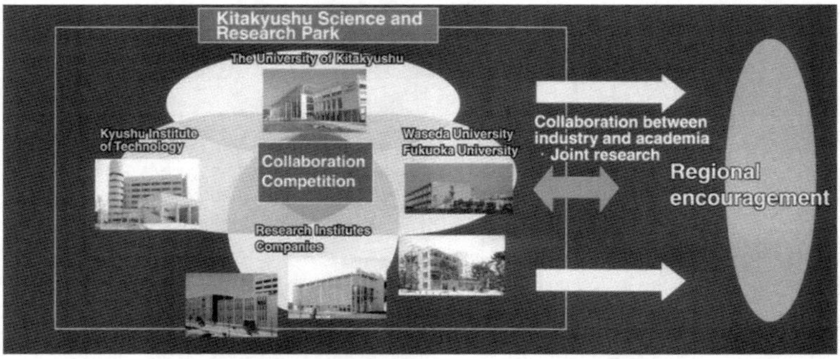


**Fig. 11** Scheme on New University of Kitakyushu. *Source* The brochure of the FAIS

and the development project was divided into three terms and promoted gradually. The first stage of the project developed 121.4 ha in the southern part of the site including the present university zone from 1995 to 2003, and Urban Development Corporation (present Urban Renaissance Agency) took charge of this project. In the second stage of the project, the municipality has charge of the project to develop 135.5 ha in the northern part of the site from 2004 to 2011. The third stage of the project is to develop the other 67.9 ha except 10 ha of the river of the site, and it is under examination now (Fig. 12).

There are 4 HEIs,[11] 16 research institutes, and 56 companies in the KSRP.[12] HEIs share their education and research ideas with each other, promote advanced education and research in leading-edge technology, encourage academia-industry collaboration and entrepreneurship, and support international student. The Campus Management Committee, which consists of the representatives of each tenant HEI for united management of the campus, plans policy on which it jointly worked and executes sharing facilities such as libraries, computer rooms, and other facilities prosperity. Various challenges for exchange and collaboration among researchers, teachers, and students in the KSRP such as joint research, credit transfer system, faculty exchange, and co-hosting congress, etc., are tried based on it.

The KSRP aims to link research and education with business and industrial field directly. For example, about 30 % of 150 faculty members in the total KSRP have business experience. They have contributed to the progress of research and education activities in the KSRP directly linked with business and industrial fields. In addition, such a directivity of the research and education activities to the real business and industrial fields has been strengthened through various research

[11] Faculty of Environmental Engineering, University of Kitakyushu (municipal), the Graduate School of Life Science and Systems Engineering, Kyushu Institute of Technology (national), the Graduate School of Information, Production and Systems, Waseda University (private), and the Graduate School of Engineering, Fukuoka University (private).

[12] As of June, 2011.

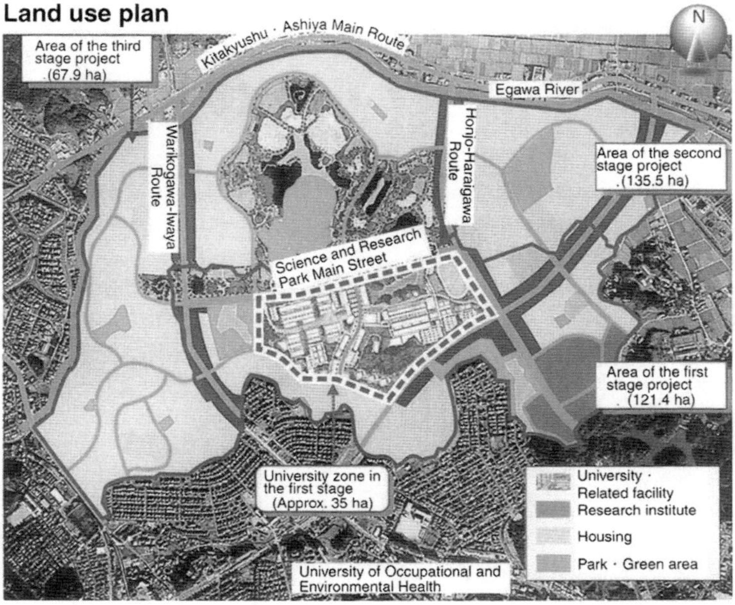


**Fig. 12** Land-use plan for KSRP development. *source* The brochurse of the FAIS

institutes in the KSRP.[13] Only, although various actors in industry, academia, and government had gathered, the KSRP is specialized for two technology fields: environment and information which are the most useful technology fields in terms of the purpose of the KSRP development because the Kitakyushu Industrial Area is still one of the biggest industrial agglomerations of environment and semiconductor businesses in Western Japan.

The KSRP was developed as an innovative milieu which promotes creation of new knowledge, technology, and value in such specialized fields. It is paid attention that the creation of the environment where the competition coexists with cooperation has been put from the policymaking stage on the mind, and the environment for "race" with competitive cooperation has been formed from the intercollegiate relations among the universities in the specialized fields. It seems the clearest feature of the innovative milieu in the KSRP.

[13] For instance, the Kyushu Laboratory of Advanced Research Institute for Science and Engineering, Waseda University aims at the academia-industry collaboration while doing a wide approach from basic theory to practical business research as a Kyushu hub of the research institute with the tradition to have contributed to the development of science, technology and industry of Japan in the field of environment, construction, telecommunication, system LSI, and robotics, etc. And, the Cranfield University at Kitakyushu has been trying to research jointly by industry-university cooperation in the field of the aeronautical engineering as a foreign research institute. Moreover, the Fukuoka Research Center for Recycling Systems has tried to match a recycling technology with the social system for the development of resource recycling society.

### 3.3 Boosting Triple Helix Innovation: Role of FAIS

By the way, in the KSRP development, the role of the Kitakyushu Foundation for the Advancement of Industry and Science and Technology (FAIS) is interesting. The FAIS aims to contribute to the creation of new businesses and the sophistication of industrial technologies through encouraging academic research and R&D based on the industry-government-academia cooperation in Kitakyushu City.

To achieve this aim, the FAIS consists of the executive board, six subordinate organizations: the Campus Administration Center, the SME Support Center, the Industry-Academia Cooperation Center, the Semiconductor Technology Center (the former SoC Design Center), the Car Electronics Center (the former Human Technology Based Cluster Development Center), and the Bureau for Robot Development Support (the former Robotics Research Institute).

In terms of policymaking in the initial stage of the KSRP development, the executive board structure is interesting. In 2001 when the FAIS was established together with the opening of the KSRP, the executive board consisted of 19 members from the industrial world, academic community, and administration.[14] The president of the executive board was the former minister of Education[15] (MEXT) and the former president of the University of Tokyo. It means that the FAIS obtained the political negotiation power in the national level through him and got a chance to synchronize its strategies with national policies such as the Knowledge Cluster Initiative by MEXT in the initial stage. Moreover, the other members who are representatives of industrial world, universities, and civil servants in the locality also contribute to improving the political negotiation power of the FAIS at the local or region level. The FAIS could improve the practice possibility of projects based on its network of networks which consists of the vertical network among nation, region, and locality and the horizontal network among industry, university, and administration owing to this political negotiation power of the executive board. In a local city such as Kitakyushu City where resources for policymaking are relatively limited, it is effective to build the network of networks in multi-dimension and various fields above-mentioned for concentrating regional competence.

On the other hand, six subordinate organizations are specialized separately for a specific role or project. To put it concretely, the Campus Administration Center carries out operations of the FAIS, administers the common-use facilities, and promotes collaboration and exchange among universities. The SME Support Center supports R&D at SMEs by providing research funding and SOHO office, promoting collaboration projects between industry and academia, and providing on-site consultation service. The Industry-Academia Cooperation Center supports improvement of local industries and creation of new industries through

[14] In 2011, the executive board of the FAIS consisted of 18 members.

[15] The Ministry of Education, Culture, Sports, Science, and Technology at present.

coordinating collaboration projects between industry and academia, gaining patents for technological seeds of universities and transferring technologies of companies.

Based on it mentioned above, the FAIS has played the role of a "coupler"[16] that unites two or more different organizations and functions like one organization to boost triple helix innovation (H. Etzkowits 2008) in the KSRP and promote the creation of new industry in the region (Cha 2011). While various organizations in the KSRP are synchronized with each other through the FAIS, the KSRP can play the role of an open platform which supports the role of the FAIS as a coupler. In this way, the FAIS has contributed to strengthening the innovation competence of the KSRP. For instance, the number of joint researches supported by external funds has persistently increased since 2001 though it slightly came down after the climax of 2005, and the number of patent application and license agreement in the running total has also increased every year since 2001 (Figs. 13; 14).

### *3.4 Synchronism, Synergism, and Synthesis: Using Cluster Projects*

The KSRP has been designated as a project area of some cluster projects initiated by the central government. It means that local projects can be synchronized, synergized, and synthesized with national projects in the KSRP.

For example, on April, 2002, the KSRP was designated as a project area of the Knowledge Cluster Initiative named the Kitakyushu Human Technology Cluster. This project aimed to create new environmental industries and its cluster. To create new environmental industries, a lot of joint researches between academia and industry were conducted by this project based on the knowledge foundation of the KSRP and the technology base accumulated by local industries in the field of environment, life/safety, and health. The system for promoting this cluster project was as follows (Fig. 15).

On the other hand, the FAIS participated in the Industrial Cluster Project named the Kyushu Silicon Cluster which was another national project started from 2001 in semiconductor/FPD[17]-related fields. The 138 companies, 10 universities, 11 local governments, 13 industry support organizations, and 64 individuals participated in the Kyushu Semiconductor Industries and Technology Innovation Association (SIIQ) which is the promotion organization of this project. The FAIS, of course, is also one of them (Fig. 16).

As the FAIS participated in the SIIQ, the members of the Kitakyushu Human Technology Cluster got more detailed information about the participants of the

---

[16] A coupler means a device that unites two circuits related with electricity or electron, and that automatically synchronizes one keyboard with the other keyboards, keyboards with pedals, or one note with another note related to music.

[17] Flat Panel Display.

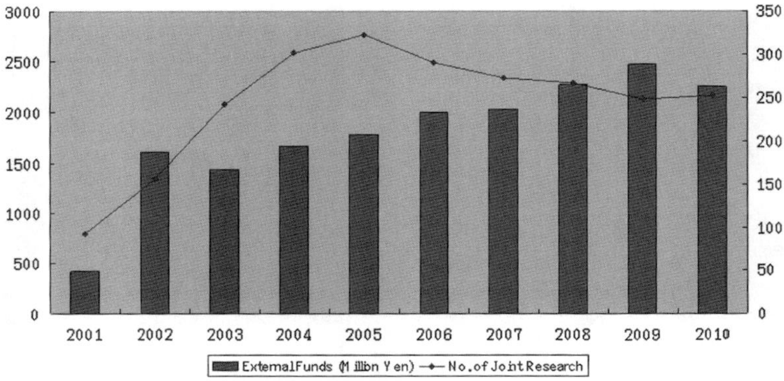


**Fig. 13** Change of external funds and number of joint researches in the KSRP. *Source* Author based on data from the KSRP's webpage (http://www.ksrp.or.jp)

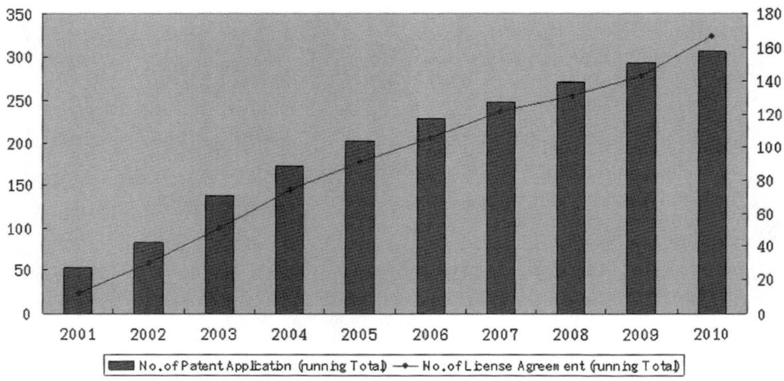


**Fig. 14** Number of patent application and license agreement by the FAIS. *Source* Author based on data from the KSRP's webpage (http://www.ksrp.or.jp)

Kyushu Silicon Cluster. It means that they got a chance to synchronize their research activities in the KSRP with the development activities of the Kyushu Silicon Cluster project in Kyushu and synergize with each other for achieving their own goal. In this context, they organized the Kyushu Wide Area Cluster Combined Headquarters Committee cooperating with members of the Fukuoka Project of a Cluster for System LSI Design and Development which was the other project of the Knowledge Cluster Initiative in Kyushu, and held a joint conference named Synergy Kyushu 2003.

The vision of the Kyushu Wide Area Cluster that the so-called "Silicon Island, Kyushu" become a hub of system LSI development in Asia was succeeded by the Fukuoka Cluster for Advanced System LSI Technology Development as a new project of the Knowledge Cluster Initiative (the Second Stage) in Fukuoka, Kitakyushu, and Iizuka area from 2009. The part related with SoC technology of

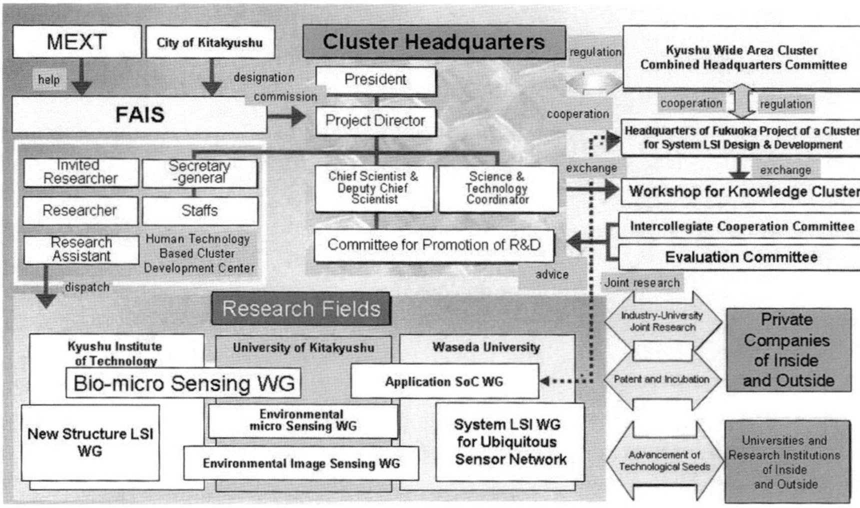


**Fig. 15** System for promoting the Kitakyushu Human Technology Cluster. *Source* Author based on a figure in Japanese from the brochure of the FAIS

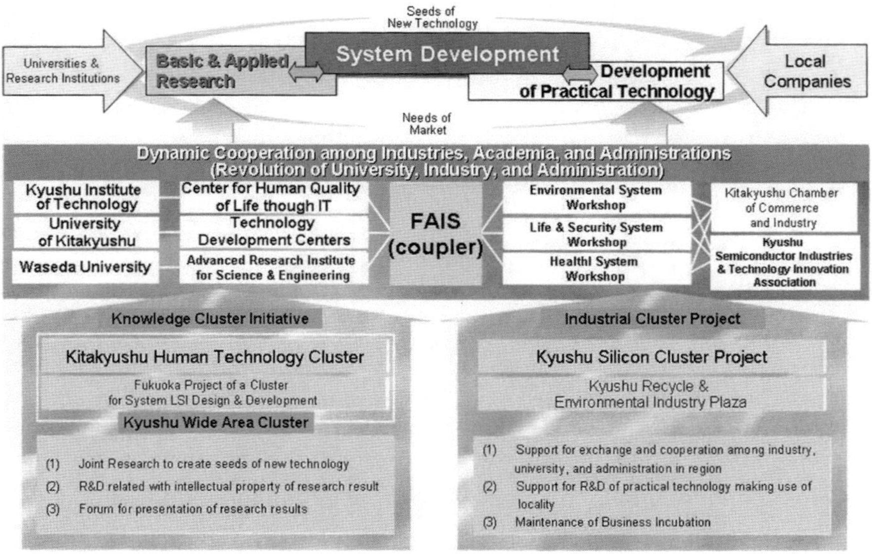


**Fig. 16** Synthesis of cluster policies and role of FAIS based on the KSRP infrastructure. *Source* Author based on a figure in Japanese from the brochure of the FAIS

the Kitakyushu Human Technology Cluster project was synthesized with this project and keeps on their research under the new framework of national policy titled the Regional Innovation Cluster Program (Global Type) from 2011.

### 3.5 Self-Directed S&T Promotion for Green Growth: from Eco-Town to Smart Community

"Environment" is one of the keywords to understand the Kitakyushu approach on building STP's ecosystem because Kitakyushu City is a symbolic place to overcome environmental pollution occurred in the process of modern industrialization. As mentioned above, although the Kitakyushu Industrial Area has supported high growth of Japan as one of the Japanese four Great-industrial Areas, it has brought about the serious environmental pollution in the 1960s in this area. In particular, the extent of the pollution was very severe in the industrial area of Tobata, Wakamatsu, and Yahata near the Dokai Bay. Then, local women's association began a movement to regain blue sky and sea from the environmental pollution symbolized "Smoke with seven colors" and "Sea of the death," and as the local government joined in this movement based on the resident network, the technology, know-how, talent, ethos, and climate concerning environment were cultivated from the process of overcoming the pollution in the City.

Based on the experiences to overcome industrial pollution, the City of Kitakyushu strategically chose an environmental approach for regional renaissance of the city visioning a "Resource Recycling-based Economic Society" involving some strategic conceptions such as:

- Strategic use of the large-scale innings in the coastal industrial area near the Dokai Bay where the land remained because of the hollowing out by the reorganization and rationalization in the industrial structure.
- Positive use of the regional experience for overcoming the pollution centered on the Dokai Bay and the regional tradition for the international exchange centered on Moji Port.
- Regional development that uses the green engineering, know-how, talent, and citizen network obtained from pollution overcoming process.
- Recycling of all kinds of wastes from the heavy consumption of energy and resource.
- Cooperation of the green engineering with developing countries and leading of the eco-business in Japan toward the cleanup production.
- Development of the recycling industry as a national venous industry.
- Development as a hub port for the distribution of the recycling industry.

The Kitakyushu Eco-town Project was started based on such strategic conceptions, and it is a starting point of green innovation based on the KSRP development. The project aims to develop eco-business totally from education and basic research in the KSRP to technological and practical research in the Practical Research Area, and industrialization in the Comprehensive Environmental Complex and the Hibiki Recycle Area (Fig. 17).

The results of basic research in the KSRP can be "seeds" for various empirical researches as sub-projects of the eco-town project in the Practical Research Area where there are sixteen facilities for industry-academia research partnership.

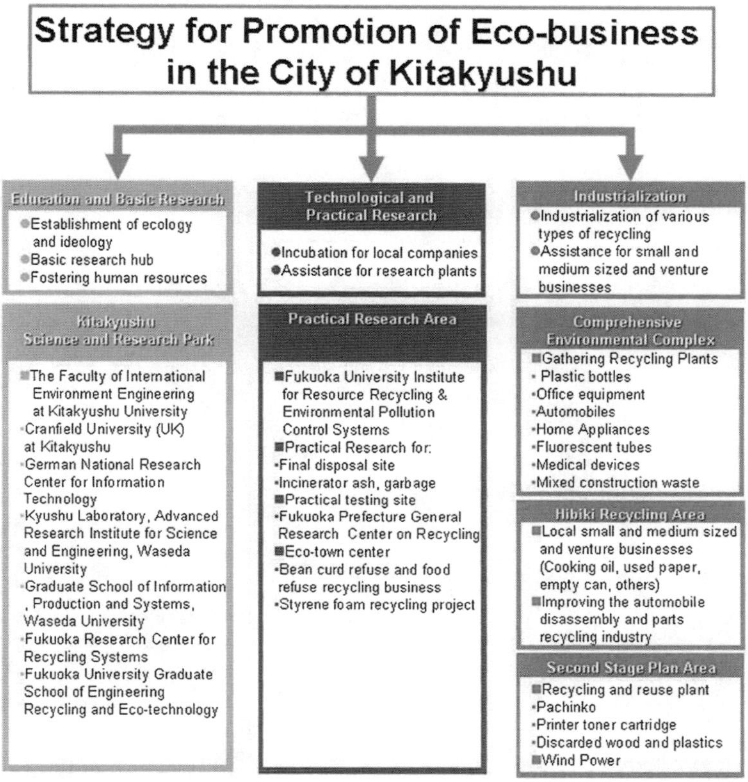


**Fig. 17** The strategy for promotion of eco-business in the City of Kitakyushu. *Source* Author based on the brochure of the City of Kitakyushu

Through the empirical researches, some of technological seeds from the KSRP evolved in the Comprehensive Environmental Industrial Complex for industrialization of the environmental technology and zero emission. There are various recycling companies such as PET bottles, an office equipments, cars, home appliances, fluorescent tubes, medical devices, and mixed construction wastes in there, which companies are trying to commercialize recycling technology.

The Kitakyushu Eco-town Project is the first eco-town project approved by the government in Japan (GEC 2005). And also, the Eco-town Center, the Hibiki Recycling Complex, and the Comprehensive Environmental Industrial Complex were the first only targeted areas for the eco-town project. This kind of strategic nature as an environmental model city is also demonstrated in the Advanced Low Carbon Technology Research and Development Strategic Guidelines which released on January 2011. The FAIS got a chance to manage the future direction for low carbon technology research and development in the City.

The guidelines have been formulated to build a research hub for low carbon technologies and promote local industries featuring the low carbon technologies by

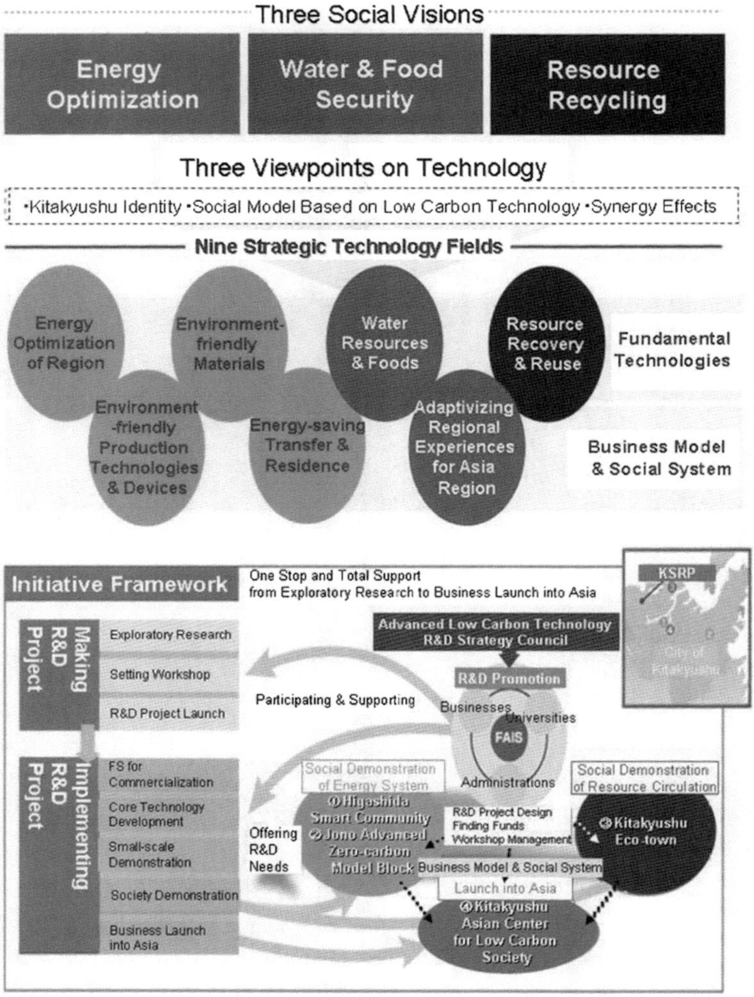


**Fig. 18** Advanced low carbon technology research and development strategic guidelines. *Source* Author based on the data from the KSRP's webpage (http://www.ksrp.or.jp)

providing developmental support. Setting up three social visions (energy optimization, water and food security, and resource recycling), the guidelines will tackle nine strategic technology fields from three viewpoints, including utilization of local potentialities. The municipality will establish ten study groups aiming to create 75 projects for the research and development within 5 years. By establishing the Project Planning Office (tentative name) based on industry-academia-administration collaboration, these projects will be operated. It goes without saying that the main place for the project operation will be the KSRP and the STP development will be extended through it into the City (Fig. 18).

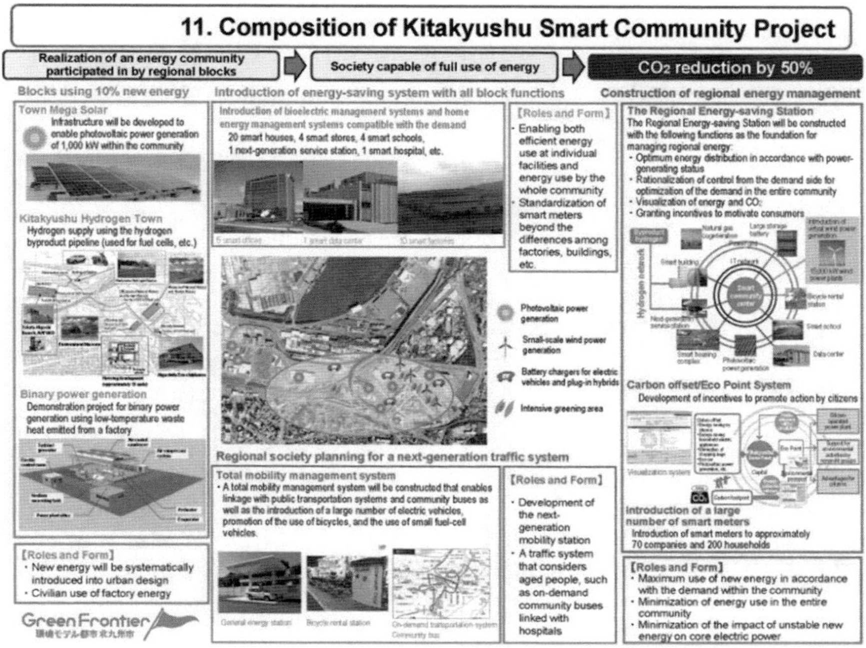


**Fig. 19** Kitakyushu smart community creation project. *Source* http://www.challenge25.go.jp (2011. 9. 30)

The Kitakyushu Smart Community Creation Project from August 2010 is also a green innovation project on the framework of the Guidelines. This project implements various demonstrations to 2014, and the demonstration results are used in the Higashida district, the Yahata-higashi ward, as stepping-stones to expanding the project to all areas of the city. The Higashida district (approximately 120 ha) is investing ¥16.3 billion over the 5-year period, which is founded on a basic plan to (1) adopt renewable energies such as solar power generation on a large scale; (2) establish a consumer-participatory energy system by means of a smart grid; (3) reduce energy consumption in places such as homes and buildings; (4) introduce electric vehicles and fuel-cell cars on a large scale; and (5) engage in highly comfortable and convenient town development through these activities.[18] In the district, moreover, the Kitakyushu Hydrogen Town Project as a part of the Kitakyushu Smart Community Creation Project was newly launched in January 2011. The project marks a world-first attempt to use a pipeline to supply the hydrogen generated in the iron manufacturing process at iron foundries to operate fuel cells (Fig. 19; 20).

[18] http://www.gov-online.go.jp/eng/publicity/book/hlj/html/201107/201107_05.html

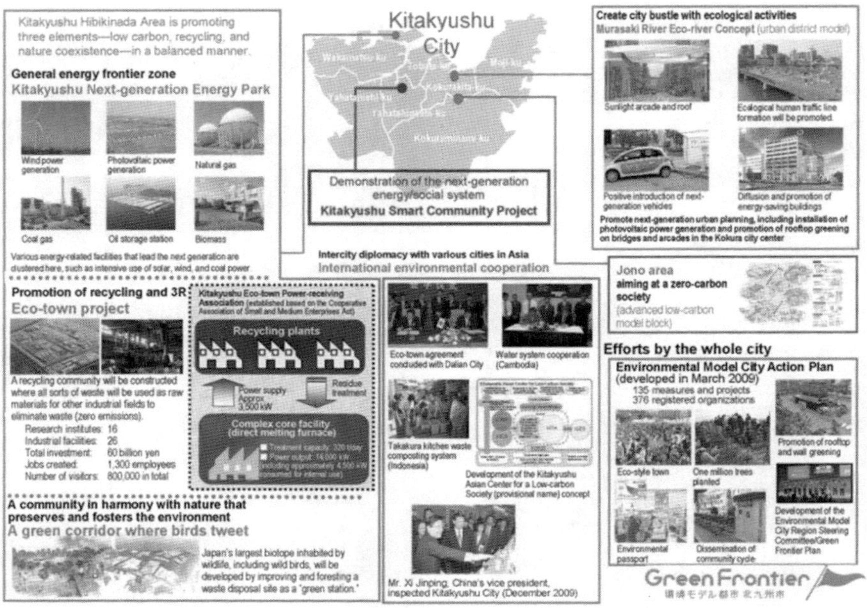


**Fig. 20** Kitakyushu City's efforts toward a low-carbon society. *Source* http://www.challenge25.go.jp (2011. 9. 30)

By the way, the implementation body of the project is the Kitakyushu Smart Community Council which consists of 46 companies and organizations including Nippon Steel Corp., the University of Kitakyushu, municipality, and the FAIS. It is interesting that many of the KSRP's members such as the FAIS participate in the council. While the council, as a local council for S&T promotion and triple helix innovation, participates in planning and promoting all of the sub-projects for building the Kitakyushu Smart Community, R&D activities in the KSRP will be synchronized, synergized, and synthesized with this project toward green innovation by the "participants" from the KSRP (Fig. 21).

# 4 Conclusion

## *4.1 Lessons*

It is clear that STP is a spatial being: there is a scale of space that various buildings or facilities related with R&D, practical education, and test production occasionally are located. Therefore, it may be necessary to make a good physical environment with charming design luring its constituents in the development of STP. However, it is not just a physical environment for gathering such buildings or

| | | |
|---|---|---|
| IDC Frontier Inc. | Sharp Corporation | NIPPON TELEGRAPH AND TELEPHONE WEST CORPORATION |
| iTest Co., Ltd. | NIPPON STEEL ENGINEERING CO., LTD. | NITTETSU ELEX Co., Ltd |
| Iwatani Corporation | NS Solutions Corporation | NISSAN MOTOR CO., LTD. |
| UCHIDA YOKO CO., LTD. | Nippon Steel City Produce, Inc. | IBM Japan, Ltd. |
| Area Service Co., Ltd. | Nippon Steel Corporation | Japan Telecom Information Service Co., Ltd. |
| OMRON Corporation | ZENRIN CO., LTD. | FUJI Corporation |
| Environmental Technology Service Co., Ltd. | SOFTBANK TELECOM Corp. | Fuji Electric Systems Co., Ltd. |
| Kitakyushu National College of Technology | SOFTBANK BB Corp. | Furukawa Electric Co., Ltd. |
| Kitakyushu Foundation for the Advancement of Industry, Science and Technology | NPO., Town Mobile Network Kitakyushu | The Furukawa Battery CO., LTD. |
| City of Kitakyushu | Electric Power Development Co.,Ltd. | Mitsubishi Chemical Corporation |
| University of Kitakyushu | Toshiba Corporation | YASKAWA INFORMATION SYSTEMS Corporation |
| Kitakyushu Chamber of Commerce and Industry | Toshiba Lighting & Technology Corporation | YASKAWA Electric Corporation |
| Human Media Creation Center/KYUSHU | TOTO Ltd. | Waseda University |
| NPO SATOYAMA | Toyota Motor Corporation | Yoshikawa Kikai Kogyo Co., Ltd. |
| SANKI Co.,Ltd. | Nano-Optonics Energy, Inc. | |
| JX Nippon Oil & Energy Corporation | West-Japan Auto Recycle Co.,Ltd (WARC) | |

**Fig. 21** Members of the Kitakyushu smart community council. *Source* http://www.challenge25.go.jp (2011. 9. 30)

facilities up. Rather it can be said as a specific place that innovative ethos to create new knowledge carrying industrial or economic value is blown up through various interactions under the science initiative. From this point of view, networking on the context of building STP's ecosystem can be said as a work to change individual and separate actions of STP citizenry into interactions among them and make an innovative community that occupies space of STP and changes it into the locus of innovation. For that reason, it is argued that networking is the crucial factor to complete building STP's ecosystem as such.

In the Kitakyushu approach, networking is the essence of building STP's ecosystem from the planning stage. First of all, the plan put emphasis on networking of academia for the knowledge creation based on competitive cooperation that is expected as the mainspring of making a new industrial city that the plan aimed. To make and activate networking of academia, many kinds of motives are given to universities and research institutes in the KSRP: joint research, credit transfer system, exchange of faculty, co-hosting of congress, university extension, and campus management committee that composed of the representative of the member HEIs. And then, STP as a common physical environment of its members plays the role of catalyst to precipitate the networking through some common-use facilities. Within these facilities, besides, the Collaboration Center, the Semiconductor Center, the IT Advancement Center have some rental labs and project rooms for private companies, and the Annex of Collaboration Center is a core building for joint research among the tenant companies. Networking in the KSRP is more expanded through these facilities to industries.

Then, the roles of two organizations are very important for networking, especially networking of networks: the Campus Management Committee as an "interface" and the FAIS as a "coupler." As mentioned above, the Campus Management Committee consisted of the member HEIs in the KSRP plans policies on which it jointly worked and decides the contents offered through the facilities though the municipality bore the expense for the facilities construction. It means that the Campus Management Committee is the point that the member HEIs meet and interact together and it makes a kind of "protocol" which is a set of rules governing the interaction among them. And the FAIS, as a coupler, connects two different constituents with each other and makes them just like one system through promoting various projects such as joint research, study group, technology transfer, etc., with the Campus Management Committee.

If we can say that the essence of innovation is "new combination," networking is, on the context of STP's ecosystem, a kind of social innovation for technological innovation because the various participants of STP development are interconnected with one another and interacted with some creative new ways to create new technology through new combination of knowledge. That is why industry-government-academia relationship has attracted considerable attention as the key to innovation in increasingly knowledge-based societies. But, the important thing is not the relationship itself but the way of interaction. In other words, an industry-government-academia relationship itself is not new one. The real new one that we have to find out to create innovation is the way of interaction among industry, government, and academia. A triple helix concept of industry-government-academia interactions, though a little complicated, is a good example from this point of view (Etzkowits 2008). Therefore, building STP's ecosystem without some strategic considerations for networking as the way of interaction among its participants cannot be a tool to promote innovation and national development. The Kitakyushu approach explains the matter clearly.

## *4.2 Clustering of Networking Sequences*

It is not easy to find more suitable paths of networking on building STP's ecosystem because it is very variable according to the conditions and circumstances. In this chapter, therefore, we try to elucidate its process with clustering of networking sequences from the point of view of regional system of innovation.

#### 4.2.1 Sequence #1: Sectoral Networking for Synchronism

In the planning stage of STP, first, specify some future constituent sectors of the park. In general, it may consist of academia (university) sector, business (industry) sector, and administration (government) sector, but it is possible to include research institute sector, consultancy sector, finance sector, or credit guarantee sector, etc., on occasion. Second, have the concept of STP jointly with all of sectors and set the role and vision of sector on STP's ecosystem through debates in each sector. Then, the concept of STP's ecosystem should be not only suitable the objective of STP development but also simple to understand. In debates, arrange or find out some opinion leaders who understand the concept and the objective very well. They will be trained and should be trained as main players who speaks for their sector and cooperate with other sectors in knotworking. Third, give a collaborative project to each sector on the assumption of the role and vision of sector on building STP's ecosystem. It is necessary that the project period is short to get the result within the period time. Of course, let make a report on the result when the project is finished, and it need to be reflected to the scheme for knotworking of networks, the next sequence.

#### 4.2.2 Sequence #2: Knotworking of Networks for Synergism

"Knotworking" means a way to organize and carry out productive activities when many players need to adjust their own actions mutually in a situation that they influence each other having the object of their actions jointly (Yamazumi and Engeström 2008). The way of thinking based on the concept of "knot" point out that whereas players or their action systems make just weakly networking, the performance of their cooperative activities begins to pulsate in haste and has it jointly, and a kind of "improvised echo" appears from it. It is impossible that knot is changed with something like specific individuals or fixed organizations as centers of control even though knotworking, in the works through cooperation, is tied up or untied some times. The type and characteristics of the initiative is going to change gradually in a series of knotworking because the dynamic horizontal movement that breaks and crosses the boundary of partition and isolation which is drawn historically and practically between some different organizations or works or cultures has come true.

Knotworking of networks is a useful approach in the construction stage of STP. Until the construction of STP is finished, we may challenge to make knotworking of networks through joint researches, study groups, and co-hosting of congress and so on. Then, mentioned above, the main players of knotworking are the opinion leaders in the planning stage. Based on the understanding about the concept and objective of building STP's ecosystem, they may try to make knotworking of networks with others from another sectors (networks) ascertaining and reflecting the role and vision of sector on building STP's ecosystem. But, because the linkage of knotworking is very flexible, variable, and weakly, it is difficult to structuralize it with any hierarchy. The important thing is knotworking itself: we can anticipate and define networking of networks in the near future observing the type and characteristic of knotworking. Of course, it is possible to adjust the number or the role of participants. And then, participants until the last in this stage are very important humanware that have the potential to be the staffs of organization that functions as an interface or a coupler for the networking of networks, the next stage.

### 4.2.3 Sequence #3: Networking of Networks for Synthesis

In the development and management stage of STP, we need to make more broadened networks to enhance the feasibility of innovation and national development through STP. To be the catalyst of innovation and growth in the national level, STP needs to make lots of network linkages and higher level integration for gaining more effective learning progress. However, using the so-called 'one by one approach' to make network linkages and integration expected is less effective. Based on the networks gaining through knotworking of networks, we can structuralize it first, and expand networking of networks secondly. Networking of networks means to make network linkages with existing networks by interconnecting and interacting with main players of it. Networking of networks is useful for propelling economy of scale, scope, speed, and status at the same time. But, because various levels of players and networks may be linked in a short time, the gap between the existing and the new is a serious threat to the stability for effective learning process in the STP. Therefore, it is very critical to make a coupler organization to play a role of auto stabilizer for building STP's ecosystem and regional system of innovation.

## References

Baba, Y. (1985). Japanese colour TV firms: Decision-making from the 1950s to the 1980s. DPhil dissertation. University of Sussex, Brighton.

Cha, S. R. (2004). Policy for development of research and Industrial Park in Tsukuba Science City. *The Korea Spatial Planning Review, 43* (in Korean).

Cha, S. R. (2008). Towards creative growth of innovation cluster: Experience and vista gained from clustered development of IT Industry in two Japanese Peripheral Cities. In *Proceeding on 2008 UNESCO-WTA International Training Workshop. UNESCO-WTA Cooperative Project Steering Committee*. Daejeon pp. 177–209.

Cha, S. R. (2011). *Regional innovation system and policy for high-tech industrial area in Japan and Korea*. Fukuoka: Hanashoin (in Japanese).

Etzkowits, H. (2008). *The triple helix: University-industry-government innovation in action*. New York: Routledge.

Freeman, C., & Soete, L. (1997). *The economics of industrial innovation, 3rd edn*. Cambridge: The MIT Press.

GEC. (2005). Eco-towns in Japan-implications and lessons for developing countries and cities-, Osaka: Global Environment Centre Foundation (GEC). Retrieved Sept 30, 2001 from http://www.ksrp.or.jp/e/info/index.html

List, F. (1841). *The national system of political economy* (English ed.). London: Longman.
MEXT. (2009). *Cluster; knowledge cluster initiative*. Tokyo: MEXT.
METI. (2008). *Industrial cluster project*. METI: Tokyo.
METI. (2009). *Industrial cluster project*. METI: Tokyo.
MEXT. (2010). *White paper on science and technology*. Tokyo: MEXT.
MEXT. (2011). *The 4th science and technology basic plan*. Tokyo: MEXT (in Japanese).
Miyakawa, Y. (2006). Regional Renaissance and sustainable development of science city in Japan on high tech clusters in global context. In *Proceeding of International Workshop on High-tech Clusters in Global Context*. UNESCO-WTA Cooperative Project, Daejeon, pp. 99–144.
Scalise P. J. (2001, January 12). Sisyphus, the Japan Specialist. Retrieved from http://www.Japantoday.com
Tatsuno, S. (1986). *The technopolis strategy: Japan, high technology, and the control of the technology of the twenty-first century*. New York: Prentice Hall Press.
Yamazumi, K., & Engeström, Y. (2008). *Knotworking: To creation of human activities knotwored*. Tokyo: Shinyosha (in Japanese).
Yazawa, S. (1990). The technopolis program in Japan. *Hitotsubashi Journal of Social Studies, 22*(1), 7–18.

# Science, Technology and Industry Co-operation in Finland

**Hannu Juuso**

**Abstract** The innovation co-operation among the universities, enterprises and other involved organisations like the science parks seems to be more and more important. The same is with the networking with all of the players in the whole value chain of innovation. Through some selected funding and operation schemes and examples from Finland I hope to emphasis elements, which could be adapted to other countries in other science parks among the WTA members. The global economy is driving the whole innovation environment towards open innovation. The global competition is not only present in the business life, but it is more and more influencing the academic and research world. The keyword in open innovation is networking. The science and technology parks have an important role—building the networks between research, business and public authorities. The trend in one sentence: *The more **R&D** (Research and Development) is changing to **R&B** (Research and Business) the more **Know—How** is changing to **Know—Who**.*

## 1 Finland

Finland is a European country, situated in the northern part of the continent. Spread over an area of approximately 338,144 km$^2$(130,558 sq miles), it boasts the distinction of being the seventh largest country in Europe. The population of

H. Juuso (✉)
Tekes, the Finnish Funding Agency for Technology and Innovation,
P.O.Box 69 FI-00101 Helsinki, Finland
e-mail: Hannu.Juuso@tekes.fi

D.-S. Oh and F. Phillips (eds.), *Technopolis*,
DOI: 10.1007/978-1-4471-5508-9_23, © Springer-Verlag London 2014

Finland is 5.3 million and Finland has two official languages, Finnish (92 %) and Swedish (6 %). Finland is a constitutional republic and became a member of the European Union in the year 1995. The capital of Finland is Helsinki. A quarter of the country's total area lies north of the Arctic Circle. Forest covers about 75 % of Finland and water bodies, around 188,000 lakes, make about 10 % of the country. Finland has been ranked the best or one of the best at the World Economic Forum's Global Competitiveness Report Index Ranking during the last 5 years. The innovation system has shown to be one of the best in the world. The situation is not static—in the global competition continuous change and strive to improvement is a necessity.

## 2 Innovation Environment

Finland suffered from a severe economic recession with high levels of unemployment at the end of the 1980s and beginning of the 1990s, resulting from among others the limited competitiveness of the more traditional sectors dominating the Finnish economy by that time, such as pulp and paper and wood products. In order to address the severe economic recession, caused by the structural problems of the Finnish innovation system, the government identified research and innovation as an important driver for future economic growth. Since the 1980s, the Finnish government has therefore focussed its policy and instruments on improving R&D intensity, thereby formulating ambitious targets for Gross Domestic Expenditure on R&D (GERD) and Business Enterprise Expenditure on R&D (BERD).

Finland's strategy has proven to be effective: the economic growth in the 1990s outpaced most of its competitors. Although the burst of the ICT bubble slowed down development at the beginning of the new century, the current economic growth of 2.9 % (2005) lies above the EU 25 average. The emphasis on R&D and innovation in its policy has made the Finnish innovation system one of the best performing in the world. A specific characteristic is its high level of expenditure on research and innovation. Total R&D spending amounted to 3.5 % of Finland's GDP in 2004, well above the overall 3 % EU target for 2010. About 70 % of R&D spending is financed by the private sector. In the NRP, the Finnish government announces a target of raising total R&D spending to 4 % of GDP by the end of the decade. Public spending on R&D is projected to increase by 5–7 % each year over the same period. The increasing levels of R&D expenditure by the government since the 1980s have been accompanied by increasing levels of BERD, especially from the emerging ICT sector in Finland (De Heide 2007).

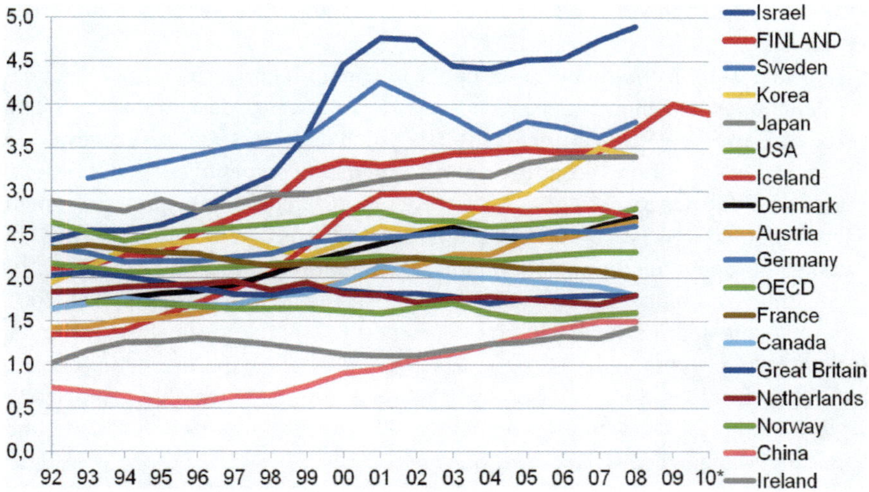


* Estimate based on queries and calculations

*sources* OECD, Main Science and Technology indicators AND Statistics Finland

## 3 Recent Changes in Innovation Environment

A series of changes in governmental structure and policy in 2007 and 2008 reflect a rethinking of innovation policy in Finland. It is too early to judge the results of the new initiatives, but two departures are worthnoting: first, a new focus on industrial sectors, or clusters, and second, increased attention to the development of regional capabilities. In addition, a powerful new **Ministry of Employment and the Economy** (**MEE**) was created in 2008 with responsibility for employment, regional development, industrial policy, innovation and technology policy, energy policy and competition policy. While these changes indicate recognition of the need for change, and some movement towards decentralisation, we do not believe whether they are sufficient to forestall the economic shocks to the large established Finnish corporations in forest products and ICT in the coming years.

The Science and Technology Policy Council authorised the formation of centres in the five areas: energy and environment, metal products and mechanical engineering, forest cluster, health and wellbeing and information and communication industry and services.

In 2007 Tekes announced the launch of Forest Cluster, Ltd—the first **Strategic Centre of Excellence in Science, Technology and Innovation** (**STI**). Forest Cluster Ltd. is a consortium aimed at coordinating top-level, longer term research programmes that combine funds and expertise from private enterprises, universities and research institutes, with public support—primarily from Tekes and the

Academy of Finland, as well as from the EU under the Seventh Framework Programme.

The other significant change is a public recommitment to local and regional level economies and innovation systems—a move towards decentralisation. **The Centre of Expertise Programme (OSKE)**, which dates to 1994, was renewed in 2007 for a 6-year period in order to "*improve regional competitiveness in line with national and European policies*" . The programme traditionally provided small amounts of funding and high-level status in the Finnish innovation strategy to encourage co-operation among universities, research institutes,

companies and municipal actors in 21 regional Centres of Expertise. The 2007 programme created 13 new Clusters of Expertise—or Competence Clusters—selected through a competitive process to represent the top expertise in their fields, which range from Food Development, Clean Tech, Energy Technology and Health and Wellbeing to Ubiquitous Computing, Nanotechnology and Tourism and Experience Management (Sabel and Saxenian 2008).

## 4 Co-operation in Innovation Activity

According to the study done by The Statistics Finland, the key co-operation partners in innovation activity are customers and equipment and material suppliers and for group enterprises the other enterprises from their own group. Around one-third of those with innovation activity connected to product or process innovations in 2006–2008 reported about co-operation with equipment and material suppliers. Similarly, around one-third said they had had collaboration related to innovation activity with customers. Forty percent of the innovating enterprises belonging to groups had co-operated with enterprises in their own group. In all, 37 % of those with innovation activity said they had co-operation related to innovations. Nearly all of these enterprises had co-operation partners in Finland, around 70 % elsewhere in Europe. Of those having co-operated, nearly one-third had partners in the United States and close on one-fifth said they had partners in China. 6 % of those having co-operated reported partners in India and one-fifth elsewhere not mentioned above. In relative terms, the least collaboration was reported with public and private non-profit research institutes (Statistics Finland 2010).

European Commission makes regularly surveys on the development of Innovation Environment in the member countries in Europe. Finland has been one of the top performing countries. One method of comparison is the amount of Small and Medium Size Enterprises (SMEs), which co-operate with universities, research centres and other organisations in research and development. Chart below presents the comparison between member countries in 2006.

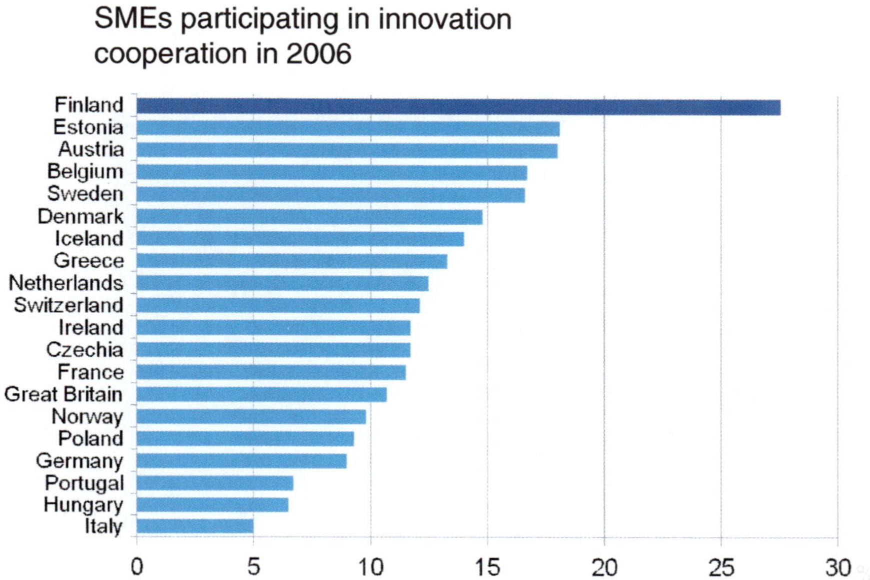

*Sources* European Innovation Scoreboard 2008

## 5 Resources and Main Actors in the Innovation Environment

There is no ideal environment for innovation and there is not a single phase, where the external resources are needed. Innovation needs all kinds of support actions throughout the whole process starting from basic research and ending to entry of the market. There are several actors in the innovation environment and the variety of the different financial and business support is outstanding. Following chart describes the resources in the Finnish Innovation Environment (European Innovation Scoreboard 2008).

Resources in the innovation environment

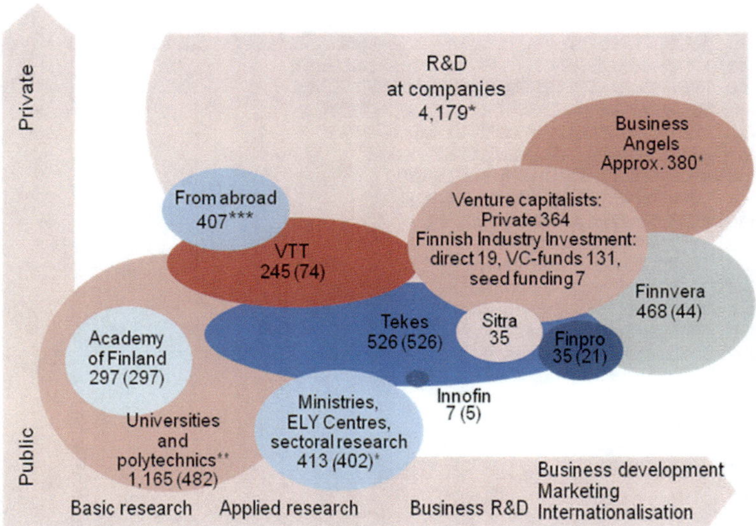


The figures represent the total extent of each organisation in million euros in 2008, those marked with star are earlier. In parenthesis the share that is funded from the State budget.
**includes polytechnics ***includes R&D costs of corporations foreign units

The above figure provides an insight view in the expenditure in R&D and innovation by actor. The figures represent the total extent of each organisation (in million Euros, €) in 2008. In parenthesis the share that is funded from the State budget. The funds of Tekes, the Academy of Finland and Innofin are funded entirely from the State budget.

The main organisations involved in innovation environment:

**The Academy of Finland**, which includes four national research councils, is responsible for the financing and strategy formulation of the basic research, research training and science policy. The Academy of Finland has a range of different funding instruments for different purposes: it provides funding for research projects, research programmes, centres of excellence in research, researcher training, international co-operation, as well as research posts for Academy Professors and Academy Research Fellows. The Academy has four Research Councils that decide on the allocation of funding within their respective fields. In 2006, approximately 15 % (€257 million) of all Government research funding will be channelled through the Academy. The Academy's responsibilities also include the advancement of scientific research and the encouragement of its exploration, and the development of international scientific co-operation.

**Tekes, the Finnish Funding Agency for Technology and Innovation**, is administered by the Ministry of Trade and Industry. Its primary objective is to promote the competitiveness of Finnish industry and the service sector by assisting in the creation of technology and innovation. The main instruments of Tekes are industrial R&D grants and loans to firms and grants for applied research for public organisations (universities, public research organisations and polytechnics) along with various expert services for business development and internationalisation. Tekes has an annual budget of about €633 million (2010), a source of funding for 1,896 projects. Tekes' funding focuses on SMEs; in 2010, 61 % of the funding for companies' projects was allocated SMEs and 70 % to companies with less than 500 employees.

**Centre for Economic Development, Transport and the Environment** (ELY Centres) The Centres for Economic Development, Transport and the Environment manage the regional implementation and development tasks of the state administration. There are 15 Centres for Economic Development, Transport and the Environment. These Centres are tasked with promoting regional competitiveness, well-being and sustainable development, as well as curbing climate change. Centres for Economic Development, Transport and the Environment operate within the administrative sector of the Ministry of Employment and the Economy. In addition to the Ministry of Employment and the Economy, their operations are steered by the Ministry of the Interior, the Ministry of Agriculture and Forestry, the Ministry of the Environment, the Ministry of Transport and Communications and the Ministry of Education and Culture. Tekes services can be obtained via these ELY Centres (14 in number) along with other public services for SMEs.

**VTT, Technical Research Centre of Finland** is the biggest multi-technological applied research organisation in Northern Europe. VTT provides high-end technology solutions and innovation services. Through its international scientific and technology network, VTT can produce information, upgrade technology knowledge, create business intelligence and value added to its stakeholders. VTT is a part of the Finnish innovation system under the domain of the Ministry of Employment and the Economy. VTT is a not-for-profit organisation and it's turnover is 292 M€(31.12.2010) and personnel 3,167 (1.1.2011).

**Sitra**, the Finnish Innovation Fund, was established as a division of the Bank of Finland in 1967 to mark the 50th anniversary of Finnish independence. Sitra's responsibilities are stipulated in law. It came under the auspices of the Finnish Parliament in 1991. Sitra's activities are financed by the yield from its endowment capital and the return of its venture capital investments.

**Finpro** is a globally operating organisation that helps Finnish companies in export activities. Finpro was founded in 1919. Finpro's partners include some 550 Finnish companies, the Confederation of Finnish Industries, the Federation of Finnish Enterprises and the Federation of Finnish Technology Industries. Finpro as a public–private partner belongs to the Ministry of Employment and the Economy Group and works in close co-operation with actors in the Finnish innovation ecosystem such as: ELY -centres, Tekes—the Finnish Funding Agency for Technology and Innovation and the Ministry for Foreign Affairs of Finland.

**Finnvera** is a specialised financing company governed by the Finnish State. Finnvera has official Export Credit Agency (ECA) status and it provides businesses with loans, guarantees, venture capital investment and export credit guarantees.

It is important to note that there are active operating links between the organisations. The ELY Centres, for example, provide regional access points for

Tekes services. Tekes has active co-operation processes with Finpro and Sitra. There is an extensive network of intermediary organisations such as technology and science parks, local or regional business development companies and business incubators. The association of Finnish Science parks—TEKEL—has 30 members, and additionally there are other innovation centres (40), local or regional business development companies (100–160, including 60 incubators), business incubators (100) and university technology transfer offices or companies (in 12–14 universities). These organisations operate as intermediaries between the producers and users/appliers of new knowledge, research results and technology. They are either private or public entities, and they perform a variety of tasks on a regional basis and according to their funding base.

**TEKEL, Finnish Science Park Association**

The first science park in Finland started in 1982 in Oulu and it was also the first in the Nordic countries. Short after that the first incubator was established 1985. The Finnish Science Park Association (TEKEL) was founded in 1988. At the moment TEKEL has 32 members, which call themselves Science Parks or Technology Centres. The TEKEL members have 1,300,000 m$^2$ of space, 780 employees, 2.200 customer companies and organisations with 14.100 business contacts. Most of the member centres are located in Finnish university cities. The TEKEL members provide companies and other organisations with premises, assistance and support. Building up common technology projects between universities and companies is especially encouraged.

Benefits for the client companies of the TEKEL members:

- Business incubator services for start-ups, business expertise for commercialising, training and consultancy services, etc.
- Networking services: domestic and international contacts.
- Facilities for innovative companies.
- Implementation of national programmes. E.g., Centre of Expertise programme (OSKE).

Technology transfer and support in creating new high-tech companies is one of the key actions of the TEKEL members. TEKEL belongs to the EEN network (Enterprise Europe Network), which is the widest business and technology transfer network in the world. The EEN network has its technology transfer roots started in 1995 in Europe as an Innovation Relay Centre (IRC) network. The activity has been co-funded since the beginning by the European Commission and nowadays it has spread widely around the world.

## 6 Tekes

Tekes, the Finnish Funding Agency for Technology and Innovation, is the main public organisation for research, development and innovation in Finland. Tekes provides funding for innovative projects aimed at generating new know-how and new kinds of products, processes and service or business concepts. The aim is to renew industries, increase value added and productivity, improve the quality of working life as well as boost exports and generate employment and wellbeing. Tekes works with the top innovative companies and research units in Finland. Tekes participates actively in the planning and implementation of technology and innovation policy in Finland. More than half of corporate R&D and innovation funding granted by Tekes goes to small- and medium-sized enterprises.

Tekes employs over 380 people. Along its Helsinki-based headquarters Tekes works in close co-operation with the 15 regional Centres for Economic Development, Transport and the Environment (ELY Centres) throughout Finland and six offices abroad in Beijing, Brussels, Tokyo, Shanghai, Silicon Valley and Washington D.C. Tekes funds its target projects through low-interest loans or grants, depending on the stage of the innovation and the nature of the proposed project.

Tekes also awards financing to foreign-owned companies registered in Finland. Foreign companies with R&D activity in Finland are not required to have a Finnish partner to be eligible for funding. The financed project should, however, contribute to the Finnish economy. In 2010, Tekes was actively funding nearly 1,900 projects with €633 million altogether and had a total of 1,535 completed projects.

SHOKs or Strategic Centres for Science, Technology and Innovation carry out long-term co-operation in fields most crucial for the future. The results are breakthrough innovations of global importance, which can be agilely transformed into growth in business life and wellbeing in society (Tekes http://www.tekes.fi).

The Strategic Centres for Science, Technology and Innovation established in Finland are new public–private partnerships. Their main goal is to thoroughly renew industry clusters and to create radical innovations.

- Energy and the environment (CLEEN Ltd).
- Metal products and mechanical engineering (FIMECC Ltd).
- The forest cluster (Forestcluster Ltd).
- Built environment innovations (RYM Oy).
- Health and wellbeing (SalWe Ltd).
- Information and communication industry and services (Tieto- ja viestintäteollisuuden tutkimus TIVIT Oy).

## 7 Tekes Programmes for Start-ups

Growth companies are the backbone of economic wellbeing. Finns must be encouraged to become entrepreneurs, and successful entrepreneurs encouraged to become investors. Finland may very well become the European centre for growth companies in the technology industry. Tekes has improved its services further to meet the needs of start-ups and growing SMEs. These services encourage companies to focus on growth and aim for global markets from the start. Success is evident in, for instance, the games industry and in the recent Deloitte and Wired listings.

Tekes offers numerous services for growth companies. Services such as Funding for young innovative growth companies and the Vigo-acceleration programme have proved to be effective especially with regard to ICT companies (Tekes Homepage).

## 8 Tuli

Tuli is a programme funded by Tekes that helps researchers and research communities to evaluate the commercial potential of a research-based inventions or ideas and aids in the process of their commercialisation. The programme was launched first time in 1993 and reshaped into a 4-year national programme in 2002. The programme was operated via a network of local technology parks and the total budget of the programme in 2005 was approximately €2.5 million. Annually there were about 500–650 business ideas identified.

The latest Tuli programme runs from 2008–2012. In 2008-2010, 14 universities, 21 polytechnics and five research institutes—all in all, 40 research organisation—are involved. The overall budget for the whole programme is approximately €50 million. Tuli aims to give birth to new research-based business with the aid of licence agreements, know-how transfers and new companies.

In other words, Tuli aims to commercialise research results in Finnish universities and research institutes.

Tuli is intended to be flexible, fast and able to take risks in a healthy manner. The programme is developed in co-operation with the R&D and innovation services of universities and research institutes that have the mission of finding and developing research-based inventions and business ideas. With Tuli's help, researchers, students and research groups are able to receive funding and versatile expert services in all fields of commercialisation.

Results from the Tuli programme.

- In year 2008 there were in total 693 projects that received Tuli-funding. 77 % of those projects came from invention reports. There were started 16 new companies from the projects funded by Tuli. There were 26 projects that were commercialised through technology transfer which brought income of €920,000. This is 34 % of the incomes reported for licencing and technology sales of Universities in 2008.
- During 2007, 389 research-based inventions or business ideas were evaluated in the Tuli programme. 181 of the projects started from invention notices. During the year, eight inventions or ideas lead to finding of a new company and two lead to a licencing deal.
- From 2002 to 2006, over 1500 research-based inventions or business ideas were evaluated. 99 projects advanced into licencing deals and 184 into finding a new business. The annual funding volume of the Tuli programme was approximately €2.5 million during 2002–2006.

Tekes is prepared to fund the 46 co-operative universities and research institutes with €12 million from 1st April 2008 to 30th April, 2011. (Tuli http://www.tuli.info).

## 9 Vigo

Vigo is a new type of acceleration programme designed to complement the internationally acclaimed Finnish innovation ecosystem. The programme bridges the gap between early stage technology firms and international venture funding. Vigo is an accelerator programme for dynamic start-up companies with potential for global growth. The programme exposes these companies to experienced entrepreneurs and executives and facilitates raising both public and private growth funding (Vigo homepages).

The backbone of the programme is formed by the Vigo Accelerators, carefully selected independent companies run by internationally proven entrepreneurs and executives. These Accelerators help the best and the brightest start-ups to grow faster, smarter and safer into the global market. The Accelerators are not consultants—they are co-entrepreneurs who invest in the companies they work with to guarantee common goals and passionate development effort.

The key objectives of the Vigo Programme are:

- Incentivising the best business developers to help the most promising start-ups grow into successful companies.
- Ensuring early stage funding for the target companies, increase their shareholder value and make them attractive targets for venture investors.
- Raising significant venture capital investments for continued expansion of the target companies after the acceleration stage.
- Invigorating the Finnish venture capital market and bring more international acceleration and venture capital players into Finland.

The Finnish Ministry of Employment and Economy (MEE) launched the Vigo Programme in 2009 in co-operation with Tekes and Veraventure. Tekes is responsible for the co-ordination on programme and Finnvera participates in the programme by making investments into the target companies.

The currently active Accelerator companies are:

- Cleantech Invest.
- Food Process Innovations.
- Koppi Catch.
- Lifeline Ventures.
- Lots.
- Veturi Venture Accelerator.

In addition to Tekes and Veraventure also Finnvera, MEE and a cluster of private investors participate in the programme. (Tekes Homepage).

The first 2 years operation has shown that the concept has been successful in raising capital for the fund. Also remarkable is that majority of the fund (now about €60 M) is coming from private sector (60 %).

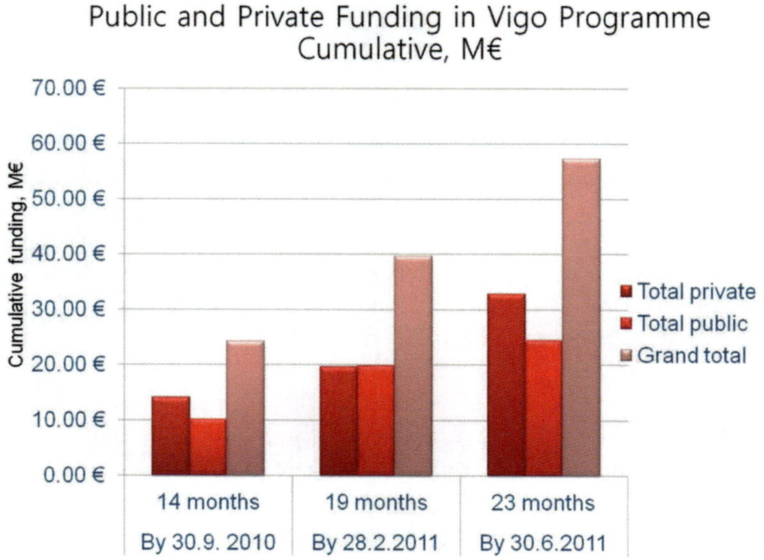

## 10 Funding of Innovative Growth Companies

Tekes established a new funding concept having the objective to generate new high growth enterprises to Finland. Tekes funding accelerates the development of enterprises which pursue for rapid international growth. The funding is adopted to the fact that only few enterprises can develop to a real success story. To ensure the process there must be more start-ups getting the possibility to grow and based on the concrete results the companies can obtain more funding even up to €1 million. This "funnel" model can be seen in the structure of the funding.

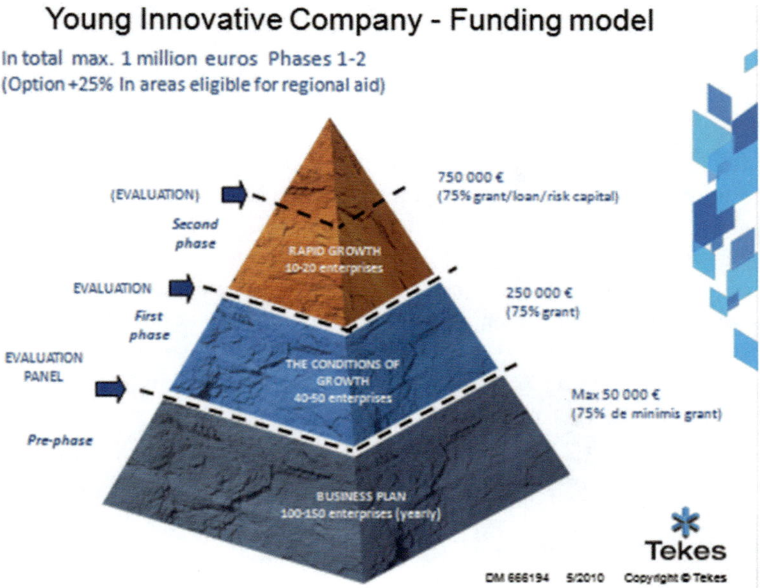


The requirements for funding are:

- a small company
- has been of existence for less than 6 years at the time when the aid of the first phase is granted
- is innovative on the basis that R&D expenses represent at least 15 % of its total operating expenses
- is fully committed to the necessary development activities
- has sufficient resources and credible chances to achieve its' business goals including fast growth and internationalisation.

Until April 2011 there were 104 companies receiving funding and from which 4 companies reached full €1 million funding. The total funding from Tekes was €74 million.

Expertise and innovation are increasingly important competitive factors for Finland. Finnish innovation is based on solid expertise, cross-sector innovation policy and on strengthening of our international presence. Finnish innovation actors share the goal of translating the creativity of individuals and communities into international success stories. For this purpose, we co-operate with progressive organisations regardless of whether they are public, private or non-profit. The innovation system is an interaction network.

# 11 Some Examples of Innovations Starting from Universities and Research Centres with the Help of Innovation Environment Actor

## *11.1 Eniram Oy*

Eniram delivers decision support systems to seafarers and shipping companies. The main product of Eniram is the Dynamic Trimming Assistant (DTA) system, which collects real-time data on the status of the vessel and displays it in a user-friendly graphical format. The system helps the ship's officers to achieve optimum trim. When the water resistance of the hull is as small as possible, the energy of the vessel is directed into movement, reducing fuel consumption and the environmental load. For a large cruise ship, the savings in fuel costs can be up to €200,000 per year. The corresponding environmental impact is a reduction in $CO_2$ emissions totalling more than 1,000 t per year.

The company was founded by three students from Helsinki University of Technology (now Aalto University) and one from Helsinki Institute of Technology. The initial evaluation of the technology was done in 2004 as a feasibility study by the Innovation Centre of the Helsinki University of Technology. The further development of business plan, establishment of the company and raise of capital was done through the National Technology Incubator Development Programme (Yrke) by Technopolis Plc, which is the largest Science and Technology Park incubator in Finland. The company succeeded to get the first product to the market in 2008. One-year later the company's turnover was one million Euros and the company had 30 employees. In the 2010 respective values were €3.3 million and 50 employees. This year (2011) the company will have a turnover of €7 million and employees 60. Currently the company has installations on ca. 100 vessels (cruise, container, tanker, bulker, ropax). Total annual savings in fuel corresponds ca. $20 million and 100,000 t of $CO_2$.

For further information, please contact:

Aki Luukkainen, VP, Marketing and Operations
Eniram Oy, Tel. +358 40 517 8387,
aki.luukkainen(at)eniram.fi
www.eniram.fi

### 11.1.1 Canatu Oy

**Nanobud revolutionises car control**

Car dashboards will become touch screens in the future. Today this may sound like utopia, but the Finnish company Canatu Oy has already developing technology that will make it possible. In 2006, a new kind of carbon structure, the nanobud, was discovered in the Helsinki University of Technology (now Aalto University Schools of Technology). It combines the properties of fullerenes and carbon nanotubes providing new interesting functionalities. Canatu makes carbon nanomaterials which enable highly transparent, conductive and flexible films.

"The original invention was a kind of diamond, so we further developed the material and product properties". "One potential use is touch sensors, because this nanobud film is flexible and thus allows us to design sensors in a more innovative manner than at present. For example, the car industry is very interested in using this method to develop complete touch sensing dashboards," says Mikko Kärkkäinen, CFO at Canatu Oy.

In practice, Canatu produces thin, transparent films that contain carbon nanobuds or nanotubes. The thickness and electrical conductivity of the film can be adjusted very precisely. "We make the world's thinnest nanotube films," continues Kärkkäinen.

No one else has produced similar films industrially, which is why Canatu joined the Tekes programme—to show that the method can be scaled from laboratory to industrial production. The pilot plant is now under construction at their facilities in Helsinki, and production is scheduled to begin in 2012. "In practice, our line is just a reactor, in which the nanostructures are synthesised, located above a printing machine that deposits them onto the film in the form of aerosol. Competing methods are much more complicated having as many as 18–20 different phases," explains Kärkkäinen. Canatu participated in the Functional Materials Programme and is particularly grateful for the flexibility shown by Tekes. "Things happen rapidly in our field. When research results forced us to change direction, Tekes was able to react quickly," states Kärkkäinen.

The work has also received international recognition. Canatu was named to the respected Red Herring 100 Global list in 2010. This prize is awarded annually to the top 100 technology companies in the world.

More information:

Chief Financial Officer, Mikko Kärkkäinen, Canatu Oy
Tel. +358 40 533 7872, mikko.karkkainen(at)canatu.com
www.canatu.com

## References

European Innovation Scoreboard 2008.

de Heide, M. (2007). Monitoring and analysis of policies and public financing instruments conducive to higher levels of R&D investments, The "POLICY MIX" project, country review Finland, Technopolis, March 2007.

Sabel, C., Saxenian, A. L. (2008). A fugitive success, Finland's economic future, Sitra Reports 80.

Statistics Finland, Innovation (2008). Innovation survey final report. Accessed 10 June 2010.

Statistics Finland, http://www.stat.fi/index_en.html.

Tekes, the finnish funding agency for technology and innovation, www.tekes.fi. www.tuli.info & TULI-ohjelman väliraportti, Teknologiaohjelmaraportti 8/2004, Tekes. http://www.vigo.fi/program1.

# Websites

Finnish Science and Technology information service: www.research.fi/en
Virtual Finland: http://virtual.finland.fi/
Study in Finland (Information on education, training and youth Issues) http://finland.cimo.fi/
Finland in Figures (Statistics Finland): http://www.stat.fi/tk/tp/tasku/suomilukuina_en.html

# Development of an Innovation Cluster in the Region: Experience of Gwanggyo Technovalley in Korea

**Deok Soon Yim**

**Abstract** Gyeonggi Province is the core area of Korea not only in terms of population, politics, and culture but also in manufacturing and industrial technologies. All the major IT companies and automakers are headquarterd in Gyeonggi Province. Recently, it has started to invest at science and technology to enhance its technological competitiveness. The Gwanggyo Technovalley (GTV) is one of the major S&T policy initiatives in Gyeonggi Province, which is owned and run by the regional government. In this chapter, the theory of innovation cluster was briefly reviewed and GTV situation in 2010 is evaluated in the context of innovation cluster. It is evaluated that GTV has excellent locational advantages but there are many weak points and policy issues for the future development. For instance, it is necessary to make detailed comprehensive master plan including marketing and globalization plans. In addition, it is required to identify the technology areas for the specialization. Finally, it is concluded that GTV case shows typical (or some specific to Gyeonggi Province) characteristics of the development of an innovation cluster in the region. From the GTV case, it can be learned that more thoughts and plans are necessary on management plan as well as hardware infrastructure development in the development of innovation clusters in regional level.

## 1 Introduction

It is acknowledged that the "innovation cluster" is one of the effective policy tools to boost technological innovation in the region. For this reason, many countries are trying to establish and use innovation cluster (science and technology park).

D. S. Yim (✉)
CEO & President, International Innopolis Research Center, Seoul, Korea
e-mail: yimdeoks@gmail.com

D.-S. Oh and F. Phillips (eds.), *Technopolis*,
DOI: 10.1007/978-1-4471-5508-9_24, © Springer-Verlag London 2014

Gyeonggi Province of Korea is no exception. It has been developing various innovation clusters by locating science and technology organizations at close distance. Among them, Gwanggyo Technovalley (GTV) and Pangyo Technovalley (PTV) are the two leading technological complexes of Gyeonggi Province. In particular, GTV has emerged as a kind of innovation cluster, where research institutes, universities, and industries are at close distance and interact together for better innovation, when Advanced Institutes of Convergence Technology (AICT) was opened in March 2008.

However, there seem to be many operational issues such as networking among innovation actors, marketing of GTV, and so on. Currently, GTV does not have a dedicated managerial body and therefore has some difficulties to coordinate activities among the tenant companies and research institutes. Since GTV is established recently, it may be too early to evaluate its performance. However, it is worthwhile to evaluate in the context of innovation cluster and draw some policy lessons. As it is the regional government-driven innovation cluster of Korea, it will give some insight to look at the typical issues of innovation cluster development in the regional innovation system.

In order to evaluate the GTV in the context of innovation cluster, the S&T policy of Gyeonggi Province was reviewed. In addition, a survey was conducted to find out the direction of GTV for the tenant companies and research institutes in GTV. Finally the conclusion was drawn to suggest the direction of the regional innovation policy in Gyeonggi province.

## 2 Concepts of Innovation Cluster

First of all, it has to be mentioned that innovation cluster is different from other similar concepts such as technopark, science park, technolopolis, technovalley and so on. Since there are many similar concepts for innovation cluster, it is necessary to compare and clarify the related concepts. Generally speaking, the science park was established to commercialize the R&D output of the university. That is why it is often located at or nearby the university campus. After the science park, similar complex came out in various names. It seems that science town refers to bigger complex than science park. Later technology park came out and it focuses more on industrial technology or applied science than science park. Then the name of valley started to be widely used after imitating the Silicon Valley. Now, people widely call those science and technology focused complexes, regardless of the size and specialization focus, as the Science and Technology Park (STP).

OECD (1999) argued that the cluster approach can be very effective in promoting industrial development. It uses the terminology innovative cluster, which can cover from production to science-oriented cluster. Yim (2000) instead started to use the innovation cluster, which mainly covers the science and technology park. The innovation cluster can be defined as a networked group of innovation actors and location(s), where the actors are creating economic and technological values by interacting, competing, and collaborating with other actors in innovation processes, which functions as the source of innovative activities for the region/nation, and has global competitiveness (Yim 2002, 2008).

In addition to OECD (1999) argument, we can say cluster approach is useful because of following reasons. First, the innovation cluster rather than whole nation is the unit of competition in the real world. The national innovation system has somewhat complex characteristics and it is often difficult to make policy to promote national innovation system itself. Second, due to the close distance among the innovation actors, the innovation cluster has various advantages in S&T knowledge production, transfer, and utilization.

Since innovation process includes not only science and technology but also the actual use in the business, the innovation cluster can be thought of the concept beyond science and technology park. Like other complex or cluster, innovation cluster has similar characteristics of system. Innovation cluster, as a sort of system, is composed of many elements; actors, processes, and cultures. The actors are normally located very closely and have organic relationship, which means they share the technological innovation process to some extent. The objects of sharing can be input, process, or output according to their needs. Often the actors share similar culture and can communicate well among. The interaction normally is about money, information, scientific or technological knowledge, business opportunities, and human resources which is shown in Fig. 1.

An innovation cluster is not only naturally formed by some factors such as regional demand and abundant resource but also artificially created by government policy. Silicon Valley is a typical example of naturally formed innovation cluster whereas Zhongghancun in Chian, Hsinchu Science Park in Taiwan, Kista region in

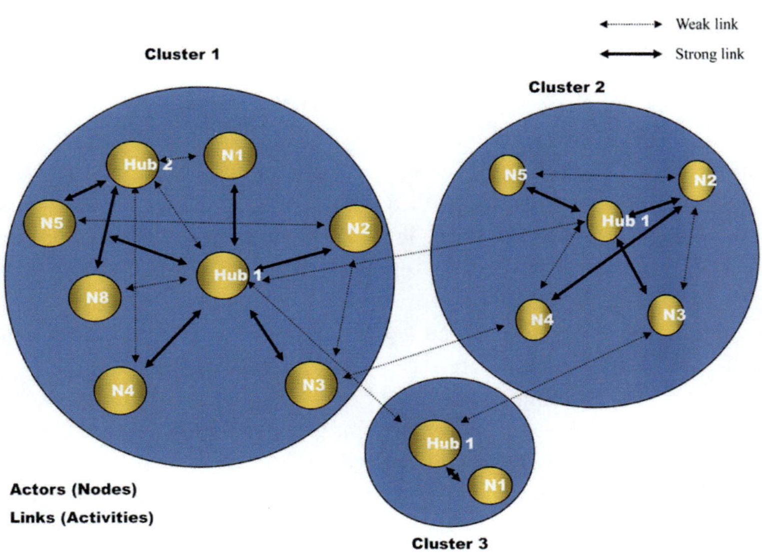


**Fig. 1** Elements of innovation cluster. *Source* Yim (2008)

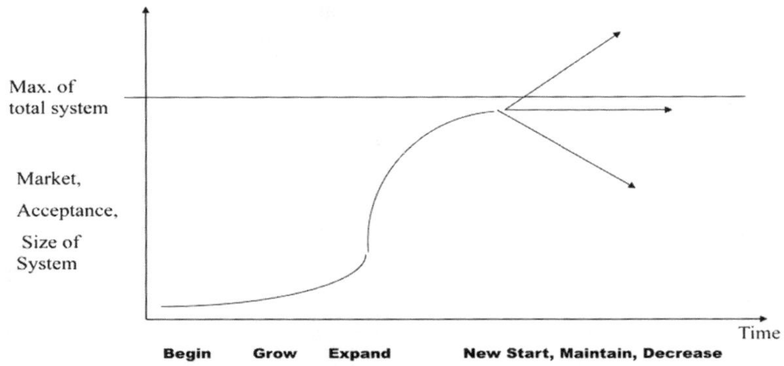


**Fig. 2** Life cycle of innovation cluster. *Source* Yim (2008)

Sweden are the ones created by government. And it is obvious that such innovation clusters are leading their national competitiveness.

The innovation cluster has a life cycle, which means it is borne but also can die away as the environment changes. The developing stage of a cluster can be classified as Pre-cluster, Emerging-cluster, Expanding-cluster, and Restructuring-cluster. First, Pre-cluster stage stands for a period that few firms have very limited links among them. There is little economic impact and Anchor companies emerge. Second, in Emerging-cluster stage, firms create links and organize among themselves to form industry associations and alliances. Third, in Expanding-cluster period, linkages and critical mass grow and the economic impact is expanded. Fourth, under Restructuring-cluster stage, there are high inter-firm links. Moreover cluster spawns new clusters and begins to adapt in changing market (Sadik 2001).

The life cycle of an innovation cluster can be drawn using time, market variables. Typical pattern of growth is that it takes a shape of exponential growth. Normally, it takes a long period of time to reach a certain tipping point for the exponential growth. However, when it reach the saturation point where the innovation cluster meet the maximum capacity of total system, it either grow again, or remain at the same level, or die away Fig. 2.

For the development paths of an innovation cluster, Yim (2002) has conceptualized two different paths, in the context of value chain process. The growth of innovation cluster can be classified into two subcategories of clustering. One is "downstream clustering" where, the high science and technology level leads and expands to production and market. The other "upstream clustering," where the strong basis of the marketing capability or production capability leads to the R&D function at the end. One example of the first type is "Silicon Valley" and that of the second type is "Dongdaemun market (Korea)" driven to subcontract production in the beginning, then with the emergence of shopping centers, synergy effects were generated from interactions between market and productive function which finally leads to the development of design (R&D) function Fig. 3.

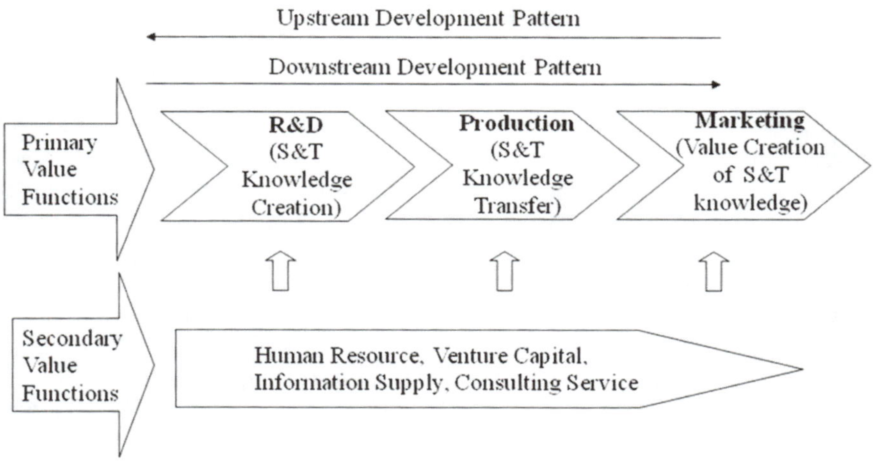


**Fig. 3** Evolution or development pattern of an innovation cluster. *Source* Yim (2002)

The life cycle and development pattern imply that innovation cluster can be successful only when certain initial conditions are met. Therefore, it is necessary to identify such factors. Generally speaking, there are many success factors for innovation cluster development. First of all, the location is important. It represents the natural conditions like land, water, weather, and transportation. Then there comes factors like excellent technological capability, high quality people, access to market, innovative culture, venture capital, and so on. In the case of government initiated cluster, the government plays important role in the development of a cluster. So, good government policy and support are often the most important factors in the initial development period Fig. 4.

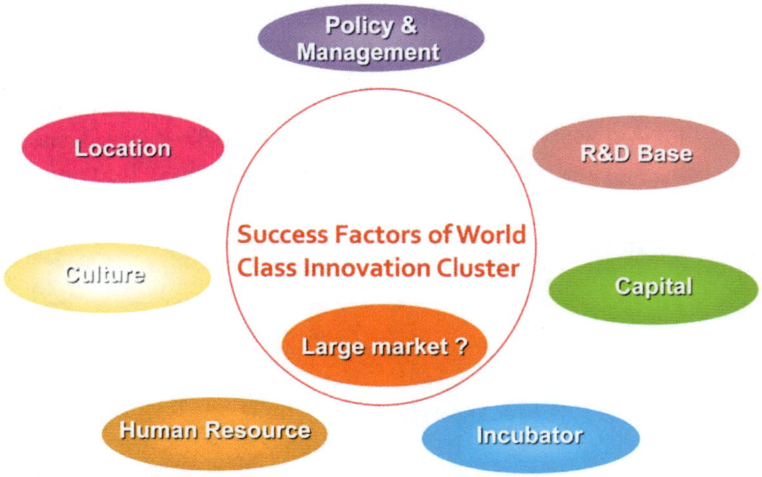


**Fig. 4** Success factors of innovation cluster. *Source* Yim (2002)

## 3 Evaluation of Gwanggyo Technovalley

Gwanggyo Technovalley (GTV) is located at Suwon City, the capital of Gyeonggi Province, which is the surrounding area of Seoul, Korea. In Gyeonggi, there are much resources available in terms of researchers, research centers, and industries. However, it is relatively neglected in terms of central government's support because of its policy, emphasizing the balanced development of nation between metropolitan area, and other regions. Recognizing lack of central government, Gyeonggi province tries to have its own regional innovation cluster. GTV has been started as the part of new Gwanggyo Region and it is expected to play a sort of innovation hub in the region Table 1.

Gwanggyo Technovalley has become a kind of innovation cluster in the structure when Advanced Institute of Convergence Technology (AICT) was opened in March 2008. Currently there are research institutes, companies, and universities at GTV campus. Korea Advanced Nano Fab Center (KANC) was established in 2008 to promote the development of the Nano-technology (NT) and related businesses in Korea. Gyeonggi Bio Center (GGBC) was established in 2007 to strengthen global competitive power and support the bio-industries industries in Gyeonggi Province. Gyeonggi Small & Medium Business Center (GSBC) was established in 2002 to provide customized supporting service to its small and medium-sized enterprises. Gyeonggi R&DB Center (GR&DBC) was established in 2007 to build knowledge network among innovation actors by attracting internal and external excellent research institutes and companies. Finally, The AICT was established jointly with the Seoul National University. It has 2 graduate programs and 9 research centers. AICT pursues collaborative and convergent researches in the application of the latest technologies in IT, BT, and NT with the objective of commercialization Table 2.

Gwanggyo Technovalley as close to Seoul, enjoys good location advantage and it can have some synergy with other innovation clusters in Gyeonggi Province. In addition, the support of regional government is quite strong. As AICT has been established, the hardware structure is almost completed.

It is too early to say which development stage the GTV is at. Judging from the interviews with the people in GTV, it seems to be approaching into an intermediate stage between 'Pre-Cluster' and 'Emerging-Cluster' in terms of development stage

**Table 1** Status of GTV

| Location | Iui-dong, Yeongtong-gu, Suwon city, Gyeonggi province |
|---|---|
| Size | 269,404 square meters |
| Construction Period | 45 months (June 2004–February 2008) |
| Investment | $ 476,250,000 |
| Tenant companies | About 230 |
| Major fields | IT, BT, NT, CT/Original technology, and technology commercialization fostering science experts |

**Table 2** Overview of 5 main institutes of GTV

| | | KANC | GGBC | GR&DBC | GSBC | AICT |
|---|---|---|---|---|---|---|
| Construction period | | 2003–2006 | 2004–2007 | 2005–2007 | 1997–2002. | 2004–2008 |
| Total building cost ($ mill.) | Total | $198 mill. | $95.8 mill. | $47.8 mill. | $85.9 mill. | $144 mill. |
| | Central gov't expenditure | $54.7 mill. | $4.7 mill. | – | $5 mill. | – |
| | Local gov't expenditure | $98.1 mill. | $91.1 mill. | $47.8 mill. | $80.9 mill. | $144 mill. |
| | etc. | $49.9 mill. | – | – | – | – |
| Size of facilities ($m^2$) | | 50,148 (16th) | 32,023 (15th) | 32,157 (7th) | 48,653 (16th) | 58,551 (16th) |
| Main field | | NT | BT | IT, BT, NT etc. | | IT, BT, NT, ET, CT |
| Function | | To provide fab service | To support joint equipment service | To provide research and business space | To support SMEs of Gyeonggi province | Education and training |
| Number of employees in each center | | 52 | 16 | 93 | 29 | 190 |
| Resident companies | Number of companies (total: 210) | 46 | 26 | 61 | 50 | 27 |
| | Number of employees (total: 2,603+) | 660 | 386 | 800 | 757 | – |
| Number of employees (Total: 2,764+) | | 712 | 402 | 800 | 850 | – |

*Source* Yim et al. (2008)

**Table 3** Problems of GTV and policy response

| Contents | Problem | Policy response |
|---|---|---|
| Master plan | H/W-oriented master plan | Comprehensive master plan including S/W sides |
| Marketing | Weak marketing & advertisement | Strengthening Global/Domestic marketing |
| Managing body | Management by government officials | Introduction of semi-public management body |
| Cluster management | H/W-oriented management for building and facilities | S/W-oriented management (network & promotion)[a] |
| Specialization | Wide and ambiguous specialization | Specialization after the review of internal capabilities |
| Critical mass | Small to reach critical mass | Need to attract more companies |

[a] Gyeonggi province officially designated the management function of Pangyo Technovalley (another innovation cluster) to the newly established Gyeonggi institute of science and technology promotion in 2012

Yim and Kim (2008), Yim (2009a), and Yim (2009b)

of an innovation cluster. Although it is being developing well, there are still some issues for the development of GTV.

First of all, the innovation cluster policy is oriented too much to the hardware building like construction of building and road. The main policy of Gyeonggi provincial government on science and technology has focused much more on the infrastructure. More than half of the science and technology budget has been spent on building up the infrastructure in Pangyo Technovalley and GTV. Although building up an infrastructure is an indispensable factor in the initial development of innovation cluster, there is no concrete policy direction or s/w plan to boot research and development activities or commercializing the technology, which are required for the successful innovation cluster. There is no master plan which stated the vision of GTV and action plan after the construction Table 3.

In order to solve those problems, we may need the following policy directions.

## 4 Conclusion and Discussion

In today's knowledge-based economy, national competitiveness largely depends on the competitiveness in science and technology. Among many factors which determine the effectiveness of national or regional innovation system, the innovation cluster is regarded as one of effective policy tools to enhance the competitiveness of science and technology in the region. There are many successful cases where the innovation clusters are leading the national or regional competitiveness in the world. However, it is also true that there are many cases, where the innovation cluster does not produce the expected result.

The GTV, as the regional government initiated innovation cluster, was completed in terms of first stage of construction and infrastructure development and shows a positive sign of growth with its location advantages. However, as it has been evaluated there are many issues and problems to be solved. First of all, the hardware-oriented policy is not enough for the innovation cluster. It needs software side policy to create technological innovation and form the network among industry, university, and research institute.

Second, the role of policy research should be emphasized. It seems that the policy research is done only prior to making the decision on the development of an innovation cluster. But the whole process of development normally takes more than a decade and constant monitoring of development and performance is a must thing to do. The lack of software side program might be explained by the fact that the policy researchers were not involved in making initial development plan.

Third, the management capabilities, including marketing function should be strengthened. Only after the building was completed, the managerial body comes in. But the marketing of an innovation cluster and attracting right tenant companies should start before the construction. With global marketing program, it can shorten the growing period of innovation cluster.

Fourth, it also needs some investment on hardware side. The people in GTV need some community space, living facilities, and easy transportation. These hardware needs also had to be included in the initial plan though.

We need more case studies to generalize the development issues and problems of regional government initiated innovation cluster. The experience of GTV gives some ideas on this matter. First, insufficient plan without proper consideration on software side program might be a typical issue in the other innovation cluster. That is partly because the regional government wants to have a visible result or does not fully understand the nature of innovation cluster. Government officials and politicians are interested mostly at hardware part of STPs like building, roads, and the software parts of innovation cluster are neglected.

Second, in many cases, little thoughts were given to the governance and management of the innovation cluster. In the beginning of policy setting, the policy researchers, community people, and the managerial body should participate.

## References

Hong, S. B., & Yim, D. S., et al. (2000). Policy Research on Overseas Emerging Innovation Clusters and Global Open Strategy. *Science and Technology Policy Institute*.

OECD (1999). Boosting Innovation: The Cluster Approach. Paris.

Sadik, R. (2001) Reaching for clusters of economic development, ED now May 2001. SRI International per Illinois Telecom Cluster website.

Yim, D. S. (2002). Innovation cluster of indian software industry: Is it evolved or developed? *Journal of Korea Technology Innovation Society*, *5*(2), 167–188.

Yim, D. S., & Kim, S. J. (2008). *Development of an innovation cluster in the region: Experience of Gwanggyo Technovalley in Korea*, UNESCO–WTA, International Workshop, October 8–11, 2008. Daejeon, Korea.

Yim, D. S. (2008). *The development and success factors of innovation cluster.* Presented at 2008 Daedeok International Conference on Innovation Cluster.
Yim, D. S. et al. (2008). The Policy Issues of Gwanggyo Technovalley and its Operational Strategy. *Gyeonggi Research Institute.*
Yim, D. S. (2009a). A plan for promoting industry. University and Research Institute, *Gyeonggi Science and Technology Center (Internal document).*
Yim, D. S. (2009b). *Regional government initiated innovation cluster: experience of Gyeonggi Province.* A paper submitted for the Korean Scientific Cooperation Network with the European Research Area.
Yim, D. S. (2012). *The role and strategy of science and technology parks for national economic development – lessons from the Korean experience.* 2012 Daedeok International Conference on Innovation Cluster.

## Websites

http://dmc.seoul.go.kr
http://www.ddinnopolis.or.kr
http://www.technopark.or.kr
http://www.mest.go.kr

# Reverse Engineered Innovation Cluster Science Parks: A Case Study for Nigeria

**Umar Buba Bindir**

**Abstract** The critical role of science, technology and innovation in evolving a vibrant and successful economy of any nation is well acknowledged. Science and Technology Parks have long been identified and applied as a veritable tool for accelerated economic development. Nigeria in recent times has conducted a series of experiments to trigger the development of its economy using technology. Although lessons on establishing S&T parks for development from many countries revealed numerous evolution options, typically implementing S&T parks have been demonstrated to be medium- to long-term projects. Furthermore, global S&T Parks development revealed that Africa is nearly totally left out in this phenomenon. To catch up, the continent requires an urgent and innovative "leap-frogging" methodology. This chapter illustrates a model to evolve viable S&T parks in Nigeria. "Reverse Engineered Innovation Clusters" from the existing networks of S&T institutions present a feasible option. Nigeria is a key country in the continent and the analysis of opportunities presented by its STI system may offer hope and relevance for the nation and for Africa.

## 1 Introduction

The critical role of science, technology and innovation in evolving a vibrant and successful economy of any nation is well acknowledged. The following facts can demonstrate this core importance of S&T for development:

*Once upon a time, not so many centuries ago, Europe was a poor continent. Then it discovered three things:*

---

U. B. Bindir (✉)
Federal Ministry of Science and Technology, Abuja, Nigeria
e-mail: ubindir@yahoo.com

D.-S. Oh and F. Phillips (eds.), *Technopolis*,
DOI: 10.1007/978-1-4471-5508-9_25, © Springer-Verlag London 2014

1. Attributes of the free market;
2. Respect and application of the rule of law; and
3. The use of science-based technology.

*Now it is rich. The same thing, with the same consequences happened in North America, and it is now happening in Asia. What about Africa? Indeed, Nigeria's position is critical.*

Nigeria in recent times has conducted a series of experiments to trigger the development of its economy using technology:

- It has attempted to transfer technologies for the development of automobile industries unsuccessfully;
- It has failed in developing an indigenous steel industry;
- Its failed attempts to mimic small-scale industrial infrastructure development are still glaringly fresh;
- Numerous failed projects on industrial estates and complexes litter the nation;
- The strategies to deploy Technology/Trade Free Zones in enhancing technology-based Foreign Direct Investment (FDI) has not been as effective as desired;
- Recently a multimillion dollar investment free zone project named Tinapa was completed and still awaits full functioning; and
- Presently, a national programme to establish and use Science and Technology Parks for development has been mounted.

Science and Technology Parks have long been identified and applied as a veritable tool for accelerated economic development. Although lessons on establishing S&T parks for development from many countries revealed numerous evolution options, typically implementing S&T parks have been demonstrated to be a medium- to long-term plan in nature. Furthermore, global S&T Parks development revealed that Africa is nearly totally left out in this phenomenon. To catch up, the continent requires an urgent and innovative "leap-frogging" methodology. This chapter illustrates a model to evolve viable S&T parks in Nigeria.

At the onset, series of indices were used to analyse data acquired through a detailed survey, to identify networks/clusters of STIs under the National S&T Parks Development Programme launched by the Federal Ministry of Science and Technology in 2006.

The programme implementation experiences so far have revealed a number of challenges including intense and costly sensitisation required to maintain the Government's understanding, keen interest and investments, resistance to the required synergy and partnerships by related sectors/institutions to minimise duplication, rivalry and enhance effectiveness of investments, weak government policy implementation on incentives such as import substitution, local content drive and consumable goods import deletion programmes, and the inability of the economy to attract non-oil/gas technology-based Foreign Direct Investments (FDIs). These suggest the need for a modified approach towards STPs to evolve in Nigeria.

A close appraisal of the programme's experiences interestingly revealed the opportunities where existing innovation clusters can be reversed engineered into viable S&T Parks. The existing cluster of STI institutions in various parts of the country presents a good foundation for strategic refocus and re-engineering to deliver viable S&T parks. Based on this, this chapter presents some potentially practical and viable S&T Parks for Nigeria.

## 2 Attributes of S&T Parks

Conventionally, S&T for development can be analysed on the basis of the interaction levels between the following three key elements:

- Science;
- Technology; and
- Innovation.

The illustration below may be useful to explain this typical thinking:

| | |
|---|---|
| A | The National Innovation System (NIS), the domain in which all activities take place |
| B | Science and Technology |
| C | Technology and Innovation |
| D | Science and Innovation |
| E | Science, Technology and Innovation |

*Domain B* typically results in the generation of new/improved/cutting-edge knowledge and methods without regard to economic benefits. This has the potential to bestow leadership upon any nation but it will not, by any means, translate automatically into tangible economic benefits.

*In domain C*, the kind of activities that would typically take place would relate to the acquisition of embodied technologies and an aggressive policy on Technology Transfer (TF) as a way to drive growth. The advantages in this domain for Africa and indeed Nigeria are limited by pace and institutional weaknesses.

The Science-Innovation link of *domain D* suggests the creation of new economically useful knowledge. With the absence of technology, it becomes a particularly difficult and nearly impracticable link, as science will seldom yield any economic benefits in the absence of technology, whether near or remote.

In *domain E*, the triad of Science, Technology and Innovation co-exists. Nigeria, and indeed every African nation, MUST aim at this domain. The joining together of Science (increasing what we know), Technology (applying what we know) and Innovation (turning our applied knowledge into economic benefits and promoting the acquisition of new knowledge through learning-by-doing and skills) is more useful than the singular contributions of any of science or technology.

For a nation to withstand competition in this era of knowledge-based economies in the globalised world there is need for it to identify its niche areas and build on it by the application of scientific methods. New technologies and industries may then be built around these areas of core competences.

## 3 The STI System in Nigeria and S&T Cluster Identification

Global distribution of S&T Parks tied to successful knowledge economies reveals the advantages of using S&T Parks for sustained economic and industrial development. Africa's current economic backwardness is clearly evident due to its weakness and inability to effectively use ST&I and its outputs to develop. Strategically, Nigeria is a key country in the continent and the analysis of opportunities presented by its STI system may indicate some hope for the future economic relevance of the nation and indeed the continent.

Elements of the growing Nigerian "formal" STI system includes 129 Universities, 125 Polytechnics, 98 Colleges of Education, over 300 R&D/policy implementation Institutions and Centres, 159 standardised Technical Colleges, 214 Vocational Schools and 1,850 Registered Open Apprenticeship Centres. The estimated STI strength of the Organised Private Sector, individual private scientists and Nigerian S&T human resources in Diaspora have generally also been assessed to be tremendous.

Despite these technology acquisition infrastructure/ingredients, clear-cut S&T Parks have not emerged in Nigeria. Recently, the Federal Government has set revolutionary goals and visions for the nation, all aimed at evolving a vibrant Nigerian economy, that can successfully compete globally. The Vision 20–2020 of the Government is aimed at raising the Nigerian economy to be among the top 20 most vibrant economies in the world by the year 2020. Indirectly, this is a resolve by the Government to revamp and modernise its Science, Technology and Innovation System to play the key role in production. This vision therefore does not leave much room for the traditional long-term evolution of S&T Parks. A creative and deliberate strategy is desired, and the "Reverse Engineered Innovation Clusters" from the existing S&T institutions presents a feasible option.

## 4 Reverse Engineered Nigerian S&T Cluster Parks

The Nigerian administrative structure simplifies the emergence of regional clusters for S&T Parks development. With 774 local governments grouped into 36 States and the Federal Capital Territory (FCT), further re-grouped into six geo-political zones, the clusters of higher institutions and research institutes on this structure have clearly indicated the opportunity of the reverse engineered S&T cluster strategy. The following illustrates the point:

(a) Universities (129)
(b) Polytechnics (125)
(c) Colleges of Education (98)

(d) Higher Education System (a + b + c)
(e) Research Institutes and Centres (over 300)
(f) The formal STI System infrastructure (d + e).

The following indices were used to analyse data acquired through a detailed survey, to identify the clusters of STIs in the six zones.

- Higher (tertiary institution);
- Course content (science and technology based);
- Number of higher degree holders (Ph.D.s) and innovators;
- Science and Technology activities (R&D results);
- Science and Technology-based infrastructure (laboratories, workshops, test facilities);
- Innovations (prototypes and products);
- Availability of infrastructure (roads, airports); and
- Proximity to local markets.

The emerging picture was used to articulate a proposal to establish S&T Parks and was presented to the Federal Government and adopted.

On this basis, the following areas have been proposed to host the government facilitated Zonal Science and Technology Parks and Technopolis:

| S/N | Zone | State | City |
|---|---|---|---|
| 1. | North East | Adamawa | Yola |
| 2. | North West | Sokoto | Sokoto |
| 3. | North Central | Benue | Makurdi |
| 4. | South East | Abia | Aba |
| 5. | South West | Osun | Ife |
| 6. | South South | Rivers | Port Harcourt |
| 7 | Federal Capital Territory | FCT | Abuja |

His Excellency, the Governor of Benue State attended the 2007 World Technopolis Association Workshop [where the present work was originally presented], as a mark of his administration's commitment to the success of the National S&T Parks programmes in Nigeria

## 5 Implementation of the Zonal STPs Programme

The Zonal S&T Parks programme commenced in 2006. The implementation process included:

- Technical surveys especially to isolate the system SWOT;
- Sensitisation programmes for the state governors and other stakeholders;
- Zonal conferences of the ST&Is in the zones;
- Identification of Parks' sectoral functions;

- Shopping for co-investing partners and anchor personalities;
- Quality space marketing and partnership with the private sector;
- Securing land and developing base infrastructure; and
- Parks' development consolidation and policy implementation.

The programme implementation experiences since 2006 has revealed a number of challenges including:

- Intense and costly sensitization required to maintain Government's keen interest and investments;
- Resistance to the required synergy and partnerships by related sectors/institutions to minimise duplication, rivalry and enhance effectiveness of investments;
- Weak government policy implementation on incentives such as import substitution, local content drive and consumable goods import deletion programmes; and
- The inability of the economy to attract non-oil/gas technology based Foreign Direct Investments (FDIs).

These suggest the need for a modified approach towards STPs to evolve in Nigeria becomes apparent.

## 6 Reverse Engineered S&T Parks

In view of the need to facilitate some quick wins, a close appraisal of the earlier identified clusters interestingly revealed areas where existing innovation centres can be re-planned and reversed engineered into clusters that would enable the immediate take off of S&T Parks. The following are only examples:

### *6.1 Sheda RIs Cluster (Agricultural S&T Park)*

Sheda is a district in the Federal Capital Territory (FCT). In this region, numerous innovation institutions exist. These include:

(a) The Sheda Science and Technology Complex (SESTCO) that houses three of the nation's advanced laboratories in Biotechnology, Physics, Mathematics and Nuclear engineering. Furthermore, the complex hosts two biotechnology spring-off companies that manufacture sickle cell anaemia drugs and HIV test kits;
(b) The Agricultural Research Stations;
(c) The University of Abuja;
(d) The National Mathematical Centre; and
(e) The National Education Research and Development Council.

## 6.2 IT Park at Lugbe

Lugbe is also another district in the Federal Capital Territory. The facilities in this area include:

(a) The headquarters of the National Space Research and Development Centre (NASRDA);
(b) The commercial office of the space command and control centre for the Nigerian Communication Satellite (NigCommSat Ltd);
(c) The headquarters of the National Biotechnology Development Agency (NABDA);
(d) The Abuja Science Village;
(e) The headquarters of the Galaxy Backbone Plc;
(f) The Abuja International Trade Fair grounds;
(g) The Abuja wonderland amusement facility; and
(h) The National Stadium complexes.
(i) Several Private Universities
(j) The Cancer Research Centre

## 6.3 Multipurpose Park at Idu

Idu is yet another district in the Federal Capital Territory. Relevant facilities in this cluster include:

(a) The headquarters of the National Agency for Science and Engineering Infrastructure (NASENI);
(b) The National Institute for Pharmaceutical Research and Development (NIPRD);
(c) The Headquarters off of the National Institute for Science Laboratory Technologists;
(d) The Idu-Karimu commercial market; and
(e) Numerous commercial S&T based companies.

## 6.4 Multipurpose Park in Zaria

Zaria is located in Kaduna State in the North Western part of Nigeria. The relevant facilities in the cluster include:

(a) The Ahmadu Bello University (ABU);
(b) The National Research Institute for Chemical Technology (NARICT);
(c) The Chemical and Leather Technology College (CHELTECH); and
(d) The National Aviation Training College.

## 6.5 University S&T Park in Ife

This cluster rotates round the institutions within the University complex. They include:

(a) The Obafemi Awolowo University (OAU);
(b) The National Centre for Technology Management (NACETEM);
(c) The Centre for Energy Research and Development (CERD);
(d) The Space Technology Research Centre and
(e) The COPINE.

## 6.6 Steel Engineering Park, Lokoja

This cluster rotates round the Ajaokuta Steel Complex facilities which include:

(a) The Steel complex;
(b) The Power generation complex; and
(c) The Engineering Training complex.

## 6.7 Oil and Gas Park in Port-Harcourt

This cluster will accommodate:

(a) The Eleme Petrochemical company;
(b) The Petroleum Refinery
(c) The Fertiliser Company
(d) The Oil and Gas Free Trade Zone
(e) The Oil Jetty

## 6.8 The Benue State Zonal S&T Parks Limited

Benue State is one of the 36 state governments in Nigeria located in the north central geo-political zone. The state government is facilitating the establishment of the zonal S&T Park located in the city of Makurdi, the state capital. To quicken the establishment process, the state government is active in facilitating the inflow of technologies and innovations from all the higher institutions in the zone. Furthermore, the governor of Benue State is providing the leadership required to encourage the other five governors in the zone to recognise and invest in the S&T Park establishment and operation. Due to this involvement of the governors and the potential support available from the states, the private sector confidence is also being increasingly boosted.

## 7 The Abuja Technopolis

Abuja, the Nigerian Capital City avails the nation the first opportunity to establish a Science City. The development of the three science and technology parks in Abuja described above will facilitate the re-engineering process and the achievement of the Technopolis objective.

## 8 The Re-Engineering Process

The objective of the re-engineering process is to re-plan and modify the productive functions of existing innovation facilities into Science and Technology Parks where the interactions of the mandates and outputs of the individual establishments is converged to deliver outputs that are more tangible and effective in improving the socio-economic system of the immediate community and indeed the nation at large.

In each case, detailed studies of the institution are being conducted to reveal their:

- Capacities;
- Capabilities;
- Outputs;
- Impacts;
- Future plans;
- Partnerships; and
- Projects/programmes.

These are then processed further to identify clearly existing overlaps, potential overlaps and hidden innovation opportunities.

The total system would naturally evolve into the desired S&T Park with clear mandates and processes, and all necessary elements such as additional research facilities, common engineering workshops, innovation products showrooms, technology incubators, technology business spin-offs, and quality service areas/facilities.

Policies and statutory functions are then proposed to fit with proposed parks and technopolis attributes.

## 9 Conclusions

The application of proper mixes and interactions between Science, Technology and Innovation is the main tool for sustainable and dynamic development of any nation. Science and Technology Parks have been demonstrated as the proper product of this mix for all countries that have successfully developed. Although broad international studies have revealed that the evolution of S&T Parks is a long-term strategy, in most cases estimated to mature in an average of 20–30 years, preliminary studies under the Nigerian Government S&T development programme have indicated that existing cluster of STI institutions in various parts of the country can be refocused to deliver viable S&T parks. The primary sample clusters presented in this chapter included:

(a) Sheda RIs Cluster that could evolve into an Agricultural S&T Park;
(b) IT Park at Lugbe;
(c) Multipurpose Park at Idu;
(d) Multipurpose Park in Zaria;
(e) University S&T Park in Ife;
(f) Steel Engineering Park, Lokoja;
(g) Oil and Gas Park in Port-Harcourt;
(h) The Benue State Zonal S&T Parks Limited and
(i) The Abuja Science City.

The main gainful product of these innovations is the evolution of Abuja Technopolis, the first of its kind in sub-saharan Africa.

This technique might be a useful tool for adoption in many developing countries particularly in Africa.

# ICT Cluster Development Strategy for High-Tech Industry Park in Mongolia

**Buyanjargal Otgonchimeg and Shagjjamba Oyunchimeg**

**Abstract** In this chapter, we review first a brief introduction on Mongolia and its development toward the knowledge-based economy. Next, we explain the Government initiative to establish a high-tech industrial park and the current situation of high-tech industry development, especially in the ICT sector and conclude the chapter with initiatives for ICT cluster development for High-tech Industry Park in Mongolia. In 2008, the Parliament of Mongolia (Resolution. No. 12) had endorsed the Millennium Development Goals (MDG)-based Comprehensive National Development Strategy (NDS) of Mongolia. It defines in a comprehensive manner that its policy aimed at promoting human development in Mongolia, in humane, civil, and democratic society, and developing intensively the country's economy, society, science, technology, culture, and civilization in strict compliance with global and regional development trends. Around the NDS, Government Action Plan 2008–2012, the list of 26 large projects approved by Government decree and ICT vision up to 2021 (draft) provides clear strategies toward the ICT sector development for the high-tech industry.

## 1 Brief Introduction on Mongolia

Mongolia is a landlocked country in northern Asia, strategically located between Russia and the People's Republic of China. Economic activity in Mongolia has traditionally been based on herding and agriculture. Besides agriculture (21.2 % of

B. Otgonchimeg (✉)
Space and Radio Communications Department, Information, Communications Technology and Post Authority, Ulaanbaatar, Mongolia
e-mail: otgonchimeg@itpta.gov.mn

S. Oyunchimeg
Policy and Planning Department, Information, Communications Technology and Post Authority, Ulaanbaatar, Mongolia
e-mail: oyunchimeg@ictpa.gov.mn

D.-S. Oh and F. Phillips (eds.), *Technopolis*, 
DOI: 10.1007/978-1-4471-5508-9_26, 

GDP), dominant industries in the composition of GDP are mining (22.7 % of GDP), trade, service, and communication (Jigjid 2001).

In 1991, Mongolia started a rapid transition from central planning to a market-oriented economy. Mongolian industry sectors comprise many traditional forms of industries, with food production and textiles being the main pillars of the industry. Mongolia's industry sector suffered a continuous decline throughout the 1990s after the country transitioned into a market economy. At the same time, progress was made in establishing the foundations for a market economy. Reforms included price and trade liberalization, the reduction of lending to state owned enterprises, the creation of a commercial banking system, and a massive program of privatization which saw the proportion of the economy in private hands rise from 4 % to over 70 % between 1990 and 2000 (Jigjid 2001).

## 2 Development Toward the Knowledge-Based Economy

At the present moment, when it is obvious that Mongolia focuses the new wave of its economic development, government agencies have been implementing a series of concrete actions to build up the foundation and ensure successful development of a knowledge-based economy. It is inevitably important to create concentration of knowledge and intellect, and build up an intellectual mine through studying the best practices of other countries in order to ensure the implementation of long-term objectives to become a country with knowledge-based economy.

### *2.1 National Development Strategy*[1]

Effective baselines for developing technology parks from national and ICT sector level are as follows:

#### 2.1.1 MDG-Based NDS for 2008–2015

The MDG-based Comprehensive NDS of Mongolia shall be implemented within the framework of the following priority areas; to achieve the MDG and provide for an all-round development of Mongolian people, to intensively develop export-oriented, private sector-led, high-technology-driven manufacturing and services, with particular focus on information, communication development, promoting bio and nanotechnology, transit transportation, logistics, financial mediation services, deeper processing of agricultural products, and create a sustainable, knowledge-

[1] The Government of Mongolia, http://www.legalinfo.mn.

based economy, to exploit mineral deposits of strategic importance, generate and accumulate savings, ensure intensive and high economic growth, and develop a modern processing industry. Moreover, policy states that Information and communication technology (ICT) is viewed as the main accelerator of Mongolia's economic and social development in the twenty-first century.

#### 2.1.2 State Policy About High-Tech Industry

In order to succeed in these strategies, the Parliament of Mongolia had given great priority on the High-tech industries and ratified the State policy about High-tech industries by its 34th resolution in 2010. Within this policy, nanotechnology, biotechnology, information technology, and electronic sectors are regarded as high-tech industries. Policy states the directions of the activities associated with the development of infrastructure, including establishment of a High-tech Industry Park, and also human resource development and encouragement of the high-tech product sales.

#### 2.1.3 Program on Developing Industry of High-Technologies

Pursuing the State Policy on the High-Tech Industry, the Government of Mongolia approved a program on developing an industry of high-technologies at the end of 2011. It is expected that the program will create legal and regulatory environments to support sustainable development of technological industries in the private sector. It also has great importance in increasing the amount of high-tech industries of the total industries contributing to the Gross Domestic Product (GDP) by creating a proper structure of financing and investment. This program states that the High-tech industry should possess more than 1.5 % of the GDP.

### *2.2 Key Organizations for National High-Tech Industry Development*

There are several organizations related to the development of an Information and Communications High-Tech Industry.

#### 2.2.1 National Development and Innovation Committee[2]

In Mongolia, the National Development and Innovation Committee (NDIC), which is currently restructured into the Ministry of Economy and Development,

[2] National Development and Innovation Committee http://www.ndic.gov.mn/.

executes strategic objectives like elaborating economic and social development policy, long- and mid-term strategies and provide guidance, developing innovation policy, long-term strategy and provide guidance, forming sector development, investment coordination policy, long- and mid-term strategy, guideline.

NDIC identified major investment projects, in the framework of the long- and medium-term, socio-economic, development policies of Mongolia, and submitted them to the Cabinet. In October 2009, the Cabinet approved a list of 26 large projects by Government Decree No. 320. These 26 large investment projects are top-priority, government-initiated projects for mineral resources processing, industrialization, those meeting infrastructure development requirements, or that otherwise hold economic or social significance. One of these projects are the "High-Tech Industrial Park" project, as given a priority Developing Intensified Agriculture and Implementing Industrialization Policy. Moreover, the "Industrial Training Complex for Information Technology" project and "Students town" projects were reflected.

### 2.2.2 Information Communications Technology and Post Authority

Information Communications Technology and Post Authority (ICPTA), the policymaking authority of Mongolia is mandated under the Prime Minister to provide primary areas of work related to the development of laws, regulations, and development policies related to information technology, post, broadcasting, telecommunications and technology development issues, development of unified registration system, regulation and implementation of policies, programs, and plans, coordination, monitoring, and evaluation.

ICTPA became a member of World Technopolis Association in 2010 and is carrying the responsibility to establish and develop Information, communications high-tech industry.

### 2.2.3 The Mongolian Academy of Science

Mongolia's first center of modern sciences came into being in 1921 when the Government of newly independent Mongolia issued a resolution declaring the establishment of "Institute of Literature and Scripts," which was later upgraded to "Institute of Sciences" and then "Institute of Sciences and Higher Education." In 1961, it was finally reorganized as "the Mongolian Academy of Sciences" (MAS). At present, under its direct supervision the MAS operates 20 research institutes and centers, including Informatics Institute which is to be a basic research facility for the future high-tech industry development.[3]

[3] Mongolian Academy of Science, http://www.mas.ac.mn/.

## 2.3 *Current Situation of High-Tech Industries*

We will briefly explain about the current development, implying the basic premises of these high-tech industry parks.

### 2.3.1 Nanotechnology and Biotechnology

"Government action plan" indicates objectives to pursue the policy to make science and technology a top-priority sector; to provide the state support in introducing and acclimatizing new technology, and to set up not less than two research institutes of biotechnology, nanotechnology, and veterinarian service equipped with world medium standards.

In Mongolia, there are three main state universities, including National University of Mongolia (NUM), Mongolian University of Science and Technology (MUST). The Mongolian State University of Agriculture prepares the human resource in the field of high-tech technologies.

NUM is the university with the highest level of research outputs in the country. The university has 22 research centers and three field stations. The Laboratory of Nanotechnology was established in 2008 at the university and focuses on education, measurements, materials, and technologies are attracting worldwide attention for nanoscience and nanotechnology Researchers of Nanotechnology Group at the NUM studied on high-efficiency organic solar cells by optimal energy levels and offsets of organic materials for the semiconductor. The present research team is implementing more than 20 projects in the frame of the National Program on Nanotechnology R&D funding during 2008–2015.[4]

The Pilot plant is established in Boroo Gold Valley in Selenge Province, 150 km north of Ulaanbataar. Productivity, trade, and market research is done by the US company 'UBIC' on the Drug anti-Cardiovascular disease and Cancer that will be produced with over 100 molecular technologies for diagnosing, neutrocetic, cosmetics, and drug manufacture. Mongolia has become a Member of International Association of Science Parks (IASP).

Main production will be focused on bio-tech drug production anti-cancer and cardiovascular disease, bio-tech blood plasma production, bio-tech egg production, bio-tech milk production. Not only this plant, but also private companies such as "Mon Chemical" LLC, "Monos" LLC operate in the bio-chemical field.

[4] The National University of Mongolia, http://www.num.edu.mn/.

**Table 1** Core indicators of the ICT sector

| | 2005 | 2006 | 2007 | 2008 | 2009 | 2010 |
|---|---|---|---|---|---|---|
| Total revenues of ICT sector (bln Tugrug) | 140.4 | 181.7 | 283.5 | 365.2 | 449.2 | 471 |
| Investment (bln Tugrug) | 34.5 | 38 | 54 | 70.6 | 73 | 89.9 |
| Tax contribution to state budget (bln Tugrug) | 54.3 | 71.4 | 108.7 | 74.4 | 76.5 | 81.8 |

### 2.3.2 Information and Communications Technology[5]

ICT serves as the main infrastructure of a modern knowledge-based economy.

ICT sector is one of the fast and intensively developing sectors and important factor that demonstrates the development of ICT in Mongolia is the contribution of the ICT development to the GDP. The ICT development of Mongolia has advanced tremendously for the past several years and made considerable progress in developing its information and communications infrastructure, particularly modern basic service and cellular services (Table 1). In order to achieve the highest sustainable economic growth, ICT should be developed and maintained in all available economic sectors and thus it could contribute to the interrelated growths of each sector.

In terms of composition of ICT sector revenues, about 74 % of total revenues are generated by mobile services and the remaining 26 % are contributed by fixed telecommunications network, VoIP, Internet, Cable TV, broadcasting, and others.

Currently, there are over 500 enterprises providing communications services with 684 special licenses and comprised around 10 % of the GDP. The number of software and development companies has reached to 100, which currently run their business in the field of IT consulting, accounting software, web design, outsourcing, and IT training.

There are educational institutions such as the Computer Science and Management School and Information, Communication and Technology School at the MUST, The School of Mathematics and Information Technology at the NUM and "Khuree," a private institute which are engaged in preparing specialists in engineer, technique, and management in the sector and activities for research and study apart from as well as the Informatics Center at the Academy of Science which runs specialized research activities.

## 3 Initiative for ICT Cluster Development in Mongolia

Mongolia having low population, insufficient market, and has landlocked location experiences a number of challenges in the process of developing an export-oriented economy while an Information technology industry is such a unique industry that

---

[5] Information, Communication Technology, and Post Authority: White Paper 2011, Ulaanbaatar (2011).

produces intangible products to shave no correlation in terms of distance or time. For, promotion of an information technology industry as a knowledge-based industry has huge potentials to serve as the remarkable stimulus for the economy of our country.

As a baseline, the provision 35 of ICT Vision 2010 indicates "Set up knowledge and education-based high-tech centers in Ulaanbaatar and the centers of the socio-economic development regions." Following the ICT Vision 2010, ICTPA has developed new policy document "ICT vision up to 2021," which is going to be approved by the parliament of Mongolia, have reflected more precise objectives toward the high-technology industry development. Moreover, E-Mongolia National Program provides several objectives toward ICT enabled economic growth, such as accelerating and promoting ICT investment and application in the economic sectors, fostering and expanding ICT application in enterprises and Small and Medium Enterprises (SME), developing domestic markets, position local ICT Companies in external markets (ICT Mongolia 2011).

## *3.1 Legal Rationale*

During the meeting with representatives of information, communications and technology sector on January 19, 2010, the Prime Minister of Mongolia has given directives to work toward building an ICT cluster to be similar format with Silicon Valley, which would support development of information technologies and other high-technological sectors in Mongolia.

Pursuing for those objectives, in June 2010, Government of Mongolia approved the resolution to develop universities and colleges in a form of a university campus and decision-related instructions were given to the Ministry of Education, Culture and Science, the Information, Post, Communications and Technology Authority, and the NDIC.

Accordingly, ICT cluster shall be developed in line with the university campus. Having location close to a university campus, i.e., to academic research and educational sector, will create the realistic benefits to coordinate activities of industry, research and development and educational sector, and to compete in international markets.

Furthermore, Parliament of Mongolia had approved the Law on Innovation in 2012 enabling legal environment for building the innovation system in Mongolia, where have been defined legal status and operation of the science park.

## *3.2 Human Resource (JTIS LLC 2011)*

All these legal rationales shows that we are just in an initial stage of developing ICT cluster in Mongolia. As mentioned before, resolution No. 320 of the Government of Mongolia issued in 2009 provides that the "Industrial Training

Complex for Information Technology" project is given in a high priority. We assume that we need to encourage human resource development of the ICT cluster, while infrastructure of the University campus is being decided and built. Also "The directives on developing Information technology outsourcing to be adhered until the year of 2015" issued by the Information, Communications and Technology Authority indicate that highly skilled professional human resources to work in outsourcing will be developed, and competitiveness of information technology companies to compete in international markets will be increased.

Following these objectives, we have entered to an agreement to cooperate with the Government of India to establish "IT & Outsourcing training center" with the purpose to train human resources capable of working in information technology outsourcing markets.

There are 44 ICT educational institutions in Mongolia training ICT professionals, out of which 17 are public institutions, and the remaining are private institutions. There are over 6,000 students studying in those institutions, specializing in 42 different areas including software engineering, network administration, information systems management, hardware engineering, telecommunications engineering, postal services, electronics engineering, optic communications, television and radio technology, satellite and wireless communications, information technology, etc. In 2010–2011 academic year, 5,871 students are studying at Bachelor's, 77 students are studying at Master's and 12 students at Doctorate level at high education institutions of Mongolia.

MUST and NUM play a major role in producing well-educated IT professionals.

There are over 7,000 people working in telecommunications, mobile communications services, Internet services, software, hardware and consulting services companies of Mongolia. Recently, The ICTPA and CRC have established a joint working group to formulate "National program on ICT Human resource development."

## *3.3 IT Park in Mongolia*

The National Information Technology Park (NITP) has been established following decree No. 107 of 2002 and officially started operation under the ICTPA since August 2002.

The main objectives of the NITP are to support the development of information technology in Mongolia, and to serve as the incubator for turning new and unique ideas of venture businesses into competitive products and services. The NITP provides incubation services for start-up companies, renting premises for exhibitions, lectures, and provides high-speed Internet connection for companies and organizations located at the premises of the NITP. Since its establishment, the

NITP has provided incubation services for about 30 software development companies, groups, and teams.[6]

Although we have a NITP, area of the operation is quiet limited as it is just around the IT service not covering the other branches of ICT sector. Therefore in order to facilitate goal to develop ICT cluster in Mongolia, we have a long way ahead.

## *3.4 ICT Cluster Development Strategy for High-Tech Industry Park*

As there are three main organizations: Ministry of Education, Culture, and Science (MESC), the Information, Post, Communications and Technology Authority, and the NDIC, following objectives were defined, from which should become clear future development strategy for high-tech industry park (Fig. 1).

Through collaboration, businesses and the public sectors can create a more dynamic local economy by ensuring key building blocks are in place. Access to a skilled workforce, quality universities, good locations, and investment capital is just one example of how this can be achieved. Individual companies contribute to the clustering effect by their relationship with suppliers and customers and by *using* their skills in networking and building alliances with other companies.

ICTPA has started the several initiatives of the establishment of "Information and Communication High-tech innovation cluster in Mongolia" with reference from international organizations and development trends of science and technology clusters, which are the holding preliminary meetings with science park experts and conducting trainings, workshops, and conferences. Furthermore, a pre-feasibility study on "High-tech innovation cluster for ICT in Mongolia" has been conducted by a local research team, since August 2011.

It became clear that the ICT industry should be developed in line with University campus of Science and technology. What is next step? In which field we should focus on? These questions will be cleared out from the result of the feasibility study to be completed in 2013.

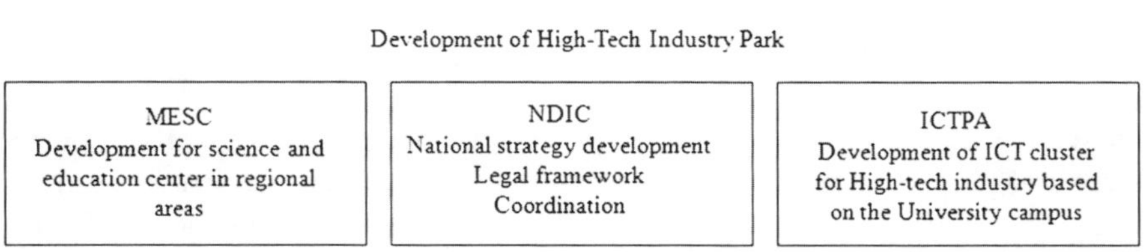


**Fig. 1** Roles of the government organizations for the development of high-tech industry park

[6] National Information Technology Park, http://www.nitp.mn/.

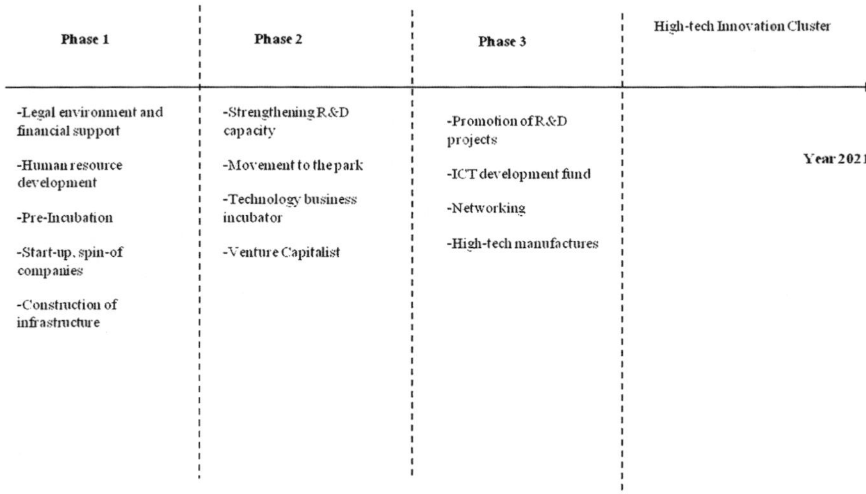


**Fig. 2** Development stages for information communications high-tech innovation cluster

Today, we expect the future development of the Information and Communications High-Tech Innovation Cluster as shown below (Fig. 2) (JTIS LLC 2011).

Up to 2021, development of the Information, Communications High-Tech Innovation Cluster can be divided into the three phases, where the first phase is more focused on Government initiatives to support construction work, human resource development, and pre-incubation. Second stage is the phase of strengthening R&D capacity, creating venture firms, developing technology business incubation. Finally, the phase 3 will be the stage of fostering the R&D growth, stabilizing financial sources, creating network between stakeholders.

# 4 Conclusion

While recognizing a benefit from successful development of knowledge-based economy, the Government of Mongolia has been taking many concrete actions to develop high-tech industries, especially for enabling legal environment. Pursuing for the state policy for developing high-tech industry, ICT sector is regarded as a high-tech industry, thus has been given a high priority, as this sector is being developed tremendously for past decades. In this circumstance, it can be seen that in a current situation, we basically established legal environment and have a certain geographic location to build ICT cluster for High-Tech industry in line with university campus. Pre-feasibility study to establish ICT cluster has been performed by local experts jointly with international organization and moreover, it is expected to have a more definite roadmap from feasibility study, which is now being under the implementation stage.

Further, we shall face big challenges in order to properly establish ICT cluster for High-Tech Industry Park. Therefore, it is crucial to have strong leadership in government level, to improve legal environment, to prepare highly trained human resources, to bring the operation of current facilities like IT Park in a new level, and most importantly to have more financial support from the government side.

## References


Information, Communication Technology and Post Authority. (2011). Policy Research on ICT Vision up to 2021. *Ulaanbaatar, p. 57–58*(92), 209–220.

Jigjid, B. (2001). *Information infrastructure and human resource development*. Ulaanbaatar http://www.ncc.gov.ph/afit/keynote/2nd/JIGJID.pdf. Accessed November 27, 2013.

Jtis, L. L. C. (2011). *Pre-feasibility study report on establishing information* (pp. 260–290). Ulaanbaatar: Communications High-Tech Innovation Cluster.


## Websites


The Government of Mongolia, http://www.legalinfo.mn
Mongolian Academy of Science, http://www.mas.ac.mn/
National Development and Innovation Committee http://www.ndic.gov.mn/
National Information Technology Park, http://www.nitp.mn/
The National University o Mongolia, http://www.num.edu.mn/

# Science and Technology Park "Technopark Gliwice" Ltd Company

**Jan Kosmol**

**Abstract** In the Silesian region, the need for applications of scientific research results to the economy seems particularly important. The Gliwice Technopark helps Silesia to achieve that objective. This chapter describes the Park's infrastructure and financing, and its challenges that stem from history of established industries and from demographics and human resources. In particular, the mentality of young people in Silesia region must be changed if we would reach a technical and economical level like that of the USA, Western Europe, and Japan.

## 1 Technopark as an Ltd Company

Technopark Gliwice was set up as an Ltd Company on April 27, 2004, four days before Polish entrance to the European Union. There were three founders:

- City Gliwice,
- Silesian University of Technology of Gliwice,
- Special Economic Zone of Katowice KSSE signed the Ltd Company contract.

Our entrance to EU was very important for the point of view of Technopark Gliwice because we were able to get money from the Structural Fund SPO 3.1 and build the infrastructure of our park. Total cost of the investment was about PLN 24 million (about €6 million). After 3 years the project was finished and at June 30, 2008 the ceremony of opening took place in Gliwice.

J. Kosmol (✉)
Technopark Gliwice, Ltd Comp, Gliwice, Poland
e-mail: jkosmol@polsl.pl

D.-S. Oh and F. Phillips (eds.), *Technopolis*,
DOI: 10.1007/978-1-4471-5508-9_27, 

Actually,

- Factory fund: 8.020.000,00 PLN (about €2 million)
- Quality System of Management ISO 9001:2001
- IASP Full Member.

## 2 Mission and Aim of Technopark Gliwice

In Poland, like in other less developed countries there exists an essential development brake caused by lack of proven methods of implementing results of scientific research into the economy. In the Silesian region, the need for applications of scientific research results to the economy seems particularly important. This results from the fact that region has been developed exactly in the opposite direction, i.e., in the direction of heavy industries mining and metallurgy. In the outcome, the delay in "High-Tech" seems to be greater than in other parts of Poland. In addition, the Silesian region belongs to the area of a very big (one of the biggest in Poland) research and development potential, i.e., a large number of universities, science and research institutes, research and development centers, and a big concentration of highly qualified technical staff (7 universities, 5 science and research institutes, 13 research and development institutes) which so far have worked for the heavy industries. It is necessary to create conditions which would enable to direct that massive intellectual potential into proinnovative conditions and incubation. The Gliwice Technopark is one of the links which enables to achieve that objective.

The need for innovative technologies is our "to be or not to be" of the post-industrial Silesian region, and the world practice has shown that it is predominantly realized by small- or medium-sized businesses. The feasibility study has shown that the Gliwice region is one of the most favorable in terms of business climate, which is an important factor for potential investors and for opening of new businesses. The system promoting innovative ideas and creation of new businesses, predominantly by young people has not been explained so far.

The creation of the "Technopark Gliwice" results as an institution which in the framework of its statutory tasks offers information on new technological achievements and access to them. The potential beneficiaries of those technologies (over 4000 per annum) know where to search for new technologies and where to realize their innovative ideas, as well as how to do it.

Generally, we support three actions:

- Incubation of new technology companies,
- Transfer of innovative technologies to SME,
- Promotion of Enterprise.

Transfer of innovative technologies and incubation processes in Technopark Gliwice may be shown in Fig. 1. There are two ways of technology transfer:

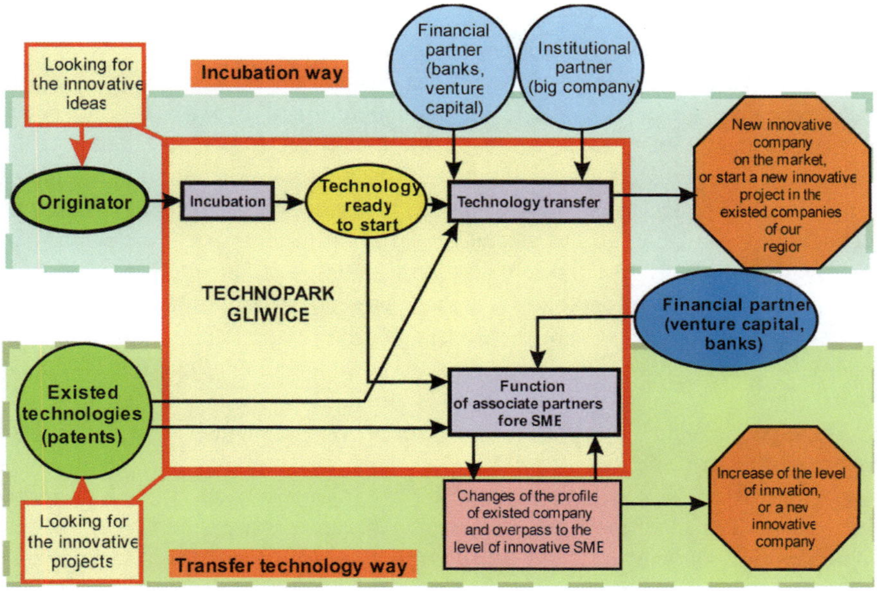


**Fig. 1** Incubation and technology transfers in Technopark Gliwice

typical way, applying existed patents or incubation way. We try to apply both methods.

The accession of Poland to the European Union has opened several new possibilities of solving difficult restructuring problems. Such a possibility offers the assistance funds, called the structural funds, for example the framework of sector operation program: The increase of (SPO-WKP) priority1 "The Development of companies' competitive behavior and the increase of innovation through strengthening the business environment to achieve step 1.3." The creation of favorable conditions for the development of companies'—i.e., provides better conditions for running the business. It was possible to create technical infrastructure in close vicinity of the Silesian Technical University. The Technopark Gliwce is a science and technology park located in the area of 8.631 $m^2$ (I stage) in the very center of the Silesian Technical University academic township in Gliwice. The strategic project objective is to create a modern investment undertaking in line with the UE standards and meeting the expectation of investors and young people who graduate from technical studies and decide to start a business in the business to innovation technology field of innovative technologies. The investment undertaking enables effective and socially favorable running of business and the exploitation of immense intellectual potential existing in the Silesian society.

To achieve that objective, the first step to do was site planning of the plot together with the existing steel structure (skeleton) on which the Technopark premises is built. The existing steel-structure was a part of the Silesian Technical

University technological engine room, the erection of which was postponed, because of financial problems at the end of the 1990.

To achieve our objective, in addition to erection of the Technopark premises and developing the adjoining land for the statutory activity, it was necessary to prepare databases on technologies available in the Technopark on "High Tech" experts. At first prepared was built a base of the technologies currently offered by the Silesian Technical University scientists. It is successively complemented with technologies available in science and research institutions of the region. The interested people are able to freely access those bases and also to formulate their needs concerning the technologies searched. The elaboration of the experts base is at first prepared by scientist of the Silesian Technical University, which enables making contacts between those searching for new technologies and those who can offer them.

The first detailed objective of the project was to intensify the technology transfer from the Silesian Technical University and research and development units from the Silesian region to small- and medium-sized enterprises by offering them favorable conditions in the Technopark area (price-attractive premises, administrative services, business consulting), assistance in access to technologies and laboratories, finding scientists supporting technology applications in small-and medium-sized enterprises, as well as the assistance in obtaining financial resources from the European Union for small- and medium-sized enterprises.

The second detailed objective was to activate incubation and creation of new innovative enterprises. The potential candidates are: 3,000 graduates from the Silesian Technical University, 2,000 research workers, also from the vicinity of the Technopark (5 science and development centers). The innovative activity covers the current development activities in the fields of: electronics, technologies connected with protective coatings technologies connected with medicine, biotechnologies, pro-ecologic technologies, as well as information technologies and teleinformatics.

The third detailed objective of the project was to create such business activity conditions which enable to invite worldwide companies representing "High-Tech" and promoting them in the Technopark area. It should be stressed that the notion of Technopark Gliwice will in the future be a certain symbiosis of several subjects operating in the area of the region and carrying out innovative activity (ARL, KSSE, technology transfer centers, e.g., in the KOMAG-Gliwice).

The fourth detailed objective was to diversify business activity in the Silesian region covered by the State programme for mining industry restructuring. The Gliwice colemine has been closed down with all the consequences. Creation of the Technopark enables business activity in the area, the more, that the area of closed-down mine is in the distance of 2 km from the Technopark seat. This is connected with the "New Gliwice" project financed by the PHARE fund and implemented by the Gliwice borough.

The additional social effect of the project was to assist the development of new jobs for over 3,000 graduates from the Silesian Technical University. This enables practical exploitation of ideas and technologies which arise every year in the form of 3,000 diplomas and ca. 200 Ph.D dissertations. It is assumed that some of those dissertations will be commercialized by means of the Technopark by placing relevant information on those dissertations in technology bases available in the Technopark and on "High-Tech" specialists.

The additional social effect of the project is supporting mainly small and medium-sized businesses in establishing collaboration with numerous institutions of the sort in UE. Within the co-operation with numerous institutions of the sort in Europe, the Technopark is able to search for co-operation with foreign the "High-Tech" companies. Such companies will be able to establish their agencies in the Technopark and co-operate with the Silesian ones. The Technopark is able to help the small and medium-sized companies in access to the UE financial assistance by developing CRAFT and other projects.

The location of the Technopark in Gliwice is justified by the fact that in the town there is an Advanced Technology Center, which concentrates over 20 research and development institutions in the region (7 universities, 5 scientific institutes, and a couple of research and development institutes) co-ordinated by the Silesian Technical University. It is obvious that every Advanced Technology Centre should have access to a science and Technology Park, where the advanced technologies may by commercialized. The location of the Technopark in Gliwice is substantiated by favorable geographical location and a dynamic expansion of transport infrastructure (A4 Highway).

The Technopark will also be an important component of very large industrial park in the Silesian region, i.e., the Katowice Special Economy Zone (KSSE S.A. which is the Technopark shareholder).

There is expectation of remarkable "suction" of innovative technologies from big companies in the Zone, directed to small innovative business acting as co-operants or suppliers. Such innovative business may be present in the Technopark in response to such needs.

The Gliwice Technopark will also be an important component of the active technology commercialization in the Silesian region. Such a net has become a subject of the whole Silesian region, and it has to be built under the supervision of the Gliwice borough. Involved in the creation of the net will be representatives of the biggest innovation centers in Europe, i.e., from Aachen (Germany), Leuwen (Belgium), and WDA (Wales).

The completion of the Technopark conception has an important impact on the role and importance of that institution in the Silesian region. It is indicated that a possibility of effective technology transfer from a university to small and medium-sized enterprises and effective incubation, particularly in the scope of innovative technologies. This will also contribute to more intense collaboration with similar institutions in Western Europe, when the conception of science and Technology Park has been approved in practice.

Figure 2 shows the location of Technopark Gliwice in the Silesian region market.

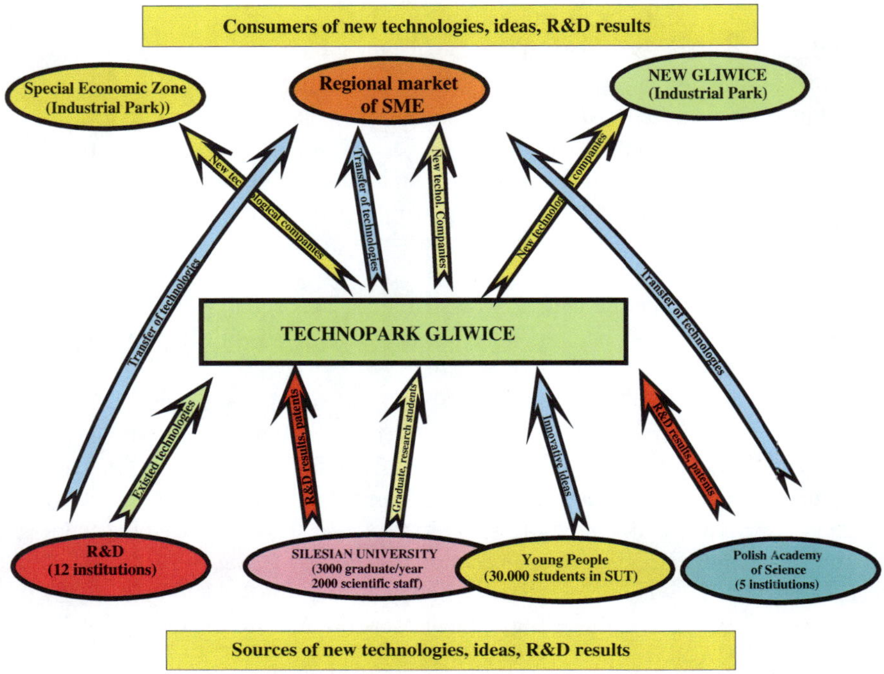


**Fig. 2** Sources and consumers of new technologies in Silesia region

# 3 Forms of Technology Transfer in Technopark Gliwice

We try to use several different methods of technology transfer like:

- Manufacturing based on specialized equipments supported prototyping,
- Project and verification services (CAD/CAM, FEM),
- Creating of innovative technology database,
- Creating of experts database,
- Creating of companies SME database,
- Offering specialized trainings,
- Organizing meetings with *z* businessmen,
- Organizing Days (Fairs) of science and industry.

## *3.1 Renting and Manufacturing Based on Specialized Equipments Supported Prototyping*

Building our Technopark we assumed that it will works as a center, supporting SME in the field of prototyping. What does it means? New (innovative)

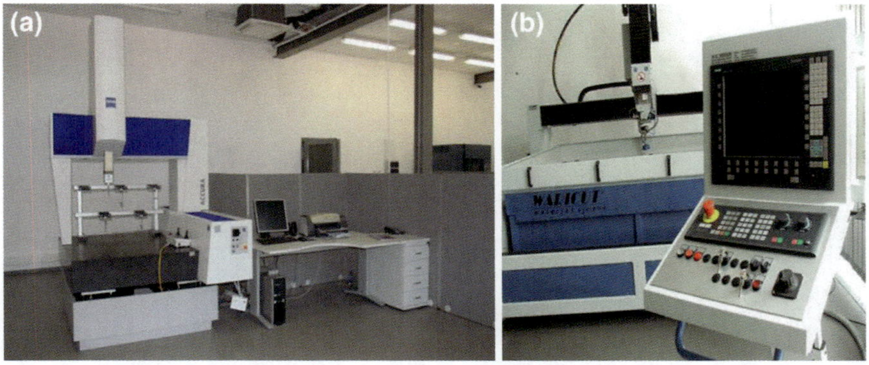


**Fig. 3** Examples of specialized machines supporting prototyping. **a** CNC measurement machine, **b** Water-jet machine tool

equipments is in many cases very expensive, for that reason small enterprises will not only be able to buy but it also operate too. We have equipped our Technopark for several specialized machines like (Fig. 3)

- Co-ordinate machine ZEISS ACCURA (Germany),
- Water-jet machine (Germany),
- 5-axis machining center from DMG (Germany),
- Rapid prototyping machine-type FDM from STRATASYS (USA).

We have created several microcompanies, incubated in Technopark and we gave them possibility of renting the machines. They may offer services for SME, for example in the field of manufacturing, supporting prototypings. The managers of microcompanies work at the university as scientific staffs and their Ph.D have been done in the same field.

## *3.2 Services in the Field of Supporting Design Processes*

Very similar looks have services offering in Technopark Gliwice in the field of High-Tech engineering software. Building our Technopark we assumed that it may works as a center, supporting SME in the field of engineering software like CAD/CAM/CAE. This kind of software supports engineering processes of designing and using of them requires high knowledge. Besides it such software is very expensive. For that reason we have equipped our Technopark with software like

- Ansys, a FEM (Finite Element Method) system,
- Catia, a CAD/CAM system,
- Solid Edge, a CAD system,
- MTS, a programming of CNC machine tool system.

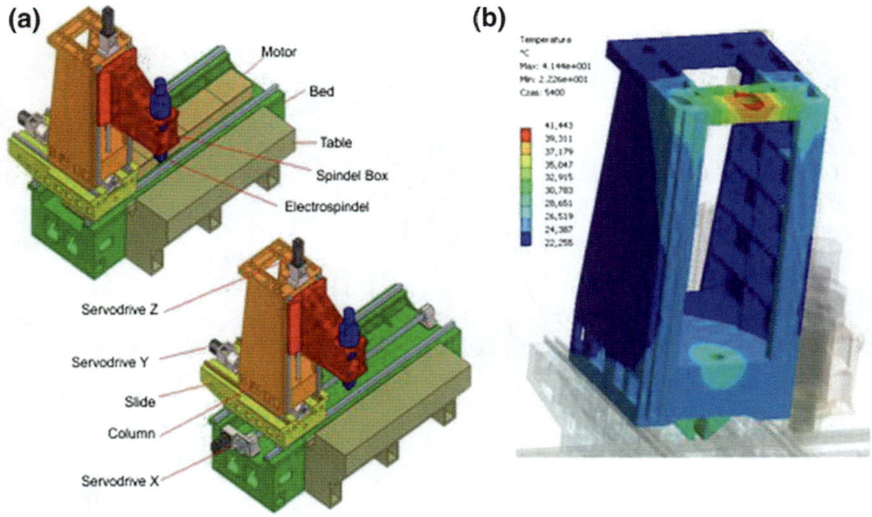


**Fig. 4** Examples of "High Tech" softwares supporting designing and manufacturing: **a** CATIA (CAD/CAM), **b** ANSYS (FEM)

Like by machines we have created several microcompanies, incubated in Technopark and we gave them possibility of renting those softwares. They may offer services for SME in the field of designing and manufacturing (Fig. 4).

## *3.3 Complex Services of Designing, Training, and Advisory*

In our opinion, transfer of knowledge is a form of technology transfer too. For that reason we offer training services, specialized and universal. Very important seems to be the specialized trainings because of processes in Technopark machines and softwares. This kind of trainings are designated first of all to the workers from Silesia region, who would like to change their jobs and for unemployed who looking for new jobs. These services are very popular and many people, young and over 45 are interested in it. Actually we offer such training services:

- Programming of CNC machine tools,
- Designing in 2D/3D environment,
- Programming of PLC controllers,
- Advanced FEM systems (ANSYS),
- Advanced CAD/CAM (CATIA) systems,
- Rapid prototyping and manufacturing,
- Water-jet machining.

The second group of trainings may be classified into general trainings in the field of business. First of all they are addressed to students, graduates, postgraduates and academic staffs, and to all other young and older habitants of our region.

Technopark Gliwice plans to implement in the years 2011–2014 of 12 projects funded mainly by:

- the European Union,
- the Ministry of Science and Higher Education.

Below mentioned examples are most interesting projects actually realized:

1. Creator of innovation-support for innovative academic entrepreneurship

- "Technoinqubation of ideas,"
- Implemented in partnership with Poznan Technology Park.

An initiative that allows students, graduates, and academics (120 in each city) make use of the 'enterprising package':

- 76 hours of entrepreneurship trainings for advanced students,
- Four subject areas (academic entrepreneurship, SME funding, creative entrepreneur, and innovation management), which are enriched with supplementary materials (recordings of experts, simulation games),

The aim of this project is primarily to prepare participants to present their idea to the investor at a joint meeting for the project partners (investment forum-2 editions).

2. An Academy of Enterprising Scientist
   Measure 8.2.1 Human Capital OP

- Implemented in partnership with the Silesian University of Technology,
- Project is addressed to academic environment and people who intend to start their own business,
- 290 participants during 2 years,
- 604 hours of training,
- Individual counseling (work on your idea),
- "Academy of Enterprising Scientist" is not only a training and consultancy, the idea is to promote academic entrepreneurship through the organization of "meetings during breakfast with the business sector and start-ups" (80 persons) and a conference (2 editions), allowing a presentation of ideas that have been successful.

3. A Scholarship Fund for graduate students to develop knowledge transfer in the regionA, Scholarship Fund for graduate students to develop knowledge transfer in the region
   Measure 8.2.1 Human Capital OP

- Project that provides a scholarship for 12 Ph.D. students from five selected faculties of the Silesian University for a period of 20 months.

- Scholars will be selected from the faculties particularly important to the development of the region (as defined in the Regional Innovation Strategy),
- In addition, scholarship holders will be subjected to an individual care of a Mentor whose task will be monitoring progress in a dissertation, and choosing the path of individual counseling,
- (20 hours for each person),
- The scholars will also undergo series of training 128 h within six cycles of training for 16 months.

4. Your own company a key to success and innovative capital for a start!
   Measure 6.2 Human Capital OP

- The projects are a continuation of a project ‘Support for those wishing to establish and maintain a business’
- So far 40 participants were trained in the field of creating a business plan and establishing an economic activity among which 21 people received financial support for their business development (up to PLN 3,700,000 each),
- A continuation of our efforts to encourage the creation of new businesses in the Silesian region is the implementation of the project’s Innovative capital for a start! The project will run from November 2011 until February 2013. For the first time the project is dedicated to innovative ventures only,
- During the project, 24 persons will extend their knowledge of creating a business plan and setting up business. 18 of them will receive financial support for development of their business activities (up to PLN 40,000) and establish a company in 2012,
- It will be also possible to apply for basic and extended bridge support combined with counseling for the newly created microenterprises.

5. Gliwice Technology Accelerator-financial and advisory support in the phase of pre-incubation of innovative solutions
   Measure 3.1 Innovative Economy OP

- A project implemented under Measure 3.1, initiating innovative activities of the Innovative Economy Operational Program,
- The amount of funding is almost PLN 6,000,000,
- The project will increase the chances of successful commercialization in the form of new businesses, through the implementation of a pre-incubation system (since the creation of an innovative idea or technology solution by consulting and support services) and then ensure financial stability in the start-up phase–it provides five capital inputs in the amount of PLN 700,000 each,
- Therefore, the Park will become a shareholder of these innovative companies.

6. An Analysis of Creativity Level and Absorption of Innovation in Enterprise
   Measure 5.2 Innovative Economy Operational Programme
   Implemented in partnership with Poznan Technology Park

- Assistance in reorganization of enterprises,
- Service aimed at using modern methods of innovation management in enterprises,

- 321 audits,
- 48 training sessions,
- Project introduces two new pro-innovative services,
- Study on innovativeness of organizations,
- Examination of problem-solving styles,
- Duration: 36 months,
- Participation in the projects is free of charge for enterprises,
- The cost of one audit which is around PLN 7,000 is paid by the EU.

7. Management, implementation, and monitoring of RIS of Silesia Region
   Measure 8.2.2 Human Capital OP
   Project in cooperation with Marshal Office of Silesia Voivodeship

- The Board of Silesia has chosen our park to carry out a system project of the Intermediary Institution of Silesia entitled 'Management, implementation, and monitoring of the Regional Innovation Strategy of Silesia,'
- Submeasure 8.2.2 'Regional Innovation Strategies,' Human Capital Operational Program,
- Our park will provide, among others technical and human resources and knowledge about available 'best practices' which will be useful for achieving the objectives of the project,
- Statistics of Technopark Gliwice projects in 2011–2014,
- The amount of financing: PLN 13,369,376,
- 8,084 h of trainings,
- 39 created companies,
- 1,582 h of consultancy,
- 2,001 participants.

## 3.4 Advisory Services

Technopark Gliwice offers several others services, first of all for SME, which helps them to get money and expert's knowledge like

- Complex application preparation of investment projects for SME,
- Creating of business outlines and feasibility study,
- Creating of technical analysis and expert's technology study.

Because of exact cooperation with the Silesian University of Gliwice we are able to offer all technical and business services because we can find experts in all fields.

## 4 Companies in Technopark Gliwice

Transfer of innovative technology to SME in Technopark Gliwice is realized first of all by the innovative companies which are situated in our technological park. Technopark Gliwice creates only the favorable conditions for development of innovative technologies and for their transferring. For that reason the companies in Technopark are so important. Because of very small area of Technopark (only 2000 $m^2$ were accessible by our clients) most of our companies belongs to IT but to architectures, CAD/CAM, chemical, and others too. They need offices only and for that reason we were able to rent such offices for 28 companies. At present we have no free offices and we cannot take new companies. Over 2/3 of all companies were incubated in Technopark Gliwice, what means that they are very young. One of them were the I3D Company situated on Polish market yet. During the first 3 years of Technopark's activities only two of 28 companies left our park. It means that terms for business in Technopark seem to be good. In 2011, the Polish Agency for Enterprise Development PARP conducted a competition among 17 best Polish Technoparks and incubators. The main criterion was satisfaction of clients situated in Technoparks and incubators. Technopark Gliwice took first place.

## 5 Pre-incubation and Incubation in Technopark Gliwce

Our Technopark works as an incubator of innovative technology too because of our location. We are situated in the center of academic village of the Silesian University of Technology of Gliwice (Fig. 5). For that reason thousands of students, graduates, and postgraduates were our potential clients and therefore one of our activities is incubation.

We have got two-stage system of incubation: pre-incubation and incubation. Forms of pre-incubation activities:

- Non-payable using of office rooms, computers, technical and business advisories, etc. (time of pre-incubation 3–12 month)
- Goal of pre-incubation stage: Creating a business study what makes possibly to decide doing next step, incubation stage.
- Forms of incubation activities,
- Using of office rooms and computers on preferential conditions,
- Non-payable using of business and technical advisories,
- Help in searching, financing the development of companies (Technopark prepared an application to Structural Fund PO IG 3.1 Initiation of economic activities"),
- Access to specialized machines and softwares available in Technopark Gliwice, help to access laboratories and scientific staff of Silesian University.

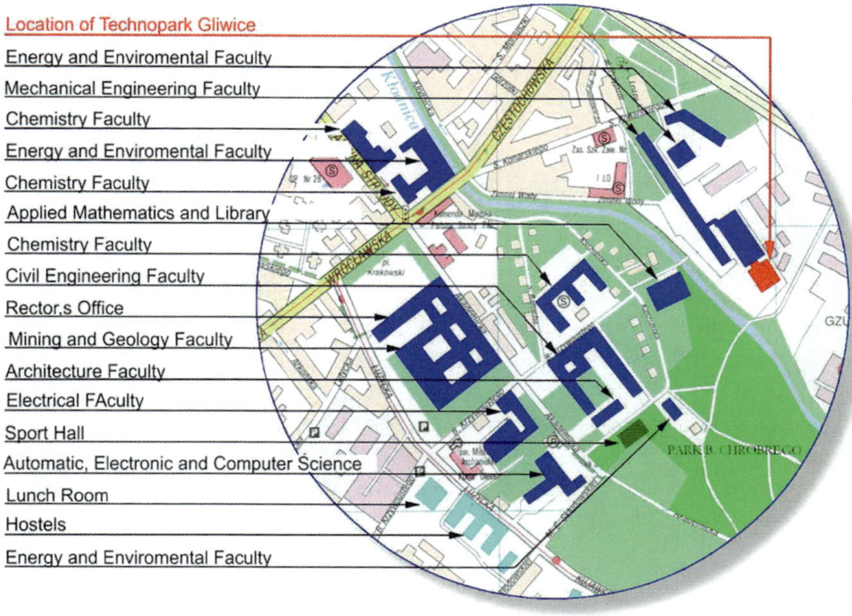


**Fig. 5** Location of Technopark Gliwice

## 6 Promotion of Enterprising

For our (Silesia region) point of view, promotion of enterprising plays very important role because of regional conditions of enterprising. The regional conditions look like follow:

- Post-industrial social surrounding (industry of mining and metallurgy) does not favor starting the individual economic activities,
- In the mentality of Silesian community dominates opinion about "superiority" working for a businessman (owner) than for "own business",
- The regional social conditions prefer education in traditional zone, it means mining and metallurgy, that is in zone of low level of innovation and enterprising,
- A much smaller mobility of Silesian community comparing to other regions of Poland.

For that reason the preferable activities of "Technopark Gliwice" are actions on promotion of enterprising, which except measurable economic results for Technopark, for example selling more High-Tech services, should change mentality, first of all of young inhabitants of Silesia agglomerate, for the point of view of taking up self-economic activities.

We try to apply several promotions of enterprising methods like:

- Trainings in the fields of individual activities in economic zone (creating business studies, registration, taxies, economic law, intellectual properties). The trainings are guided first of all to students and doctorants. All such trainings are non-payable.
- Initializing of innovative activities in the form of financial grants new companies may get up to zł 40.000 intended to start with a new enterprise.
- Action like "Open doors," guided first of all to young people from secondary schools and university students.
- Action like "Days of science and industry." The main goal is to popularize scientific achievements but with markets destination.
- Meetings with experienced businessmen and people with high authority in the field of enterprising (action like "Club of businessmen").

The mentality of young people in Silesia region have to be changed if we would like to reach technical and economical level like people in USA, Western Europe, Japan. It takes much time but such activities like incubation in Technopark Gliwice makes possibly to accelerate the process.

## 7 Summary

Technopark Gliwice belongs to one of the younger technological parks in Poland and to the smallest. But achieved results during the first 3 years of our activities shows that the way we have chosen seems to be good. We hope that cooperation with others technological parks, for example, Deajeon Technopark and with WTA help us to speed up the transformation in our region.